Environment and Plant Ecology

Second Edition

JOHN R. ETHERINGTON

Senior Lecturer in Botany,
University College of South Wales, Cardiff

with a contributed chapter by

W. ARMSTRONG

Senior Lecturer in Botany,
University of Hull

JOHN WILEY & SONS

Chichester · New York · Brisbane · Toronto · Singapore

Reprinted October 1975
Reprinted July 1976
Reprinted June 1985

Library of Congress Cataloging in Publication Data:
Etherington, John R.
Environment and plant ecology.

Bibliography: p.
Includes index.
1. Botany—Ecology. I. Title.
QK901.E85 1982 581.5 81-16167

ISBN 0 471 10136 2 (Cloth)AACR2
ISBN 0 471 10146 X (Paper)

British Library Cataloguing in Publication Data:
Etherington, John R.
Environment and plant ecology—2nd ed.
1. Botany—Ecology
I. Title
581.5 QK901

ISBN 0 471 10136 2 (Cloth)
ISBN 0 471 10146 X (Paper)

Typeset in Great Britain by Speedlith Photo Litho Ltd,
76 Great Bridgewater Street, Manchester M1 5JY.
Printed and bound in Great Britain by
Biddles Ltd, Guildford and King's Lynn

For my family and the memory of Nemo
who helped so much.

Contents

PLANT RESPONSES

PLANTS IN ECOSYSTEMS

Preface to the First Edition

The terrestrial plant, rooted on one spot and originating from seed which may be distributed almost at random, is at risk to a whole range of environmental hazards which a mobile organism can escape. From the moment that germination begins, the developing plant must cope with soil chemical and physical conditions, climatic and microclimatic effects, pathogens and herbivores. The germinating seed carries a specific blueprint of genetic data which dictates its response to a complex of environmental pressures. If the genetic scope for phenotypic plastic response is exceeded by one, or more, of these pressures, then germination or establishment will fail. Another seed of the same species, but carrying a different set of gene recombinants, may survive, but, if it cannot, it may be that the environmental conditions are entirely outside the genetic capability of the species.

Within the first few days after germination, even though the arrival of the seed was a random event, a degree of ecological pattern is generated. One species survives, another is lost; one seed dies but its genetic brother, differing in a few genes, persists, perhaps as the progenitor of a new ecotypic stock. With time, the microclimate oscillates and alters, the plants grow and change in their morphology and physiology and, as the seedlings occupy more space, so competition begins. Competitive interaction causes not only visible change but also distorts 'normal' physiological responses, sometimes exceeding their limits. By the time the first seeds have developed to adult plants, an interacting web of environmental and biological limitations has either dictated the survival of individuals or at least profoundly influenced their phenotypic development.

At any time, the ecosystem is a legacy of previous events: its component species, their morphology and physiology, the distribution of soil organic matter and inorganic nutrient elements, all of these factors and others preserve evidence of past development. As an ecosystem grows and stabilizes the wastage rate is enormous and, for every plant which survives to reproduce, perhaps many thousands are lost in failure to germinate, attack by herbivore or pathogen, killing by drought or frost, competitive exclusion or a multitude of other environmental effects. Out of this carnage emerges a group of individuals which have passed the genetic sieve of intense selection and whose offspring will, in turn, inherit combinations of attributes which best fit them to adapt to the localized environment and to withstand the competitive pressures of their neighbours. This complex species–environment interaction, coupled either with sharp geological or topographical discontinuity or with large-scale climatic gradients, produces the ecological associations so easily recognized by the field worker, in the first case as discrete communities and, in the second, as some form of continuum.

This book attempts to examine some of the details of this intricate jigsaw of plant, animal, microorganism and environmental interactions as it manifests itself in plant physiological ecology. The problem is to dissect the network of energy fluxes, elemental cycles and control systems and to present them in the light of the individual species' behaviour and competitive interaction. In a decade which has seen the increasing application of techniques for analysing continuous multiply interacting variables, combined with the introduction of systems analysis to ecological modelling, it might be thought unwise to attempt this dissection. It is the author's view, however, that the ultimate in ecosystem modelling is to describe not only general functional relationships but also the population dynamics of the species concerned.

The systems approach, at the present time, is admirably suited to the modelling of generalities such as whole ecosystem energy flow, water transport or nutrient cycling, but it is much less useful in describing species behaviour within the system. It is very easy to insert hypothetically feasible values in a general model of ecosystem nutrient cycling and to compare the computed results with the real world. It is much less easy to make the same approach at the individual species level without the backing of physiological–ecological experimentation. Unlike the whole-system situation, the requisite values for individual species modelling cannot be inferred from the general. Herein lies the need for a comparative experimental study of physiological ecology in the field and in controlled environmental conditions.

With the exception of the first chapter, this book is intended to provide a background of information concerning the environment and plant response which the undergraduate will not otherwise easily find in a single book. It has been fairly liberally laced with references which open the doors to more specialist textbooks and to some more important or interesting original papers. The first chapter is rather different, being a thumbnail sketch of the present situation in ecology. It also provides more general reference to those parts of plant ecology which are not discussed in detail in this text.

I would like to thank all of those colleagues and students who, by discussion and listening, have brought me to a fuller understanding of ecology. In particular, I cannot overstate the debt which I owe to teaching as an aid in marshalling ideas. The best learning aids are insatiable reading, note-taking and a captive audience! To countless authors, a word of thanks for the discovery that so many different viewpoints and interpretations exist, for me the greatest pleasure of literature research. I must also express my gratitude to Professor A. J. Rutter, from whom I first realized that physiological ecology would be more rewarding than a conventional 'lab. bench' study of physiology or biochemistry. As a consequence I enjoy a paid hobby; an immensely satisfying experience.

Inevitably the text must include errors and omissions for which I must be responsible. I shall be grateful for any comment or discussion from readers concerning future amendments.

University College of South Wales
Cardiff
May 1974

J. R. ETHERINGTON

Preface to the Second Edition

During the years which have so quickly passed by since the first edition was written, many more papers and texts in physiological ecology have been published. I have included new material from these sources and attempted to satisfy those critics who justly complained that the first edition gave an imbalanced treatment of the underground and aerial environments. This has generated some additional chapters and lengthened others; with regret I have compensated for this increase by omitting the introductory chapter of the first edition as its contents may be found in more detail elsewhere. It has been replaced by a much shorter introduction and a synopsis of the history of ecology with suitable reference to sources of information.

The first edition was written at the time of change to SI units with some consequent confusion. In this edition I have attempted to adopt SI throughout (HMSO, 1977; Incoll *et al.* 1977). There are a few exceptions, for example where 'light' measurements cannot meaningfully be converted, where conversion might obscure an author's original intention, and in Chapter 10 contributed by W. Armstrong. Some preferred prefixes have been avoided where their use would produce numbers which are difficult to remember, for example the use of seconds in production ecology involves problems which are avoided by using days, weeks or years. The use of symbols has been rationalized in this edition to eliminate ambiguities. I have attempted to follow previous practice (Slatyer, 1967; Worthington, 1975; Monteith, 1973) except for the use of some mnemonic symbols which may be more helpful to the student.

I am grateful to my critics, some kind and some less so, who have helped in opening the doors to at least some parts of my ignorance. To my colleagues in Cardiff, past and present, I can only say thank you for talking with so much enthusiasm about all of those parts of science which form the ingredients of this mixture which we name ecology. I have been fortunate to be surrounded by so many different interests.

University College
Cardiff
June 1981

J. R. ETHERINGTON

Symbols

Symbols used in Chapter 10 by Dr. W. Armstrong have not been rationalized with the rest of the text, and where there is ambiguity or units are not SI this is indicated.

A	Area (cm^2).
$A_{harvest}$	Number of propagules sown: species A.
A_{sown}	Number of propagules harvested: species A.
a	Root radius (m).
α	Dimensionless coefficient in equation 2.8 and 13.3.
B	Biomass Density (kg m^{-2}); subscripted respectively for total, primary, secondary, decomposer and dead organic matter components: total; 1ary; 2ary; decomposer; dead organic.
$B_{harvest}$	Numbers of propagules sown: species B.
B_{sown}	Numbers of propagules harvested: species B.
B_r	Bowen ratio.
b	Soil buffering capacity = dC_{soil}/dC_{liq}.
β	Dimensionless coefficient in equation 13.3.
C	Sensible heat exchange (W m^{-2}).
C	Concentration (mol cm^{-3}) (Chap. 10—not SI).
C_{cond}	Conduction of sensible heat (W m^{-2}).
C_{conv}	Convection of sensible heat (W m^{-2}).
C_{soil}	Concentration of a component (mol m^{-3} soil).
$C_{solution}$	Concentration of a component (mol m^{-3} solution).
c	Flux parameter (dimensionless) relating the diffusive and mass flow contributions of solute movement to a root.
C_p	Specific heat of air at constant pressure (J m^{-3} K^{-1}).
D	Leaf area duration (m^2time^{-1}).
D	Diffusion coefficient (cm^2 s^{-1}). (Chap. 10)
D_{if}	Diffusion coefficient (m^2s^{-1}).
D_{is}	Dispersal coefficient (m^2 s^{-1}).
D_{en}	Density of individuals (number m^{-2}).
d	Displacement height (m).
Δ	Slope of the saturation vapour pressure versus temperature curve for water in air (Pa K^{-1}) (on other gradient where applicable).
E	Einstein (Avogadro's number of quanta = 6.02×10^{23}).

E_A	Unit leaf rate (kg m^{-2} $time^{-1}$).
E_v	Evaporation, transpiration or vertical flux density of water vapour (kg m^{-2} s).
e	Base of natural logarithms (2.718).
e_p	Water vapour partial pressure (Pa).
e_s	Water vapour partial pressure at saturation (Pa).
ε	Emissivity.
ε	Fractional porosity (Chap. 10).
F	Leaf area ratio (m^2kg^{-1} dry weight).
F	Faraday (96,500 coulombs) (Chap. 10).
F_S	Solute flux density (mol m^{-2} s^{-1}).
$f_{x=0,t}$	Oxygen flux at zero distance and time t (ng cm^{-2}) (Chap. 10—not SI).
G.	Grazing consumption of biomass (kg dry wt m^{-2} or kg m^{-2} $time^{-1}$). Subscripted c for current year.
γ	Psychrometric constant.
H	Total vertical heat flux density through a canopy (W m^{-2}).
H_y	Hydraulic conductivity (m^2Pa^{-1} s^{-1}).
h	Height (m).
I	Current (ampere): in analogue treatment of oxygen diffusion proportional to mol O_2 s^{-1} (Chap. 10).
i_t	Diffusion current (ampere) in electrometric measurement of oxygen diffusion (Chap. 10).
J	Hydraulic flux (m^3 m^{-2} s^{-1} = ms^{-1}).
J	Joule (SI unit of work) = 1N m^{-2}.
K	Kelvin (0 °C = 273 K).
K	Dimensionless constants.
K_{ext}	Extinction coefficient.
K_C	Diffusivity for CO_2 (m^2 s^{-1}).
K_H	Diffusivity for heat (m^2 s^{-1}).
K_M	Diffusivity for momentum (m^2 s^{-1}).
K_V	Diffusivity for water vapour (m^2 s^{-1}).
K_m	Michaelis constant.
$K_{1,2 \text{ or } 3}$	Rate constants in ion-uptake kinetics.
k	von Karman's constant.
kg	Kilogramme (SI unit of mass).
L	Leaf area index (m^2 leaf m^{-2} ground).
L_A	Leaf area (m).
L_C	Litterfall of current year (kg dry wt m^{-2}).
L_f	Litterfall (kg dry wt m^{-2} or kg m^{-2} $time^{-1}$).
L_n	Latent heat of vapourisation of water (2.45 MJ kg^{-1}).
L_W	Leaf dry weight (kg).
l	Length (cm) (Chap. 10).
$\log_e$ or ln	Natural logarithm.
λ	Wavelength (μm).

λ_{max}	Wavelength of peak energy content in a continuous spectrum (μm).
M	Metabolic energy exchange (W m^{-2}).
M	Respiration rate (mol cm^{-3} s^{-1}). (Chap. 10—not SI).
M	Vertical transfer of momentum (N m^{-2} = Pa).
m	Metre (SI unit of length).
me %	Milliequivalent per 100 g (not SI but retained for convenience in ion-exchange studies).
N	Number in a population.
N	Newton = 1 kg m s^{-1} (SI unit of force).
n	Number of electrons to reduce one O_2 molecule (=4) (Chap. 10).
P	Photosynthesis (various units in text).
PAR	Photosynthetically available radiation (W m^{-2} or E m^{-2} s^{-1}).
P_A	Projection area (m^2).
Pa	Pascal = N m^{-2} (SI unit of pressure).
P_{Ev}	Potential evaporation or evapotranspiration (kg m^{-2} $time^{-1}$).
P_{gross}	Gross primary production (kg m^{-2} $time^{-1}$ or J m^{-2} $time^{-1}$).
P_{net}	Net primary production (kg m^{-2} $time^{-1}$).
P_r	Precipitation (kg m^{-2} or kg m^{-2} $time^{-1}$). Sometimes expressed as equivalent depth of water (m).
ψ	Water potential (Pa). Subscripted: p—pressure; s—solute; m—matric. Other subscripts are explained in text.
q	Specific humidity (kg water kg^{-1} air).
Q	Quantity diffusing (mol). (Chap. 10).
R	Relative growth rate (kg kg^{-1} $time^{-1}$).
R	Resistance (ohm); in analogue treatment of oxygen diffusion proportional to dx/dA s cm^{-3} (Chap. 10).
R	Universal gas constant (8.3143 J mol^{-1} K^{-1}).
R_{abs} and R_{in}	Radiant absorption and income (W m^{-2}).
R_i and R_o	Irradiances above and within a canopy (W m^{-2}).
R_{net}	Net radiation (W m^{-2}).
R_{rad}	Reradiation from a surface (W m^{-2}).
R_{sw}	Shortwave irradiance (0.3–3.0 μm) (W m^{-2}).
R^+ and R^-	Upward and downward radiant Flux density (W m^{-2}).
RRR	Relative reproductive rate (equation 13.2).
r	Diffusive resistance (s m^{-1}). Subscripted: d—external boundary layer); c—cuticular; h—heat; 1—leaf total; m—mesophyll; s—stomatal; st—bulk stomatal resistance of whole canopy; v—water vapour. Other usages are explained in the text.
r_{ep}	Reproductive rate constant.
ρ	Density of air (1.2 kg m^{-3} at 20 °C).
s	Second (SI unit of time).

Σ	'The sum of'.
σ	Stefan-Boltzman constant (5.5×10^{-8} W m^{-2} K^{-4}).
T	Time (s).
$\mathbf{T}$	Temperature (K).
T_{emp}	Temperature (°C or K as specified).
t	Tonne (= 1000 Kg).
t	Time (s) (Chap. 10).
τ	Tortuosity factor <1 (Chap. 10).
U_w and U_w^0	Chemical potential of water in a system and of free, pure water at the same temperature and elevation (J mol^{-1}).
u_h	Windspeed at the top of a vegetation canopy, height h (m s^{-1}).
$u(z)$	Windspeed at height z (m s^{-1})
u^*	Friction velocity (m s^{-1})
V and V_{max}	Rates of ion uptake (mol kg^{-1} s^{-1}) by roots and tissues.
V	Potential difference (volts): in analogue treatment of oxygen diffusion equivalent to a concentration difference (mol cm^{-3}) (Chap. 10—not SI).
W	Dry weight (kg).
x	Constant in 3/2 power law.
x	Distance (cm) (Chap. 10).
y	Year.
Z	Term in potential evapo-transpiration equation $= (r_v + r_{st})/r_h$.
z	Height (m).
z_0	Roughness length (m).

Chapter 1

The aims and development of plant ecology

Ecology may be defined as the study of the interrelationship between plants, animals, microorganisms and their environment: the word is derived from the Greek *oikos* (a dwelling or, more literally, homestead) and *logos* (study of). As a scientific discipline it is little more than 100 years old but, by the beginning of this century, Clements (1905) was already exhorting ecological workers to adopt the methods of the plant physiologist and, by 1917, Tansley had used the expression 'physiological ecology' to describe an experimental investigation of the mechanism controlling the distribution of two plant species in relation to soil chemical conditions. This book is an account of plant physiological ecology and the 'environment' of its title is intended to include not only the physical and chemical but also the biological environment provided by microorganisms and animals. Without microbiological aid plants would not exist and without animals, plant life might display very different characteristics.

Modern man (*Homo sapiens sapiens*) had evolved by the Late Palaeolithic some 40,000 years ago. Certainly, from this time, the development of language and the needs of a hunter–gatherer life must have imposed fairly sophisticated ecological concepts of the organism–environment relationship: where to look for particular animal and plant species, identification of toxic and edible materials, identification of the food-plants of different animals and, judging from modern pre-agricultural peoples, an amazingly detailed knowledge of the medicinal and other uses of the plants surrounding him. Much of this knowledge, accumulated through the millennia, is now recorded in the folk-names of plants. An interesting insight is given by Matthews' (1886) account of such names amongst North American Indians. In many ways these names have kinship with the aims of one of the most recent developments of our subject—applied ecology.

The word ecology did not become widely used until the very end of the last century. It appears in the title of Warming's classic text on ecological plant geography but he cites only one other reference to its previous titular use; in a text on ecological plant anatomy. Warming attributed the modern definition to Haekel (1866): 'the body of knowledge concerning the economy of nature—the investigation of the total relations of the animal to its inorganic and organic environment', but he also noted that Reiter (1865) had previously used the word 'oecology' though in a somewhat different sense.

Ecology is a very young science which did not commence major growth until the 1950s, in part as a delayed response to the realization that human population increase was outstripping both the food-supplying abilities and the biological resilience of the world environment. The delay was by several millennia for, throughout written history, many warnings have been sounded. Plato, in *Critias*, written about 350 B.C., mourned for the destruction of trees in Attica and complained of the subsequent soil erosion and loss of sustenance for beast and man alike (Crowther, 1953). His words might have been written yesterday, so evocative are they of our present concern for the remaining wilderness.

Over 2000 years later Malthus again prophesied disaster (1798) unless humanity realised that its population and appetite would quickly outstrip the sustaining capacity of agriculture. This warning also was unheeded, for both political and religious reasons. The consequences of the Industrial Revolution, in particular the development of communications, and the application of science to agriculture through developments in fertilizer technology and mechanization helped to confirm popular opinion that the Malthusian prophesy was fallacious and it was not until the beginning of this century that concern again began to develop, a concern which has revolutionized biological attitudes to human population growth, agriculture and the ecological basis of the conservationist movement. Unfortunately, though accepted by the scientific community, the idea that man, as any other animal, must achieve an ecologically sound equilibrium with his environment and all other life, remains politically unacceptable. In the words of Sears (1964) ecology may even by considered a 'subversive science' and Nicholson (1970) has further suggested that its delayed development as an academic discipline was not wholly fortuitous, resources devoted to the physical and chemical sciences being 'less productive of thoughts dangerous to the "system"'.

Slow to develop, ecology is now a rapid-growth sector of biological science and many more papers have been published in the last decade than during the whole history of the subject, despite the relatively early establishment of specialist organs such as *Journal of Ecology* (1913) and *Ecology* (1920). Prior to this, much work of an ecological nature was published in the established journals of natural history, general botany and zoology. Many other ecological works appeared in book-form, for example chapter three of *Origin of Species* (Darwin, 1859) has been described as the 'best ecological text ever written' (Harper, 1977). Very little ecological thought and experimentation is entirely new.

Much of the recent flood of ecological publication has been influenced by the needs of applied ecologists working in fields such as environmental pollution and agro-ecology. This is a little unfortunate as some fundamental studies are, by comparison, neglected, for example mixed-species population interactions, comparative physiological ecology and distribution studies of wild species. During the 1960s the establishment of the International Biological Programme (IBP), though promoted for reasons of human welfare (Worthington, 1975), produced many basic studies of natural ecosystems in relation to photosynthetic production, total metabolism and nutrient cycling.

In the previous edition a review of the general history and current status of ecology was given. Space does not permit such treatment here but it is hoped that the brief historical tabulation provided by Appendix Tables 1.1a, b and c will tempt the reader to consult the remarkable literature of earlier years. These fragmentary references present only the author's viewpoint of history and, as is so common, others may consider the picture both incomplete and misleading. In the later years, from the 1950s onward, it is much more difficult to distinguish seminal ideas amongst the flood of publication: the tables present titles of texts or monographs from which information may be delved.

PLANT PHYSIOLOGICAL ECOLOGY

Plants occupy almost every available earthly niche: algal cells on the surface of permanent snow, Cyanobacteria in hot saline springs, and lichen crusts on bare, arid, often hot rocks represent the extremes. Even the most hospitable environments impose some stress: photosynthesis is never maximal because carbon dioxide is rare and there is too much oxygen. Water supply oscillates from too much to too little while the cyclic course of temperature weaves above and below the optimum. Soils rarely provide ideal conditions: they are usually nutrient deficient, may be ionically unbalanced or even seriously toxic. The list of problems is almost endless and each one constrains physiological function below the optimum defined by laboratory experiments.

The incidence of stress usually follows a predictable pattern, being either continuous or following a seasonal course. Predictable stress imposes directional selection which, in recent years has been interpreted as an optimization strategy (Maynard-Smith, 1978) but, because there are so many stresses and potential responses, any such strategy is bound to be a compromise. Furthermore, any single change may cause repercussions in other parts of the ecosystem, which are not intuitively predictable. Any selected strategy will be sensitive to environmental change which we now believe to have occurred in all habitats including the tropics (Livingstone, 1975). In consequence we can draw a dynamic picture of the plant ecosystem as an ever-changing web of environmental and biological interactions in which no organism is ever fully adapted to its surroundings.

The physiological ecologist is concerned with the isolation and study of these adaptive strategies and with their interpretation in the broader context of ecosystem function and environmental change. Observation and measurement of plant species-distribution or behaviour provides the raw material of speculation concerning climatic, edaphic and biotic constraints on growth. The hypothesis derived from such speculation may be testable by species-comparative experiments ranging from environmental effects on whole-plant physiology to detailed investigations of biochemical adaptation to specific environments or to herbivore deterrence.

Before laboratory work can be undertaken the researcher needs to define the physical and chemical nature of the environment, not only by instantaneous measurement but often by continuous recording through the seasons. The development of easily portable recording instruments, starting with various direct-writing devices in the first quarter of this century, have greatly eased the data collection problem of the micrometeorologist. The post-war availability of servo instruments such as the potentiometric recorder, development of solid-state circuitry in the 1960s and the introduction of direct computer interfacing through punch- or magnetic-tape data loggers has revolutionized this field, a revolution which is continuing with the development of microprocessor controlled equipment permitting an increased scale and frequency of sampling.

The chemical environment likewise is becoming more accessible with rapid development of instrumental techniques. Many traditional time-consuming colorimetric and titrimetric analyses were largely replaced by flame emission and absorption spectrometry during the 1960s and more recently by x-ray or neutron activation methods. Some of the more difficult anionic analyses which still depend on classical methods have been hastened by the development of autoanalysers and are increasingly being replaced by electrochemical methods. Despite these advances one of the greatest problems in defining the soil environment lies in the need for extraction techniques which simulate long-term plant nutrient extractive behaviour. In agriculture this is fairly well understood as extraction techniques may be correlated with fertilizer response trials but in ecological work the assessment of nutrient availability for plant growth remains an underworked and very difficult field.

The Environment

Chapter 2

The environment above ground: energy exchange

INTRODUCTION

It has been traditional for ecologists to describe an environment in terms of its associated organisms and such factors as soil conditions, temperature, sunlight, rainfall, wind and humidity. The meteorological factors are a reflection of the pattern and utilization of solar radiation which, on a global scale, drives the atmospheric circulation with its associated cycles, and, at a microenvironmental level, provides the main long-term control which is rarely overruled by such catastrophic (but nevertheless determinant) events as extreme drought, fire, flood or frost.

Terrestrial organisms live in a thin boundary layer between the solid fabric of the earth's crust and the inhospitable environment of interplanetary space. This boundary layer, the earth's atmosphere, is strongly affected by the net flux of radiant energy between sun, earth and extraterrestrial space and it is with the multitude of energy environment effects in the boundary layer that this chapter is concerned.

Energy, the capacity to do work, is encountered in biological systems as either the kinetic energy of electromagnetic radiation ('light'), thermal motion of molecules, electric current flow, the motion of objects larger than molecules, or as potential energy stored in chemical or physical systems. These forms of energy are all interconvertable and, by a variety of trapping mechanisms during such conversions, autotrophic organisms utilize energy fluxes to build complex molecular structures and create larger scale patterns of tissue, organism and ecosystems.

The First Law of Thermodynamics states that: *Energy can neither be created nor destroyed:* this concept underpins all attempts to map and measure the pathways of energy flow in physiological and ecological systems. The Second Law is stated in various forms, for example: *Heat passes spontaneously only from a system of higher to a system of lower temperature.* This may be generalized as: *systems tend to approach a state of maximum probability and spontaneous changes are ones associated with increasing probability states.* In physical systems this implies increasingly random molecular orientation or motion and likewise in biological systems, a break-down of ordered structures. Thus any living organisms or ecosystems, unless in the suspended animation of extreme cold or desiccation, requires a basal metabolic energy expenditure to maintain organization and prevent senescence.

These thermodynamic concepts furnish an ecological 'rule-book' through which community energy flow, organization and materials-cycling can be studied. The consequence of the first law is that any ecosystem structure or function has an energy cost and implicit in the second law is the suggestion that successions and climaxes have maintenance energy requirements which, if not fulfilled, lead to ecosystem degeneration including loss of species diversity and reduction of living biomass and/or productivity.

All objects above the temperature of absolute zero emit energy by radiation and, in turn, may absorb radiant energy. The amount of energy radiated by a body is a function of the fourth power of the absolute temperature of its surface (Stefan's Law). For a perfect radiator this may be written as;

$$R_{rad} = \sigma T_{emp}^4 \tag{2.1}$$

where R_{rad} = the total energy emitted by unit area of a plane surface into an imaginary hemisphere subtending it ($W\,m^{-2}$), σ = the Stefan-Boltzmann constant ($5{\cdot}57 \times 10^{-8}\,W\,m^{-2}\,K^{-4}$), and T_{emp} = the absolute temperature (K).

A perfect radiator ('black body') has an emissivity of unity and shows the theoretical Stefan's relationship. It will also be a perfect absorber, trapping all incident radiation. Real objects do not conform to this perfect behaviour and have emissivities of less than unity, for example leaves usually range between long-wave emissivities of 0·94 and 0·99, wet soils between 0·90 and 0·95, but some highly polished metal surfaces have very low values, for example silver, 0·02. The equation must thus be rewritten for imperfect radiators as:

$$R_{rad} = \varepsilon\sigma T_{emp}^4 \tag{2.2}$$

where ε, the emissivity, is a value between 0 and 1. The radiation emitted by a body has a wavelength distribution which is related to its surface temperature, wave propagation being a function of molecular oscillation which increases in frequency with temperature rise; The peak of the wavelength distribution (λ_{max}) is related to temperature by the Wein displacement law:

$$\lambda_{max} = \frac{2897}{T_{emp}} (\mu m) \tag{2.3}$$

The distribution of radiant energy around this peak is asymmetric so that only 25 % is of shorter wavelength and 75 % is of longer wavelength. Some examples of λ_{max} are tabulated below:

Object	Approximate temperature (K)	λ_{max}	Quality
Sun's surface	6000	0·48	Blue-green
Tungsten lamp filament	3000	0·96	Short-wave infrared
Habitable, unstressed surfaces	287	10	Long-wave infrared

Within any closed system the balance of energy will be a function of the relative surface temperatures and the absorbance and emittance characteristics of all component parts of the system. It will, furthermore, depend on any other modes of energy transfer and energy-dependent changes of state. For this reason energy balance studies demand, in addition to radiation measurements, the measurement of conduction, convection, the evaporation of water or the melting of ice.

Exchanges of radiant energy in the biosphere take place within the approximate wavelength range 0·3–100 μm and for convenient reference this may be subdivided as:

0·3–0·4	0·4–0·7	0·7–3·0	3·0–100 μm
Ultraviolet	Visible	Short-wave infrared	Long-wave infrared (Thermal infrared)

Solar short-wave (Ultraviolet, Visible, Short-wave infrared); Emissions at terrestrial temperatures (Long-wave infrared)

All-wave radiation

THE EARTH'S RADIATION BALANCE

Radiant energy from the sun reaches the upper atmosphere with a flux density of 1360 W m^{-2}, the *solar constant*, and a wavelength distribution approximating that of a 6000 K black body (Figure 2.1) except for some discrete absorption bands, the Fraunhofer lines imposed by passage through the solar atmosphere. As it penetrates the earth's atmosphere the radiant flux is depleted of energy exponentially with depth and, as Figure 2.1 shows, this depletion by atmospheric absorption and scatter occurs both at discrete wavelengths and also throughout its spectrum. The well-defined absorption bands are attributable to specific

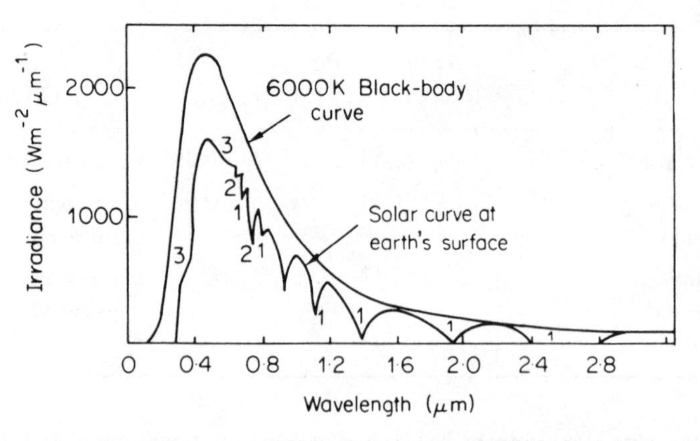

Figure 2.1 The spectral distribution of irradiance from a 6000 K black-body measured at solar distance and compared with the actual spectral distribution of solar radiation at the earth's surface. The absorption bands are due to water vapour (1), oxygen (2) and trioxygen (3). After D. M. Gates, *Energy Exchange in the Biosphere*, Harper and Row, New York, 1962, Figure 3.

molecules: most of those in the infrared are due to water vapour and the strong ultraviolet attenuation is caused by trioxygen (ozone).

Not all of the solar income which penetrates the atmosphere and reaches earth's surface is absorbed. Some is reflected, the proportion being governed by the *short-wave reflection coefficient* (albedo) of the surface. Cloud cover has a high coefficient (*c* 0·5) and reflects solar shortwave radiation back to extraterrestrial space causing the brilliant contrasts of clear and clouded regions in photographs taken from spacecraft. Fresh snow is almost totally reflective (0·8–0·95) but other natural surfaces have considerably lower values, for example bare soil *c* 0·3, broad-leaved forest 0·15–0·2 and coniferous forest 0·1–0·15. The radiant energy absorbed by any surface is given by:

$$R_{sw,abs} = R_{sw,in}(1 - r_{sw}) \tag{2.4}$$

where R_{sw} = irradiance (0·3–3·0 μm)(W m^{-2}), r_{sw} = short-wave reflection coefficient (albedo), and the subscripts abs and in mean, respectively, absorbed and surface income.

In the case of the terrestrial atmosphere and of water bodies, which are semi-transparent to short-wave radiation, the same equation is applicable if the effects of reflection are integrated with depth.

Earth's surface and atmosphere thus act as an absorbing sink for the incident solar radiation which is not reflected or scattered back to space. The absorption causes either temperature increase or latent heat conversions such as the evaporation of water which absorbs approximately 2·5 MJ kg^{-1} (Figure 2.2). This storage of sensible heat and temporary transfer of latent heat to the atmosphere must be counterbalanced if earth's surface and atmosphere is not to heat up. Figure 2.2 shows that the balance is maintained by long-wave

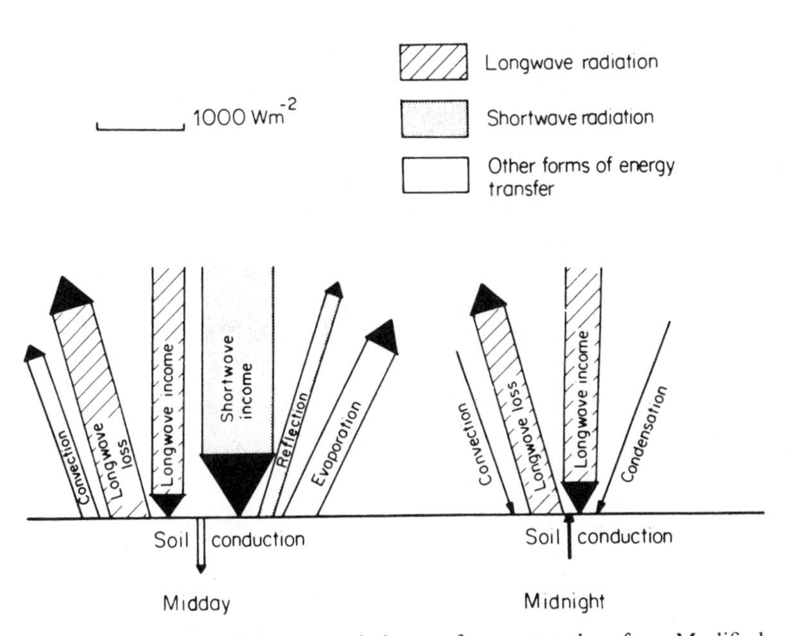

Figure 2.2 The day and night energy balance of a vegetated surface. Modified, with permission, from R. Geiger, *The Climate Near the Ground*, translated from the German 4th Edition (1961), Harvard U.P., Cambridge, Mass., 1965.

(3–100 μm) radiant loss to extraterrestrial space from the 'warm' atmosphere and terrestrial surface. Gates' (1962) speculative values used in Figure 2.2 suggest that this loss is about two-thirds of the solar income and that the remaining third is returned to extraterrestrial space by short-wave reflection and scattering. This relationship implies an exact equilibrium which is now supported by satellite measurements of solar income, reflection coefficient and long-wave loss: Rosenberg (1974) cites values for income and loss which agree to within the limits of sampling precision. The satellite-measured loss is 230 W m^{-2} and calculation through Stefan's equation (2.1) gives equivalence with a 'surface' temperature of 253 K (−20 °C) which will be some compound function of the cold upper atmosphere, its warmer lower levels and the terrestrial surface.

Figure 2.2 summarizes the situation, showing that about half of the incoming solar short-wave reaches the earth's surface and that half of this income is then dissipated in the latent heat storage of evaporating water. Thus, one quarter of the total solar income to earth becomes involved in the mass transfer processes of the water cycle. The latent heat is, of course, not lost to the system and will subsequently appear as atmospheric heating when the water vapour returns to the earth's surface as precipitation. The strong attenuation of the short-wave income by atmospheric absorption, reflection and scatter are of strong biological significance, as the maximum heat loading to which sunlit plant leaves may be exposed is barely sublethal. The selective filtering (Figure 2.1) is of equal significance as it removes most of the ultraviolet which is lethal or mutagenic to exposed cells and also removes a larger proportion of the biologically non-useful infrared than of the physiologically available visible light.

RADIATION BALANCE OF ECOSYSTEMS AND ORGANISMS

Soil, water and organism surfaces absorb downward fluxes of short-wave radiation from the sun and long-wave radiation from the atmosphere. They exchange energy amongst themselves and emit energy to the atmosphere and extraterrestrial space by long-wave radiation. These various omnidirectional radiant fluxes may be resolved into two components, downward and upward. The difference between the two is the *net radiation* of the system and is a measure of the energy available to drive such environmental processes as water evaporation:

$$R_{net} = R_{abs} + R_{rad} \tag{2.5a}$$

where R_{net} = the net radiation, R_{abs} = the absorbed irradiance (+ve), and R_{rad} = the emitted reradiation (−ve), all in $W\,m^{-2}$. For terrestrial surfaces R_{net} may be resolved into downward (+ve) and upward (−ve) fluxes.

Figure 2.3 shows the great diurnal swing of energy balance in a temperate environment: not only is the supply of solar short-wave cut-off during night but the secondary transfer processes are both slowed and reversed in direction, transpiration–evaporation being replaced by condensation of dew while the soil

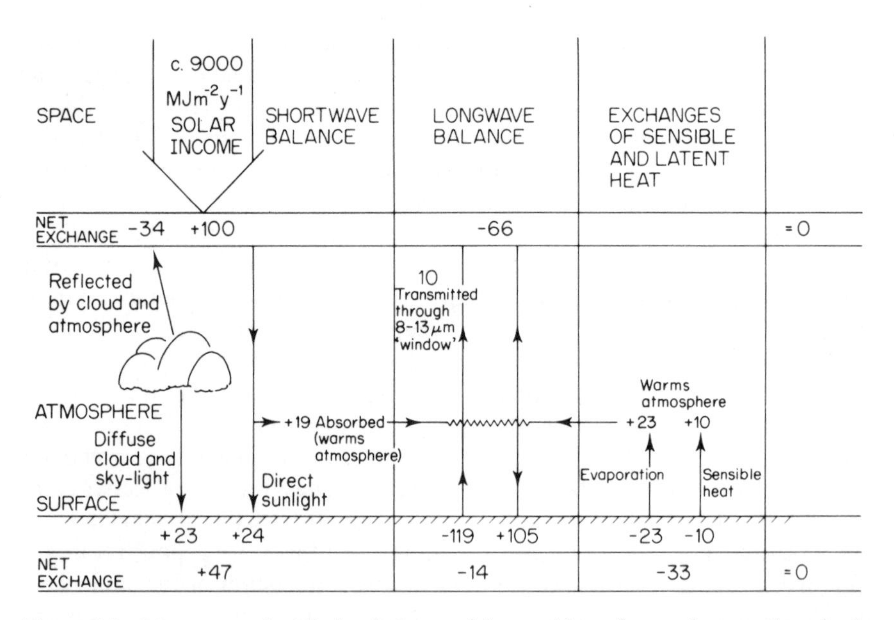

Figure 2.3 Mean annual radiation balance of the earth's surface and atmosphere in the temperate zone. ⁄\⁄\⁄\ symbolizes absorption and emission by atmospheric gases. After D. M. Gates, *Energy Exchange in the Biosphere*, Harper and Row, New York, 1962, Figure 1. Annual income from Monteith (1973). Units are percentage of total income.

is radiatively cooled instead of heated. The downward and upward components of the long-wave flux show little diurnal variation and consequently, substituting midday and midnight values in the net radiation equation:

$$R_{\text{net}} = R_+ + R_- \tag{2.5b}$$

where R_+ = the solar short-wave downward + long-wave downward, and R_- = the reflected short-wave upward + long-wave upward, in W m^{-2}.

$$\text{Midnight } R_{\text{net}} = (0 + 264) + (0 - 334) = -70\ \text{W m}^{-2}$$

$$\text{Midday } R_{\text{net}} = (598 + 246) + (-120 - 387) = 337\ \text{W m}^{-2}.$$

The components of the calculation are shown in Figure 2.4a. The data were taken from Monteith (1973, Figure 5.9) for Bergen, Norway in mid-April. Because the

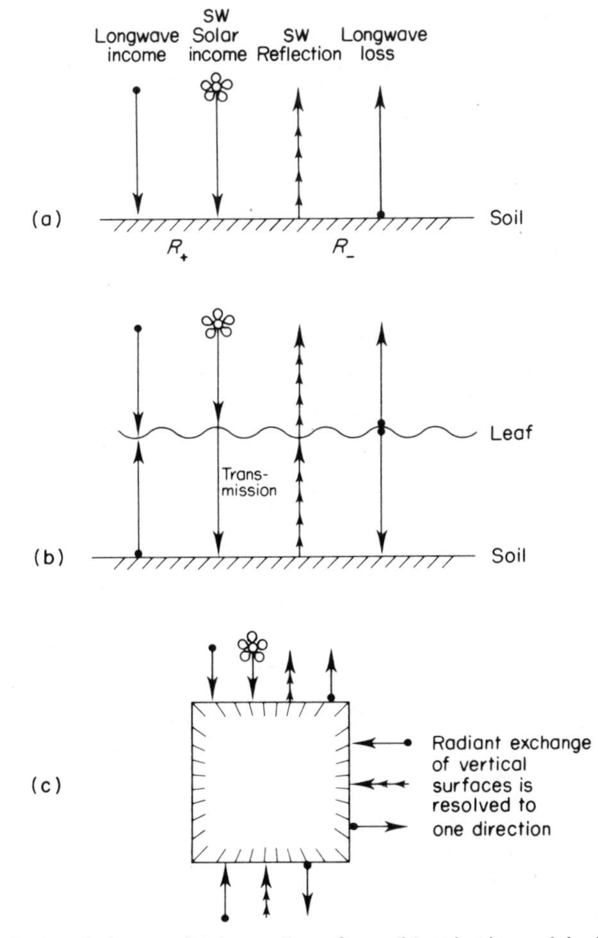

Figure 2.4 Radiation balance of (a) a soil surface; (b) a horizontal leaf; and (c) a three-dimensional object. Negligible longwave reflection is assumed.

balance between long-wave downward and long-wave upward does not have a large diurnal variation, the net radiation balance of the ecosystem surface shows a strong positive correlation with solar short-wave income, and similarly, because the long-wave components do not change greatly with season, the net radiation balance may remain negative throughout the day during winter in high latitude, for example Monteith's data for Bergen show a midday $R_{net} = -70\ W\ m^{-2}$ in January, consequent on a small solar income and high albedo of the snow covered surface.

Earth's radiation balance has been discussed without mention of seasonal and diurnal variations. Such a generalization cannot be made within the ecosystem which is affected by short-term events such as the passage of cloud shadows, by the regular cycle of day and night and by the longer cycle of the seasons. In some cases the minute to minute variations of radiation balance may exceed the magnitude of seasonal differences and most certainly may be limiting factors in the life of organisms.

The net radiation status of an individual organism or organ is also specified by equation 2.5 though calculation is complicated by the need to treat separately the upward, downward and horizontally facing surfaces of the organ (Figure 2.4b and c). In the context of plant ecology the radiation balance of the individual leaf and the whole vegetation canopy is of great interest as it will be the definitive factor governing both equilibrium temperature and also the direction and magnitude of water vapour movement.

If the net radiation balance of the leaf is positive it will reach an equilibrium temperature which is greater than that of the surrounding air unless the total dissipation of latent heat by transpiration of water plus conductive–convective transfer of heat to the surrounding air is sufficient or more than sufficient to counterbalance the radiant energy input. Leaf temperature and transpiration will thus be a compound function of radiation balance, ambient air speed, temperature and water vapour pressure in addition to the diffusive resistances of leaf and boundary-layer, of which the former is under physiological control (see pp. 165 and 184).

LEAF TEMPERATURE

The land-plant faces an insoluble dilemma of gas exchange, the need for carbon dioxide conflicting with that of water conservation. The atmospheric CO_2 concentration is very low (0·05 % w/w = 0·03 % v/v) and the plant must process a large volume of air to satisfy its photosynthetic carbon demand. Almost universally the evolutionary solution to this 'chemical engineering' problem has been the development of the thin, porous leaf with a path length of only a few tens of micrometres between the photosynthetic mesophyll cells and the external atmosphere. The gas-exchanging surfaces of the mesophyll cells are 'wet' and even when this surface water is in equilibrium with a low cell water potential (p. 156) will give a water vapour pressure in the intercellular space of more than 0·9 of saturation. At 30 °C, a common arid climate shade-temperature, this is equivalent to about 3 % w/w H_2O. The external relative vapour concentration in an arid environment may be as low as 0·25 % and the diffusion gradient for water is thus much greater than it is for CO_2, which cannot exceed 0·05 % w/w. Adequate CO_2 absorption under arid, sunlit conditions must be accompanied by massive water loss. Further discussion of these gas exchange problems and their relationship to stomatal and cuticular control appears on pp. 165.

This problem has long been appreciated and was quantified by Brown and Escombe (1905) with a very sophisticated study of leaf energy balance which showed more than 80 % of the incident radiation to be used in water evaporation. The authors described transpiration as a 'safety valve' and commented that the temperature gradient between leaf and surroundings need never be large in unsaturated air. They also realized that the xeromorphic leaf must dissipate heat by radiant emission rather than by water loss. Not all early authors were so discerning and Maximov (1929) noted that the obviously harmful consequences of excessive transpiration '... led many authors to the opinion that transpiration is nothing more than an inevitable evil, ...'. Maximov himself considered that its main function was the driving of the mineral salt transport system and secondarily the dissipation of surplus radiant heat but he also realized that more water was transported than needed for salt transfer. The water use efficiency (kg water kg^{-1} dry matter production) of crop plants ranges between 100 and 1000 (p. 130): for a *Zea mays* plant this may represent up to 4 kg day^{-1} water loss (Maximov, 1929) while a tree may lose several hundred kg day^{-1} in temperate conditions and a forest canopy between 3 and 4 kg m^{-2} day^{-1} equivalent to 3 or 4 mm of transpired water (Rutter, 1968).

The explanation for this apparently enormous loss becomes obvious if the solar radiant income to the ecosystem is considered. In the most extreme conditions, with reflected and scattered cloudlight added to direct flux, the income may exceed the solar constant (1360 W m^{-2}) and a leaf above a reflective surface such as dune-sand may receive 1·2–1·3 times this income. Without evaporative cooling a sunlit surface will become very hot and it is common experience that dark-coloured metal objects such as motor-car bodies become painfully hot in the sun, reaching temperatures which will easily denature proteins. Deely and Borden (1973) recorded soil temperatures of 50 to 60 °C, and the 94 °C in Death Valley, California must surely be a world record (Mooney *et al.* 1975).

Energy fluxes of this magnitude will evaporate 1–2 kg water m^{-2} h^{-1} and it is hardly surprising that a sunlit canopy should lose a great deal of water or that some water loss should be a necessary part of leaf temperature control.

The development of satisfactory techniques of leaf temperature measurement during the past 10 to 15 years has permitted a reassessment of the interrelationship between energy input, leaf water loss, leaf temperature and plant physiology. Provided that sufficient information is available concerning the physical characteristics of a leaf and the influence of its physiological status on such characteristics, then its energy balance may be entirely specified by the following equation (C. T. Gates, 1968):

$$R_{abs} = R_{rad} + C + L_h E_v + M \quad (2.6)$$

where R_{abs} = the radient energy absorbed by the leaf, R_{rad} = the reradiation from the leaf, C = the sensible heat exchange, M = the metabolic energy balance (photosynthesis versus respiration), all in $W m^{-2}$, and L_h = the latent heat of evaporation of water ($J kg^{-1} = 2{\cdot}5 \times 10^6$), and E_v = the transpiration rate ($kg m^{-2} s^{-1}$).

R_{rad} is related to leaf temperature through Stefan's Law (equation 2.2) while C and E_v are rather complicated functions of air-velocity, leaf–air temperature differential and water vapour pressure gradients. For leaves, C mainly comprises turbulent convective exchange: the heat storage capacity of leaves is usually so low that it may be neglected in the energy budget, as also may M, which Brown and Escombe (1905) correctly concluded for photosynthesis was less than 1 % of the incident radiant energy.

The dimensions and shape of leaves have a strong influence on the aerodynamics of convective heat and mass transfer. Small leaves have a lesser resistance to convective heat exchange than large leaves and deeply lobed leaves gain or lose heat more easily than entire leaves of equivalent size. Some ecologists have suggested that small and/or dissected leaves are an adaptive advantage in arid or high-light habitats, permitting increased convective and evaporative cooling. Lewis (1972) for example presents evidence that increased leaf dissection both genotypic and phenotypic occurs in populations of *Geranium sanguineum* (bloody cranesbill) in less shaded habitats. The preponderance of small sclerophyllous leaves in mediterranean vegetation has been similarly interpreted.

Gates and Papian (1971) used an expanded form of equation 2.5 to compute leaf temperature and transpiration rate for a wide range of environmental conditions. Their tabulations of energy budgets reveal a number of ecologically relevant and occasionally surprising relationships.

Damage may accrue when air temperature is high and windspeed low if transpiration is limited by either high humidity or stomatal closure. Under these conditions the energy budget predicts leaf temperatures as much as 20 °C above air temperature (Figure 2.5*C*) and if the air temperature is extreme it would be possible for leaves to reach *c*. 60 °C. Salisbury and Spomer (1964) measured leaf over-temperatures of 22 °C in an alpine environment while Turner and Tranquillini (1961) cited in Tranquillini (1964) showed that *Pinus cembra* (arolla pine) needle temperature could fluctuate by as much as 20 °C within a few seconds and, in the spring, could cross and recross the freezing point range (−4 to −8 °C) several times during the day.

The effect of leaf size is such that the over-temperature of 23 °C, shown in Figure 2.5*C* for 1000 $W m^{-2}$ irradiance, would be reduced to 13 °C if the leaf size was 1 × 1 cm rather than 5 × 5 cm. This is because the convective and evaporative cooling is increased by the consequent reduction in boundary layer resistance, an effect also caused by increasing windspeed (Figure 2.5*A* and *B*). The

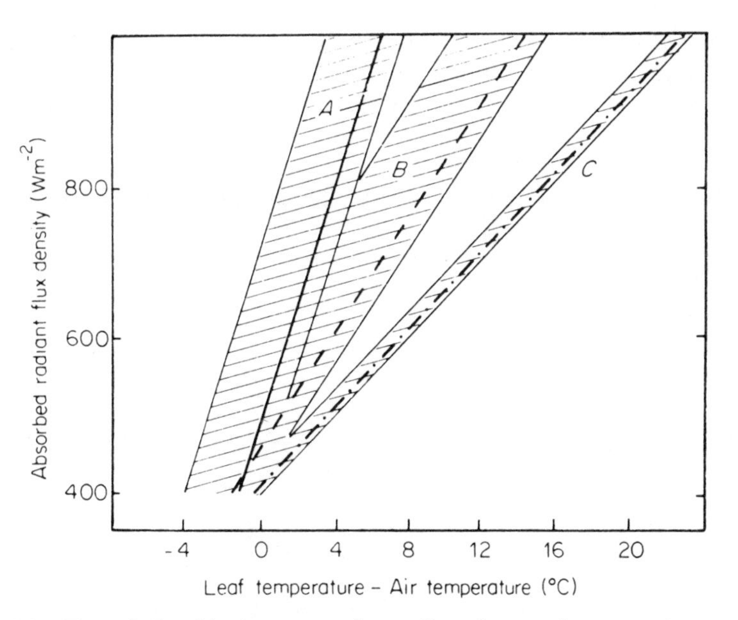

Figure 2.5 The relationship between solar radiant input, air temperature and leaf temperature. The plotted lines are for relative humidity 75 % while the hatched bands represent the range from 0 to 100 %. All values are for an air temperature of 20 °C and a 5 × 5 cm (length × breadth) leaf. *A*: airspeed 100 cm s^{-1}; leaf internal diffusive resistance 1.0 s cm^{-1}. *B*: airspeed 10 cm s^{-1}; resistance 1.0 s cm^{-1}. *C*: airspeed 10 cm s^{-1}; resistance 10 s cm^{-1}. Plotted from Gates and Papian (1971).

leaf cooling caused by wind and by evaporation may be harmful if the air temperature is near freezing, an under-temperature of 4 to 5 °C being sufficient to frost-damage sensitive species. It is partly for this reason that snow-cover may give protection from cold and desiccation damage.

If the windspeed is low, radiant heating in bright sunlight may be sufficient to permit photosynthesis even though air-temperature may be below the photosynthetic minimum. Some arctic and alpine flowers rely on this effect, their parabolic structure and heliotropism focusing energy into the centre of the flower and giving over-temperatures which reach +6 °C and, in the case of tubular and other enclosed floral structures, a greenhouse effect results in a temperature increase of up to 25 °C (Lewis and Callaghan, 1975). Such temperatures most certainly encourage insect pollinators as well as improving the plant-physiological status.

Solar heating may occasionally cause severe damage if transpiration is limited: both fruits and woody branches may suffer sunscald injury (Treshow, 1970) and unusually exposed shade leaves may be similarly injured. This has been noted for *Arum maculatum* (cuckoo pint) in a South Wales woodland after wind damage to the tree canopy (Etherington, 1976 unpublished). Rackham (1975) suggests that similar damage to *Mercurialis perennis* (dog's mercury) was caused by leaf overheating in sunflecks.

Acute injury of this kind is not common: leaf overheating is more often reflected in reduced photosynthesis, and Gates (1965) predicted from calculated leaf temperatures and the temperature dependence relationship of photosynthesis, that on days of high insolation, the photosynthesis of *Zea mays* (sweetcorn) would decline during the mid-day period and recover in the afternoon though never sufficiently to equal the mid-morning rate. Figure 2.6

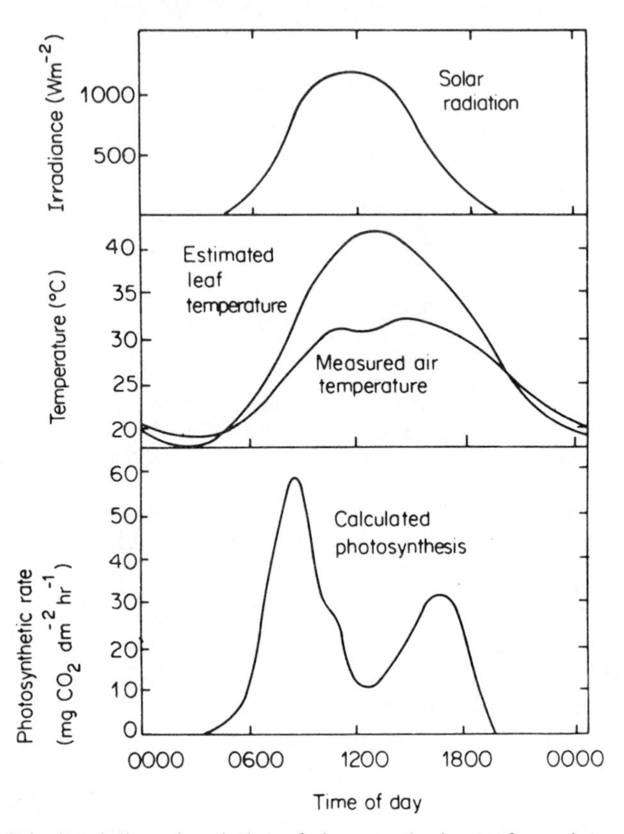

Figure 2.6 Calculated diurnal variation of photosynthetic rate for maize corresponding to diurnal course of solar radiation and air temperature. Reproduced with permission from D. M. Gates, Energy, plants and ecology, *Ecology*, **46**, 1–10, Figure 7 (1965).

shows his calculated curves which were based on field data of Thut and Loomis (1944) and temperature dependence curves from Waggoner *et al.* (1963). Field measurements in full sunlight support Gates' suggestion and Figure 2.7 shows the evidence of a mid-day slump in photosynthesis, mid-day stomatal closure and an associated slight recovery of declining leaf water potential.

The interpretation of these data is difficult because the photosynthetic slump could be caused either by direct temperature-induced damage to a photosystem, or by stomatal closure which would limit CO_2 uptake. Mid-day stomatal closure

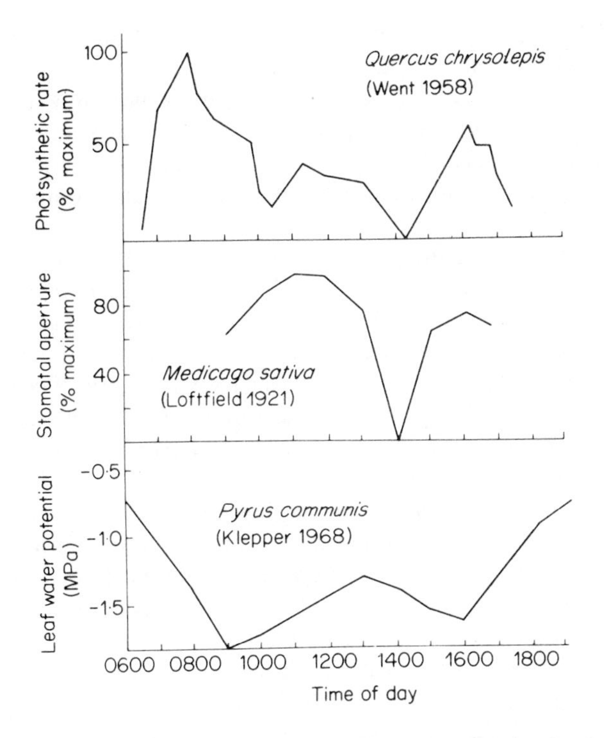

Figure 2.7 The diurnal course of photosynthesis, stomatal aperture and leaf water potential of three plant species growing in natural or semi-natural conditions.

has been known-of for many years and might be caused by water deficit or by temperature induced increase in intercellular space CO_2 concentration which would trigger closure (Heath and Orchard, 1957). Further discussion appears on p. 166. Irrespective of mechanism, the optimal leaf temperature for photosynthesis is in the range 15–30 °C for C-3 plants and 35–45 °C for C-4 plants (see p. 130). In either case predictable and measured leaf temperatures may quite often exceed these optima and reduce total daily carbon fixation.

The role of transpiration in controlling leaf temperature may be appreciated from the difference between Figure 2.5*B* and *C* at 1000 W m^{-2} irradiance. The increase of leaf diffusive resistance from 1 to 10 s cm^{-1} raises the leaf temperature by more than 8 °C to well over 20 °C above air temperature if the windspeed is only 10 cm s^{-1}. Such values have been shown experimentally, for example Lange (1959) found that *Citrullus colocynthis* (bitter apple) leaves remained 10 to 15 °C below the air temperature of 40 °C but cutting the petiole to sever the water supply caused temperature to increase by 22 °C. Similarly, *Phragmites communis* (reedgrass), growing in one of the world's hottest environments, Death Valley, California, with an air temperature of 43–46 °C, but copious soil water supply, maintained a leaf temperature 8 °C below ambient (Pearcy *et al.* 1974).

Gates (1980) describes three distinct strategies of leaf temperature control only one of which relies on water evaporation and consequently would fail in dry soil. A second strategy is vertical leaf posture which not only reduces radiant interception but also improves convective cooling: a few plants such as *Lactuca serriola* (prickly lettuce) are 'compass plants' and alter their leaf posture during the day, intercepting most radiation in the morning and evening and avoiding insolation during the mid-day period.

A third strategy is reliance on very small leaves which are more efficiently cooled by convection, for example desert shrubs such as *Prosopis* spp. (mesquites) and *Larrea divaricata* (creosote bush). Gates (1980) notes that such leaves appear to be particularly water-use-efficient in high radiation environments, contrasting with the broad thin blades of shade leaves which are more efficient in low light conditions. The leaf-rolling habit of the Gramineae has been interpreted as limiting radiant interception rather than functioning only in a water conservative role (Ripley and Redmond, 1975).

It is easy to be oversimplistic concerning control of transpiration and leaf temperature as the internal resistance of the leaf is a physiological variable which may change as rapidly as the environmental factors. Landsberg and Butler (1980), for example, describe the control of stomatal opening in *Malus pumilla* by the gradient of water vapour pressure from leaf to air; in this case transpiration remains relatively constant as leaf net radiation balance increases.

THE LEAF CANOPY AND ITS MICROCLIMATE

Discussion of leaf temperature has so far been confined to the individual leaf. In the context of plant ecology it is the behaviour of the whole canopy which is important and, as the subsequent discussion will show, a multilayered leaf system may impose conditions on the individual leaf ranging from intense sunlight, low humidity and high windspeed at the top of the canopy to very low light intensity and still, moist air at the base, not to mention the differences in light quality (spectral distribution) and the physiological differences between leaves which have developed in different light and humidity regimes.

A bare soil or rock surface is a site of abrupt vertical change in meteorological conditions and is also subject to very rapid changes with time. Surface temperatures may be extreme in bright sunlight, ranging from 50 °C, or so, in temperate conditions (Deely and Borden, 1973) to 94 °C in the subtropics (Mooney *et al.* 1975), but the air, a few centimetres above, may be more than 40 °C cooler than this, because of wind-forced convection (see p. 25). A vegetation canopy 'spreads' vertical profiles by absorbing solar energy, not at a single surface, but at a succession of surfaces which may span tens of metres in the deepest canopies and also acts as a 'porous' layer through the whole depth of which the momentum of moving air may be absorbed. Again this spreads the vertical windspeed profile and reduces the rate at which materials are transferred to or from the ecosystem surface. The physiological activity of the component plants also plays an important part in the transfer of materials within the system and, for this reason, Etherington (1978) has described the terrestrial plant as '... a biological bridge between two very contrasting parts of the environment.'

Radiant energy

In general, short-wave solar radiation penetrating canopy is depleted exponentially by successive leaf layers, approximately obeying the Bouguer–Lambert Law (Monsi and Saeki, 1953):

$$R_i = R_o e^{-K_{ext} L} \tag{2.7}$$

where R_i and R_o are, respectively, the irradiances within and above the canopy ($W m^{-2}$); L = leaf area index (leaf area per unit ground area; $m^2 m^{-2}$); K_{ext} = extinction coefficient (dimensionless); e = base of natural logarithms (2·718).

The relationship of radiant absorption to canopy characteristics produces a theoretical profile in which each layer (unit of leaf area index) absorbs a constant proportion of the incident radiation (Figure 2.8b) and conforms to the exponential depletion curve of equation 2.7. This is often plotted semi-logarithmically to give the straight line relationship of Figure 2.8c. It would be simplistic to accept that all canopies behave in the same way, but Newton and Blackman (1970) investigated attenuation of radiation in different canopy structures ranging from the vertically disposed sword-like leaves of *Gladiolus* sp. to the horizontal, and uniform with depth, leaves of *Agrostemma githago* (corn cockle). They concluded that interception of light, with respect to leaf area is close to exponential irrespective of leaf orientation. The extinction coefficient (K_{ext}) was particularly low for *Gladiolus*, supporting previously suggested values of K_{ext} = 0·6–0·9 for broad-leaved species and K_{ext} = 0·3–0·5 for grasses (Monsi and Saeki 1953).

No attention has been given to the effect of wind on the overall optical characteristics of canopies but, as it effects both vertical and horizontal disposition of leaves, it is likely to homogenize the time-averaged system causing it to behave as a uniform suspension of chlorophyllous elements showing exponential attenuation with depth. The variable interaction of canopy architecture with solar angle and cloud diffusion of incoming radiation, further adds to this time-averaged homogenization.

Species with high K_{ext} values and large leaf area indices (L) may absorb most of the incoming light, for example Stern and Donald (1962) showed up to 33% interception by 3·5 cm of a *Trifolium subterraneum* (subterranean clover) sward compared with only 16% by the same depth of *Lolium rigidum* (a grass). At the base of grass–clover swards (L = 6·7–14·7) the light intensity was reduced to substantially less than 5% of full daylight (plotted as zero in their graphs). Both Stern and Donald and Newton and Blackman measured 'light' intensity with conventional photocells (either selenium or silicon). This raises problems of changing wavelength distribution (quality) which are further discussed below.

The 'average' values of radiant flux density at different depths in the canopy may, with the photosynthetic capacities of the different leaf layers be used to predict canopy potential photosynthesis (p. 346). The distribution of short-wave radiation in the canopy is also instrumental in creating the daytime profiles of temperature, water vapour and carbon dioxide. The presence of 'holes' in the real canopy results in the projection of the full solar beam to various depths, producing *sunflecks* which move systematically in response to solar angle. They also oscillate and flicker at high frequency as the canopy is tossed by wind. Sunflecks may significantly increase whole canopy photosynthesis (Kriedmann *et al.* 1973), or in shade habitats, decrease the photosynthesis of individual leaves.

The penetration of the canopy by short-wave radiation raises several problems of measurement (p. 36) and physiological interpretation, related firstly to the

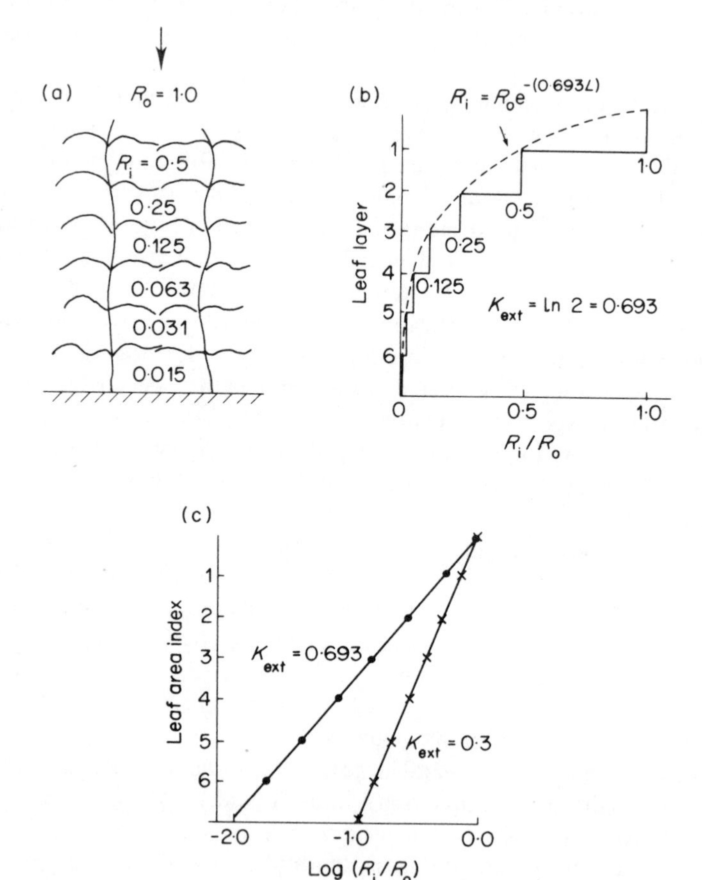

Figure 2.8 Attenuation of incident radiation by passage through a leaf canopy: (a) a model canopy attenuating radiation by a factor of 0.5 at each complete leaf layer; (b) the relationship of irradiance to depth of the model canopy and its interpretation by the Monsi and Saeki (1953) equation; (c) log-reciprocal plots of irradiance with canopy depth for the model canopy ($K_{ext} = 0.693$) and for a more upright leaf architecture ($K_{ext} = 0.3$).

spatial and temporal heterogeneity of the system and secondly to the change of spectral distribution with depth. Transmission through leaves selectively removes the photosynthetically active wavelengths and the visible spectrum becomes relatively green-enriched. The near infrared (0·7–3·0 μm) is less strongly attenuated than the visible, and the shade spectrum is consequently enriched with the morphogenetically active far-red. Ross (1975) gives some ratios for energy content of sun- and shade-light at different wavelengths which illustrates these two changes:

	green (0·54 μm) / red (0·66 μm)	red (0·66 μm) / far-red (0·73 μm)
Sun	1·2	1·2
Shade (Leaf area index 5)	2·1	0·1

Bjorkman and Ludlow (1972) measured the irradiance on a rainforest floor as 2·5 % of the above-canopy clear sky value but the PAR quantum flux was only 0·41 % of the exterior value, indicating the large change in spectral quality which makes the sub-canopy light less photosynthetically useful and differentially active in phytochrome controlled processes such as germination, development and flowering (see p. 223).

The underwater radiation environment has often been compared with terrestrial shade; attenuation of radiation penetrating non-turbid water closely follows equation 2.6, and submerged canopies are analogous to their terrestrial counterparts. There is, however, one invariable difference. Far-red is strongly absorbed by water and the red/far-red ratio is greater than that of full sunlight. Spence (1975) gives r/fr = 3·6 to 4·7 at 1 m depth and 500 at 4 m. Little is known of phytochrome mechanisms in aquatic plants: most available data concerns floating or shallow water species (Sculthorpe, 1967): red algae have been shown to possess phytochromes (Dring, 1967) and a recent study suggests that heterophylly in *Hippuris vulgaris* (mares-tail) is controlled by r/fr ratio (Bodkin *et al.* 1980).

Wind

The movement of air over a surface is retarded by friction which results in a constant downward transfer of energy (momentum). This creates a turbulent boundary layer of which the depth is a function of surface irregularity. Matter and thermal energy are carried toward or away from the surface in rotating eddies and consequently reduce the gradients in vertical profiles of temperature, water vapour and carbon dioxide concentration.

Air flow over a uniformly rough, level surface and without complication by convective eddying, results in a simple profile of windspeed variation which is generated by frictional eddies (Figure 2.9a). The windspeed (u) profile with height (z) is logarithmic through most of the depth of the turbulent boundary layer (Figure 2.9b and c). At some considerable height above the surface, air flow is laminar and speed no longer changes strongly with height: laminar flow is also maintained very close to the surface but here, change of speed with height causes strong shearing (Figure 2.9a).

The aerodynamic properties of a relatively smooth surface are described by the *roughness length* z_0, a measure of the frictional ability of the surface to absorb momentum from air moving over it. It is the height at which the hypothetical

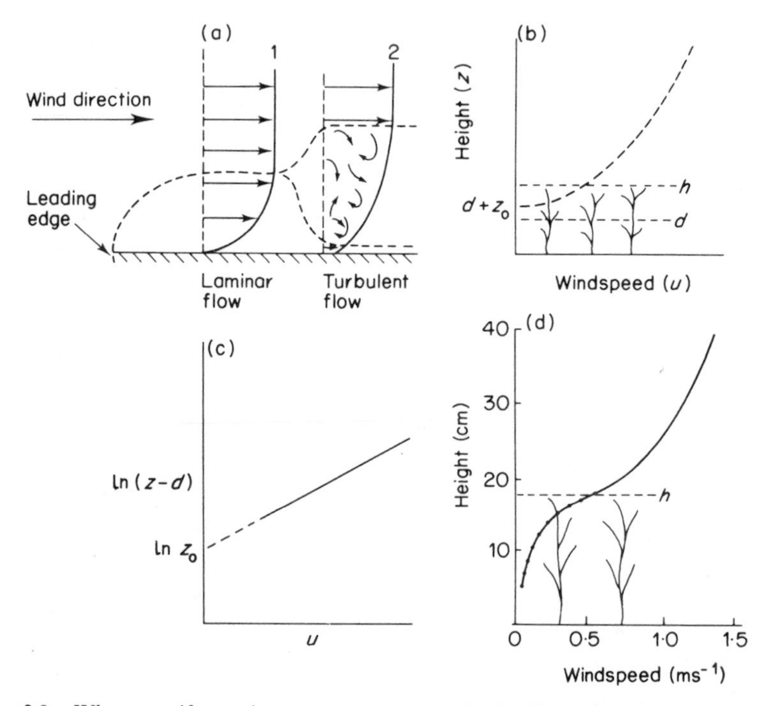

Figure 2.9 When a uniform air-stream encounters the leading edge of a solid surface its initially laminar-flow profile, (a) curve 1, is disrupted into turbulent flow, (a) curve 2, though a thin laminar-flow layer is maintained close to the surface. Broken lines indicate the top of the laminar flow layer and the top and bottom of the turbulent flow profiles. Arrow lengths are proportional to windspeed. (b) The windspeed profile described by equation 2.8, showing the displacement height (d) and roughness length (z_0) with respect to vegetation height (h). (c) The logarithmic plot of the windspeed profile shown in (b). The abcissa is plotted as difference from the displacement height ($z - d$) so that extrapolation to zero windspeed gives $\ln_e z_0$. (d) The windspeed profile above and within a canopy calculated by equation 2.8 and equation 2.9 ········ using empirical values abstracted from Ripley and Redmond (1975), Figure 18. Within the canopy, windspeed does not decline so abruptly as extrapolation of the exterior logarithmic profile would suggest.

extrapolation of the logarithmic wind speed profile becomes zero (Figure 2.9c). In the presence of a vegetation canopy with fairly close-packed roughness elements the effective surface is raised above the soil. The height at which the arithmetic windspeed profile extrapolates to zero is $(d + z_o)$ where d is the *displacement height* and represents the effective plane at which momentum is being absorbed (Figure 2.9c). For rigid surfaces z_0 and d are constant but, in vegetation stands, both values may be changed by distortion of the canopy by the wind. A reduction of d and increase of z_o have been recorded for forest canopies (Rauner, 1975) and it has also been suggested that streamlining of leaves in the wind reduces z_o (Grace, 1977).

The wind profile above the canopy may be described in the form:

$$u(z) = \frac{u^*}{k} \log_e \left(\frac{z - d}{z_o} \right) \tag{2.8}$$

where $u(z)$ = the mean windspeed at height z, u^* = the friction velocity, (both in $\mathrm{m\,s^{-1}}$), k = von Karman's constant, and z, d and z_o are defined as above (m).

The friction velocity is related to the tangential rate of rotation of frictional eddies and von Karman's constant has been experimentally established for air as *c.* 0·4, more or less irrespective of surface type. The value of u^*/k may, in any case, be found from the slope of the windspeed versus log height profile. The equation contains three constants, k, d and z_o, which must be obtained from experiment because there is no theory from which they can be calculated. Businger (1975) notes that this is 'not quite satisfactory' and gives some discussion of the relationship between d, z_o and the height, cross-section and porosity of the roughness elements.

The interpretation of airspeed profiles is complicated by sensible heat flux effects, momentum and matter transfer in the boundary layer being modified by the interaction of convective buoyancy with frictional turbulence. If the surface is warmer than the air, convection causes vertical elongation of the frictional eddies thus increasing the vertical transport of heat and matter. By contrast, atmospheric temperature inversion will suppress eddying and reduce vertical transport rates.

Equation 2.8 describes the logarithmic nature of airspeed variation in the turbulent layer above the canopy. For mathematical convenience these profiles are extrapolated to their hypothetical zero speeds but, in reality there is a relatively weak but complex air-flow within the canopy. In the upper, more leafy region, windspeed declines exponentially with depth but at lower levels, if leaf development is sparse, does not decrease so much as the extrapolation of this curve would suggest (Figure 2.9d) and may actually increase if the undercanopy is very open as in a forest.

The canopy windspeed profile may be described using the equation:

$$\frac{u(z)}{u_h} = \left[1 + \alpha \left(1 - \frac{z}{h} \right) \right]^{-2} \tag{2.9}$$

where $u(z)/u_h$ = ratio of windspeed at height z to that at the height of the canopy (h), α is an empirical coefficient lying between 1·0 and 5·0 and is a function of mean leaf area index, the mean drag coefficient of leaves at height z and the eddy diffusivity of momentum in air (see below). The drag coefficient is a measure of the ability of a body to absorb momentum from an airstream. (A streamlined body has a low coefficient.)

This oversimplified account of air-flow above and within canopies illustrates the complexity of the system of which much more extensive accounts may be found in Monteith (1973 and 1975); Businger (1975) and Grace (1977).

Temperature

Air temperature within the canopy is governed by radiation balance, latent heat dissipation and eddy-diffusive transfer of sensible heat. There is consequently a strong diurnal migration of the temperature profiles (Figure 2.10)

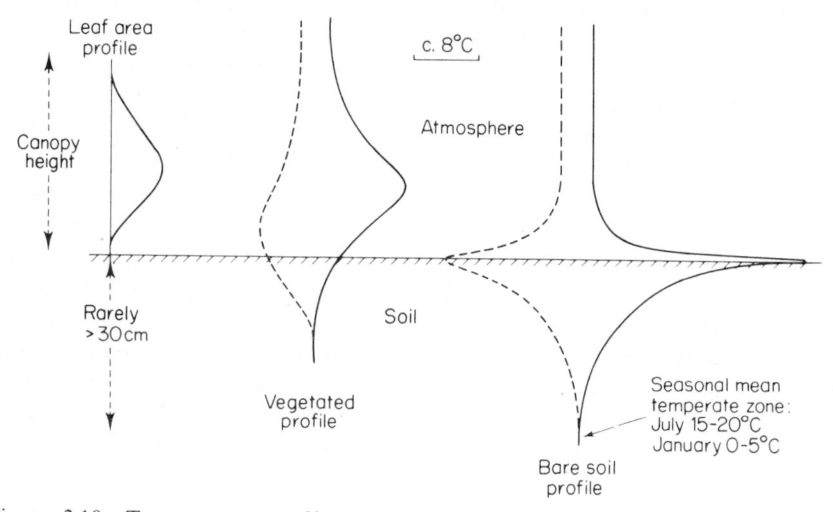

Figure 2.10 Temperature profiles through vegetated and bare-soil surfaces; night ----- and day ——. Note that the below ground profile must be considered at a constant scale, irrespective of vegetation type, whereas that of the above ground profile must be varied according to the depth of the vegetation canopy.

which, over bare soil show abrupt vertical gradients, large day–night differentials and considerable conduction of heat to or from the surface few cm of soil. A vegetation canopy effectively insulates the soil from these effects and vertically elongates the above-ground profiles. With a clear sky the bare soil surface is strongly heated during the day and radiatively cooled at night: if the surface cools sufficiently during the night the adjacent air may exceed its water saturation vapour pressure and dew may form. With low air temperature the same conditions will cause a radiation frost. These effects are modified by cloud which will limit daytime heating and reduce night-time radiative loss. Rising windspeed truncates the profiles by increasing forced convection. Soil temperature profiles show a relatively shallow day–night differential, as heat transfer is by conduction in a medium of relatively low heat conductivity. Sandy soils are more conductive than clay or peat, and consequently the diurnal migration of the profiles is more marked in dry sand than any other soil. Wet soils behave rather differently as large amounts of heat may be transferred by evaporation and condensation of water in the pore space (Monteith, 1973). Below the depth of diurnal temperature change the soil is in a longer-term thermal equilibrium with the mean seasonal air-temperature, while at great depth there is no seasonal shift and the temperature is some function of the annual mean. The low thermal conductivity of soil, particularly when peat-covered, is the reason why boreal permafrost soils thaw and erode if the protective vegetation mat and peat cover is destroyed.

Water vapour

The capacity of air to carry water vapour is very strongly temperature dependent, for example saturated air at 1 °C contains 5·2 g m^{-3} H_2O (saturation vapour pressure 0·65 kPa) but at 30 °C this has increased to 30·4 g m^{-3} H_2O (SVP 4·2 kPa). Even small temperature changes have a great effect: increasing the temperature from 20 to 21 °C increases the saturated water content from 17·3 to 18·3 g m^{-3} (data from Slavik, 1974). It should be born in mind that water potential is logarithmically related to water vapour pressure, thus relatively small diffusion gradients for water will be in equilibrium with very large negative potentials. It is consequently predictable that the diurnal changes of temperature profile will have an enormous effect on that of water vapour pressure with the added complication, in a vegetation canopy, that the leaf surfaces present a source of water vapour which is strongly physiologically controlled; thus canopy profiles are consequently a complex function of physical and physiological conditions (Figure 2.11).

The day and night profiles of saturation vapour pressure (SVP) almost parallel the temperature profiles: the actual vapour pressure (AVP) profile at night is very little different from the SVP profile as the night-time cooling of the atmosphere may often reduce the water carrying capacity to a level at which dew formation commences (SVP). This diurnal temperature change acts as a dehumidifier of the daytime atmosphere. During the day the warm atmosphere above the canopy is usually unsaturated and the difference between the unsaturated and saturated vapour pressure values represents the saturation deficit. This saturation deficit curve is not only a measure of the evaporating power of the air but also reflects the amount of water which is being transpired at different levels.

Near the top of the canopy the deficit is quite large as a consequence of convectional heating from the sun-warmed foliage. Deeper in the canopy the deficit is less, both because there is less solar heating and because the relatively still air in the canopy is being humidified by transpiration and by evaporation from the soil surface. At some point near, or just below the soil surface the actual vapour and saturation vapour pressure profiles converge.

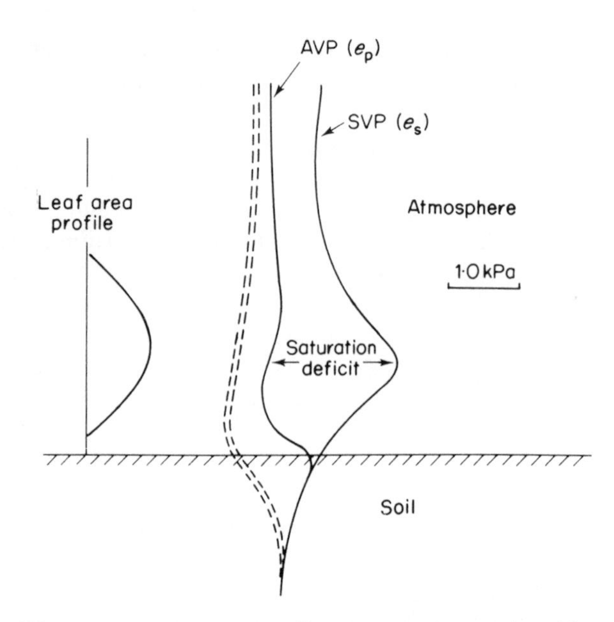

Figure 2.11 Water vapour pressure profiles through vegetated and bare soil surfaces: night ----- and day ———. At night the air may reach saturation ($e_p = e_s$) and the profiles become more or less parallel to those of air temperature. During the day the saturation vapour pressure profile remains coincident with that of temperature but the actual vapour pressure profile is a compound function of the fall in relative humidity caused by radiant heating of the air complicated by the addition of varying amounts of water vapour from transpiration. Note that a decrease of 1 kPa in vapour pressure from saturation at 20 °C represents a very large reduction of water potential approximating 60 MPa.

Carbon dioxide

Above a bare rock surface there will be no systematic gradients in the CO_2 profile but above a vegetated soil, CO_2 uptake during photosynthesis and CO_2 evolution from plants, animals and microorganism respiration establishes diurnally varying profiles of concentration. These are, however, very difficult to characterize as they represent only small changes in the very low background concentration of 320–330 $\mu l\,l^{-1}$ (*c.* 0·6 g m^{-3}).

The night-time profile (Figure 2.12a) shows an increase in concentration from above the canopy to soil-level which is a consequence of enrichment by respiratory CO_2 evolution. During the day the upper part of this profile is reversed by photosynthetic CO_2 assimilation and an equilibrium with eddy-diffusion from above is set-up in which the minimum CO_2 concentration is at some depth in the canopy. Below this inflection of the profile, the concentration rises steadily toward the soil surface reflecting the respiratory CO_2 output of the poorly illuminated lower canopy and of the soil biota. The height at which the canopy concentration equals that of the external atmosphere is related to the

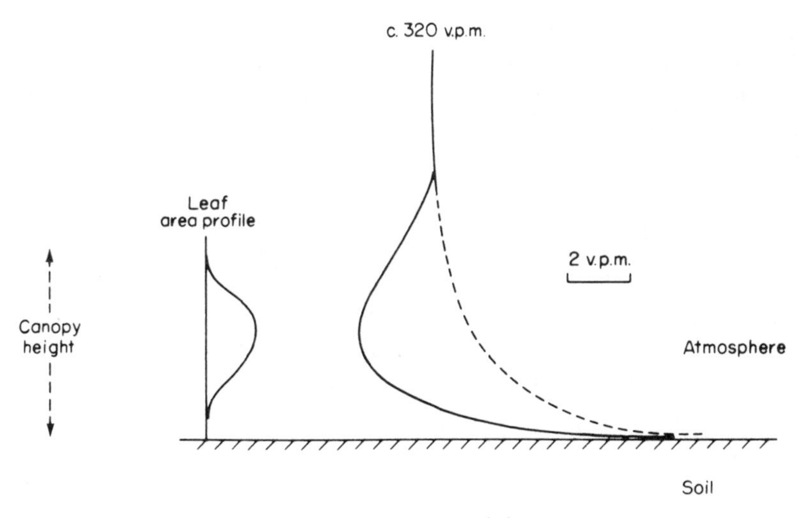

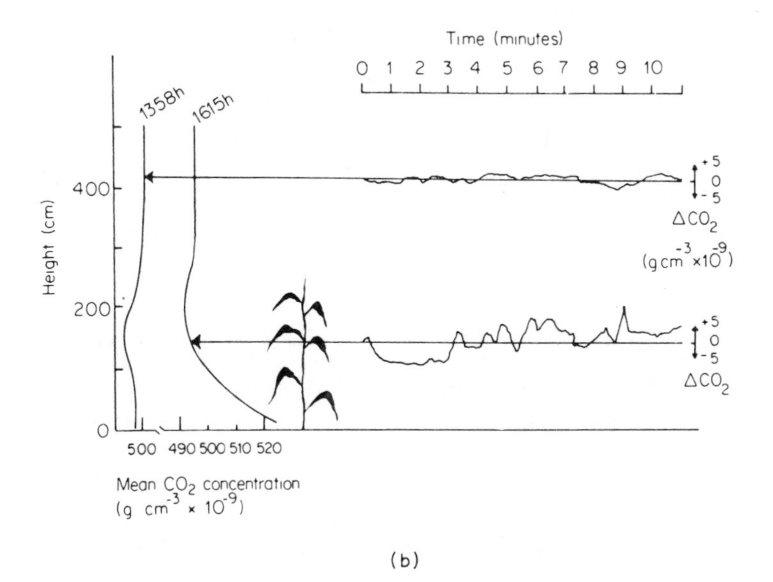

Figure 2.12 (a) Carbon dioxide concentration profiles through a vegetated surface: night ----- and day ———. The distribution of leaf area in the canopy is indicated by the diagram on the right. (b) Mean carbon dioxide concentration profiles in a *Zea mays* canopy at the indicated times. The mean values were measured over a 10 min sampling period. The variation during sampling (ΔCO_2) is shown for two indicated heights, plotted to the same CO_2 concentration scale as the profiles. These results were collected in a uniform crop canopy with clear sky insolation; steady-state conditions! Data from E. R. Lemon, *Harvesting the Sun*, Academic Press, New York, 1967, Figure 21, p. 288.

depth at which canopy light intensity has fallen to the compensation point and CO_2 evolution and uptake are equal. The concentration changes caused by photosynthesis are rarely more than 5 $\mu l\,l^{-1}$ but respiratory enrichment near the soil may be much greater since the soil atmosphere, in which diffusive exchange is limited, may contain 10,000 $\mu l\,l^{-1}$ or more.

Because the total concentration is so low, and utilization by photosynthesis in the canopy so inhomogenous, profiles must be based upon mean values from fairly long sampling periods. Figure 2.12b shows that the variations at any one sampling point may be as large as the differences measured in the mean profiles. This imposes a serious limitation on the technique as a method of measuring gross community photosynthesis and for this reason it can only be used under ideal, uniform conditions (p. 32).

Conclusion

The micrometeorological characteristics of a canopy are complex and strongly inter-correlated, having strongly developed vertical properties and a less systematic horizontal variation according to the distribution of individual plants, branches and leaves. The regular diurnal and seasonal cycles bear a superimposed 'noise' of short-term effects such as the passage of cloud shadows, turbulence induced flickering of sunflecks with associated variation of air and leaf temperature, humidity and CO_2 mixing. This four-dimensional variability is very different from the constant, often low light intensity conditions of the environmental chamber but it must be a significant factor in the physiology and life-cycle of the plant in the field. The measurement and simulation of these characteristics and the assessment of their affects is a relatively new, difficult but potentially rewarding field of ecological endeavour (Tibbitts and Kozlowski, 1979).

TRANSFER OF MASS, MOMENTUM AND HEAT

The information provided by wind profiles may be used in the *aerodynamic method* of calculating vertical fluxes in the vegetation canopy if it is assumed that water vapour, carbon dioxide and heat move vertically in a manner analgous to the vertical transfer of momentum:

$$M_z = K_M \rho \left(\frac{du}{dz}\right) \tag{2.10}$$

where M_z = the vertical transfer of momentum ($N\,m^{-2}$), K_M = the eddy diffusivity of momentum (turbulent transfer coefficient) ($m^2\,s^{-1}$), and ρ = the density of air at ambient temperature ($kg\,m^{-3}$).

The eddy diffusivity of momentum is a measure of the ability of turbulent air in the boundary layer to transfer momentum to the surface. It is a function of eddy size and rotational velocity and, hence, is also proportional to the transfer coefficients for other air properties. If the air temperature and concentrations of water vapour and carbon dioxide are measured at different heights in the canopy, their vertical turbulent transfers may be calculated by means of the similarity hypothesis:

$$K_M = K_H = K_V = K_C$$

where K, the eddy diffusivity, is subscripted respectively, H, V and C for heat, water vapour and CO_2. It is necessary also to have information on the vertical distribution of leaf area so that the shape of the windspeed profile can be determined from Equation 2.9.

With fully forced convection the eddy structure is circular, the boundary layer is said to be in a neutral stability state and the wind profile above the surface is exactly logarithmic. If the surface is heated and strong convection occurs, the buoyancy effect vertically distorts the eddies enhancing vertical transfers and creating an unstable condition. Similarly a temperature inversion may damp the vertical component of the eddies, creating stable conditions in which vertical transfer is suppressed. Calculations of canopy fluxes must, if necessary, be corrected for these departures from neutral stability transfer conditions. Detailed discussion is given by Thom (1975).

An alternative method of whole canopy flux measurement is the *Bowen ratio* or energy balance technique. Total heat flux H($\mathrm{W\,m^{-2}}$) through the canopy is specified entirely by:

$$H = C + L_h E_v \tag{2.11}$$

where C = the total sensible heat flux density in air ($\mathrm{W\,m^{-2}}$), L_h is the latent heat of vaporization of water ($2{\cdot}45\ \mathrm{MJ\,kg^{-1}}$), and E_v is the vertical flux density of water vapour per unit area ($\mathrm{kg\,m^{-2}\,s^{-1}}$).

During the daytime the total heat flux may be approximately equated to the net radiation (R_{net}) which may be measured with a net radiometer above the canopy (p. 34). Conduction to or from the soil may need to be included in H and, in critical work, advection of energy by a horizontal air-stream and storage of heat in biological and physical systems must be considered. The vertical heat fluxes may be calculated as follows:

$$L_h E_v = \frac{(R_{net} - C_{cond})}{1 + B_r} \tag{2.12}$$

where R_{net} = the net radiation flux density ($\mathrm{W\,m^{-2}}$), C_{cond} = the conduction of sensible heat to or from the soil ($\mathrm{W\,m^{-2}}$), and B_r = the Bowen ratio. The Bowen ratio is defined as:

$$B_r = \frac{C_{conv}}{L_h E_V} \simeq \gamma \frac{dT_{emp}}{dq} \tag{2.13}$$

where C_{conv} = the vertical flux density of sensible heat in air ($\mathrm{W\,m^{-2}}$), γ = the psychrometer constant ($\simeq 4{\cdot}2\ \mathrm{kg\,H_2O\,kg\ air^{-1}\,°C^{-1}}$), dT_{emp} = the gradient of temperature through the canopy (°C), and dq = the gradient of specific humidity through canopy ($\mathrm{kg\,H_2O\,kg\ air^{-1}}$).

The similarity hypothesis ($K_H = K_{other\ properties}$) is again invoked to formulate equations for the vertical fluxes of CO_2 and water vapour if profiles of concentration are known. Much more detailed information concerning calculations of aerodynamic and Bowen ratio transfers is available in Monteith (1973), Thom (1975) and other authors in Monteith (1975).

A major problem of any micrometeorological study of canopy fluxes is the advection of energy into or out of the ecosystem by moving air. As an air-stream passes from one system to another it takes both time and distance to reach a new equilibrium windspeed profile, temperature and water content. Profile measurements made close to a boundary may not be representitive of conditions within an extensive stand. If a small area of vegetation is surrounded by sun-heated dry soil, the advection of warm dry air will enhance transpiration rates and modify other physiological functions: various workers have referred to this as the 'oasis effect'. On a larger scale, Hogg (1971) has noted that easterly advection of warm, moist air into western Britain produces a January temperature anomaly of +15 °C compared with, for example, central U.S.A. at the same latitude.

Satisfactory measurements of canopy fluxes thus demand very large upwind 'fetch' of uniformly similar vegetation so that profiles may develop. Monteith (1973) cites a fetch:height ratio of 200; about 10 km for a tall forest! Such demands can rarely be satisfied except in crop, grass, tundra and occasional forest stands and it is for this reason that they feature prominently in the micrometeorological literature. Advection of energy must be the rule rather than the exception in most terrestrial ecosystems.

POTENTIAL EVAPO-TRANSPIRATION

Micrometeorological treatments of canopy fluxes require information from two or more levels above the surface but the combination of aerodynamic and energy balance techniques allows the derivation of equations which require measurements at one level only. The Penman equation (1948) for potential evapo-transpiration is of this form. Evaporation is calculated as a function of the heat budget plus the total air movement over the system times its mean saturation deficit: it assumes a closed canopy, ample water supply and no physiological restriction of water loss. Thom (1975) presents a form of this equation originally derived by Monteith (1965), which also includes the physiological effect of 'bulk stomatal resistance':

$$L_h E_v = \frac{\Delta H + (\rho C_p [e_s - e_p]/r_h)}{\Delta + \gamma Z} \tag{2.14}$$

where Δ = the slope of vapour pressure versus temperature curve at ambient temperature (Pa °C^{-1}), ρ = the density of air at ambient temperature (kg m^{-3}), C_p = the specific heat of water (1010 J kg^{-1} °C^{-1}), e_s = the saturation vapour pressure of water at ambient temperature (Pa), e_p = the partial vapour pressure of water at ambient temperature, γ = the psychrometer constant (66 Pa °C^{-1}),

$Z = (r_v + r_{st})/r_h$, and r = resistance, subscripted h for convective heat transfer; v for water vapour and st for 'bulk stomatal resistance' ($s\,m^{-1}$).

The values of Δ, ρ, C_p, e_s, γ and L_h may be looked-up in standard tables (e.g. Monteith, 1973). If stomata are fully open and, as is often the case in mid latitudes $r_h = r_v = r_{st} = c.\ 40\,s\,m^{-1}$ and if H, e and temperature are measured, the solution of the equation gives the potential evapo-transpiration, the maximum rate of water loss in the absence of physiological limitation. Substituting values for $H = 84\,W\,m^{-2}$, air temperature of 20 °C and daily mean saturation deficit of 20 % (2340 − 1872 = 468 Pa) the equation predicts an evapo-transpiration rate of $0{\cdot}14\,kg\,m^{-2}\,h^{-1}$. These values are probably representative summer means for temperate conditions. With stomatal closure r_{st} may reach $1000\,s\,m^{-1}$ and the calculated loss is reduced to $0{\cdot}02\,kg\,m^{-2}\,h^{-1}$. These two arbitrarily calculated values not only illustrate the substantial effect which stomatal closure is likely to have on canopy water loss but also closely approximate actual measurements of canopy water loss with maximum values of perhaps $4\,kg\,m^{-2}\,d^{-1}$ (Rutter, 1968).

The measurement of r_v and r_h is discussed in some detail by Thom (1975) and the relationship of r_{st} to individual leaf resistance is dealt with in chapter 7.

MEASURING TECHNIQUES FOR THE AERIAL ENVIRONMENT

The foregoing discussion has revealed the great complexity of the above-ground energy environment and turbulent atmospheric boundary layer. A bewildering variety of techniques and instruments are available for the definition of this system but their use, and the interpretation of results, is fraught with difficulty. Some attempt at clarification will be made in this section but, in no way is it possible to present a full manual of methods in so short a space.

Radiant energy

Measurements are required in all of the wavebands conveniently subdivided at the beginning of the chapter (p. 9). To construct an energy budget the solar short-wave income (0·3–3·0 μm) must be balanced against the low temperature long-wave loss (3·0–100 μm) while measurements for photosynthetic studies must be made in the visible 'light' part of the spectrum (0·4–0·7 μm) and expressed either as energy flux or quantum supply. Interest in environmental simulation, and in control by light quality, has demanded spectroradiometric instruments which can measure the distribution of radiant energy content with wavelength.

All instruments measuring irradiance are termed *radiometers*. Total solar radiation received by a horizontal surface is measured with a *solarimeter* and a *net radiometer* measures the difference between upward and downward all-wave fluxes. A solarimeter may also be used to measure irradiance from artificial lights. PAR is measured with specially filtered and calibrated *quantum flux meters* or much less satisfactorily, with various *photometers* (light meters) of which the spectral sensitivity roughly matches that of the human eye.

The operating principle of all radiant sensors depends on either thermal effects caused by radiant absorption or on photo-electric effects in which quantum absorption disturbs the electronic configuration in the sensor, generating a voltage or changing a resistance. One exception to this generalization is photochemical integration using, for example, photographic emulsions, uranyl oxalate or anthracene as photoreceptors (Maggs and Alexander, 1970; Kubin, 1971).

Thermal radiant sensors are most commonly thermo-electric, a bank of thermocouples (thermopile), or thermistors, with a low emissivity coating (e.g. optical black) being electrically balanced against a set of shielded or low emissivity (e.g. MgO coated) elements. Radiant heating of the blackened sensor gives an out-of-balance voltage which may be suitable read or recorded. A few instruments, now not much used, sense the temperature differential with liquid in glass thermometers, bimetallic strip or pneumatic detectors and distillation gauges.

The sensor must give linear output with irradiance, be insensitive to wind and temperature, should have zero-stability and constancy of response. Cosine correction should be such that the instrument registers the correct equivalent horizontal irradiance when the incident beam is not vertical (Anderson, 1971; Kubin, 1971; and Szeicz, 1975).

Solarimeters may be equipped with a glass dome, restricting transmission to 0·3–3·0 μm, so measuring total short-wave irradiance. It may be filtered to cut off wavelengths below 0·7 μm (e.g. Schott RG 695 optical glass dome). One filtered and one unfiltered solarimeter then give total short-wave, short-wave infrared and, by difference, visible light (the UV component is usually a very small proportion of the 0·3–0·7 μm radiant flux).

Long-wave infrared may be measured with similar thermoelectric elements but the dome must be of thin polyethylene film which has negligible absorption between 0·3 and 100 μm, or the sensor must be unshielded, in which case it is sensitive to wind cooling unless fan-ventilated at a rate which swamps wind effects. All-wave radiometers of this type register the complete spectrum from 0·3–100 μm and are usually exposed as net radiometers with one unit facing upward and one downward, connected electrically 'back to back' so that a zero reading represents zero net radiation, positive output is a net downward flux and negative, a net upward flux of all-wave radiation. If necessary the short-wave component can be subtracted from this by measuring with a pair of solarimeters, one inverted. This inverted instrument gives a measure of short-wave reflection from the surface (albedo). Details of instruments, their exposure, recording and other problems may be found in Anderson (1971), Kubin (1971), Monteith (1972), Szeicz (1975) and Fritschen and Gay (1979).

Investigation of spectral quality of light below a canopy or underwater demands spectroradiometric recording in which the energy flux is sensed not in total but at discrete wavelengths. Several such instruments are available and measure spectral energy distribution between 0·3 and 3·0 μm using suitable filters or a diffraction grating to scan the spectrum. Their use is of particular interest in specifying the change in photomorphogenic activity of light below a canopy or defining the quality of light available for shade-plant photosynthetic systems (Edwards and Evans, 1975).

Photoelectric sensors are now very freely available as selenium barrier layer cells, silicon and germanium photodiodes, vacuum tube photocells, photomultipliers and photoconductive cells of indium antimonide, lead or cadmium sulphide. All of these cells differ in their spectral response curves and are unsatisfactory as sensors of light for plant ecological purposes unless this limitation is understood. The cells may, however, be fitted with suitable filters and cosine correctors adapting them to many types of measurement because of their cheapness and sensitivity. Norman *et al.* (1969) described a PAR or quantum sensor which is fitted with interference filters which match the spectral sensitivity of a photodiode to the general photosynthetic response curve of the green plant. McCree (1973) makes the helpful observation that, though the bimodal spectrum of energy absorption by chlorophyll is well known, it is less often appreciated that it is 'topped-up' by the absorption of spectra of other pigments, mainly yellow. If the action spectrum is plotted *per unit of quantum flux absorbed* it is almost rectangular in the spectral band 0·4–0·7 μm. He suggested that measurements of quantum (photon) flux between these limits should represent PAR for all practical purposes. The quantum sensor of Norman *et al.* (1969) is filtered to give this response. Subsequently in this book the term PAR will be defined in this way and the word 'light' used synonymously with it unless otherwise qualified.

The calibration of radiation and PAR sensors poses many problems, not least because such instruments require frequent recalibration. Details are given in Kubin (1971) and Fritschen and Gay (1979). Radiation is usually expressed as irradiance with units of $W\,m^{-2}$ or its time integral of $J\,m^{-2}$. These energy flux expressions are directly compatible with measured changes of stored or latent heat and, *via* the Stefan's law relationship (equation 2.2) satisfactorily permit energy budgeting and water loss calculation. PAR may also be expressed as $W\,m^{-2}$ (0·4–0·7 μm) which is satisfactory for the relatively imprecise needs of production ecology (chapter 11) but, for critical photosynthetic work PAR is now more often measured as a photon flux density (Einstein $m^{-2}\,s^{-1}$) where 1 Einstein (E) = $6{\cdot}02 \times 10^{23}$ quanta (Avogadro's number of quanta). Incoll *et al.* (1977) note that this is not an SI compatible unit and suggest its replacement by mol $m^{-2}\,s^{-1}$ where mol = mole of photons: the numerical value is identical to $E\,m^{-2}\,s^{-1}$.

It is useful to have some information on the relationship between energy flux, photon flux and conventional photometric illuminance measurements expressed in lux but it is impossible to give simple conversions as they are wavelength dependent ($E\,m^{-2}\,s^{-1} = 1{\cdot}2 \times 10^{9}\lambda^{-1}\,W\,m^{-2}$ if λ is in cm). Approximate conversions are possible from the nomogram of Figure 2.13 for light of the same spectral composition as sunlight. These conversions cannot be used for light of other spectral quality and are intended only to aid in the interpretation of irradiance, illuminance or photon flux data cited elsewhere in the book.

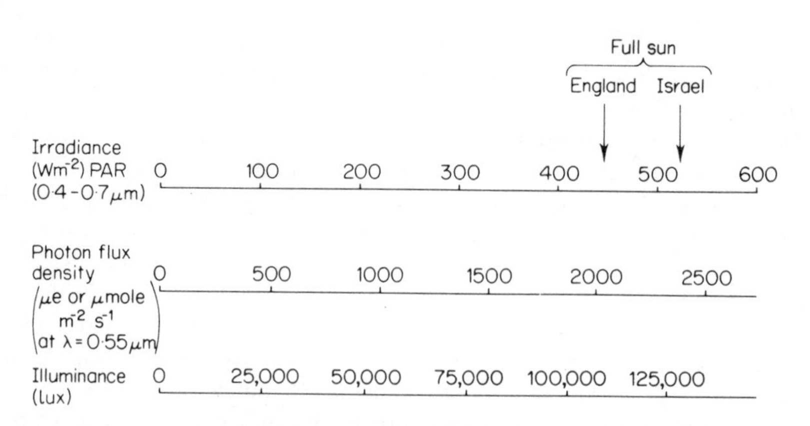

Figure 2.13 A nomogram permitting approximate conversions between irradiance, photon flux density, and illuminance measurements *in sunlight*. The photon flux density is given for a wavelength of 0.55 μm, the mid-point of the PAR spectrum. Irradiance from an incandescent (tungsten filament) lamp will not differ very much from these values in conversion but the 'cool' light of a fluorescent tube lamp may have up to 25% more lux per watt.

Wind

The most widely used methods are the standard meteorological rotating cup anemometer and the heated sensor anemometer. The cup instrument may give integrated records on a mechanical counter, or produce a continuous record if fitted with suitable electrical contacts. Two main disadvantages are its frictional resistance preventing the measurement of low airspeeds and its failure to operate accurately in a turbulent environment such as a crop canopy. Heated sensor instruments rely on forced convective cooling of an electrically heated element which may be a wire, a thermocouple or a thermistor. Accurate instruments read down to airspeeds of 0–1 m s^{-1} with a response time of 0·5–1·0 s. One disadvantage is that they respond to the scalar wind, that is the sum of both horizontal (vector) and vertical turbulence components and because of this need careful interpretation. Details are given in Monteith (1972), Szeicz (1975), S. B. Chapman (1976) and Fritschen and Gay (1979).

Temperature

Air temperature may be measured with mercury-in-glass thermometers if spot readings are adequate and if precaution is taken to shield the thermometer from direct radiation. Remote reading and recording methods are more commonly needed: suitable sensors are thermocouples, wire resistance thermometers, temperature sensitive diodes or thermistors. The first, very easily home-made, is a wire junction of two dissimilar metals, commonly copper–constantan or chromel–constantan. A potential difference is created across the junction, the magnitude of which is related to temperature, and is read by reference to a second junction maintained at a known temperature; often melting ice in a vacuum flask. The resultant voltage (39 μV $°C^{-1}$ for copper–constantan) is registered by a suitable potentiometric recorder, microvoltmeter or data logger. The major problem with thermocouples is the very low voltage which must be measured and the difficulty of avoiding spurious thermal voltages if there are dissimilar metal junctions elsewhere in the system. Cheapness and convenience often outweigh these problems.

Resistance thermometers made of platinum give about 0·4 % change per °C but a relatively low resistance. Thermistors may have a much higher resistance (kilo- to megohms) and a high temperature coefficient; but they are available as small beads or flakes and are consequently very useful for environmental measurement. Both sensors suffer the disadvantage that they must be carefully matched before use and, though thermistors are relatively inexpensive, resistance thermometers are of high cost. Now that silicon and germanium diodes are so cheap, their linear temperature response and ease of recording make them an attractive proposition for thermometry. Methods are reviewed by S. B. Chapman (1976), Monteith (1972), and Fritschen and Gay (1979).

Surface temperatures may be measured with thermocouples or thermistors touching the leaf, soil or other surface but considerable problems of thermal contact may arise. It is much more satisfactory to monitor leaf and other surface temperatures using an infrared thermometer for contactless measurement. The instrument responds to the long-wave radiation from the surface, usually in the spectral band 8–13 μm, which can be calibrated to surface temperature using Stefan's law (equation 2.2). Very expensive infrared thermal imaging devices have also been used for large scale aerial or satellite measurements of vegetation temperature. An extensive review is given by Perrier (1971).

Water vapour

Water vapour is present in air at a partial pressure e_p which is a variable fraction of the saturated vapour pressure e_s. Water will not evaporate into a saturated atmosphere ($e_p = e_s$) but if a saturation deficit exists, it will, and in doing so stores latent heat. This is the basis of the wet and dry-bulb psychometric method of measuring humidity. Air is aspirated over a pair of temperature sensors, one of which is fitted with a capillary sleeve dipping in a water reservoir. Evaporative latent heat demand depresses the wet-bulb temperature (e.g. 6·2 °C at 20 °C if $e_p/e_s = 0{\cdot}5$). This is the most robust and satisfactory instrument for field recording but it must be adequately radiation shielded, the sensor should be small to minimize any residual radiant absorption and it must be aspirated at a rate exceeding 3·6 m s^{-1}. Long (1968) describes the construction of a suitable instrument. Recording is as for temperature sensors.

Water vapour may be monitored with other instruments such as the infrared gas analyser (IRGA), resistance hygrometers which measure the changing surface resistance of a sulphonated polystyrene element as it equilibrates with water vapour pressure, lithium chloride resistance meters which operate similarly, dew-point meters which cool a sensor until the temperature of $e_p = e_s$ is reached, and lithium chloride dew cells which measure the equilibrium temperature of a saturated solution of LiCl which is evaporating water. None of these methods, with the exception of IRGA, has been extensively used in field recording systems but several find application in diffusion porometers, environmental control systems and portable humidity meters (Monteith, 1972; S. B. Chapman, 1976; Fritschen and Gay, 1979).

Carbon dioxide

Most routine measurement of atmospheric CO_2 now relies on IRGA techniques in which the strong absorption of infrared at *c.* 4·3 μm is used to compare the concentration of CO_2 in a sample cell with either a standard CO_2 concentration or differentially with another flowing sample. Modern instruments, operated differentially, will give a full scale deflection for 10–15 v.p.m. difference with atmospheric background of 300 + v.p.m. Less sensitive than IRGA is the conductiometric technique in which deionized water is equilibrated with a bubbling air sample and its conductivity monitored (Sestak *et al.* 1971; Szeicz, 1975).

Problems of sampling and measuring CO_2 arise from the CO_2 permeability of many plastics used in pipework: metal is best but difficult to install. Water vapour may give interference absorption unless the samples are dried or suitable optical filters are installed on the analyser. Some drying agents such as silica gel give problems of CO_2 adsorption. There are many difficulties in the operation and calibration of IRGA systems (Sestak *et al.* 1971).

Environmental measurement in the field poses serious problems of data logging which, in the past, could only be solved labour intensively. The last decade has seen rapid developments in the digitizing of analogue signals and their interfacing to magnetic tape recording systems which have replaced punched tape. Recording and processing of environmental data is now itself a specialist topic and references may be found in Monteith (1972), Sceicz (1975) and Fritschen and Gay (1979).

Chapter 3

Soils

INTRODUCTION

Almost any part of the earth's surface which supports vegetation also bears a covering of soil. It is thus possible to define a soil as 'the material in which plants root' though this is a limited definition as soil is formed by the long-term modification of parent geological material through a combination of biological, climatic and topographic effects, the multifactorial interactions of which lead to a great variety of potential soil types. A further strong source of variation is the feedback between organism and environment. Plant species are often habitat-specific but their presence and activity gradually produce alterations in the environment which may influence their own vitality and allow other species to invade and supersede them.

Earlier views of pedogenesis suggested that the sequence of invasions by organisms contributed to a monocyclic situation in which a soil would gradually evolve, with its associated vegetation, until it reached an equilibrium with the prevailing climate, but it is now accepted that many soils are polycyclic, their pedogenesis having been interrupted and diverted by repeated intervention of external factors. Soil formation must, then, be looked upon as a long-term process of complex interactions leading to the production of a mineral matrix in intimate association with interstitial organic material both living and dead.

The earth's crust is composed of a relatively few rock-forming minerals mainly based upon various alumino-silicates, metal silicates and variable quantities of free silica (SiO_2). In their primary form these minerals arose by the solidification of molten magma and are referred to as igneous rocks. Weathering and denudation leads to their mechanical and chemical breakdown and redistribution; the secondary products of this process are the sedimentary rocks which form from the deposition of weathering products in ocean basins (water transport) or on terrestrial surfaces (aeolian, glacial or solifluion transport). Some water-deposited sediments may be biogenic or of chemical origin; for example, most limestone rocks. A third category of rocks arises by the influence of heat and/or pressure on pre-existing igneous or sedimentary rocks: these are named metamorphic. Table 3.1 gives some examples of these varied rock types.

The alumino-silicate minerals are the most widespread constituents of igneous rocks and may be subdivided into two main groups: the feldspars (potassium,

Table 3.1 Soil-forming rocks and their composition

	Rock-type	Example	Composition
IGNEOUS	QUARTZ RICH (acid magmas)	GRANITE	Mica, alkali feldspar, quartz in coarse, discrete crystals
IGNEOUS	QUARTZ POOR (basic)	BASALT	Fine crystalline structure of plagioclase feldspar and augite
SEDIMENTARY	Coarse texture Water-deposited	SANDSTONE SAND	Quartz grains cemented with Fe_2O_3, $CaCO_3$ or loose.
SEDIMENTARY	Fine texture Water-deposited	MUDSTONES SHALES CLAYS	Predominately finely particulate alumino-silicates with various cementation such as Fe_2O_3, $CaCO_3$
BIOGENIC AND CHEMOGENIC		LIMESTONE	$CaCO_3$ as organic remains or as chemically precipitated material. May be very pure as in chalk
AEOLIAN		LOESS	Wind-transported deposits of medium texture. Often $CaCO_3$-rich
GLACIAL		DRIFT TILL BOULDER CLAY	Often heterogenous mixture of diverse geological material. Often clay-rich and impermeable
METAMORPHIC	Metamorphosed sedimentary	MARBLE SLATE	Metamorphosed limestone Metamorphosed shale
METAMORPHIC	Metamorphosed igneous	GNEISS	Metamorphosed quartzites and granitic rocks

calcium–sodium and sodium alumino-silicates) and the micas (complex potassium alumino-silicates with included hydroxyl groups, often containing magnesium, iron, sodium and fluorine).

These minerals are all comparatively easily weathered and it is their breakdown products which form the fine-grained mineral matter of most soils and, because of their rather complex constitution, provide most of the mineral nutrients required by plants. The small particle size of this clay mineral gives the soil alumino-silicate fraction a large specific surface which is physicochemically very active in the nutritional and water relationships of the soil.

The coarse-grained particles of most soils are derived from free silica in the parent rocks. Silica is very resistant to both chemical and physical weathering processes, at least under temperate climatic conditions, and tends to accumulate in the soil mantle as a mechanical matrix of sand and gravel particles amongst which the alumino-silicate clays are distributed. It is the denudation of this material which leads to the formation of the sedimentary rocks and these, having already passed through one cycle of weathering process, tend to be softer and more rapidly reweathered than the hard, primeval igneous rocks. For similar reasons the metamorphic rocks have more physical and chemical coherence than their igneous or sedimentary progenitors and so resist the weathering process more successfully. Table 3.2 summarizes the weathering characteristics of the various rock-forming minerals.

THE WEATHERING PROCESS

Bare rock surfaces and developing soils are exposed to a range of physical, chemical and biological processes which lead to mechanical and chemical disruption of their components (Table 3.3). Chemical and physical processes alone give rise to abiotic crusts of weathering products which are only the raw material of soil formation. However, careful investigation shows that most bare rock surfaces do not remain free of life for very long and the physicochemical weathering processes are soon reinforced by the often potent effects of numerous microorganisms. Jacks (1965) stresses the role of the lichen symbiosis in this early phase, suggesting that lichens are able to extract nutrients which would be unavailable to higher plants. Retention of water by the thin layer of lichen, fungal and bacterial organisms on rock surfaces prolongs the period during which chemical processes can proceed, splitting the rock alumino-silicates by hydrolysis and carbonation into the simpler clay alumino-silicates. Carbon dioxide released by respiratory processes must further accelerate this type of weathering. Photosynthetic energy fixation by the algal component of the lichen will increase the available organic matter at the surface and Rogers *et al.* (1966) have shown the appreciable contribution which the blue-green algae of some lichens may make to nitrogen fixation. Free-living blue-green algae make a very substantial contribution to nitrogen fixation, particularly in hot, neutral to alkaline environments (Stewart, 1973 and p. 235). Invasion by bryophytes increases the photosynthetic capacity of the living layer, more organic matter and weathered

Table 3.2 Rock-forming minerals, their composition and form in soil

Rock-forming mineral		Chemical constitution	Form in soil
ALUMINOSILICATES	FELDSPARS	Alumino-silicates of K, Na, Ca, Alkali feldspars are KNa forms and plagioclase feldspar NaCa	Mainly as clay-sized particles in much modified form. Very little unaltered feldspar
	MICAS	Hydroxyl-alumino-silicates of K, Na, (KMgFe), K(Mg) and K(Fe)	Similar to feldspar in producing much modified clay mineral
SILICATES	OLIVINE AUGITE GARNET PYROXENE AMPHIBOLES	Mg, Fe, Ca, Na and Mn silicates	Not usually represented in any quantity
	QUARTZ	SiO_2	Persists in soil as sand and gravel, often forming main matrix
ALUMINIUM OXIDE IRON OXIDE	(Al_2O_3) Fe_2O_3	Fairly low concentration in most rocks	Becomes concentrated by loss of silica in tropical wet conditions (ferrallitization). High concentrations in some tropical soils (oxisols)

Table 3.3 Weathering processes

Physical	Chemical
WETTING—DRYING E.g. Disruption of layer lattice minerals which swell on wetting	HYDRATION E.g. Reversible change of haematite to limonite which is accompanied by swelling and so disrupts cementation of sandstones etc. $Fe_2O_3 \leftrightarrows Fe_2O_3 3H_2O$
HEATING—COOLING E.g. Disruption of heterogeneous crystalline rocks in which inclusions have differential coefficients of thermal expansion. Surface flaking of large boulders, particularly in arid climates, due to sun heating	HYDROLYSIS E.g. Silicate breakdown $K_2Al_2Si_6O_{16} \rightarrow Al_2O_3 2SiO_2 2H_2O$ Orthoclase Kaolinite K and surplus Si are washed away in solution
FREEZING E.g. Disruption of porous, lamellar or vesicular rocks by frost shatter due to expansion of water during freezing	OXIDATION—REDUCTION E.g. $Fe^{3+} \leftrightarrows Fe^{2+}$ causes disruption of cementation as Fe^{2+} is much more soluble than Fe^{3+}
GLACIATION E.g. Physical erosion by grinding process	CARBONATION E.g. $CaCO_3 \leftrightarrows Ca(HCO_3)_2$ leads to solution loss of limestone or disruption of $CaCO_3$ cemented rocks as the hydrogen carbonate is more soluble than the carbonate
SOLUTION E.g. Removal of more mobile components such as Ca, SO_4, Cl etc.	CHELATION Essentially a consequence of biochemical activity, various metals being dissolved as chelates with organic products of plant and microorganism activity
SAND BLAST E.g. Erosion of upstanding rocks in arid, desert conditions	

rock becoming incorporated in what must now be recognized as a thin soil layer covering the surface.

Various exudates from these organisms, other than respiratory CO_2 are likely to speed the pedogenetic process; for example, some of the lichen acids have strong chelating properties; organic acids generally are potent in dissolving mineral components, while some species of bacteria directly influence the solubility of nutrient elements and cementing compounds. Mulder *et al.* (1969) cite the solubilization of phosphorus by microorganism-formed organic acids and microorganism reduction of insoluble ferric phosphates. Further examples are the bacterial reduction of manganese and iron, increasing their solubility. The cycles of sulphur, nitrogen and phosphorus are all strongly governed by microorganisms through the sizes of the organic and inorganic pools and the rates of change between soluble and insoluble, available and unavailable forms. Further discussion may be found in Campbell and Lees (1967), nitrogen; Cosgrove (1967) and Halstead and McKercher (1975), phosphorus; Freny (1967), sulphur; and Ehrlich (1971), minor elements. Many non-essential elements are also biologically cycled and concentrated, for example Uernelöv (1975) describes the formation of mono- and di-methyl mercury from relatively immobile inorganic sources.

Colonization of a juvenile soil by higher plants adds yet another complication to the soil-forming process, greatly increasing the energy-fixing capacity of the surface and increasing the supply of decaying organic matter. Soluble organic compounds also diffuse into the rhizosphere zone from the roots and wash into the soil surface from leaf-drip. Deeper penetration of roots will tend to increase the depth range of the cyclic processes involving nutrient elements, soluble elements leached downward being returned to the surface by transport through the plant.

Rock weathering is therefore, for a short time, a physicochemical process but rapidly becomes biogenic with a consequent increase in the overall rate. Pedogenesis may thus be considered as a biological phenomenon by which crusts of weathered rock debris are converted to true soils comprising a complex mineral matrix in association with a great range of organic compounds; very often carrying a rich microorganism population which is a reflection of the nature of the parent material and its interaction with climate, topography, plant cover and age.

COMPOSITION OF SOILS

The components of a soil can be classified into four categories:

(i) A matrix of mineral particles derived by varying degrees of breakdown of the parent material.

(ii) An organic component derived from long- and short-term additions of material from plants, animals and microorganisms above and below ground.

(iii) Soil solution held by capillary and adsorptive forces both between and at the surface of the soil particles, its amount varying with the balance between precipitation, evapotranspiration loss and drainage. It is a dilute solution of many different organic and inorganic compounds and forms the immediate source of plant mineral nutrients.

(iv) The soil atmosphere occupies the pore space between soil particles which, at any time, is not water-filled. Its composition differs from the above ground atmosphere as it is normally lower in oxygen and higher in carbon dioxide content. In very wet soils oxygen may be almost absent as it can only be replenished by slow diffusion in solution from the surface.

The relative proportions of the mineral and organic solids to the liquid- or gas-filled pore space are determined by the particle size distribution of the mineral matrix (the soil texture) and the binding of these fundamental particles into larger units or aggregates. The pore spaces between aggregates are much larger than those between the textural particles: the aggregated condition is often described as soil *structure*. Aggregation is related to the organic content of the soil and to its degree of microbiological activity.

Soil particle size distribution is usually expressed as frequency in arbitrary size-classes: a commonly used set of classes is that of the International Society of Soil Science:

		Equivalent diameter range (mm)
I	Coarse sand	2·0–0·2
II	Fine sand	0·2–0·02
III	Silt	0·02–0·002
IV	Clay	<0·002

This classification, though arbitrary, achieves a separation which is pedologically useful, the boundary between silt and clay roughly marking a transition in the properties of the material. Much physicochemical activity of soils is a function of the surface area: silt, fine sand and coarse sand have specific surfaces, respectively, of less than 1, 0·1, and 0·01 $m^2 g^{-1}$ compared with the clay fraction which commonly ranges between 50 and 700 $m^2 g^{-1}$ (Fripiat, 1965). Hence the clay content determined by particle-size analysis gives a very good idea of the ion-exchanging, adsorptive and capillary activities of the soil. This physical difference is also accompanied by chemical differences. The principal components of the sand/silt fractions are quartz, unmodified silicates and alumino-silicates from the parent material (mainly feldspars, micas, pyroxenes, amphiboles, olivene, etc.), iron oxides, and a few larger particles of true clay mineral. In temperate soils quartz is usually the commonest mineral, making up 90–95 % of the sane fraction in soils derived from sedimentary rocks. It may be less common in igneous soils depending on the content of the parent rock.

The clay fraction is typically different from the sand/silt, comprising an assemblage of minerals formed by weathering of alumino-silicate rock minerals. The coarsest of the clay fractions may occasionally contain a little finely particulate quartz and mica but the finer material is almost entirely true clay with some iron and aluminium oxide. The clays are layer-lattice crystalline materials based upon two subunits of silicon–oxygen and aluminium–oxygen; the details of the structure will be discussed in the next chapter. Their small size and consequent large surface area, coupled with strong ion-exchanging activity, is responsible for the importance of soil clay particles as the major source and reservoir of plant nutrients and in governing soil water relationships, aeration characteristics and association with organic macromolecules.

SOIL DEVELOPMENT—PEDOGENESIS

The comparatively recent realization that many geochemical processes involve biological activity tends to blur the distinction which was previously made between geological weathering and pedogenesis. However, the latter process is recognizable as that which develops the characteristic morphology and 'metabolism' of a soil by which it may be recognized to type in the same way that the morphology and physiology of an organism permits its specific identification. The phenotype of an organism is developed by the interaction of genetically determined raw material with the environment, while the manifested soil type analogously develops from the influence of environment on geological parent material.

Soil-forming factors are passive and active:

Passive factors

1. Parent material.
   (a) Physical constitution.
   (b) Chemical composition.

2. Topography.
   Influential on both macro- and micro-scale through its influence on drainage, soil movement and insolation.
3. Time.

Active factors

1. Rainfall.
   (a) Determines the direction of solute translocation (up or down) according to precipitation/evaporation (P/E) ratio.
   (b) In conjunction with other factors determines depth to water table and therefore capillary water and aeration status.
   (c) Indirect effect through influence on vegetation.
2. Energy balance and temperature.
   (a) Interacts with rainfall in governing P/E.
   (b) Influences rates of physicochemical processes, thus controlling rate of organic turnover in the soil.
   (c) Influences the growth rate of vegetation and other organisms.
3. Humidity/Evaporation.
   Low relative humidity gives high evaporation rates and offsets the influence of high rainfall by reducing amount of downward water movement.
4. Wind.
   Of minor importance except that high maintained wind speeds cause increased evapotranspiration and may limit vegetation growth.
5. Biosphere effects.
   The activity of living organisms must be considered definitive in pedogenesis as they speed-up and modify the physical and chemical processes which would normally occur in their absence. Their narrow ecological tolerance ranges tend to amplify the environmental differences already influential in pedogenesis.
   (i) Plants
      (a) Direct plant activities such as the secretion of organic acids and enzymes. Respiratory production of CO_2.
      (b) Input of organic matter after death.
      (c) Micrometeorological influence of vegetation cover.
   (ii) Animals.
      (a) Direct interaction between primary production and consumption by animals with consequent effect on input of organic matter to the soil.
      (b) Soil-dwelling micro-arthropods, molluscs, lumbricids, etc., have a direct influence on organic matter turnover and incorporation.
   (iii) Microorganisms.
      Bacteria and fungi play a key rôle in the geochemical, biochemical and biophysical processes of pedogenesis.

The parent material factor strongly influences soil composition both through its chemical characteristics, some of which are handed on to the derived soil, and through its physical constitution which influences, for example, leaching rate and soil aeration. Generally speaking, silicate-rich rocks with a high content of plant nutrient elements produce soils which are eutrophic or mesotrophic and highly favourable for plant growth, while silica-rich, acid rocks tend to produce oligotrophic soils with a rather poor vegetation. Physically, a high silica soil will produce a sand-rich or coarse-textured soil which is prone to leaching loss of soluble nutrients. This contrasts with silicate-rich rocks which may give a high clay content and good nutrient status but provide so little coarse pore space that the soil may be badly aerated.

Topography influences soil formation through drainage and retention of water. The soils of waterlogged hollows and, on a larger scale, of valley floors and extensive ill-drained plains or plateaux often have quite different characteristics from adjacent well-drained soils on identical parent material. Such topo-sequences have been described as *catenas* (p. 80). Most important in this context is the aeration status of the soil in relation to decomposition of soil organic matter and to the redox characteristics of the soil. Atmospheric oxygen can diffuse freely into the air-filled pore space of a soil, but when it becomes water-filled and oxygen has to diffuse in solution, it is not able to keep pace with the respiratory demand of soil microorganisms. In consequence, the soil becomes anaerobic, its redox potential falls and its microflora is replaced by a population of anaerobes. The reduced rate of oxygen entry leads to an increase of organic matter in the soil and the fall in redox potential may convert iron to the ferrous state, increasing its solubility and mobility in the profile. Other inorganic consequences of the redox change will be described later (chapter 10).

Time as a factor in any ecological context is all too often neglected. The life span of many large, woody plants lies in the range 10^2 to 10^3 years and, as they may be of prime importance in the soil formative process, no soil can be considered to be in biotic equilibrium before many tree generations, representing several thousand years, have passed by. Much of the land area in the Northern Hemisphere emerged from the late glacial period only 10,000 years ago so that it is doubtful whether its soils and vegetation can have been in any sort of equilibrium before extensive human interference became important during the Bronze Age.

The time required for soil formation appears to be widely variable as it is so strongly influenced by climatic and parent material factors. Evidence from buried soils under Bronze Age tumuli in Northern Europe suggests that podsolization has taken place in many heathland areas during the past 4000 years, probably following human destruction of the original forest cover (Dimbleby, 1962) and the same author cites other evidence which suggests that podsolization can be even more rapid. Under tropical conditions, soil formation may occupy a much longer time-scale. Mohr and van Baren (1959) wrote: 'The process of laterisation in the tropics ... must be regarded as a geological phenomenon. This is because the time factor, next to climate, is of the greatest importance in determining the final stage in the cycle of weathering'. Soil formation may thus range from being a short-term process comparable in scale to the generation time of many plants to a long-drawn-out and probably cyclic process over periods of near geological time.

Rainfall appears to play a dominant part in pedogenesis, coupled with evaporation rate and the drainage characteristics of the soil and site. With a continuous excess of precipitation over evaporation, there is uninterrupted downward movement of water in the soil profile. Initially, soluble salts such as carbonates and more mobile ions such as sodium are leached from the surface layers, a distinct front of solute loss advancing down the profile with time. With more coarsely textured soils, or a greater surplus of percolating water, material in the clay fraction may be washed downward and redeposited below (*lessivation*); in many temperate soils this is the first evidence that *illuviation* of the *B* horizon is occurring (see p. 50). With extremes of downward water movement, coupled with soil acidity and nutrient deficiency, the iron oxide coating of the soil grains may become soluble and the iron, after translocation downward, is redeposited at a deeper level in the soil. Under these circumstances there is nearly always an accompanying transport of organic materials from the surface horizons and formation of a *B* horizon organic-rich layer. Other metal sesquioxides such as manganese and aluminium may be *eluviated* from the *A* horizon along with iron to reappear as concentrations in the *B* horizon. Such soils have been named *pedalfers* in reference to the iron and aluminium translocation.

Interruption of the high P/E regime by a dry season may temporarily stop the downward leaching or, in some cases, return soluble materials to the surface. If the dry period is extended it may come to dominate the pedogenic process, which will then resemble the one which would occur with a P/E ratio constantly less than one. Soluble materials are carried upward in the soil capillary matrix and, if the water table is sufficiently high, may lead to massive enrichment of the surface horizon with soluble salts (characteristically carbonates, sulphates and chlorides of calcium and sodium) and the development of 'alkali' and saline soils in arid zones. Surface enrichment with calcium carbonate and other solutes is a characteristic of *pedocal* soils.

Temperature obviously interacts strongly with the water balance effects described above. Solar radiation is the main energy source for water evaporation, hence high radiation, high temperature and low P/E tend to be correlated together. High temperatures have a further marked effect in increasing the rates of chemical and, in particular, biochemical processes. As the photochemical step of photosynthesis is rather insensitive to temperature, a rise tends to increase respiratory breakdown of organic matter more than it increases photosynthetic production. This is generally reflected in the increased organic content of soils formed at lower temperatures. Figure 3.1 shows the strong interaction between temperature, soil wetness and organic matter accumulation. In cold climates photosynthetic production exceeds the rate of decomposition in the soil but, as temperature increases, so bacteria take over from the fungi as the major decomposers and the rate increases. At about 25 °C with good soil aeration there

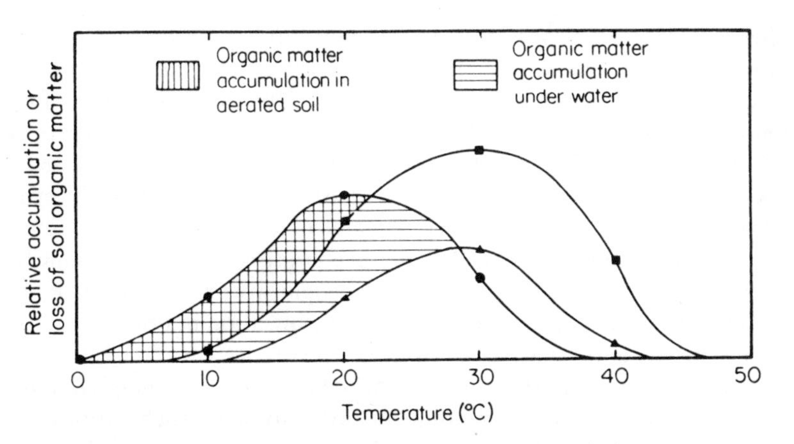

Figure 3.1 The relationship of soil organic matter accumulation to wetness and temperature: net photosynthetic input of organic matter (●); net respiratory loss in dry (■) and in wet (▲) conditions. Reproduced with permission from E. C. J. Mohr and F. A. van Bahren, *Tropical Soils*, van Hoeve, The Hague, 1959, *Figure* 56, *p*. 280.

is a balance between production and breakdown, and above this temperature organic matter cannot accumulate.

If the soil is flooded, the reduced oxygen supply permits only anaerobic bacteria to function and organic matter continues to accumulate in waterlogged soils up to about 35 °C. These relationships partly account for the high organic content of peat soils formed under conditions of waterlogging (temperate and tropical valley peats), high rainfall (blanket peats) or cold (tundra and alpine soils).

SOIL PROFILES

The interaction of these pedogenic factors permits the differentiation of a variety of soil types which may be defined by the nature of the mineral matrix, the vertical distribution or organic matter and the movement and redeposition of various inorganic constituents. Soils are described and identified by reference to their *profiles*—the sequence and nature of the horizons (layers) exposed in a pit-section dug through the soil mantle. The dimensions of a definitive profile vary with soil type as some soils show little horizontal variability, whereas others may show recurrent variation over horizontal distances of several metres. The smallest three-dimensional volume of a soil needed to give full representation to such features is termed a *pedon* (Soil Survey Staff, 1975).

'A soil horizon may be defined as a layer which is approximately parallel to the soil surface and that has properties produced by soil forming processes but that are unlike those of adjoining layers' (Soil Survey Staff, 1960). Horizons may usually be identified visually in the field but they also have chemical and physical properties which can be measured in the laboratory to confirm the diagnosis.

Horizons may be classified by reference to both position and constitution. Historically, the soil horizons of the Russian steppes were the first to be classified by the *ABC* terminology (Dokuchayev, 1900). The *A* horizons were the dark-coloured surface layers and the *C* horizons the underlying layers largely unaffected by the soil-forming process. The *B* horizons were transitional in properties and position. Adoption of this terminology in western Europe led to the concept of the *B* horizon as one into which materials were washed (illuviated) by downward water movement. This, however, raises difficulty with some tropical soils (latosols) in which the *A* horizon is recognizable but the horizon in the *B* position is enriched with sesquioxides, not by illuviation but by the leaching loss of other components (silicates and silica).

Five main groups of horizons may be defined and are designated *O*, *A*, *B*, *C* and *R*. Their definitions are fairly broad in order to solve such problems as the terminology of sesquioxide-enriched horizons described above (Soil Survey Staff, 1975).

O horizons

Organic horizons forming above the surface of the mineral matrix and dominated by fresh or partially decomposed organic matter.

A horizons

Mineral horizons formed either at or adjacent to the surface and characterized by enrichment with organic matter and/or downward loss (eluviation) of soluble salts, clay, iron or aluminium and consequent enrichment with silica or other resistant minerals.

B horizons

Mineral horizons forming below the surface and having one or more of the following features. (i) Enrichment with inwashed clay (lessivation), iron, aluminium, manganese or organic matter. (ii) Residual enrichment with sesquioxides or silicate clays which has occurred other than by the removal of carbonates or easily soluble salts. (iii) Sesquioxide coatings of mineral grains sufficient to give a more intense colour than horizons above or below. (iv) Alteration of the original rock material to give silicate clays or oxides in conditions where (i), (ii) and (iii) do not apply.

C horizons

Mineral horizons below the *B* layer but excluding true bedrock and lacking any characteristics of the *A* or *B* horizons. Formed under conditions little affected by the pedogenic process. May or may not be the material from which the *A* and *B* horizons formed.

R horizons

Bedrock. May or may not be the material from which the *A*, *B* and *C* horizons formed (soil may develop in drift overlying bedrock).

It must be stressed that the *O*, *A* and *B* horizons are entirely pedogenic and are unrelated to geological layering; the process of horizon formation may in fact continue uninterrupted downward through a geological discontinuity such as the boundary of superficial drift and bedrock. The number and relationship of the soil horizons is governed, in particular, by the direction and quantity of water movement in the profile. With a high ratio of precipitation to evaporation (>1) downward washing of materials predominates leading to the eluviation of surface (*A*) horizons and illuviation of *B* horizons. If the P/E ratio falls below one then capillary rise of water may occur, transporting soluble materials upward towards the surface.

Profiles may be further subdivided by numbering from the surface downward, e.g. O_1, O_2, A_1, A_2, A_3, etc. Each subdivision is an integral unit and needs its own separate definition as a horizon. Transitions between horizons may be shown by the use of both respective symbols; thus *A/B* includes features of both an *A* and a *B* horizon. A number of other horizon characteristics may be indicated by the use of a lower case suffix: b indicates a buried horizon, ca a calcium carbonate enrichment, g gleying, h an illuvial enrichment with organic matter and p a horizon modified by ploughing. Gleyed horizons are grey, blue-grey and green-grey due to the reduction of iron from the ferric to the ferrous condition. They may often be mottled with red-brown patches where ferric compounds have been redeposited in soil voids, root channels or around living roots (see p. 313). The system of soil profile description used in soil classification by the Soil Survey of England and Wales is described by Hodgson (1974). Ragg and Clayden (1973) have used the U.S. system to classify British soils; they review and discuss the comparison with the British system outlining difficulties and criticisms.

SOIL CLASSIFICATION

Classifications of soil types are inevitably a little arbitrary as they constrain a natural continuum in a linear or branching system. The majority of soil classifications have been based on a taxonomic and/or genetic approach, though purely artificial classifications are possible. The earliest attempts were made in Russia at the end of the last century and were type-defined by a combination of climatic/vegetation data and soil morphology. During the past half-century the U.S. Department of Agriculture Soil Survey has been responsible for the continuous development and refining of a system much influenced by Dokuchayev's (1900) final classification of Russian soils which defined a group of zonal soils, roughly corresponding geographically with the great climatic regions of the earth's surface, and two other groups which depart from the nature of zonal soils by reason of their unusual water relationships or parent material. The U.S. system in its revised form (Thorp and Smith, 1949) was widely used in the mapping of American soils and, in modified form has been adopted elsewhere. It does suffer defects imposed by the use of genetic characteristics for soil identification which may prevent a soil from being placed until its pedogenesis is understood. A European system utilizing binomial naming was published at about this time (Kubiena, 1953) and many of its soil types resemble those of the U.S. system.

The most recent advance is the current U.S.D.A. taxonomic scheme (Soil Survey Staff, 1975) which recognizes ten *Orders* identified by manifested characteristics such as colour, morphology, chemical constitution etc., and permits immediate classification of any newly observed soil type. Table 3.4 gives an outline of the taxonomic groups and an indication of their relationship with previously identified soil types. The diagnostic characteristics omit any which are seriously affected by the intensity of genetic processes, thus avoiding the problems of classifying immature and ill-understood soils. The orders, nevertheless, tend to represent the various zonal and associated azonal soils of the 1949 system.

Further subdivisions are made: into suborders; great groups; subgroups; families and series. The major subdivisions are indicated by prefixing, for example a peaty podzol with iron pan belongs to the Order Spodosol, because of wetness the Suborder Aquod and because of pan formation Great Group Placaquod. Further details are given in Soil Survey Staff (1975) and Ragg and Clayden (1973). A simplified account of the classification is given by Buol *et al.* (1973).

SOILS—THE GEOGRAPHY OF THE GREAT GROUPS

The earliest classification of soil-types showed an awareness that their distribution was a generalized reflection of latitudinal variations in climate. In this awareness, pedology was closely akin to plant ecology in the Clementsian view of the climatic climax (Clements, 1916; 1936). In both fields the subsequent collection of more detailed information has shown that climatic determinism may be misleading if it is accepted on too local a basis: in both soil and plant distribution there are other strongly acting factors such as topographic water regime, physicochemical status of the parent material, grazing intensity, and anthropic effects such as extensive forest clearance for hunting and agriculture.

Current classifications make varying degrees of allowance for these interactions but there is still sufficient evidence of a zonal distribution of soils for Figure 3.2 to give a crude picture of the relationship between climatic temperature and wetness as a determining factor in the worldwide pattern of soils. In the following sections a brief description of the more widespread zonal soil-types is given.

Tundra soils

The tundra climate is exceptionally harsh with extremes of low temperature, high windspeed and, often, aridity. As a result, many soils of this zone are pedologically immature with little or no true horizon differentiation. On stony

Table 3.4 Soil Classification: the Comprehensive System of the U.S. Dept. Agric. (Soil Survey Staff, 1975)

Order	Description	Nature of soil and previous names cited later in the text
1. Entisol	Soils on recent deposits; immature; little or no horizon formation	Soils on unconsolidated parent materials (*Regosols*) such as stabilized dune soils and alluvia. *Lithosols* on rocky parent material
2. Vertisols	Soils with large proportion of swelling clay Montmorillonite. Widely cracking to great depth when dry and consequently 'self-ploughing'. Mainly tropical	*Grumosols, Regur, Black Cotton Soils*. Often with marked micro-relief features (gilgai) related to soil movement
3. Inceptisols	Young unweathered soils on a variety of parent materials	Alluvial soils with horizon differentiation. Some *Brown Forest Soils, Brown Earths* (*Sols bruns acides*), calcareous soils and Gleys
4. Aridosols	Soils of arid climates (low precipitation/evaporation ratio)	Desert soils, *Sierosems, Solonchak*, some *Solonetz* and some *Brown Soils* and *Reddish Brown Soils* of arid grasslands
5. Mollisols	Soil with dark mollic epipedon* of high base saturation. No leaching is apparent	(a) Various *Brown Forest Soils, Brown Earths* and their associated gley soils (b) Soils on limestones: Rendzinas and associated gleys (c) *Chernozems* and other associated prairie and steppe soils of arid climates
6. Spodosols	Strongly leached soils with a subsurface *Spodic* horizon recognized by deposition of organic material and amorphous aluminium compounds with or without iron	*Podosols* of various types, *Peaty Podsols* and some *Peaty Gley Podsols*
7. Alfisols	Soils with an *Argillic* B horizon, that is, containing clay which has been washed in from above and with moderate base saturation	Some gley soils. *Brown Earths* with strong downward movement of clay (*Sols bruns lessivés*). Some *Grey-Brown Podsolic* soils
8. Ultisols	Strongly leached soils with lessivation to form an argillic B horizon but with base saturation too low to qualify as alfisols. Mainly (sub) tropical	*Red-Yellow Podsolic* soils, some *Laterites* and *Latosols*
9. Oxisols	Soils with an *Oxic* horizon in which the clay mineral has a very low exchange capacity and consists mainly of kaolinite, iron oxides and quartz fragments which remain as a weathering residue after removal of silica. Tropical	*Latosols, Laterites*
10. Histosols	Organic soils with a substantial depth of peat. (*Histic* epipedon)	Bog soils of ombrogenous and soligenous type, some *Peaty Gleys*

* Epipedons include all horizons which form at the surface and are darkened by organic matter.

substrata such soils are named *Lithosols*, and on sands, gravels and other unconsolidated materials, *Regosols*. Soils of this description are not limited only to the tundra but may occur in other parts of the world if the climate and parent material are unfavourable to soil development. For this reason they may be considered *azonal* soils.

Many ill-drained tundra soils (Figure 3.3) have a peaty or organic-rich *A* horizon resting on a pale coloured gley horizon which, during the summer

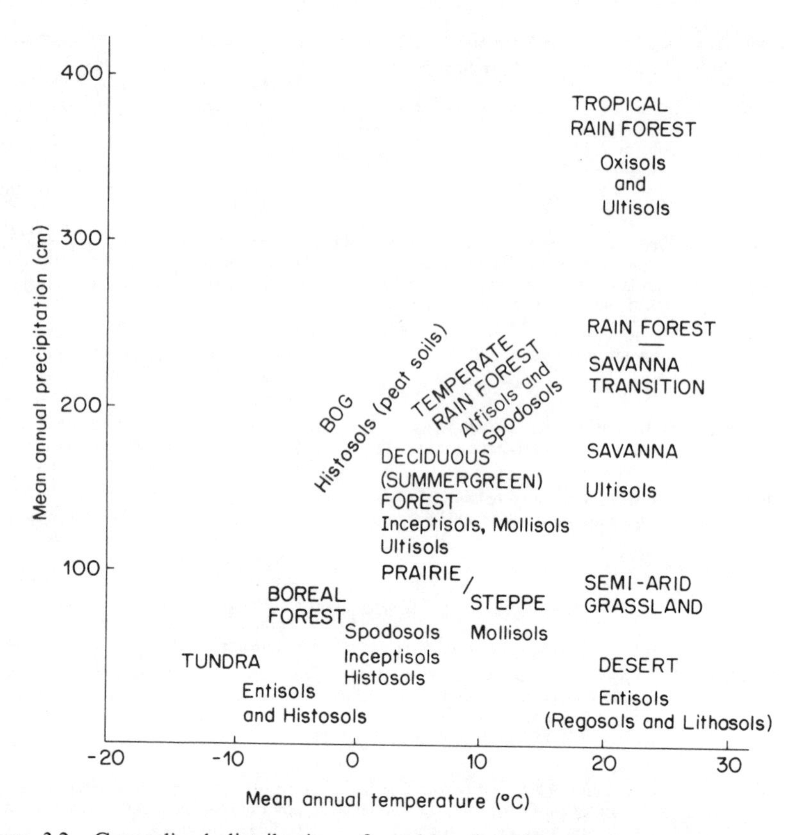

Figure 3.2 Generalized distribution of world soil and vegetation types. No precise boundaries are shown as transitions are very variable and independently related to other factors such as geology, hydrology and land-use history.

months, may be almost liquid, its drainage impeded by the permanently frozen soil (permafrost) beneath. The soil is strongly seasonal in its biological activity, being completely frozen for the greater part of the year. The expansion and contraction of the soil water on freezing and thawing results in the enlargement of soil voids and a general mixing of the soil (cryoturbation) which leads to portions of the *A* horizon becoming incorporated at other levels in the profile. The same process often causes migration of clay mineral which forms coatings on stones or an indurated zone at the base of the profile where it is squeezed between the permafrost and the overlying soil on freezing.

Cryoturbation leads to the patterning of tundra soils which is so obviously seen from the air. The surface is dissected into regular polygons of which the centres are raised and the borders depressed. One cause of this is mud-explosion in which semi-liquid mud is forced from the central regions of the polygons by the lateral pressure of ice forming in wedge-shaped cavities along the sunken borders. Such morphological features remain as 'fossil' features of soils which formed in the northern hemisphere in the periglacial climate following the last glaciation (Fitzpatrick, 1956).

Ill-drained soils of this type belong to the *Inceptisols* (*Cryaquepts*) or, if a deep peat layer accumulates under very wet conditions they may be *Histosols*. With free drainage various *Spodosols* (*Podzols*) are able to form under dwarf-shrub heath as frost-heaving does not disrupt the soil horizons. Freely-drained base-rich materials have a richer herbaceous vegetation on poorly developed *Brown Earths* which may be *Mollisols* or *Inceptisols*. Rieger (1974) gives a clear review of the nature and classification of Arctic soils and some information on the generally more extreme climate Antarctic soils by Ugolini (1970).

Tundra vegetation ranges from isolated patches of lichen, mosses and occasional higher plants of Rock Tundra to the more continuous cover of moss, lichen, *Carex* (sedge) or dwarf-shrub Tundra. Almost any depression may accumulate water over the permafrost, giving rise to Tundra Moor with peat-forming mosses. The whole vegetational physiognomy is markedly xeromorphic, the frequently frozen soil and shallow rooting imposed by waterlogging interacting with extremes of wind exposure and low humidity to cause potentially serious water loss. The majority of the plants are drought-tolerant (poikilohydrous mosses and lichens) or avoid water loss by their low stature, xeromorphy and short growing season. Most of the vascular plants either belong to the geophyte and hemicryptophyte life-forms or are dwarf chamaephytes with xeromorphic leaves.

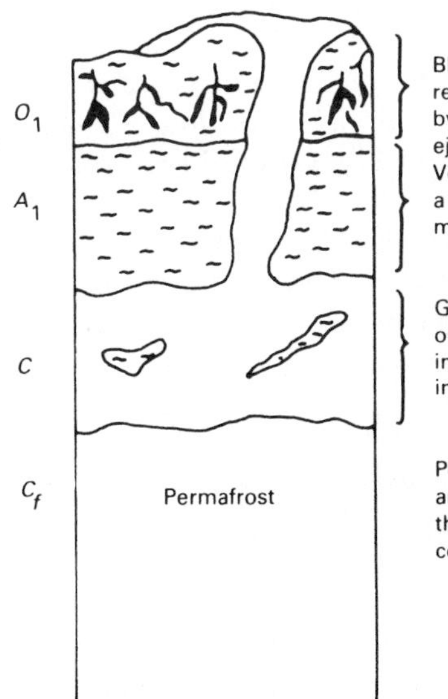

Figure 3.3 A tundra soil profile with permafrost-impeded drainage (Cryaquept). During the winter months the whole profile is frozen. Summer thawing is rarely to more than one metre. Vegetation may be dominated by cyperaceous species, replaced by *Sphagnum* spp and other mosses in more oligotrophic conditions.

The harsh climate seriously limits species diversity and some areas may be extensively dominated by a single species; thus moss tundra may be dominated by *Polytrichum* species and lichen tundra, for example, by *Cetraria* or *Cladonia* species. *Carex* tundra has a grassy aspect due more to the inclusion of *Carex* and *Eriophorum* (sedge and cotton grass) species than to an abundance of grasses. In high summer the meadow-like appearance is accentuated by profuse flowering of various circumpolar species of prominent-flowered plants such as *Ranunculus* and *Potentilla* (buttercups and cinque foils). Other flowering plants characteristic of high latitudes are numerous species of *Gentiana* and *Saxifraga*. *Dryas octapetala* (mountain avens) is so typical of this habitat that its remains, in lake deposits, may be used diagnostically of past climate.

The very limited chamaephytic flora include the dwarf birches and willows: *Betula nana*, *Salix herbacea*, *S. reticulata* and *S. arctica*. In more southerly areas and slopes of warmer aspect *Betula* and *Salix* species may form a continuous thicket cover. The more heath-like tundra may be dominated by one of several ericaceous plants, examples being *Empetrum nigrum* and *Cassiope tetragona* (crowberry and mountain heather).

Boreal forest soils

To the south of the tundra regions of northern Europe, North America and Asia the higher annual temperatures permit an enlargement of the life-form spectrum. The dwarf-shrub cover of the southernmost tundra gives way to Boreal or Taiga forest characteristically dominated by needle-leaved conifers. The soil most frequently associated with this formation is the *Podzol*, though, as a soil type, it may also extend southward into the broad-leaved forests and heathlands of lower latitudes, a zone usually dominated by various brown and grey forest soils.

The fully developed podzol shows an extreme degree of horizon differentiation which is best seen in the *Iron-humus Podzol*. This is most often found under heathland with ericaceous dominants; the more typical conifer forest podzol is the *Iron Podzol*. Figure 3.4 shows the interrelationship of a range of podzol types which may form according to degree of wetness, parent material, etc.

The most prominent feature of the podzol is the intense eluviation of the surface horizons with mobilization of organic matter and its redeposition at lower levels. The process is so potent that all easily soluble inorganic materials are leached downward and the cation exchange complex is desaturated (p. 88) with a corresponding increase in soil acidity. This drop in pH and the removal of calcium cause a change in the microbiological pattern of organic matter decomposition leading to the accumulation of a peaty, fibrous organic layer near the surface. This material has been named raw-humus or *mor* (p. 100).

The fully developed podzol profile (Figure 3.4a) has a surface O_1 horizon of undecomposed plant litter (A_{00} in other terminologies) which grades into the $O_2(A_0)$ horizon of peaty mor which is considerably decomposed but still contains microscopically recognizable plant fragments. The bottom of this layer rests on the first horizon of true mineral soil, the A_1, which is stained black or grey by organic matter. Its individual sand and larger silt particles can be seen to be bleached due to the mobilization of the ferric sesquioxide which normally coats such grains, giving them a brown coloration. Below the A_1 is the A_2 horizon

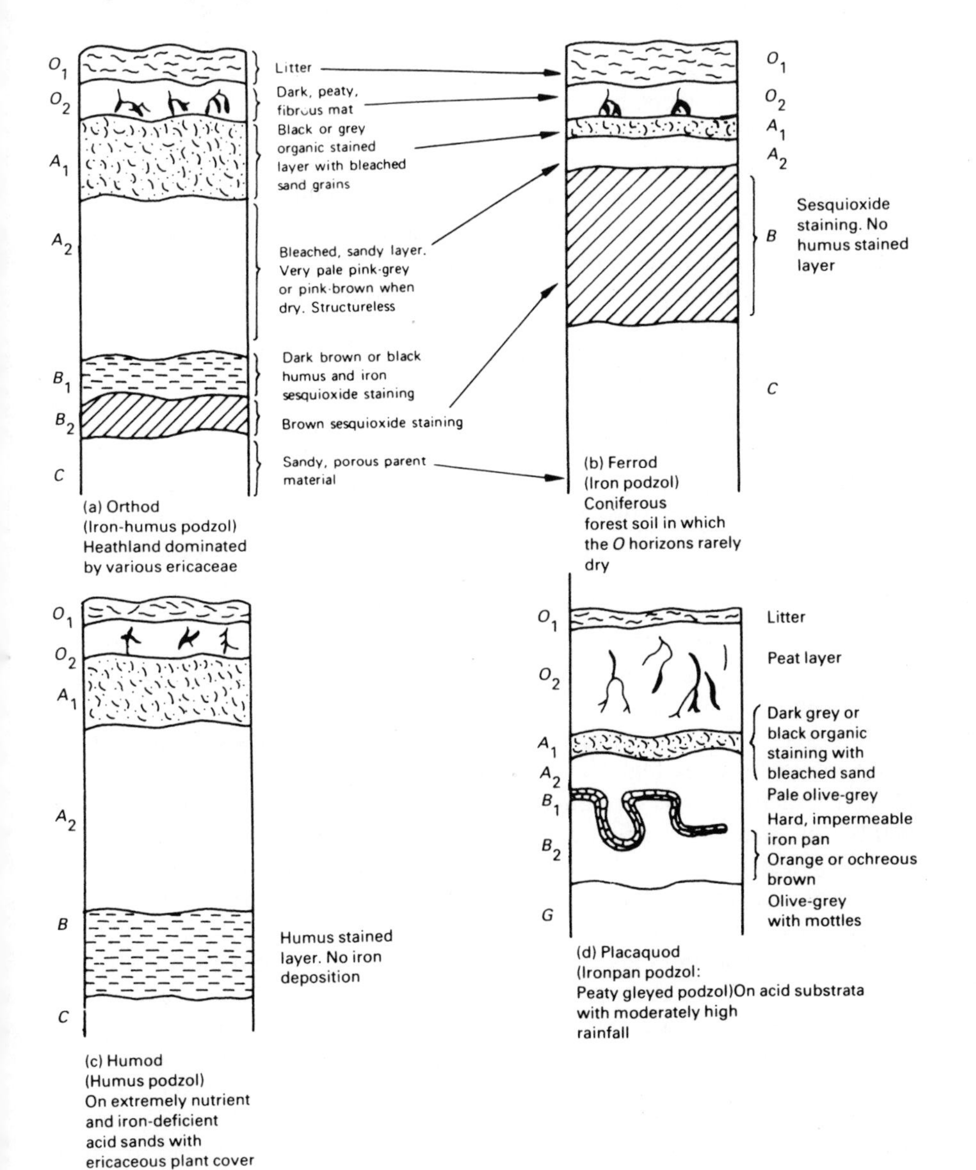

Figure 3.4 Profiles of four Spodosol (podzol) types (see also Figure 3.11). The name, podzol, derives from a Russian word meaning ash and referring to the wood-ash colouration of the bleached A_2 horizons.

which shows the same extensive bleaching of its larger particles but no organic staining. The horizon has a whitish to pale grey, ashen appearance and is a characteristic of all well-developed podzols. Bleaching due to removal of ferric sesquioxide is a diagnostic feature of *albic* horizons. With such potent eluviation it is not surprising to find that the A_1 and A_2 horizons are also often depleted of manganese and aluminium sesquioxides, of clay mineral and of more soluble compounds and ions such as carbonates, calcium, magnesium and potassium.

Below the *A* horizons there is a sharp transition to the illuvial zone of the *B* horizons in which percolating water has redeposited various materials to form the B_1 horizon, which is enriched with organic matter and ferric/aluminium sesquioxide, and the B_2 horizon enriched predominantly with ferric/aluminium sesquioxide. These illuvial horizons are diagnostically the *spodic* horizons of the spodosol. The humic and sesquioxide coatings of the mineral grains may cause cementing of the particles and 'pan' formation. Some iron pans have a rock-like consistency and are impervious to water, thus causing a superficial waterlogging of the *A* horizons (Figure 3.4d).

The *C* horizon of the podzol, if it is also the parent material, is most likely to be a sand, the rock-rubble of a sandstone or acidic igneous rock such as granite. All of these parent materials weather to form coarse-textured soils which are much more likely to show full podzolization than fine-textured parent materials which impede free drainage in the profile. Podzol formation is thus strongly influenced by parent material and also by topogenic waterlogging, vegetation type and human interference. The boreal forest consequently has a mixture of soil types of which the podzol is dominant only when other factors permit. There is evidence in northern Europe that much podzolization and soil acidification is anthropic following the removal of the primeval forest (Dimbleby, 1962) which accounts for the occurrence of podzols in the broad-leaf forest areas to the south of the climatic podzol zone.

The vegetation of the podzol, characteristically considered to be needle-leaved forest, may also be heathland in these more southerly areas. In northern Europe, where heathland is most prominent, they are usually dominated by ericaceous plants; for example, the central species of much European lowland heath is *Calluna vulgaris* (ling). Under rather dry conditions *Erica cinerea* (bell heather) may become codominant or take over full dominance. Schimper (1898) long ago recognized this association as an edaphic one and it has been more recently realized that it is strongly anthropic, being maintained by burning and tree felling (Tansley, 1939). In the absence of human control the instability of the association is manifested in the ease with which it is invaded by, for example, *Pteridium aquilinum* (bracken) and subsequently colonized by *Betula* spp. or conifers. With increasing wetness there is a transition to various types of wet heath and extensive peat formation. Under high rainfall conditions ombrogenous peat may form from *Sphagnum* spp., *Eriophorum* spp. or *Molinea caerulea.* Extensive tracts of country may become covered with blanket peat which, pedogenetically, is the *O* horizon of a peaty podzol or peaty gley. Similar peat formations may occur in valley bottoms but, under these soligenous conditions, the peat is more nutrient-rich. Tansley (1939) gives an extensive account of the relationships of the various upland and lowland heath formations.

Boreal forest, for geographical reasons, like the tundra, is mainly confined to the northern hemisphere. It forms a prominent belt across North America, northern Europe and Asia and includes a large range of locally dominant conifer species, mainly of the genera *Picea, Pinus, Abies* and *Larix* (spruce, pine, fir and larch). The diversity is exemplified by contrasting the range of dominants in these continents. In the east of North America *Picea glauca, P. mariana, Larix laricina, Abies balsamina* and *Pinus banksiana* may variously be dominant, but further west the climatic change results in their replacement by *Pinus contorta* and *Abies lasiocarpa*. In Western Europe different species occur, *Pinus sylvestris* and *Picea abies* commonly dominating. Further east, and in Asia, *Larix sibirica, Abies sibirica* and *Pinus obovata* are prominent.

The processes involved in podzolization commence with leaching and decalcification and proceed to the formation and migration of clay mineral; formation of free oxides and destruction of clay mineral; mobilization of sesquioxides and their deposition in the *B* horizon (Bunting, 1965). The fully developed podzol is usually very acid and its cation exchange complex is strongly desaturated (p. 88). It is difficult to specify precise pH ranges but Pearsall (1952) identifies pH 3·8–4·0 as the upper limit for mor humus formation and most podzol horizons are near or below this pH.

In the northern hemisphere the processes leading to podzolization probably started when the land-surface emerged from periglacial conditions at the end of the last glaciation and accelerated when man began extensive deforestation in the late Neolithic or early Bronze Age (Dimbleby, 1962). There is now evidence to suggest that certain plant species hasten the podzolization process; for example, the ericaceous heath which so often succeeds cleared forest on acid brown soils (Grubb *et al.*, 1969). Other plants recognized as soil acidifiers are some conifers and, amongst the broad-leaved tree species, *Fagus sylvatica* (beech) and *Quercus* spp. (oaks) in Britain (Mackney, 1961) and *Populus trichocarpa* (poplar) in Alaska (Crocker and Dickinson, 1957).

Summer deciduous forest soils

The *Brown Forest Soils* are characteristically those of the broad-leaved summer forest formation, the most productive of temperate zone ecosystems. The high productivity is reflected in the great biological activity of these soils in which microorganisms and the soil fauna play a very considerable part in pedogenesis.

Like the podzol, the brown forest soil is as much a function of parent material as of climate. Most typically it forms on silicate-rich parent materials: of the sedimentary rocks, the shales, mudstones, clays, finer sandstones and also impure limestones. On igneous substrata it is most often found on base-rich materials such as basalt and other deep-seated lavas.

The brown forest soil is maintained in a constant state of horizon mixing by the activity of its prolific earthworm fauna. As a result the *B* horizon rarely shows sharp delimitation from the *A*, other than in its lower organic content and difference in colour and structure. It is often designated (*B*) to indicate its indeterminate nature. Any leaching which occurs is usually counterbalanced by biological mixing but in some cases (*Sols bruns lessives*—leached brown soils) some downward washing of clay has occurred and, on more siliceous parent materials, leaching of carbonates may have lowered the pH into the acid range (*Sols bruns acides*—acid brown soils).

The central group of brown forest soils is of high cation saturation (above 50 %) and has a pH near 6. These are, however, variable characteristics since calcareous brown soils, on limestones, may be of pH 7 or above, and have high calcium saturation of the exchange complex, while some leached brown soils may be well below pH 6 and have a cation saturation of less than 35 %. The acid brown soils, forming on siliceous parent materials, podzolize easily and have a base saturation and pH range overlapping with that of the podzols. Pearsall (1952) suggested that pH 4·8–5·0 was the lower limit for the mull humus formation (p. 99) characteristic of brown forest soils but their pH range extends upwards above pH 6·5, which he considered the lower limit for the distribution of calcicole plants.

The *A* horizon of the nutrient-rich brown forest soil is markedly structured as a result of earthworm activity. The worm casts, particularly those voided underground, contain organic matter which has been comminuted in the animal's alimentary tract and forms a rich substrate for fungal and bacterial growth. The binding effect of the hyphae coupled with the influence of bacterial gums stabilizes the soil in the casts into persistent structural aggregates which, in some particularly rich soils, may form a large proportion of the *A* and (*B*) horizons, giving them a soft, spongy and well-aerated nature.

The action of the worm population keeps the surface of the soil more or less clear of organic debris so that O_1 and O_2 horizons are minimal or non-existent. In some of the more acid soils there is formation of *O* horizons (Figure 3.5) as the worm population falls with increasing acidity. Worms appear to be absent below a pH of *c.* 4·8 (their activity is instrumental in mull humus formation so that this limitation is further evidence for Pearsall's suggestion noted above). Soil Survey Staff (1975) attribute the nature of the upper horizon of the brown forest soil to the activity of earthworms in pulling leaf litter into their burrows, thus incorporating it uniformly into a considerable depth of mineral soil. This is comparable with the effect of a grass sward in which the roots are responsible for returning a large proportion of the photosynthesate into the soil.

The vegetation of the brown forest soils is typically a broad-leaved deciduous forest, though many areas have been taken by man for agriculture. Under arable management large amounts of organic matter must be added if the soil is to retain its originally well-structured condition while heavy inorganic fertilization is needed to compensate the loss to the crop. There is a lesser problem in using these soils for grass production as much organic matter is returned to the soil by the root system, but inorganic fertilization is required and some care has to be taken

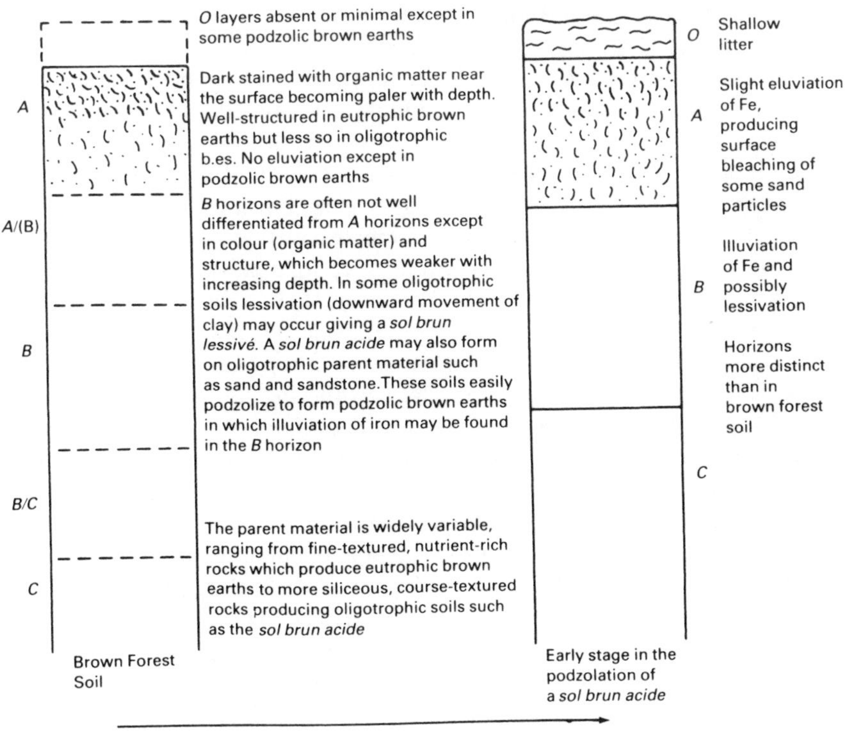

Figure 3.5 Brown Forest Soils or Brown Earths: their profiles and relationships. In the USDA Taxonomy, base-rich brown earths may be *Mollisols*: leached *sol bruns acides* are *Inceptisols* and brown earths with downwashing of clay (lessivation) are *Alfisols*. If the clay-enriched B-horizons are base-desaturated there may be overlap with the *Ultisols* of warmer climates (See Table 3.4).

to avoid damaging the soil structure by overstocking and 'poaching' by cattle trampling, or by the use of heavy machinery.

The natural forests of these soils are physiognomically similar in all parts of the world. There is usually a single stratum of dominant trees and, below this, sparse strata of undershrubs and herbs. Few climbers or epiphytes are present so that the general aspect is much less luxuriant than that of tropical forest.

Five general types are recognized (Polunin, 1960).

1. Oakwoods of west and central Europe. These are dominated by *Quercus robur* and/or *Q. petraea* with a number of associated species such as *Fraxinus excelsior*, *Populus* spp., *Betula* spp., *Ulmus* spp. and *Alnus glutinosa* (Ash, poplars, birches, elms and alder). These vary in abundance according to soil wetness, pH and nutrient status. A number of shrub species such as *Corylus avellana* and *Crataegus monogyna* (hazel and hawthorn) are common and the ground flora may vary from a luxuriant cover of herbaceous plants on rich soils to tussock grass cover of species such as *Deschampsia caespitosa* on moist or heavy clay soils or to a

Pteridium aquilinum–Rubus fruticosus (bracken-blackberry) association on light sandy soils.

2. Forests of North America, East Asia and south-east Europe. These are more varied in their species, the wide range including species of *Quercus, Fagus, Betula, Carya* (Hickories), *Juglans*, (walnut), *Acer* (maple), *Tilia* (Basswood), *Fraxinus, Ulmus, Liriodendron* (Tulip tree), *Castanea* (chestnut) and *Carpinus* (hornbeam). The underflora of these forests resembles its western European counterpart in its wide variety of herbaceous, grass or Pteridophyte cover. In some areas bordering the boreal forest region conifers may occur as codominants with the deciduous species.

3. Beech forest. Widespread forest areas in Europe may be dominated by *Fagus sylvatica* which forms such a dense, shade-casting canopy that the shrub and herb layer is usually either very sparse or absent while the soil is characteristically covered with a thick brown mat of the slow-decaying beech leaves.

4. Southern Beech. An association of the southern hemisphere dominated by *Nothofagus* spp. closely crowded in the tree layer with relatively few undershrubs having a luxuriant carpet of Bryophytes and ferns.

5. Variants due to topographic waterlogging, containing *Salix* or *Alnus* spp. (alder or willow) as dominants and having a rich underflora of hygrophyllous herbs. According to soil nutrient status and acidity the underflora may range from a wet heath *Sphagnum* spp. or *Molinia caerulea* (purple moor grass) association to a lush, rich fen flora under eutrophic conditions. Under these circumstances the soil will be of a humic gley type (p. 75) rather than the brown forest soil common to the rest of the formation.

Subhumid and arid grassland soils

In central continental areas the gradient of decreasing precipitation in the deciduous forest region leads to a greater tree spacing and the development of a park-like vegetation. Further reduction in rainfall prevents tree survival and the forest vegetation gives way to the subhumid plainland grass associations: the prairies of North America, the steppes of eastern Europe and Asia, the pampas of South America and the grasslands of southern Africa and Australasia.

These grasslands are essentially midcontinental in their distribution and their associated soils are much influenced by climatic conditions. The annual precipitation range lies between 1000 mm and less than 350 mm. At the lower end of this range there is a transition to semi-desert conditions and, also, rainfall effectiveness may be reduced by the high intensity of summer storms. The upper limit overlaps the precipitation region of summer deciduous forest but the continental climate with continuous sub-freezing conditions during the winter months and the high evapotranspiration in the midsummer months is unfavourable to tree growth.

Leaching is not extreme as the winter-frozen soil prevents continuous downward water movement for a large proportion of the year while high evaporation rates induce an upward capillary movement during the summer.

The climatic range for grassland as a climatic/biotic climax is thus fairly wide and includes soils which range from near relatives of the podzolic type to poor arid soils in which biological activity is limited by persistent drought. The soils thus span the range from highly leached soils with spodic horizons of R_2O_3 and organic carbon enrichment to soils in which a past or present low P/E regime has caused secondary accumulation of calcium carbonate to form a calcic horizon at some level in the profile.

The range of soil type, while reflecting latitudinal climate, is also governed by parent material and soil hydrology; consequently, sandy materials may become leached to thin, red podzolic soils while adjacent lime-rich loess, limestone, basalts and dolerite form the *Chernozemic* soils so typical of the region. In river valleys, strips of forest cover may penetrate deep into the open grasslands and have wetter varieties of mollisols, ranging from brown forest soils to various gleys (see pp. 59, 75). This characteristic geographical pattern has been almost entirely eliminated in North America and Eurasia as the majority of rich grassland soils have been cultivated for cereal growth, and the former grassland persists only in relict sites.

The modal *Chernozem* profile (Figure 3.6) is dominated by a dark-coloured *A* horizon of extreme depth (to 1·5 m) and high organic content. The horizon may

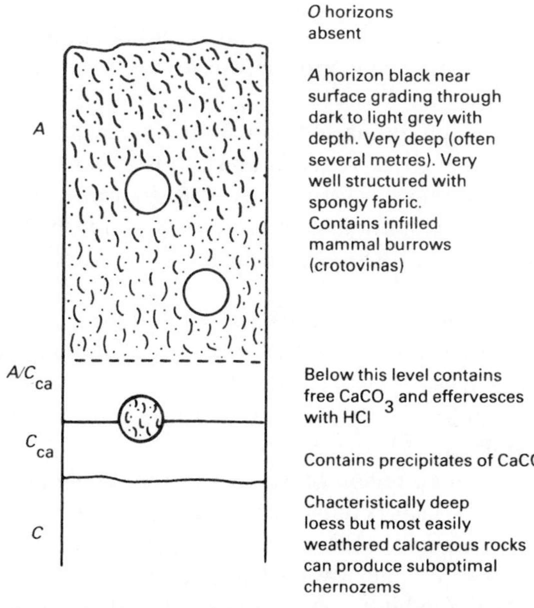

Figure 3.6 Grassland soils: a *Vermaltoll* (Chernozem) profile. The formation of this soil requires a calcareous parent material, characteristically loess, and a low P_r/E_v water regime. As P_r/E_v increases so the surface layers become $CaCO_3$ desaturated and the depth to a $CaCO_3$ horizon is a function of climatic wetness (see Figure 4.3). The name, chernozem, is of Russian derivation and means 'black soil'.

be calcium carbonate saturated and has a pH range from 6·5 to 8·0 or more and rests directly on a highly calcareous *C* horizon. At some level in the profile there is a zone of calcium carbonate enrichment (C_{ca} or B_{ca}) the depth depending on annual P/E balance. In North America the depth of this horizon increases with precipitation on a west–east gradient.

The chernozem is extremely biologically active with a high earthworm population and an accompanying collection of soil-dwelling mammals whose burrows are so extensive as to form a diagnostic part of the profile with their infilled remnants. In the dark *A* horizon they often appear as 'white eyes' containing the contrasting material of the calcic *C* horizon. Chernozems belong to the Order *Mollisols* because, like some base-rich Brown Forest Soils, they have a *Mollic* epipedon (surface horizon) which diagnostically contains over 1% organic matter, is dark-coloured, of high base saturation and has strong structural stability. The active mixing and structure formation by earthworm and other animal burrowing is the reason for the Great Group name: *Vermaltoll.*

Under rather more humid conditions the black chernozem soils are replaced by those which have been named *Brunizems* (*Prairie soils*) and *Reddish Prairie soils* in North America and *Degraded* and *Leached Chernozems* in Eurasia. In all of these soils some downward translocation of materials may be detected in the formation of a *B* horizon and a brown coloration formed by weathering production of iron oxides. (If the *B* horizon is enriched with translocated clay the soils may be transitional to *Alfisols.* It is possible that these soils may formerly have carried a forest cover before its removal by human activity. On coarser parent materials under these conditions *Thin Red Podzolic* soils (*Ultisols*) may also be formed.

With an annual precipitation below 350–400 mm the Chernozems give way to the *Chestnut soils* of Northern America and Eurasia and with further reduced precipitation these intergrade with various *Arid Brown Desert* and *Semi-desert soils.* Under these drier conditions there is a transition from full grass cover to sparse grass or desert scrub with the inclusion of more xeromorphic species.

In Eurasia the Chernozems are dominated by grasses including the genera *Festuca*, *Koeleria* and *Stipa* and, in North America, by *Boutoloua*, *Agropyron*, *Andropogon*, *Sorghastrum*, *Koeleria*, *Stipa*, *Poa*, *Panicum* and others. Associated with these is a large range of flowering herbs which, in certain aspects, may appear to dominate the grassland.

As a major vegetational region of the earth, these grasslands were not only the reflection of climatic factors but also showed the strong biotic pressure of the grazing mammals formerly so abundant on the plainlands of the midcontinents. The role of fire in these habitats was probably equally important in maintaining the open nature of the habitat and it may be that these grasslands could be regarded as deflected climax (plagioclimax) associations which were prevented from achieving the full climatic status for the region. Man has, however, now destroyed much of the vegetation but the soils remain as identifiable entities, albeit modified by cultivation.

Mediterranean and subtropical soils

Soils of these regions are much influenced by the length of the dry season and the general aridity of the environment. In this they resemble the grassland soils but, of course, do not experience a cold winter climate in which biological and chemical weathering come to a standstill. For this reason most of the varied soils occurring in the region show considerable weathering effects compared with the relatively unweathered state of the modal chernozem.

Mediterranean soils

In the drier regions of the mediterranean type climate *Terra rossa* (red soils) are characteristically developed on limestone or dolomite karst and on some other basic rocks. The *Terra rossa* have a constitutional, but not genetic, affinity with the tropical latosols, their red colours being caused by the enrichment of the surface layers with iron oxides consequent on decalcification of the iron oxides during the intensely dry summer months produces a vivid red coloration of the soil which often contrasts with the whiteness of the fairly pure limestone from which it forms.

The profile (Figure 3.7) constitutes a poorly developed *A* horizon of low organic content, grading through an *A*/*B* transition to a red, clay enriched, *B* horizon which then shows a sharply defined junction with the parent limestone below. The texture may vary from clayey to sandy according to the nature of the impurities in the limestone or to the ingress of wind-blown material.

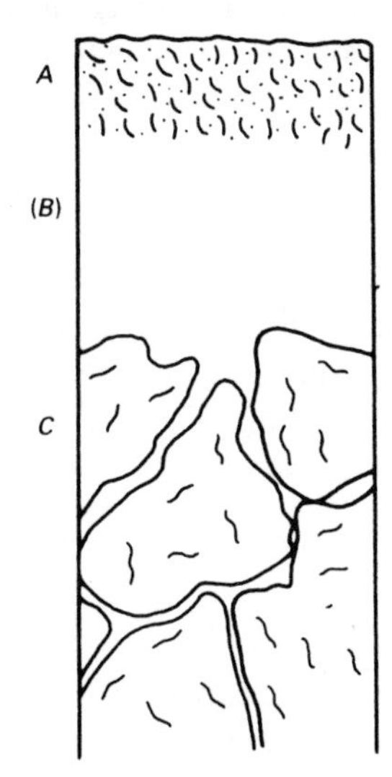

Figure 3.7 *Terra rossa* and *terra fusca* profiles. The *A* horizons are often shallow (< 10 cm) and deficient in organic matter. The vivid red of the terra rossa and the ochreous colour of the terra fusca are more strongly shown in the (*B*) horizons. Even on pure limestones the *A* and upper (*B*) horizons may be decalcified, being neutral or slightly acid and contrasting with the comparable temperate rendzinas.

Morphologically the profile resembles the limestone rendzinas of more temperate climates and Stace (1956) suggests that there is a continuum of characteristics between these two soil types.

Siliceous rocks or wetter areas have a more biologically active brown soil covering of *Terra fusca* with a greater organic content than the terra rossa and in which earthworm activity produces a greater mixing. On limestone parent materials the two soil types appear to be interconvertible, the removal of forest cover promoting the change to *terra rossa* while reestablishment of a closed vegetation cover protects the soil surface from intense summer inolation and may permit a terra rossa to revert to the terra fusca condition (Bunting, 1965). The clay enrichment and relatively high base status of their *B* horizons places both soil types in the Order *Alfisols*. Most *terra rossa* belong to the Great Group *Rhodustalfs* while the brown *terra fusca* are *Hapludalfs* which are genetically related to the brown woodland soils of cooler climates (Bridges, 1978).

In Europe, *terra rossa* carries a low scrub or heath-like vegetation, in France termed *garigue*. It represents the remains of a formerly more luxurient forest cover of *Quercus ilex* (evergreen oak) and *Pinus halpensis* (aleppo pine) which has been removed by man and his voracious companion, the goat. The consequent soil erosion has produced the characteristic karst scenery of the limestone country of the north Mediterranean. Many *terra rossa* are formed on the eroded remains of *terra fusca* soils and the *garigue* vegetation may be considered a degeneration product of the varied and diverse *maquis* thicket woodlands of the biologically richer *terra fusca*.

The sclerophyll *maquis* vegetation of the terra fusca is much richer than the garigue and contains a wide variety of shrubs which were probably the subdominants of former forest cover. They include such plants as *Olea europea* (Olive), *Cystus* spp., *Myrtus communis* (myrtle), *Rosemarinus officinalis* (rosemary) and many others. The original dominants in the Mediterranean area were *Quercus suber* (cork-oak) and *Pinus pinaster* (maratime pine). Similar sclerophyllous vegetation and soils may be found in the Cape region of South Africa, in western California (chaparall), in central Chile and in parts of South Australia. In all cases these vegetation types are characteristically exposed to repeated burning and often to heavy grazing. The convergent evolution of unrelated species to a very similar physiognomy in these habitats is very striking (Cody and Mooney, 1978).

Subtropical soils

The sub-tropics embrace a range of warm climatypes from wet to continuously arid. Warm wet conditions favour soil-formative processes which have some resemblence to those of the full tropics and it is not surprising that the diversity of soil types found in this climatic zone show affinities with both temperate and tropical types and also form geographical continua into higher and lower latitudes where local conditions of geology, topography or drainage are favourable.

In warm, moist environments ranging from the sub-tropics to warm temperate areas the predominate soils are *Red* and *Yellow Podzolic* types which, in the U.S. classification belong to the *Ultisols* characterized by a base-desaturated argillic *B* horizon into which clay has been lessivated from above. Iron compounds have been eluviated from the *A* horizons which thus show some degree of bleaching. In warmer, well drained sites the *B* horizon iron oxides are dehydrated giving a pronounced red colour and the yellow colours of hydrated oxides are manifested in lower-lying sites hence the naming of the two types.

Profiles are usually well drained and acid with thin O_1 and O_2 layers, a dark A_1 horizon containing organic matter and grading into a bleached A_2 which then shows a sharp boundary with the red or yellow *B* horizon (Figure 3.8). Their former vegetation was usually a summergreen forest but in many cases this has been removed by man. Richards (1952) describes the transition between tropical rain forest and drier habitats: evergreen forest giving way first to summergreen forest followed by savanna forest, thorn forest and finally arid scrub. The red/yellow podzolic soils belong usually to the wetter part of this transition but Richards describes a great diversity of vegetation and soils for the region on a worldwide basis.

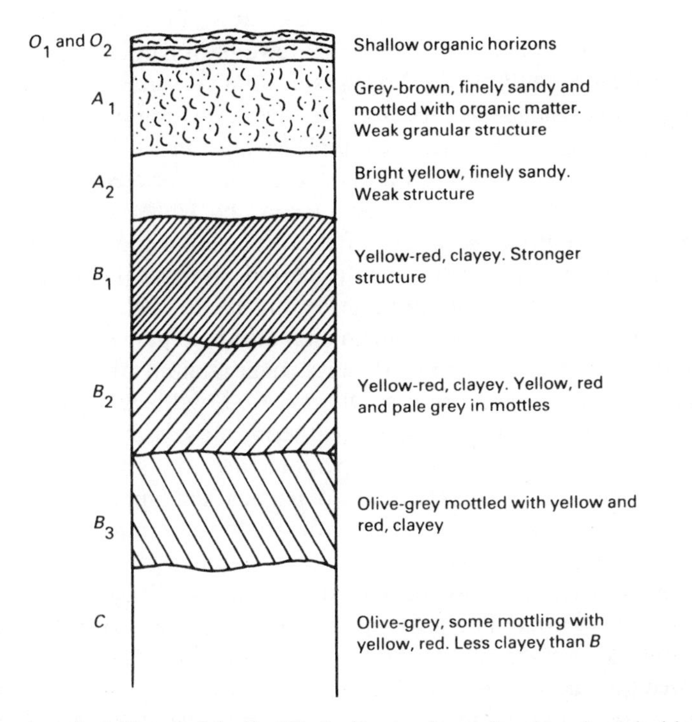

Figure 3.8 A typical *Ultisol* of the Red Podzolic type formed under mixed deciduous and pine forest. Most of the profile is below pH 5, is strongly unsaturated and the *B* horizons are rich in Fe and Al oxides. The *B* horizons are typically clay-enriched by lessivation.

Under base-rich conditions of parent material or topogeny, many soils show a higher base saturation of the argillic *B* horizon and little or no eluviation of the *A* horizon. These soils are *Alfisols* and form part of a continuum linking red-yellow podzolic soils to the mediterranean soil types and to the Alfisols of some temperate forests. Most of these alfisols, like the red and yellow podzolic soils, formed under an original forest cover which may no longer exist.

With a gradient of increasing environmental dryness there is a continuum link to temperate grassland through to desert soils and also to many savanna soils of hotter climates. The more biologically active of these grassland soils may have mollic horizons and belong to the Mollisols but, once drought is seriously limiting to plant growth, little or no horizon formation occurs and the soils may be various *Aridosols* or, in the most extreme conditions of drought, or parent material limitation, azonal *Entisols*: skeletal *lithosols* forming on bare rock and *regosols* on unconsolidated substrata.

Tropical soils

Since the writing of the first edition, the general neglect of tropical soils has been to some extent remedied by the appearance of several monographs. Burnham (1975) and Sanchez (1976) give accounts of soil types and properties, Young (1976) discusses tropical soil survey and McFarlane (1976) re-examines the controversial topic of the soils which have been named *laterites.*

Richards, as recently as 1952, suggested that in some areas of the tropics vegetation had existed in its present form since 'a very remote geological period'. This view is no longer tenable, for example Livingstone (1975) presents evidence of climatic change and forest advance or recession associated with post-Tertiary wetter and drier periods in Africa. Simpson and Haffer (1978) make similar suggestions concerning the origin of the Amazonian biota from its Tertiary origins *via* a series of refugia during dry periods. Neither the soils nor the great species diversity of the tropics can now be directly attributed to the great age or unchanging nature of the ecosystem and it is also becoming plain that the impact of man has been more substantial than was once realized (Whitmore, 1975). The geological history of rain forest is fully reviewed by Flenley (1979).

In the humid tropics deep crusts of weathering may reach a maximum of 200 m depth (Young, 1976). On base-rich parent materials this weathering process favours the genesis of the most extreme of tropical soil types, the *Oxisol* in which large quantities of easily weatherable silicates have been lost, causing residual enrichment with sesquioxides particularly of aluminium and iron, stable 1:1 lattice kaolinitic clays (p. 84) and sometimes native quartz grains. Soils on more siliceous rocks become kaolinite-enriched by reaction of dissolved silica with aluminium sesquioxide thus permitting the formation of a base-poor argillic subsurface horizon characteristic of the order *Ultisols.*

Superimposed upon these differences of parent material, the effects of relief and age of surface substantially effect soil type so that, under relatively similar Rain Forest canopies there may be localized catenary mosaics of Oxisols on terraces and plateaux, Ultisols on shallow slopes and tropical variants of temperate Brown Forest Soils (*Inceptisols*) on steep, unstable slopes (Figure 3.9). If the parent material is highly siliceous and coarse grained, these soils may be replaced by very deep *Spodosols*: tropical podzols which generally carry a less rich rainforest variant: Heath Forest (Whitmore, 1975).

Increasing base-richness of the parent material associated with a seasonal dry period may permit the development of *Alfisols* in place of Ultisols, differentiated by the much higher base-saturation of the argillic horizon. Climates and parent materials which favour the development of Alfisols often also produce a catenary sequence (p. 80) in which the Alfisols of gentle slopes and plateaux are replaced by *Vertisols* in low-lying areas where the swelling clay, montmorillonite, may be synthesized by resilication of Al_2O_3 giving these soils their characteristic 'self-ploughing' nature (Figure 3.9).

The seasonally dry climate, with increasing intensity of periodic drought, favours the replacement of full rain forest firstly with Monsoon Forest and then a sequence of Savanna Forest, Thorn Forest, Tropical Grassland (savanna) and finally Tropical Desert. Richards (1952) and Whitmore (1975) suggest burning as a strong factor influencing the rain forest–savanna boundary and Richards also attributes all tropical grasslands to the effects of fire. If this is the case it is not surprising that many savanna soils have affinities with those of rain forest areas, thus Mohr and van Baren (1959) note that fossil cuirass' of hardened laterite are now usually under grassland, and Bawden (1965) described a diverse range of sands (Tropical Podzols?) and Entisols on savanna-clad slopes with black clay soils (Vertisols) in low-lying areas and on plains of base-rich material. Kowal and Kassam (1978) give a detailed account of the soils, ecology and agriculture of savanna.

Oxisols

The diagnostic feature of these soils is an *oxic* horizon within 2 m of the surface or a continuous layer of *plinthite* within 30 cm of the surface and not overlain by a

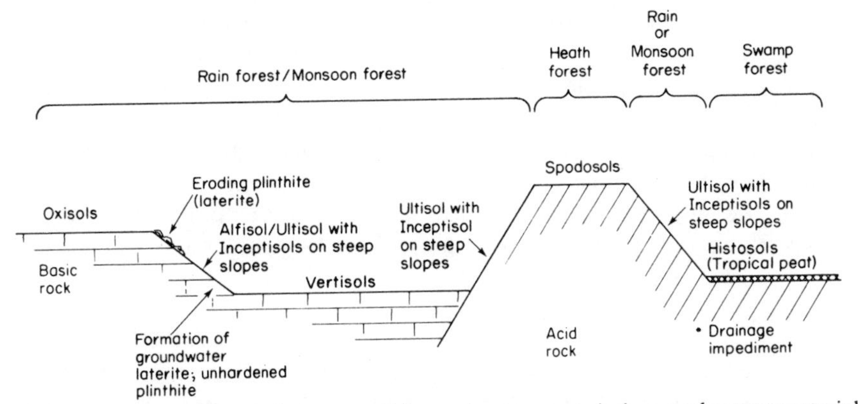

Figure 3.9 The relationship between soil formation, geomorphology and parent material in equatorial climates with slight to pronounced dry season.

spodic or an argillic horizon. Oxic horizons are made up of low-charge kaolinitic clays and sesquioxides of aluminium and/or iron and having a cation exchange capacity (p. 85) of less than 10 me 100 g^{-1} clay. It is essentially a weathering residue left by solution removal of silica from silicate minerals and in fluctuating water table conditions may be almost entirely Al_2O_3 and kaolinite, almost white in colour, but in freely drained conditions dehydrated Fe_2O_3 is very stable and gives the oxic horizon a dark red or dark brown colour.

Plinthite is a sesquioxide kaolinite layer formed by segregation of iron from other horizons or from higher adjoining areas. It may be soft and exhibits irreversible hardening when repeatedly wetted and dried or it may already be hardened *in situ.* It most nearly corresponds with the 'soil' originally defined by Buchanan from the Latin *later* (= brick) as *Laterite* (McFarlane, 1976).

The high-temperature humid climatic processes which produce residual enrichment soils also cause concentration of several characteristic metals such as Ti, Cr, Ni, Zn and Co. Base-rich alumino-silicate parent materials which give more sesquioxides on weathering are more likely to form Oxisols than more siliceous rocks. When Al_2O_3 meets silica-rich ground water it may resilicate to form kaolinitic clays or other clays.

Plinthite-containing soils (laterites) include both Oxisols and Ultisols. They are usually associated with a seasonally drier climate (Burnham, 1975) which is in agreement with McFarlane's (1976) model of laterite formation which involves iron segregation associated with an oscillating water table during down-wasting of the land surface. At the end of this process, which may be cyclically repeated, disafforestation causes induration and loss of permeability of the plinthite. The early stages of plinthite formation may occur beneath the argillic horizon of an Ultisol and thus a common tropical landscape feature is a catena of eroding 'fossil' laterite in raised areas merging into lower-lying Ultisols with unhardened plinthite (e.g. Burnham, 1975).

Oxisols show considerable diversity relating to parent material, relief and environmental history. They may range from the erosion surface laterites which are common in the savanna environments of Brazil, Zaire and Western Uganda (Figure 3.10a) to soils of various rain forest and seasonal forest environments typified by Figure 3.10b. Oxisols are the commonest of tropical soils, covering 22 % of the surface (Sanchez, 1976). Agriculturally, because of their extreme state of weathering and leaching, they pose a great problem of nutrient element limitation of plant growth and difficulty of retention of cationic fertilizers related to their low cation exchange capacity (p. 85). In some areas, though relatively limited, the presence of unhardened plinthite in the profiles makes them susceptible to surface cementation on exposure to sunlight and air.

Alfisols and Ultisols

With the exception of the azonal Aridosols, Alfisols form the second commonest tropical soil type (16 % area) and Ultisols the third (11 % area). Both of these Orders are of highly weathered and leached materials which are characterized by illuviation of clay to argillic, *B* horizons. Both soil types show a catenary sequence (p. 80), from well drained upper slopes to poorly drained footslopes, of increasing plinthite (groundwater laterite) concentrations in the subsoil. The separation of Oxisols with groundwater plinthite from these plinthitic Alfisols and Ultisols is not entirely clear as the relationship is of a continuum nature but any soil with illuviated clay overlying a continuous plinthite sheet is excluded from the Oxisols. Alfisols are formed generally on younger or more base-rich parent materials and the catena through plinthitic soils is often completed by transition to Vertisols in which montmorillonite clay has been synthesized by resilication and base status increased by transport from above (Figure 3.9).

A typical drainage impeded Ultisol profile is shown in Figure 3.10e. It differs little from an Alfisol in similar catenary position except for the lower base saturation of the argillic horizon. Upper slope Ultisols with good drainage often resemble the soils of the sub-tropics already described (Figure 3.8) though horizon differentiation is often less marked than it is in these Red-Yellow Podzolic Ultisols of cooler climates. Nye (1954) discusses the formation of an Alfisol catena over granite–gneiss in Nigeria: the soil profiles are shown in Figure 3.12.

Alfisol and Ultisols are typically formed in seasonally dry climates and consequently associated with seasonal forest, savanna and grassland rather than full rain forest. Agriculturally the Ultisols resemble the Oxisols in their extreme degree of weathering and nutrient element leaching. Enrichment with sesquioxides and kaolinitic clay results in a very low cation exchange capacity, hence artificial fertilization is difficult, leaching-loss being very rapid in wetter climates. Despite their higher base-status, Sanchez (1976) notes that many Alfisols pose an erosion problem on slopes and once the gravelly plinthite concretions of the *B* horizons are exposed may be very difficult to till. He also notes, however, that plinthite-containing Ultisols have been cultivated in the U.S.A. for over 200 years.

Spodosols

Extremely deep Tropical Podzols occur on unconsolidated alluvial and coastal sands and similar but shallower soils on very coarse-grained siliceous rocks such as granite or sandstone. The profiles are very similar to temperate zone podzols (Figure 3.10c) and their extreme acidity and nutrient deficiency permits superficial accumulation of a mor humus layer despite the climatic warmth. Many tropical podzols carry rain forest but, presumably because of nutrient limitation, it may be rather species-poor, of less than normal height and belongs to the heath forest vegetation type. Whitmore (1975) suggests that burning may convert such vegetation to savanna on podzols.

These tropical podzols belong to various thermic suborders of the Spodosols. In wet conditions Thermaquods (Ground Water Podzols) are formed on highly siliceous, almost pure quartz materials. In drier areas the profiles resemble

Humus Podzols, a thermic group of the Humods (Soil Survey Staff 1975). Burnham (1975) describes a mosaic of these wetter and drier podzol types on the ridges and depressions of old beach deposits in Malaya with heath forest on the ridges and a transition to swamp forest in the depressions.

Rivers draining podzol areas are typically 'blackwaters' consequent on dissolved organic matter and iron sols. The *Rio Negro* is so named for this reason.

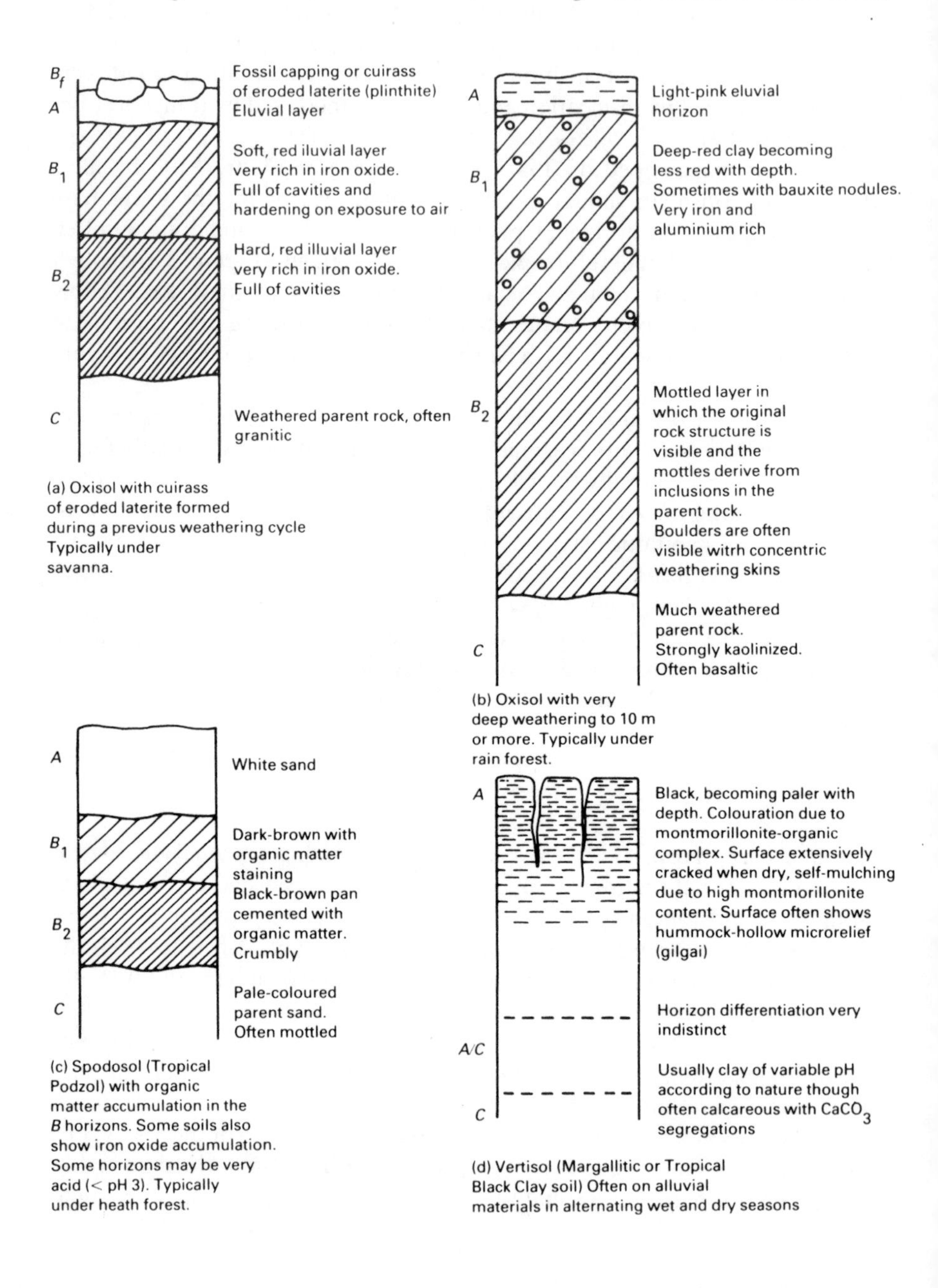

(a) Oxisol with cuirass of eroded laterite formed during a previous weathering cycle Typically under savanna.

(b) Oxisol with very deep weathering to 10 m or more. Typically under rain forest.

(c) Spodosol (Tropical Podzol) with organic matter accumulation in the *B* horizons. Some soils also show iron oxide accumulation. Some horizons may be very acid (< pH 3). Typically under heath forest.

(d) Vertisol (Margallitic or Tropical Black Clay soil) Often on alluvial materials in alternating wet and dry seasons

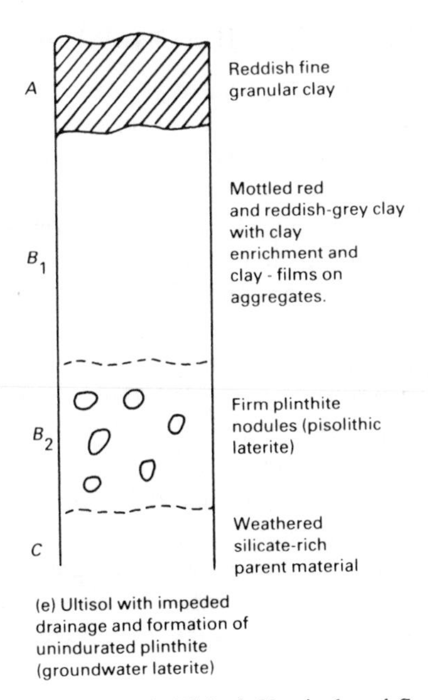

Figure 3.10 Tropical Oxisol, Ultisol, Vertisol and Spodosol profiles.

The acid, nutrient deficient drainage waters, in areas of impeded drainage may promote *Histosol* (Tropical Peat) formation carrying a swamp forest variant of the heath forest (p. 78).

Vertisols

Agriculturally attractive, these Tropical Black Clays, Black Cotton Soils, *Regur* or Margallites, are formed in low-lying areas or plateau surfaces of base-rich materials in seasonally dry climates. They are enriched with 2:1 lattice montmorillonitic clays which swell on wetting and produce the 'self-ploughing' characteristic for which the Order is named. Cracks which open in dry weather become infilled by falling soil crumbs eventually *inverting* the profile. Vertisols are frequently formed at the base of drainage catenas and the montmorillonite synthesized by resilication of Al_2O_3 by inwash of dissolved silica from above. For the same reason the profiles are typically base-rich and often contain free-calcium carbonate at least in the deeper parts of the profile (Figure 3.10d). The swelling phenomenon often leads to the formation of micro-relief hummock–hollow patterns or *gilgai* ranging from a few centimetres to several metres in width.

Because they are agriculturally tractable, most natural vegetation of vertisols has been destroyed and they now bear, in semi-natural conditions, secondary forest or savanna but it seems likely that their former vegetation must have been some form of seasonal forest.

VEGETATION OF THE ZONAL SOIL TYPES

Descriptions of the vegetation associated with the various soil types have, of necessity, been limited in this text. Early source-descriptions may be found in Schimper's classic *Plant Geography upon a Physiological Basis* (1898) while more recent plant geographies are those of Dansereau (1957), Polunin (1960), Good (1974) and Walter (1979). Oosting (1956) gives a brief general description of North American vegetation which is dealt with more extensively and with beautiful illustrations by Gleason and Cronquist (1964). Ellenberg (1963) describes the vegetation of Central Europe and gives a great deal of useful environmental and physiological information. Richard's classic, *The Tropical Rainforest* (1952) is now supplemented by Whitmore's (1975) account of the far-eastern rain forests and by Flenley's (1979) geological history of the formation, while Walter (1971) gives a general account of tropical vegetation with substantial physiological and environmental background. Specialist texts on various aspects of specific vegetation zones are those of Mooney and di Castri (1973) on mediterranean vegetation, and Weigolaski (1975) on tundra ecosystems. A number of introductory texts have recently appeared, for example Longman and Jenik (1974) and Janzen (1975) on tropical ecology and Pruitt (1978) on boreal ecosystems.

AZONAL SOILS

The azonal soils or *Entisols* are incipient soils in which pedogenesis has hardly begun. They show no obvious profile differentiation and thus have no apparent affinity with the neighbouring zonal soils. They are classified, according to parent material, into *regosols*, *lithosols* and *alluvial* soils. Regosols are formed on soft, unconsolidated materials such as dune sand, loess and glacial till, while the lithosols are associated with stony substrata and include many desert and tundra soils and soils of unstable slopes. Some thin rendzinas (p. 79) also fall into the lithosol category as do *rankers* with an OA_1 horizon resting directly on nutrient-deficient, siliceous parent rocks.

Alluvial soils are very variable and are classified by texture, ranging from the finest clay/silt soils through sand and gravel to boulder deposits. Many *marsh soils* are of clay/silt texture of either freshwater or marine (saltmarsh) origin. The waterside meadow or *warp* soils of many temperate zone river valleys are often rich, having a strong biological activity and being maintained in a pedogenetically young condition by repeated input of silt. Another characteristic alluvial form is the *gyttja* soil, initially formed under water in eutrophic conditions, and consisting of clay with much organic matter, coloured black by the precipitation of ferrous sulphide.

INTRAZONAL SOILS

These are soils in which a local factor dominates over the zonal climate and vegetation in pedogenesis. There are three normal variants of hydromorphic, calcimorphic and halomorphic types. Many classifications have considered these to be distinct soil types but the USDA taxonomy presents most of them as deviants from the adjacent zonal soil forms.

Hydromorphic soils

The most widespread variant in humid climates is the *Gley*, which has a part of the profile waterlogged and consequently in a reducing rather than an oxidizing condition. The detailed consequences of waterlogging are discussed in Chapter 10 but the immediate morphological effect is seen in soil colour. Iron in the ferric condition gives most soils their yellow, brown or red colorations which contrast strongly with the blue-grey and grey colours of the ferrous compounds in waterlogged soils. This subdued coloration is often intermingled with a mottling of yellow or red-brown where local oxidation occurs adjacent to air-filled voids or to living roots.

Another consequence of the lowered oxygen tension and reduced oxygen diffusion rates in wet soils is that organic matter decays rather slowly, either causing organic enrichment of the *A* horizon or build-up of superficial peat.

The origin of the gley characteristic may be a high ground water level or the retention of surface water by a superficial impervious layer in the soil, the latter situation forming *surface water gleys* (*pseudogleys*). Schlichting and Schwertmann (1973) have edited an extensive symposium volume on these hydromorphic soils.

The gley soils show enormous variation in response to degree of wetness and nutrient/acidity status. Under eutrophic, high pH conditions *mull gleys* occur and have an *A* horizon not dissimilar to those of adjacent brown forest soils. Increasing wetness brings waterlogging nearer to the surface, causing fen-like conditions to develop and the accumulation of a surface layer of nutrient-rich fen peat (Figure 3.11). Increasing acidity and nutrient deficiency reduce rates of organic decomposition and produce a range of soils having affinities with the podzols: these are the *humic gleys*, *peaty gley podzols* (*half-bog soils*) and *bog gleys* with a thick peat cover (Figure 3.11).

The acid gley soils best show their relationship to the podzols in the sequence of soils commencing with the peaty gley podzols which have a thin iron pan in the upper part of the *B* horizon, but otherwise are similar in profile to the normal podzol (Figure 3.11). Their A_2 horizon is, however, grey or greenish-grey in colour with none of the pale pink tinge associated with well-drained podzols. The iron pan, which is hard and black or brown in colour, underlies this layer but, below it, the remainder of the *B* horizon is bright reddish-brown and well-aerated. Crompton (1956) suggests that superficial waterlogging mobilizes iron which reprecipitates as a pan, in the *B* horizon, when it comes into contact with this better aerated layer. The peaty gleys are consequently typical of areas with high rainfall and low evaporation in which the surface soil remains very wet during most of the year. As the pan thickens it may increase this effect by impeding drainage.

The origin of peat-forming ecosystems (mires) is controlled by the interaction of hydrological, climatic and geochemical factors. Three distinct categories can be classified (Moore and Bellamy, 1973). *Primary peats* depend on flow of drainage water into depressions and their formation reduces the water-holding capacity of the basin. *Secondary peats* extend beyond the confines of the basin, the peat retaining water by capillarity and increasing the water retention of the

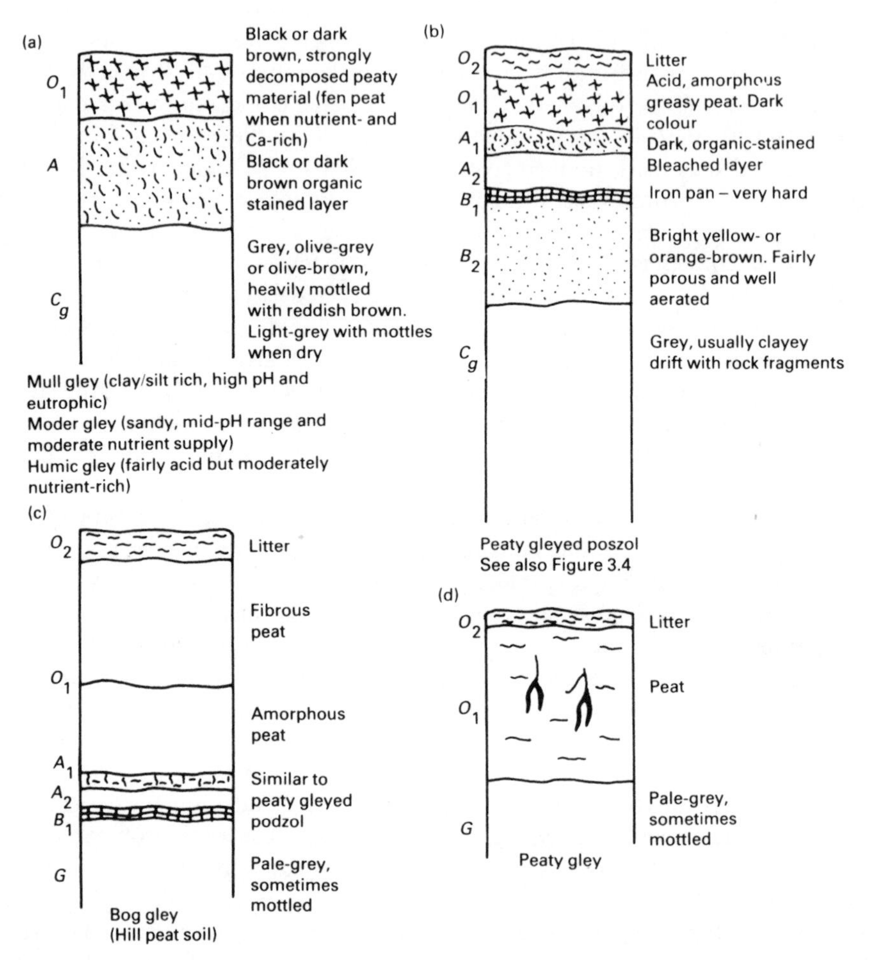

Figure 3.11 The influence of wetness, acidity and nutrient status on the formation of gley soils. (a) Under nutrient-rich conditions a mull-gley is formed. With declining nutrient status and increasing acidity either moder-gley or humic-gley soils develop. These three types have similar profile and differ mainly in chemical characteristics, humus formation and vegetation. (b) In circumstances conducive to podzolization moderately high rainfall causes formation of peaty gleyed podzols while very high rainfall initiates ombrotrophic peat (blanket peat) formation giving (c) bog gley soils often overlying a thin iron pan similar to (b). Shallow peat soils often form directly overlying a gley horizon as peaty gley soils (d).

surface but the water remains in hydrological contact with the groundwater. *Tertiary peats* are similar but their perched water table is out of contact with ground-water, fed only by rain (Figure 3.12).

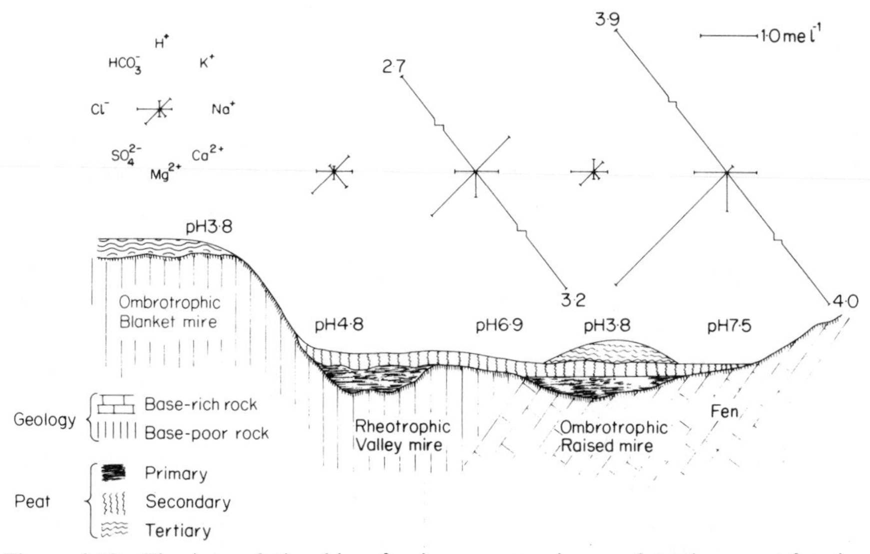

Figure 3.12 The interrelationship of primary, secondary and tertiary peat-forming environments. The rose-diagrams show the relative concentrations of important ion species in the interstitial water of the surface peat.

Primary mires are *rheotrophic* (Greek *rheos*—a stream) or, alternatively, *soligenous* while the secondary and tertiary rainfed systems are *ombrotrophic* (Greek *ombros*—a rainstorm) or *ombrogenous.* They vary from each other in nutrient content: if the groundwater is calcareous and nutrient-rich, rheotrophic mires may form *fen peat* but even when the drainage water is acid and oligotrophic, primary *basin peats* are substantially more nutrient rich and less acid than secondary and tertiary ombrotrophic *raised-bog* and *blanket peats* which depend on rain-dissolved materials and blown dust for nutrition.

Primary peats are found throughout the world, though more commonly at high latitudes their formation requiring only the accumulation of drainage water and not the high P_r/E_v ratios which engender secondary and tertiary mires. Despite this, extensive lowland mires in the tropics are more or less confined to riverine areas and coastal sites where they merge with mangrove swamps some of which also form peat. Ombrotrophic peats do occur in the tropics with sufficiently wet conditions, for example, Richards (1952) describes very deep raised bog peats comprising the very acid, porridge-like remains of swamp-forest trees and overlying bleached clay soils which strongly resemble the 'seat-earths' of coal-seams.

Secondary peat grows ombrotrophically from a primary basin mire (Figure 3.12) to form the classically dome-shaped *raised bog* and, as it becomes older and higher, the peat is progressively more oligotrophic until, when it loses contact with ground water, the transition to tertiary peat is completed. In temperate and boreal environments, many flat or gently sloping surfaces become covered by ombrotrophic *blanket peat* without any sequence of primary and secondary peats beneath.

If these soils are engendered only by high P_r/E_v conditions they must be considered zonal rather than azonal in their geography. There is some suspicion that temperate blanket peats and iron-pan peaty gley podzols may be anthropic, having developed after man removed the original tree-cover reducing the transpiration demand in already rather wet climates (Crampton, 1968; Moore, 1975).

Insight to the relationship between mire types is given by their water chemistry (Moore and Bellamy, 1973): rheotrophic peats are dominated by high concentrations of elements such as calcium and sulphate-sulphur derived from inorganic soil (Figure 3.12). The transition to the ombrotrophic peats of the raised- and blanket-mires is accompanied by a great reduction in the concentration of all elements and a shift in their relative proportions to resemble those of the incoming rainwater. For this reason, coastal ombrotrophic peats may be enriched with sodium, chloride and potassium from seaspray but without major effects on vegetation. Alkaline waters such as those of fen peats also contain substantial quantities of bicarbonate.

The various gley soils belong to aquic suborders of the U.S. Comprehensive system unless the peat layer (histic epipedon) exceeds 40–60 cm, in which case they are classified in a separate order, the *Histosols*. Some soils satisfying other specialized criteria are also included as histosols (see Soil Survey Staff, 1975; and Buol *et al.*, 1973). Mull, Moder and Humic Gleys may be Aquolls or Aquepts, Peaty Gley Podzols and Bog Gleys are either Aquods or Fibrists and Peaty Gleys are Aquepts or Fibrists. Under tropical conditions with very acid, calcium deficient drainage water, the tropical peats may be Fibrists or Saprists. Aquic variants of all other tropical soils may be encountered, thus Aquox, Aquults and Aqualfs are either Groundwater Laterite soils or Low Humic Gley soils while gleyed Vertisols are Aquerts. The wetter forms of all these soils may bear tropical swamp forest, often species-poor and resembling the heath forest of tropical spodosols.

The vegetation of acid gleys in Northern Europe varies according to wetness and oligotrophy. The humic gleys often have a swamp woodland of *Salix* and *Alnus* spp (willow and alder) or a poor marsh vegetation of herbaceous hydrophytes. The peaty gleys usually have a surface mat of peat-forming mosses, including *Sphagnum* spp., with a cover of grasses such as *Molinia caerulea* (purple moor grass). Under slightly drier conditions the vegetation of the iron pan podzols may be of Ericeae such as *Calluna vulgaris* (ling) and *Vaccinium myrtillus* (bilberry). The bog gleys, much wetter soils, generally support peat-formers such as *Sphagnum* spp., *Molinia caerulea* and *Eriophorum* spp. (cotton grass). In basin peat areas some indicators of nutrient enrichment such as *Juncus* (rushes) spp. may appear, but generally the vegetation of valley bog, raised bog and blanket bog is superficially similar, conditioned by nutrient deficiency, acidity and waterlogging. Because these characteristics are continuous and partially independent variables, the range of intermediate soil types and vegetations is very large (Moore and Bellamy, 1973).

The variation of soil type with topographic wetness is expressed in the concept of the *catena*. First described in relation to tropical soils, it is encountered wherever topography generates a pattern of repeating soil types. Such toposequences may arise on the same or on differing parent materials and reflect not only soil wetness but also the influence of evaporation, insolation and erosion. Figure 3.13 shows a number of catenary soil sequences of various types.

Calcimorphic soils

Soils formed on relatively pure limestone in the temperate zone are generally *Rendzinas*. They are usually rather immature, thin soils with an *A* horizon resting directly on the *C* horizon limestone, or perhaps on a C_a horizon under drier conditions (Figure 3.14).

The *A* horizon is typically black, dark grey or dark brown in colour and ranges from a few cm to *c.* 35 cm in depth. The dark coloration is caused by organic matter in the presence of high concentrations of calcium carbonate; there are often free fragments of limestone parent material in this horizon which have been incorporated by earthworm activity.

Those in which the A horizon is mollic belong, like the chernozems and brown forest soils, to the *Mollisols* of which a separate suborder, the *Rendoll* is reserved for the rendzina (Soil Survey Staff, 1975).

The rendzinas are most often associated with pure limestone and, on clay-rich calcareous materials, deeper calcimorphic brown soils may form. With increasing temperature there is a transition between the rendzinas and the terra rossa/terra fusca complex (Stace, 1956).

Rendzinas may be highly alkaline, their pH values ranging between *c.* 7·5 and 8·3, reflecting the high calcium carbonate content. Within this pH range the solubility of many nutrients is much reduced, for example, iron deficiency and lime-induced chlorosis are typical of such soils. As a consequence, many plant species of these soils are *calcicoles*, either obligate or facultative, while *calcifuge* plants are usually excluded from the association. Further discussion of this problem is presented in Chapter 9.

The native vegetation of most rendzinas is woodland; in northern Europe probably of *Fagus sylvatica* (beech) or, with increasing rainfall and harder parent limestones, *Fraxinus excelsior* (ash). In central Europe alpine rendzinas are formed under a vegetation of cushion grass or grass heath, but in the more oceanic climate of Britain and north-west Europe the extensive chalk and limestone grasslands dominated by *Festuca rubra* and *F. ovina*, or by taller grasses such as *Zerna erecta* or *Brachypodium pinnatum*, are biotic plagioclimaxes maintained by rabbit- or sheep-grazing. With lessened grazing pressure there is a subsere through a limestone scrub of *Swida sanguinea* (dogwood) and *Crataegus monogyna* (hawthorn) to a woodland dominated by *Fraxinus excelsior* and/or *Fagus sylvatica.* The limestone grasslands are particularly rich in herbaceous rosette and creeping species which give them a characteristic colourful appearance during spring and summer and comprise a range of calcicoles of varying degrees of exclusiveness and a few more ubiquitous plants.

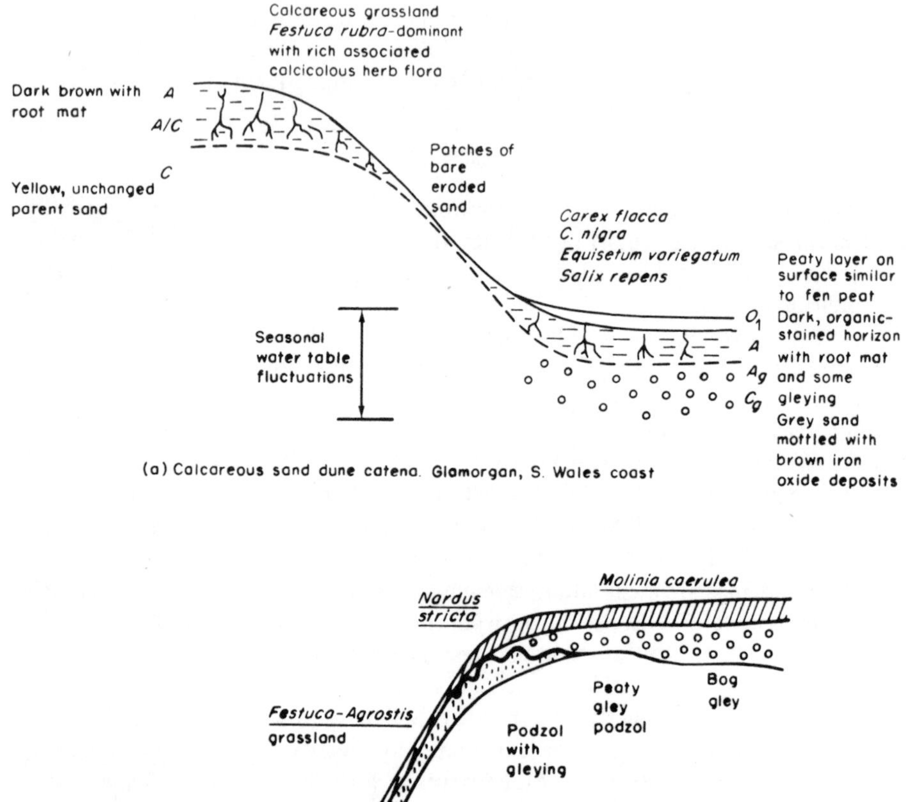

(a) Calcareous sand dune catena. Glamorgan, S. Wales coast

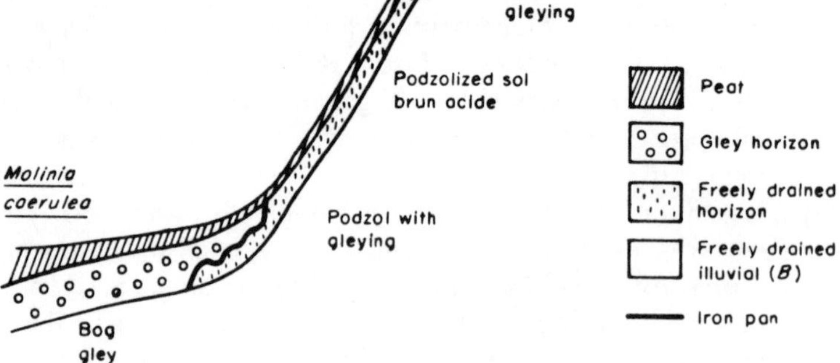

(b) British upland catena characteristic of high rainfall with oligotrophic parent materials

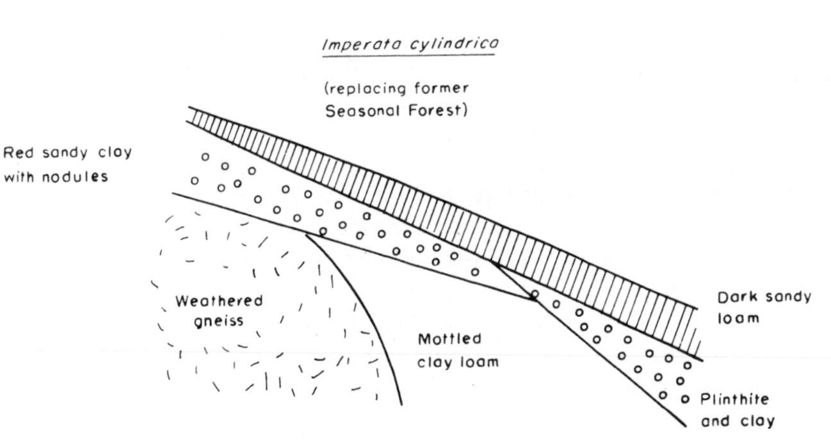

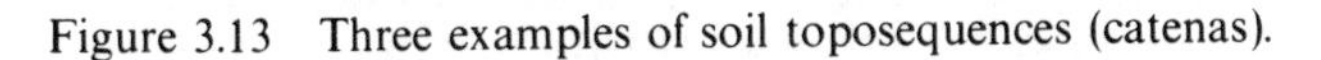

Figure 3.13 Three examples of soil toposequences (catenas).

Halomorphic soils

Most halomorphic soils are formed in arid zones by surface accumulation of soluble salts related to high ground water levels, impermeable substrata and high evaporation rates, thus having an originally hydromorphic character. Kubiena (1953) describes primary salt soils, which are gleys in which a range of salts have crystallized, at the surface, by evaporation and capillary rise under a low P/E regime. Secondary salt soils occur under the same climatic conditions by superimposed waterlogging, for example as a result of overirrigation or leaking irrigation canals and have caused serious agricultural loss in areas of extensive irrigation such as the Indus Basin in the Punjab.

The most salt-rich saline soil is the *Solonchak* which has a salt-saturated pale grey *A* horizon overlying a mottled gley layer. In the spring the soil may be

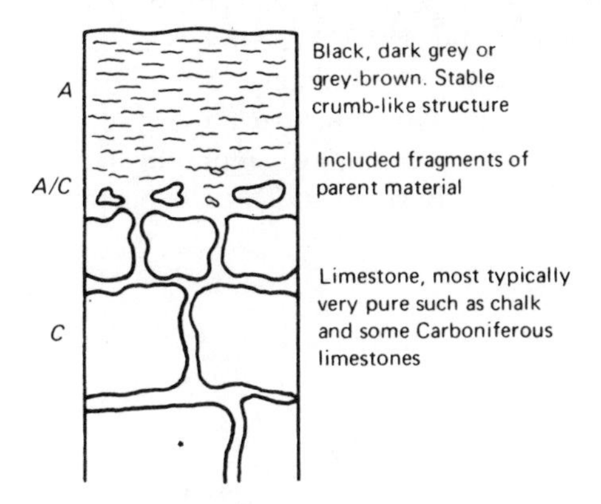

Figure 3.14 A *Rendoll* (rendzina) profile.

flooded but, as the season advances and the soil dries, it turns white, and patchy efflorescences of salt crystals may occur at the surface. This effect and the pale colour of the soil are responsible for the alternative name of *White Alkali soil*. The salts are usually chlorides and sulphates of sodium, calcium and magnesium with a soil pH of above 8·0 due to the presence of some carbonates. If very large amounts of sodium carbonate are present, the pH may rise to 10·0 or more.

The saline soils contrast with the carbonate soils or *Solonetz* in which the predominate salt is sodium carbonate. The soil organic matter is very dark in colour, causing these soils to be called *Black Alkali soils*. They are not salt-enriched at the surface, as is the solonchak, the sodium in the upper part of the profile being in exchangeable form and only appearing as a carbonate accumulation lower in the profile. Most of these soils are saline members of the order Aridosols, mainly Salorthids and Natrargids.

The native vegetation of these soils is strongly halophytic and includes a high proportion of species drawn from the family Chenapodiaceae; for example, the Great Basin Desert of the Western U.S.A. has extensive areas of salt desert which may be dominated by Shadscale (*Atriplex* spp.) and Winter Fat (*Eurotia lantana*). With higher salt content *Salicornia* spp. (glass wort) become dominant. Species of *Atriplex*, *Salicornia*, *Halimione*, *Suaeda* and *Salsola* may be found on saline soils whether of salt deserts or of temperate zone salt marshes in many parts of the world.

The coastal regions of the temperate zone are fringed by salt-marshes which, depending on current precipitation–evaporation balance may have a soil solution more saline than seawater to relatively non-saline. These marshes carry a specialized halophyte vegetation (V. J. Chapman, 1976, 1977) of which some physiological characteristics are discussed in chapter 9. The soils, usually of alluvial material, are often strongly reducing as sodium-deflocculated clay and silt seal their pore structure against ingress of air. The deeper horizons are thus either grey-blue with reduced iron compounds or black with ferrous sulphide deposits (see p. 304).

In tropical climates the salt marsh fringes are replaced by mangrove swamps (Lugo and Snedaker, 1974) which may range from pioneer mudflat colonization with daily tidal flooding to much less saline soils inland. New alluvium trapped by the trees is green-grey, reducing and alkaline to neutral: further from the sea black sulphide muds occur and, with drainage these give extremely acid (pH 2–3) sulphate soils (cat-clays) containing free sulphuric acid, coloured bright yellow with basic ferric sulphates (Burnham, 1975; Chapman, 1977; Bridges 1978).

Chapter 4

Chemical and physical properties of soils. The root environment

INTRODUCTION

The soil is a three-phase system of solid, liquid and gaseous components, each of which has its own physical and chemical properties, in an equilibrium, or transient-state, relationship with the others. The liquid and gaseous phases are, in small soil volumes, fairly homogeneous. By contrast the solid phase is heterogeneous, comprising a range of different sized inorganic particles of silica, silicate clay, metal oxides and other minor components, all in varying degrees of association with different types of organic matter. The nature of this association and the characteristics of the organic matter vary greatly with soil type.

In addition to these components, each soil has a distinctive flora and fauna of Bacteria, Fungi, Blue-green algae, Algae, Protozoa, Rotiferae, Nematoda, Oligochaetae, Mollusca and Arthropoda. This assemblage of organisms not only reflects the present status of a soil, but also affects the pedogenetic process by modifying the course of organic decomposition, by influencing chemical processes and by altering the physical structure of the soil. The interaction of an earthworm population with soil fungi and bacteria, for example, causes the strong aggregation of many mollic soil horizons.

THE MINERAL MATRIX AND CLAY CHEMISTRY

The soil fabric consists of a mixture of sand, silt and clay-sized mineral particles (p. 111). In cool climates the clay fraction is predominately of 2:1 and some 1:1 layer alumino-silicate clays (see below), while highly weathered soils in hot climates more commonly contain 1:1 layer clays mixed with various amorphous and crystalline species of iron and aluminium oxides and hydroxides.

The clay-sized fraction provides a major source of ion-exchanging activity to which must be added the effect of soil organic matter. The clay particles, with their large specific surface (p. 45), are readily decomposed by weathering processes and so provide the primary source of replenishment for the soil-plant mobile nutrient element pool.

Grim (1953) gives a detailed account of clay mineral structure and attributes the basic generalizations to Pauling (1930). Electron micrographs show that clays are made up of flat plate-like or needle-like crystals in which X-ray diffraction studies have shown atomic layer lattices involving two types of structural subunits (Figures 4.1 and 4.2). One consists of two sheets of close-packed oxygen atoms or hydroxyl units in which aluminium (or magnesium or iron) atoms are embedded in octahedral coordination, each aluminium atom thus being

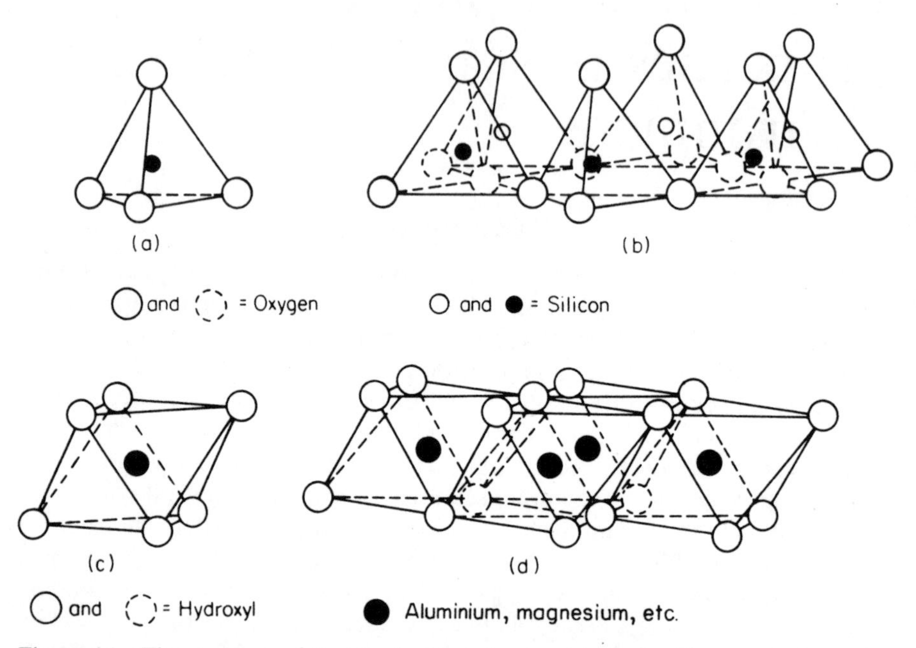

Figure 4.1 The structure of the clay lattice subunits. (a) Single silica tetrahedron; (b) sheet structure of silica tetrahedra; (c) single octahedral unit; (d) sheet structure of octahedral units. Reproduced with permission from R. E. Grim, *Clay Mineralogy*, McGraw-Hill, London, 1953. Figure 1, p. 43.

equidistant from six O— or OH— positions. The other subunit is a sheet of silica tetrahedra in which each silicon atom is equidistant from four oxygens.

Soils contain a wide variety of silicate clays but a few common examples may be cited. Highly weathered soils such as the oxisols and ultisols usually contain *kaolinite* while soils of less extreme weathering, for example, the mollisols, often contain *vermiculite* and *montmorillonite*. Many immature soils have *mica* clays which are usually direct derivates of mica in the parent material.

The silicate clays mentioned above have layer lattices which are based on the aluminium octahedral sheets and silicon tetrahedral sheets bonded together to form a 1:1 lattice with one aluminium and one silicon sheet or a 2:1 lattice with a central aluminium sheet sandwiched between two silicon sheets. Kaolinite is a 1:1 clay in which the unit cells of the lattice are hydrogen bonded together

(Figure 4.2) and do not swell on hydration as water molecules cannot enter between the lattice layers. By contrast, the 2:1 lattices of vermiculite and montmorillonite show strong swelling on hydration as they have, respectively, magnesium and calcium ions in the interlayers, associated with varying amounts of water of hydration. The thickness of the unit cell may vary from 1·0 to 1·5 nanometers according to the degree of hydration. The 2:1 mica clays have potassium ions in the interlayers, without water of hydration, and do not swell on wetting. Soils such as the vertisols which show very marked volume changes and crack formation during the wetting–drying cycle have a high proportion of swelling clay, usually montmorillonite.

Highly weathered oxidols or ultisols may, in addition, contain the amorphous aluminium silicates *allophane* and *halloysite* often with considerable amounts of clay-sized iron and aluminium (hydr)oxide. These may be amorphous precipitates of $R(OH)_3$ or crystallized to form the primary hydrated oxides *goethite* and *gibbsite* ($FeOOH$ and $Al(OH)_3$). These may dehydrate to haematite and bohemite (Fe_2O_3 and $AlOOH$).

CATION EXCHANGE

If a solution of a neutral salt such as potassium chloride is flushed through a soil the outflowing liquid is usually found to be depleted of potassium, enriched with other metal cations and often reduced in pH due to enrichment with hydrogen ions. These changes have been wrought by the *cation exchange complex* of the soil, consisting mainly of the clay fraction and the organic matter. The cation exchanging property is caused by the presence of unsatisfied negative charges at the surface of the clay and organic particles; metal cations and hydrogen ions are bound at these sites, sufficiently loosely to show exchange with ions in the bathing solution. Table 4.1 shows the *cation exchange capacity* (CEC) of various clays and Table 4.2 that of different soils. It may be seen that the CEC of sandy soils is associated mainly with their organic content but that of finer textured soils is a function of clay content. The ion-exchanging behaviour of organic matter is discussed later (p. 108).

Silicate clays have ion-exchanging properties for three reasons: (i) broken bonds at the edges of aluminium/silicon units produce unsatisfied charges; (ii) some exposed hydroxyl units may lose hydrogen ions in exchange for metal cations under certain pH conditions and (iii), usually most important, individual atoms in the lattice may be isomorphously substituted; trivalent aluminium for quadrivalent silicon and divalent iron, calcium and magnesium for aluminium. The substitution does not disrupt the lattice sufficiently to alter the crystal structure but it does result in a surplus negative charge which causes cation binding at the crystal surface. Montmorillonite has a high cation exchange capacity which is mainly due to considerable isomorphous substitution of magnesium for aluminium. By contrast, kaolinite has a much lower CEC, deriving only from broken bonds and hydroxyls, there being no isomorphous substitution within the lattice.

Table 4.1 The cation exchange capacity of some clay minerals

Clay	C.E.C. (me %)
Kaolinite	3–15
Montmorillonite	80–150
Illite (mica-type clays)	10–40

Table 4.2 The cation exchange capacity of different soils*

Soil	Horizon	% Organic C	% Clay	C.E.C. (me %)
Spodosol (low clay content)	A_1	1.1	1	5
	A_2	0.03	1	1
	B	2.5	7	28
	C	0.02	5	1.4
Mollisol (high clay content)	A	2.0	36	27
	B_1	1.1	38	29
	B_2	0.6	34	26
	C	0.1	29	25
Oxisol	A	2.7	51	11
	B	0.4	51	5
	C	0.2	36	3
Soil organic carbon	—	—	—	c. 100–400

* Data mainly from Soil Survey Staff (1975).

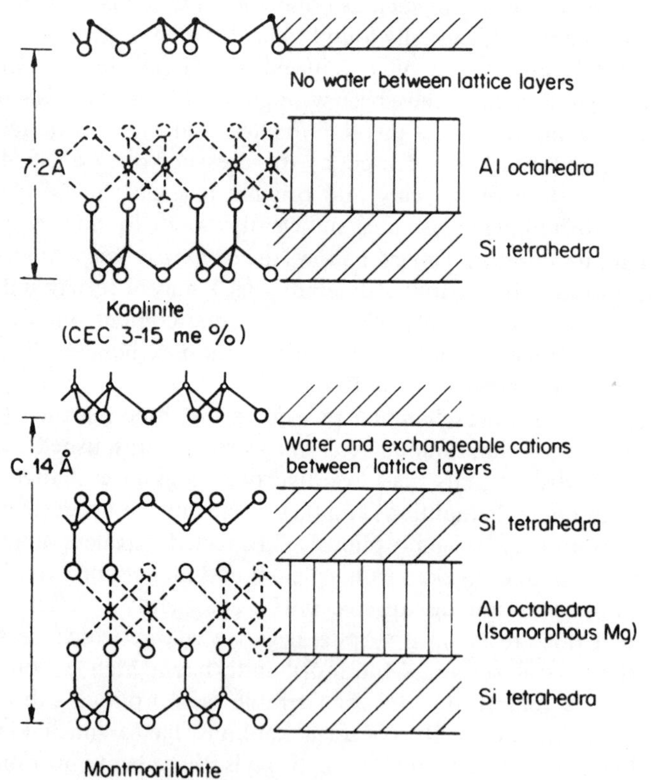

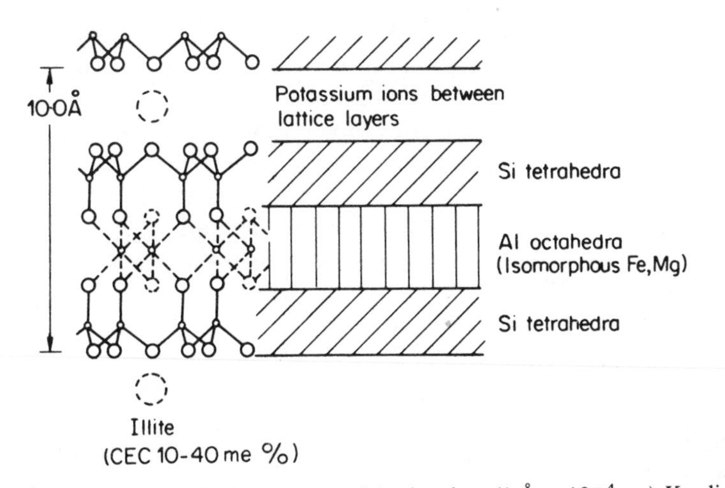

Figure 4.2 The crystal structure of 1:1 and 2:1 lattice clays (1 Å = $10^{-4}\mu$m). Kaolinite is a non-swelling 1:1 clay which does not take in water molecules between the lattice layers. It has no isomorphous replacement and, consequently, a low CEC. Montmorillonite has a 2:1 lattice which, in the expanded form, holds a layer of water between the lattice layers. In the dehydrated condition the lattice closes to a unit spacing of about 9.3 Å. The cation exchange capacity of montmorillonite is high because aluminium is isomorphously replaced by magnesium and a large exchange surface is available between the lattice layers. Illite is a micaceous, non-swelling clay which structurally resembles montmorillonite but has its lattice layers bound together by non-exchangeable potassium ions and cannot expand by absorption of water. Though it shows extensive isomorphous replacement of aluminium by iron and magnesium it has a low exchange capacity as the interlattice layers are not available for exchange. After R. E. Grim, *Clay Mineralogy*, McGraw-Hill, London, 1953, Figure 2, p. 44.

The presence of ion exchangers in the soil is of great importance both in pedogenesis and in the soil–plant nutritional relationship. Most metallic elements which are taken up by growing plants are absorbed as cations but they exist in three forms in the soil: (i) as sparingly soluble components of mineral or organic material; (ii) adsorbed onto the cation exchange complex and (iii) in small quantities in soil solution. Free cations in solution are easily leached from the soil if the P/E ratio is high but the exchange complex forms a reservoir of nutrients which are not so easily lost in this way, though maintaining a continuous supply to the soil solution by slow equilibration. Hence soils may exist, under high rainfall conditions, which maintain a steady supply of nutrients to the plant cover without becoming rapidly depleted of nutrients by the leaching process. The plants also act as bio-cyclers in this relationship, the root systems extracting nutrients from deeper horizons and thence returning them to the soil surface in litter. As decomposition proceeds the liberated cations return to the exchange complex of the surface layers.

Despite their nutrient-retaining behaviour many soils ultimately do become acidified by desaturation of the exchange complex; metal cations being replaced by *exchangeable hydrogen*. The acidification of a neutral salt solution on passage through a soil is caused by the flushing of exchangeable hydrogen ions from the exchange complex. There is, however, some difficulty in interpreting this observation as the leachate from an acid soil almost invariably contains aluminium amongst the displaced cations. Aluminium ions rapidly react with water to form insoluble aluminium hydroxide and hydrogen ions; it is thus difficult to know whether the increased acidity of the leachate derives from exchangeable hydrogen or exchangeable aluminium in the soil. It is generally assumed (Rorison, 1973) that, however leached a soil, the predominant exchangeable cation is aluminium and hydrogen ions are only generated secondarily on equilibrium with solution:

$$\text{CEC–Al} \rightleftharpoons \text{Al}^{3+} \underset{+3\text{H}_2\text{O}}{\rightleftharpoons} \text{Al(OH)}_3 + 3\,\text{H}^+.$$

It is conventional to express the *cation saturation* of a soil as the percentage of the cation exchange capacity occupied by metal cations. Cation exchange measurements are usually made at a standardized pH as the behaviour of some ion binding sites is pH dependent.

Most techniques for measuring cation exchange capacity, individual exchangeable cations, cation saturation and exchangeable hydrogen have depended on various exchange displacement methods in which a high concentration of a single salt is leached through a soil sample so that the exchange complex becomes totally saturated with a single cation species and the originally adsorbed cations appear in the leachate. Neutral ammonium acetate has been most widely used; metal cations may then be measured in the extract and cation exchange capacity determined by distilling the adsorbed ammonia from the soil sample and measuring it either volumetrically or colorimetrically. Exchangeable hydrogen may be estimated from the difference between the total metal cation value and the cation exchange capacity.

Exchangeable hydrogen or exchange acidity may also be measured directly, for example by leaching with a buffer solution of barium chloride and triethanolamine, followed by titration to estimate the exchange acidity resulting from displacement of both hydrogen and aluminium ions. Details of these techniques may be found in Jackson (1958), Chapman and Pratt (1961), Black (1965) and Hesse (1971). The origin of cation exchange activity in a number of different mechanisms, depending on dissociation of different components of the system, leads to a pH sensitivity; thus cation exchange capacity, exchangeable hydrogen and exchangeable metal cations should all be measured at a specified pH (often pH 7). It might be more desirable in plant nutritional studies to measure these values at the prevailing soil pH.

ANION EXCHANGE

Materials which have free positive charges on their surfaces act as anion exchangers. Silicate clays and organic matter both have some such charges which are sufficiently spatially removed from negative charges not to be neutralized, hence kaolinite has about 2 me % and soil organic matter 5–10 me % anion exchanging activity. These values are determined by exchange saturation with anions, for example phosphate, in acid solution (e.g. Hesse, 1971). Many of the clay minerals and metal oxides of highly weathered tropical soils show substantial anion exchange capacity, for example halloysite 15, allophane 17 and gibbsite 5 me % (Sanchez, 1976). This effect is compounded by the occlusion of silicate clay particles in a sheath of iron oxide in such soils. These strongly anion exchanging minerals also have a substantial portion of their cation exchange capacity variable with pH thus, in acid soils dominated by oxide and 1:1 lattice clays, there may be a 'negative' cation exchange capacity. When leached with neutral potassium chloride these soils produce an alkaline leachate by displacement of OH^- ions (−ve ΔpH) compared with the more usual +ve ΔpH of 2:1 lattice clays. Most soils, even those which have oxic horizons, manifest a +ve ΔpH but a few oxisol subsoils show the negative relationship and are particularly easily leached of cations but show retention of nitrate, phosphate and sulphate by anion exchange.

SOIL SOLUTION

The soil solution is the aqueous component of a soil at field moisture content (Adams, 1974). Its dissolved electrolyte content is a function of the exchange equilibria of cations with the exchange complex, solution equilibria of soluble inorganic materials and microbiological mineralization equilibria of such materials as nitrogen and sulphur containing organic compounds. The majority of techniques for extracting soluble electrolytes involve either the addition of water to the soil or some delay between field sampling and the withdrawal of the solution, time in which nitrification and denitrification may alter the status of the solution. The techniques which have been used to overcome these problems as far as possible are reviewed by Adams (1974). One of the two most widely used is displacement, in which a suitable solution is added to the top of a carefully packed soil column: moving downward, it displaces the interstitial solution which is collected at the bottom of the column. Suction plate or pressure plate extraction is the second technique (p. 155), but in this case the 'sieving' action of the ceramic plate or the effect of altered gas pressure may have modified the solution composition. Some soil solution analyses are given in Chapter 9, Table 9.3.

Soil solution components may move through the soil by mass flow with soil water or by diffusion. Movement by diffusion alone is defined by Fick's first law:

$$F_s = -D_{if}\frac{dC_{soil}}{dx} \tag{4.1}$$

where F_s = the flux of solute (mol m^{-2} s^{-1}), D_{if} = the diffusion coefficient (m^2 s^{-1}), C_{soil} = the concentration of diffusible solute in soil (mol m^{-3} soil), and x = distance (m).

If water movement is involved, mass flow effects must be added:

$$F_s = -D_{is}\frac{dC_{soil}}{dx} + JC_{liq} \tag{4.2}$$

where D_{is} = the dispersion coefficient ($m^2 s^{-1}$), which differs from the diffusion coefficient because water movement itself causes some dispersion of molecules, J = the water flux ($m^3 m^{-2} s^{-1}$), and C_{liq} = the concentration of solutes in soil solution ($mol\, m^{-3}$ solution).

These equations define the movement of solutes in one dimension in response to water movement and diffusion gradients. Nye and Tinker (1977) develop the more complex equations of the cylindrical geometry of the rooting cylinder or the spherical geometry of systems such as soil aggregates.

Solute movement in mass flow may occur downward as a result of infiltrating rainfall, upward with capillary rise to satisfy surface evaporation or in any direction to meet root absorption demand. In each case it is possible that soil solution may be carried into regions of the soil where it is no longer in equilibrium with ion-exchanging, solution or organic complexing activity. In these cases the composition of the solution may change, thus incoming rainwater may remove solutes from the exchange complex in exchange for hydrogen ions causing downward leaching. If the rainfall income contains solutes which have a high affinity for soil organic matter, they may be scavenged on the way down the profile. Several of the heavy metals, for example lead, behave in this manner and may become concentrated near the soil surface as a result of input from exhaust emissions (p. 369).

SOIL ACIDITY

Soil acidity is associated with the presence of hydrogen and aluminium ions on the exchange complex and the existence of an equilibrium solution of hydrogen ions in the interstitial water of the soil. As an intensity factor it may be defined by the conventional physical chemical concept of hydrogen ion activity expressed as pH.

pH is defined as the negative logarithm of hydrogen ion activity where activity is understood to mean effective concentration. The product of hydrogen ion and hydroxyl ion activity is constant for dilute aqueous solutions and equals 10^{-14}. In pure water the hydrogen and hydroxyl ion concentrations are equal and have the value 10^{-7} g ions/litre, hence the pH of pure water is 7. Increasing acidity raises the H^+ ion concentration, lowers the OH^- ion concentration and lowers the pH value. Increasing alkalinity raises the OH^- ion concentration with a corresponding reduction in H^+ concentration and thus increases pH value. Soil pH is measured electrometrically using a glass electrode referred to a calomel half-cell or, less accurately, using indicator solutions or papers.

The chemical definition of pH is only valid for simple, aqueous solvent–solute systems. Measurements of pH in suspensions of surface-charged solids, though giving repeatable results, must be interpreted with caution. Readings taken with the electrodes in the supernatant liquid sometimes differ by more than 0·5 pH units from readings with electrodes in the sediment. Further differences appear if the electrolyte concentration of the soil solution is altered, for example the addition of a neutral salt such as potassium chloride usually reduces pH, presumably by displacement of hydrogen ions from the exchange complex. The electrode positional effect is probably due to a junction potential arising at the boundary between the calomel electrode and the suspension: potassium ions may enter the cationic atmosphere of individual soil particles while the chloride ions are largely excluded and move only through the external solution. The potassium ions thus move forward, as a front, faster than the chloride ions. Olsen and Robbins (1971), using suspensions of ion exchange resin, obtained results which support this theory. With the further complication that potassium may exchange with ions of different valency from adjacent soil particles, the junction effect leads to aberration in the pH reading compared with that in simple solution (Black, 1968).

The indicated soil pH may thus be strongly affected by the position of the glass electrode but little by the calomel electrode. The consensus of opinion is that the pH reading obtained is a reflection of the mean pH of the liquid surrounding the soil particles. This pH is susceptible to dilution effects and therefore, for consistency of reading, various standardized soil : water ratios have been used, for example 1:2·5 or 1:1, while some workers have attempted to simulate the field situation more closely by using a stiff paste of soil and water. After adjusting the water content of a soil sample it is necessary to allow some time for pH equilibration as fairly slow processes such as cation exchange and/or carbonate dissolution are involved.

Addition of water to solutions of electrolytes causes a pH shift toward neutrality; alkaline solutions decrease and acid solutions increase in pH. In soils, the consequences of dilution are much more unpredictable as the addition of water reduces the electrolyte concentration in the soil solution, thus causing further dissociation of cations from the exchange complex. It also, temporarily, alters the carbon dioxide concentration in solution which, in calcareous soils, may strongly influence pH. Addition of water also increases the thickness of water films on soil particles, so modifying the local influence of adsorbed ions on pH and also altering the junction potential effect described above.

Despite the difficulties of interpreting soil pH values they show strong correlations with soil type, vegetation type, profile horizon and, agriculturally, with crop growth, lime requirement and mineral nutrition. Natural soils usually have pH values between about pH 3·0 and 8·4, the upper value being the calcium carbonate equilibrium with atmospheric carbon dioxide concentration and the lower value, the soil solution equilibrium with a highly hydrogen-saturated soil. More extreme values do occur in unusual soil types: some alkali soils with high sodium carbonate content reach values of pH 10·0–10·5 while drained gleys may produce sulphuric acid by oxidation of ferrous sulphide, their pH falling to 2·0 or below. Such low pHs may also be found in spoil heap soils derived from sulphide ores or from coal spoil with a high ferrous sulphide content.

Many years ago Salisbury (1925) showed that almost all of the calcium carbonate in dune sand was lost by leaching within 250 years in the prevailing rainfall climate of 80 cm/year and, during this period, the pH fell from 7·8 to 5·0. This rate of change may be lessened on finer textured parent materials but, even so, with an excess of precipitation over evaporation, decalcification is a pedological fact.

Jenny and Leonard (1934) investigated soils of the loess zone of the central U.S.A. on a transect spanning a precipitation gradient from 30 to 100 cm/ann and showed that the depth to free carbonates was strongly correlated with rainfall (Figure 4.3). After decalcification, further leaching leads to the desaturation of the exchange complex and an increase in the acidity of the soil solution, due to the increased ratio of hydrogen to metal cations ionizing from the exchange sites. Figure 4.3 shows that titratable hydrogen was first detected in this study at an annual precipitation level of about 60 cm and with a soil pH of about 7. Increasing rainfall to 90–100 cm/year is accompanied by a fall in soil pH to 5·0–5·5 and a rise in titratable hydrogen to 6–8 *me*/100 g. The main source of titratable hydrogen is likely to be dissociation from an unsaturated exchange complex.

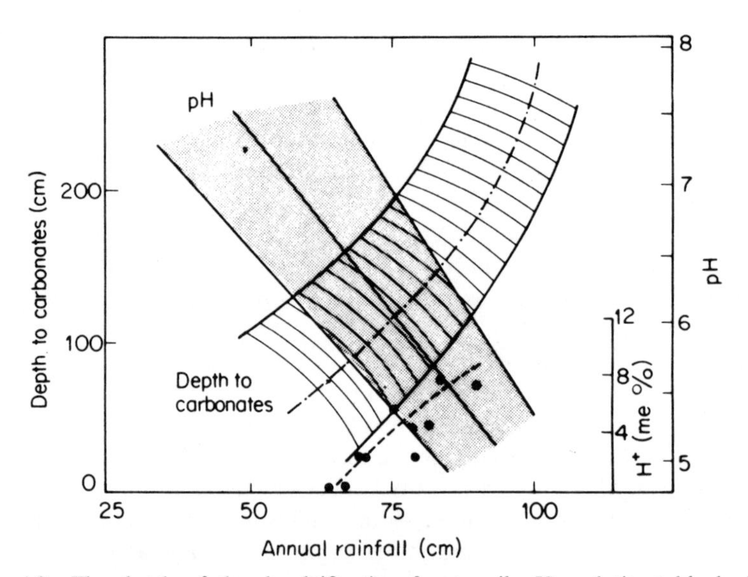

Figure 4.3 The depth of the decalcification front, soil pH and titratable hydrogen (exchange acidity) in relation to annual rainfall in central USA. ● = titratable hydrogen. Data of Jenny and Leonard (1934).

Leaching, in pedogenesis, may therefore be considered as a process in which an advancing front of decalcification or cation desaturation passes down the profile, accompanied by a fall in soil pH. Natural soil profiles are not always, however, more acid near the surface than in the deeper horizons as other factors are confounded with the leaching process. Biocycling of elements returns cations to the soil surface in litterfall and delays or even reverses the leaching loss; faunal soil mixing limits downward movement while annual P/E and seasonal distribution of precipitation have a very strong influence on leaching. High evaporation rates may actually enrich the surface soil with solutes carried upward in capillary water and produce pedocal soils of very high pH. Figure 4.4 indicates the range of pH

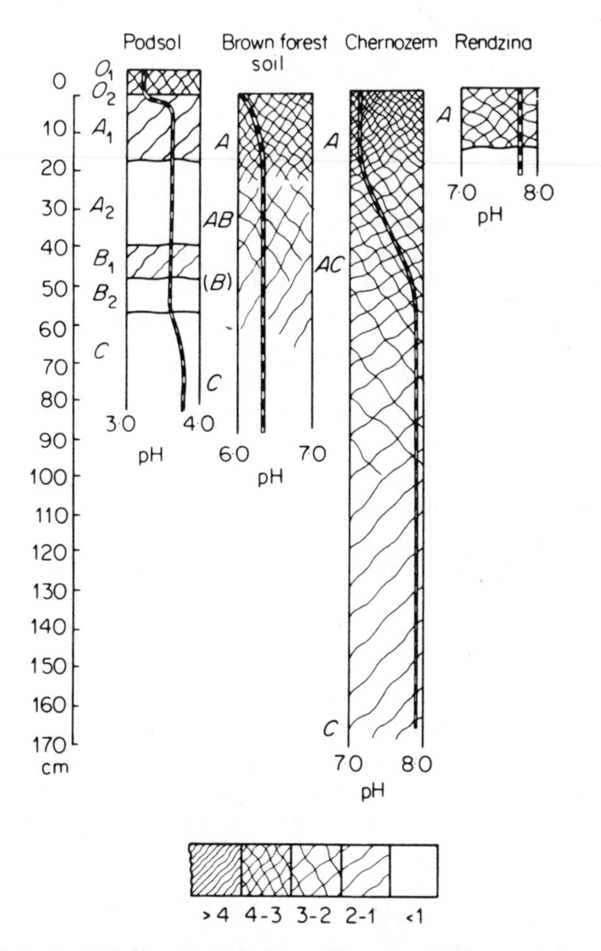

Figure 4.4 Organic matter and pH profiles of several contrasting soil types.

values which may be encountered in natural soil profiles and also shows the very steep profile pH gradients which may be found when there is a marked build-up of organic surface layers. The steepness of these gradients is perhaps superficially concealed by the pH notation in which one unit represents a tenfold change in H^+ ion concentration (Rennie 1966).

Pearsall (1952) delimited various boundary pH values which are ecologically significant. Plants regarded as calcicoles usually occur above pH 6·5 and contrast with the extreme *calcifuges* of heath and moorland soils below pH 3·8–4·0. Soils above pH 6·5 are generally cation-saturated and have a subclass of *calcareous* soils containing free calcium carbonate, while soils below pH 3·8–4·0 are strongly desaturated and have a considerable content of exchangeable hydrogen (aluminium).

These pH limits are also reflected in the nature of the soil organic matter: raw humus or *mor* is associated with soils of below pH 3·8 while *mull* is characteristic of the more cation-saturated soils of pH 4·8–5·0 and above. Between 3·8 and 4·8 intermediate organic matter forms occur. Soil organic matter is discussed in more detail later in this chapter.

Many differences between soils of differing pH are due to the processes of 'soil metabolism' which vary strongly with soil pH. Between pH 5·0 and 8·0+ both bacterial and fungal decomposition proceed rapidly, but below about pH 5·0 bacterial activity is reduced and fewer fungal species are found. There is also a change in fauna, the numerous worms and snails of higher pH soils being replaced with a less diverse population of arthropods: mainly mites and springtails.

Limitation of bacterial activity by acidity and accompanying calcium deficiency slows the rate of organic decomposition so that acid soils tend to accumulate a thick, superficial mat of undecomposed organic matter: the O_1 and O_2 horizons so characteristic of spodosols. Increasing wetness also inhibits oxygen diffusion in the soil, encouraging anaerobiosis and slowing decomposition. Hence the histosols or bog soils may accumulate layers of acid peat many metres in depth.

Long-term waterlogging also interacts with soil pH, tending to reduce the pH of alkaline soils and increase that of acid soils so that most anaerobic soils are usually in the range pH 5·0–7·0. In alkaline soils the reduction of pH may be caused by accumulation of carbon dioxide and, possibly, by release of organic acids from microorganism metabolism. Acid soils tend to be increased in pH probably by the conversion of inert ferric sesquioxide to the more basic ferrous hydroxide (Greene, 1963). Wetness and soil pH are thus inextricably bound together and may be subject to seasonal oscillations, peat bogs, for example, becoming more acid as they dry out and oxidation replaces the prevailing reducing condition. Calcareous soils, by contrast, increase in pH as they dry and reach a pH equilibrium with a lower carbon dioxide concentration. These effects may also occur on a microscale in the structural aggregates of the soil which may be oxidized on their surfaces but reduced within the bodies of the aggregates (Greenwood, 1961; Crampton, 1963).

Figure 4.5 presents some of the generalizations which may be made, relating pH wetness, soil processes and vegetation. The accumulation of organic matter associated with both low pH and increased wetness is also reflected in a greater C/N ratio, while the inhibition of bacterial activity limits the nitrifying process so that acid or waterlogged soils contain little or no free nitrate. A strong pattern of vegetational distribution is also imposed by acidity and wetness. As soil parameters these two factors and their interaction are possibly the most important in defining the nature of the ecosystem. An ecologist, armed with the

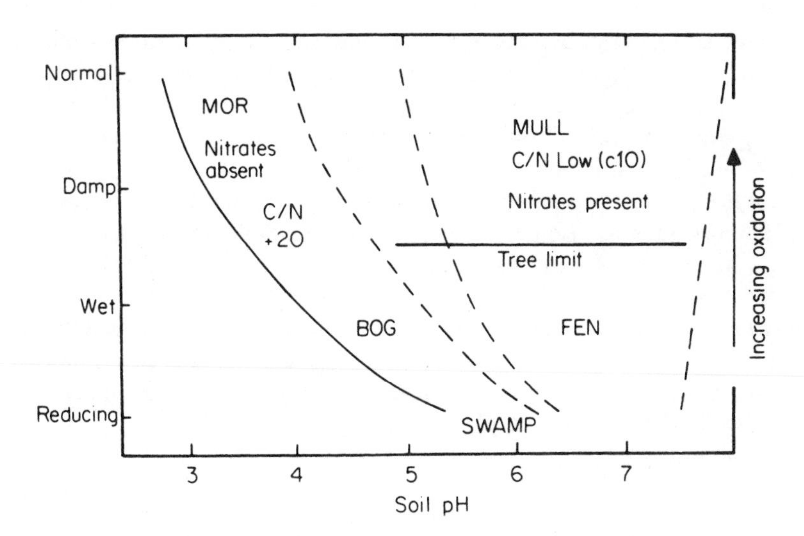

Figure 4.5 Soil characteristics in relation to wetness and pH. Reproduced by permission from W. H. Pearsall, The pH of Natural Soils, *J. Soil Sci.*, **3**, 48, Figure 4 (1952).

knowledge of soil pH and long-term wetness, could essay a fairly accurate prediction of soil type, vegetation and fauna for a given geographic area.

Some of the reasons for the overriding importance of these two factors are discussed by Greene (1963). The value of soil pH can, in most normal soils, be considered as an index of its exchangeable cation saturation as most soils between about pH 5·0 and 6·0 pass from approximately 25 % to 75 % saturation. Under waterlogging conditions there is a further relationship between pH and redox potential which is governed by the equilibrium between ferric sesquioxide and ferrous hydroxide. $Fe(OH)_2$ acts as a mild oxidant and, while it remains present in the soil, redox potential does not fall to a level at which the soil can become seriously toxic by production of sulphide or excessively high ferrous iron concentrations. During the reduction process, ferrous hydroxide is produced, which is weakly basic and causes a rise in soil pH.

Soil pH and soil chemistry

Soil pH is easily measured but interpreted with difficulty due to its relationship with a large number of complex inorganic equilibria which may also be under the control of redox conditions and biological activity.

The balance of H^+ and OH^- ions in soil solution, expressed by pH and pOH, is directly controlled through the degree of saturation of the cation exchange complex: low levels of saturation will result in a large equilibrium concentration of H^+ (low pH) and will also be associated with low concentrations of metals such as calcium, magnesium and potassium which are normally supplied to plants via cation exchange equilibria.

The concentration of cations with sparingly soluble hydroxides is related to pOH through the common ion effect (Figure 4.6) and there may be various other phase equilibrium processes which are active in specific pH ranges, for example the precipitation of phosphorus by iron and aluminium at low pH.

The biological consequences which are discussed fully in chapter 9, may be a simple reflection of the solubility-availability of particular elements at different soil pH values but may also be related to above-threshold limits of toxic elements such as aluminium iron and sometimes magnanese. Low pH also has a strong secondary impact on plant growth through impeded nitrogen fixation, nitrification and denitrification. The organisms concerned have a high pH optimum, bacterial fixation may be limited by calcium deficiency and both nitrogenase and nitrate reductase enzymes have a requirement for molybdenum which may be deficient at low pH (Figure 4.6).

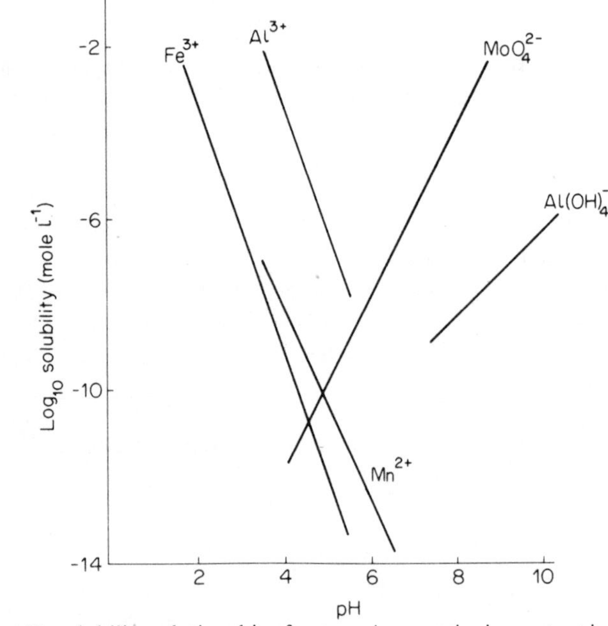

Figure 4.6 pH–solubility relationships for some ion-species important in pedogenesis or plant nutrition. Calculations of Sillen and Martell (1974): data of Lindsay (1972) and Bohn, McNeal and O'Connor (1979).

The phase equilibria of several biologically significant cation–hydroxide systems are shown in Figure 4.6 together with that of $CaMoO_4$–MoO_4^{2-}. It is perhaps simplistic to consider that these hydroxide equilibria are the sole soil-solution controllers of ion concentrations but the diagrams nevertheless spotlight some important relationships.

Iron, in well oxidized soils, occurs almost entirely in the trivalent form and, at soil pH 3.0, or less Fe^{3+} may be sufficiently concentrated to be toxic. Between pH 3 and pH 4 it declines in concentration by $\times 10^4$ and, at pH 6 to pH 7+ which is common in limed agricultural and natural limestone soils the activity of Fe^{3+} is reduced to less than 0·001 pM. Thus it is hardly surprising that pH induced iron deficiency and plant adaptations for gaining iron from high pH soils are frequently encountered (p. 268).

Manganese shows similar pH related solubility changes and problems of deficiency may be encountered in alkaline soils. It is less certain that toxic concentrations may arise at low pH but iron and manganese also show redox related solubility changes and both elements may reach very high concentrations in poorly aerated, acid soil solutions (p. 311).

Aluminium, an inessential element, is phytotoxic at very low concentrations (less than 0·04 μM) and may influence plant growth in any soil more acid than about pH 5·5. Because of its amphoteric behaviour it also reappears in soil solution in alkaline soils above pH 8·0 and toxicity has been reported in soils derived from pulverized fuel ash (PFA) produced by coal-fired electrical power generation (Jones, 1961).

Molybdenum, essential for the nitrogenase and nitrate reductase enzymes, is available to plants as the molybdate anion which also shows a dramatic pH related solubility change of $\times 10^6$ between pH 7·0 and pH 4·0 with the consequence that deficiency may be encountered in acid soil.

Further ionic interactions may take place, related to pH, for example between iron, aluminium and phosphorus in acid soils and calcium and phosphorus in neutral to alkaline soils. Near-neutral soil solutions often contain the phosphate anion ($H_2PO_4^-$) at between 10^{-5} and 10^{-6} M. Aluminium and iron may appear in similar concentrations at soil pH 3·0 to 4·0 and, with log solubility products of, respectively, −22 and −18 for orthophosphate formation, this would scavenge the orthophosphate anion to less than 10^{-12} M so explaining the extreme phosphorus deficiency of many acid soils. In alkaline soils the concentration of soluble phosphate is again forced to a low level by the insolubility of the calcium phosphates.

The effect of soil pH and its chemical correlates on plant growth and distribution is discussed extensively in chapter 9.

Soil pH, as a criterion of hydrogen ion concentration, is an intensity factor reflecting degree of acidity, but soils also vary quantitatively in acidity according to the degree of unsaturation of the exchange complex. This is important in considering the agricultural implications of soil acidity, particularly lime requirement.

For agricultural purposes it is normal to adjust soil pH to about pH 6·5. Addition of lime ($Ca(OH)_2$) or ground limestone ($CaCO_3$) initially raises the pH of a soil by neutralizing the free hydrogen ions in the soil solution. This, however, promotes further ionization of hydrogen from the unsaturated exchange complex and the pH cannot rise to 6·5 until all of the exchangeable hydrogen has been neutralized and replaced by metal cations. Thus the lime requirement is not only a function of original soil pH but also of total exchangeable hydrogen. Acid soils with a high clay or organic matter content have a large reservoir of exchangeable hydrogen compared with sandy soils with a low cation exchange capacity. The sandy soil has a low *buffering capacity* and is easily changed in pH by addition of lime compared with the well-buffered clay soil (Figure 4.7).

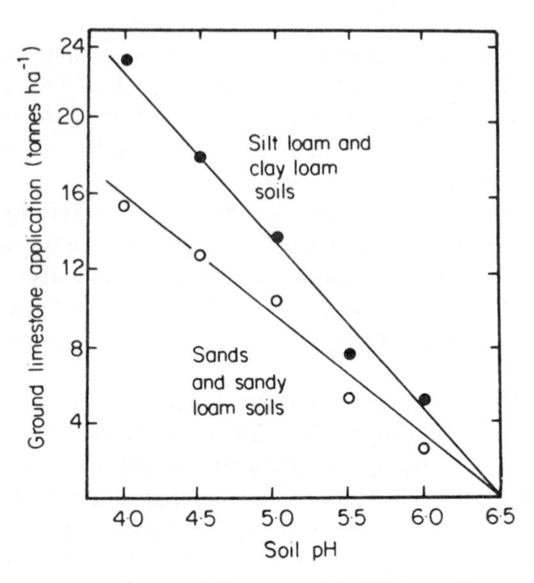

Figure 4.7 The requirement of ground limestone ($CaCO_3$) to adjust soil acidity to pH 6.5. Data of Jackson (1958).

SOIL ORGANIC MATTER

The organic matter derived from surface-living plants and animals and from soil-dwelling micro- and macroorganisms, is a key factor differentiating the mineral 'crust of weathering' from true soils. The nature of the organic matter is governed both by input and by 'soil metabolism', consequently vegetation, climate, parent material and topography all have a strong influence.

The initial input is in the form of surface litter and subterranean dead root material but, after encountering various soil processes, this organic matter may be grossly modified and has, in the past, been referred to as *humus*. Current practice is, generally, to replace this term by 'soil organic matter'.

The first step in the complex process of its formation is the comminution and transport of the original plant fragments, usually by the soil fauna. The earthworm, *Lumbricus terrestris*, is an example of an animal which is spectacularly successful in transporting and incorporating litter which it pulls into its deep burrows (1–2m) and devours on the spot (van der Drift, 1965). The remains of digested litter are incorporated in casts which are voided both at the surface and at all depths in the burrow system. The very rapid disappearance of litter in deciduous forests and its uniform incorporation throughout the soil profile pays tribute to the great efficiency of the earthworm population in this process.

Many other less mobile organisms also assist in this primary comminution of litter but leave their faecal products in the surface layers. They include many Annelida (both Lumbricid and Enchytraeid worms), Myriapoda (centipedes and millipedes), Isopoda (woodlice) larvae of Diptera (flies), Mollusca (slugs and snails) with many other Arthropoda. These animals also carry fungal and bacterial inocula amongst the litter and rapidly initiate a succession in which phylloplane organisms are sequentially replaced by species characteristic of fresh litter and later by old litter and then soil-dwelling species. This successional change is accompanied by an increase in microorganism-feeding animals such as Protozoa, some Nematoda (eelworms), some Enchytraeid worms and the ubiquitous microarthropods: Collembola (springtails) and Acari (mites).

The ecology of these animals is complex and closely related to soil and vegetation-type (Wallwork, 1970, 1976; Vanek, 1975). Earthworms, for example, which are responsible for major soil mixing in high pH soils, become less frequent and less species-diverse in more acid conditions. The deep-burrowing Lumbricids disappear at a threshold of about pH 4·5 leaving only surface-dwelling worms such as some Enchytraeids, thus permitting the accumulation of a persistent surface organic layer often of *mor* humus type (p. 100) (Barley, 1961; Satchell, 1967).

Many acid soils contain coprogenic elements which are well-shown in Babel's excellent photomicrographs (1975a). The frequency of enchytraeid droppings may be used as a quantitative characteristic diagnostic of some humus forms (Babel, 1975b), and Kubiena (1953) describes an 'insect mull' containing prominent arthropod droppings and associated with the *moder* humus form of the pH range intermediate between *mull* and *mor* (p. 100).

The importance of the joint activities of the fauna and the decomposer organisms cannot be overstressed. Detailed and extensive accounts are given by Dickinson and Pugh (1974), Kilbertus *et al.* (1971) and Swift *et al.* (1979). The composition of the soil organic components which are in a long-term steady-state relationship with input and decomposition is not only a reflection of litter chemistry but also of the breakdown and resynthesis capabilities of the local faunal-microorganism association. Examples are soil polysaccharides, the bulk of which are derived from bacterial cell wall and capsule materials and the more stable, persistent components of soil organic matter which are based on microbiologically resynthesized polyphenolic structures.

Some of the organic matter is remarkably resistant to microorganism attack and recent radiocarbon dating studies have revealed residence times for soil organic carbon which are much longer than was once thought. Paul *et al.* (1964) and Scharpenseel (1971) give mean ages ranging from a few hundred to several thousand years. Predictably, age increases with depth and Jenkinson (1975) and Jenkinson and Rayner (1977) investigating a deep Brown Earth soil from the classic Broadbalk experiment at Rothamsted, England, found an age of 1400 years in the 0–23 cm depth range, increasing to over 12,000 years below 2 metres.

The soil has been unmanured and under almost continuous wheat culture since 1843 and the soil samples were collected before 1955 from which date nuclear weapon testing has contaminated surfaces with non-natural ^{14}C. The more stable components of soil organic matter such as humic acids (p. 107) usually give older dates than the less stable fulvic acids and water-soluble low molecular weight compounds (Scharpenseel and Schiffmann, 1977) thus giving some problem in the interpretation of dates. Stout *et al.* (1980, in press) give an extensive review of the ^{14}C technique as a means of studying the rate of soil organic carbon turnover.

Humus forms and pedogenesis

In temperate climates the range of organic matter types reflects the gross differences between soil groups; moderate rain coupled with a nutrient-deficient, well-drained parent material tends to produce podsolization and *mor* humus while finer textured, nutrient-rich rocks with broad-leaf forest cover develop brown forest soils and *mull* humus forms. These two terms were originally used in the late 1800s by Muller (Howard, 1969) to describe the organic matter of beechwoods on, respectively, oligotrophic and eutrophic parent materials. Muller's classification is still widely used though problems do arise in applying it to other than cold temperate habitats.

Mor, or raw humus, is characteristic of the true podsol and forms at a low pH (below 3·8–4·0) in a nutrient-deficient milieu. Acidity and lack of calcium limit bacterial activity, thus slowing decomposition and causing the accumulation of a deep O_1 litter layer. Earthworms are absent, thus preventing the litter from being mixed with the mineral soil. It decays slowly, *in situ*, to form the O_2 layer of true mor. This is a compact, rather tough, black or brown peaty material in which the individual fragments are microscopically recognizable as plant material. Most of the plant roots are confined to this horizon or to the surface of the A_1 and it also contains large quantities of dark-coloured fungal mycelium. The discontinuous distribution of organic matter in the profile is quite characteristic, the majority occurring in the O horizon and a small amount of finely particulate material washing into the A_1 horizon. The remainder of the profile has little organic content except for a layer in the B_1 horizon which, probably, is carried down and redeposited in a colloidal form, perhaps associated with iron.

The formation of mor, though partially a function of climate and parent material, is also related to plant cover and human or other biotic interference. There is, in fact, a whole syndrome of soil acidification which has been interpreted as soil deterioration caused by specific types of plant cover and, at least in western Europe, attributable to human destruction of broad-leaf forest and its replacement by heathland (Dimbleby, 1962).

Mor is microbiologically poorer than mull, supporting a smaller bacterial population and a fungal population which, though prominent in biomass, is often limited in species numbers. Figure 4.8 shows the distribution of fungi and bacteria in soil samples taken from a sand dune sere and ranging, with soil age, from pH 6·8 to pH 4·3. In the early stages of colonization there is a marked increase in

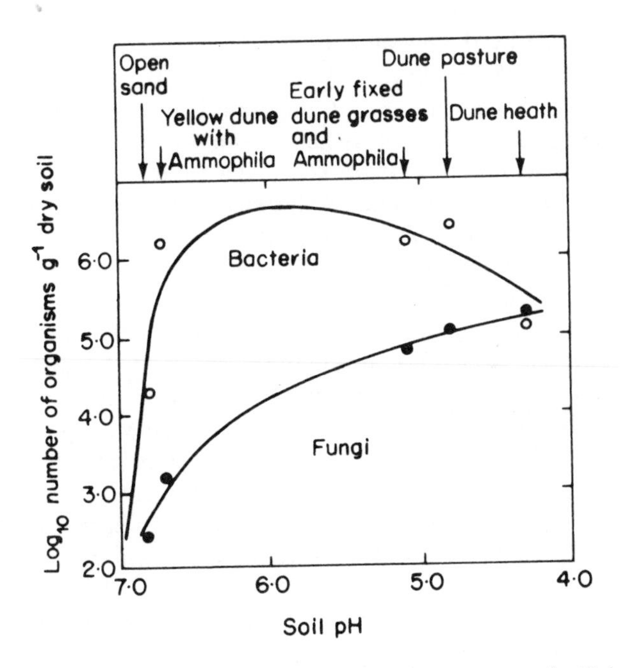

Figure 4.8 The numbers of soil bacteria and fungi in relation to soil pH during a sand-dune succession. Note that organism numbers are plotted on a logarithmic scale. Soil samples from 5–15 cm depth. Data of Webley, Eastwood and Gimingham (1952).

the populations of both groups, but, once the succession has reached acid dune heath, bacterial numbers decline sharply (Webley *et al.* 1952). Warcup (1951) also found that numbers of soil fungi were similarly increased by falling soil pH and that different species tended to be represented in various parts of the pH range. A useful review is given by Hattori (1973).

The process of podsolization and its apparent relationship to certain plant species has attracted the attention of numerous workers who have tried to show that various organic compounds derived either from the living vegetation, the litter layer or the O_2 mor layer, are responsible for the intense eluviation of the *A* horizon. Deb (1949) and Bloomfield (1953a) reviewed the previous literature and noted that organo-iron may move, as a humus-protected sol or as a complex organic ion. Bloomfield, in this and in a number of other papers (1953b; 1954a, b, c) showed that soil sesquioxides could be mobilized by aqueous extracts of the leaves of many conifers and some broad-leaved species. He also noted (1965) that the activity of leaf extracts in mobilizing sesquioxides correlated with their content of polyphenols but, since both broad-leaved species and conifers showed this activity, postulated that the residence time of polyphenols in the litter layer may be crucial and might be affected both by enzymatic effects in the decay process and by the palatability of the litter to the soil fauna. The problem is also further complicated by the possible implication of soil microorganisms in the mobilizing process (Aristovskaya and Zavarzin, 1971). Flaig *et al.* (1975) have extensively reviewed the metal complexing ability of humic acids.

Mull humus characterizes the richer brown forest soils and some other mollisols of the temperate zone. It is associated with soil pH values above 5·0, an abundance of divalent cations and a well-developed earthworm fauna. A grey, brown-grey or blackish material, it is diffusively incorporated amongst the soil mineral particles by biological mixing. Microbial decomposition is so rapid that no plant or microorganism remains are recognizable; it has an amorphous, colloidal nature. The association with soil particles is so intimate that it cannot be separated by normal, mechanical means and is believed to be physicochemically bound as a 'clay–humus complex'. Kubiena (1953) suggests that the complex is microbiologically stable and is responsible for the persistent good structure of agricultural soils derived from broad-leaved forest or grassland areas. The natural vegetation of mull soils is a luxuriant reflection of the eutrophic nature of the material, is rich in species and, by contrast with the vegetation of mor soils, has many deep-rooting plants which are active in nutrient cycling and produce easily decomposable litter.

The rapid disappearance of litter and its incorporation into both *A* and (*B*) horizons of mollisols is associated with intense earthworm activity. Many years ago Darwin (1881) noted that earthworms in pasture soils required only 30 years to produce a surface layer of worm casts 6 inches (15 cm) or more in depth; an annual turnover of some 20 tons/acre (45t ha^{-1}). This represents less than half of the material ingested and voided by worms as many of the casts are produced below ground. These casts contain the ground-up remains of plant litter, already attacked by cellulase enzymes, homogeneously mixed to a paste-like consistency with soil to form a perfect substrate for microorganism growth. As a result, fungal hyphae and bacterial gums bind the casts into stable aggregates which persist for long periods and form the bulk of the material in the *A* and (*B*) horizons. Such soils are very porous and have a spongy consistency underfoot. The earthworm mixing process is almost certainly responsible for the formation of the clay–humus complex.

The rate of removal of litter and its relationship to mull/mor formation is affected by the differential palatability of various plant species to earthworms. Satchell (1967) noted that the well-recognized podsolizers, the conifers and the beech, are the least palatable, the oak which can occur on podsols is intermediate, while the well-recognized mull formers are the most palatable. This observation has considerable significance in the light of Bloomfield's (1965) suggestions that leaf litter with a long residence time will be most active in mobilizing iron with polyphenolic leachates. Satchell (1967) also notes that destruction of worm populations with toxins leads to surface accumulation or organic matter in normally mull soils. The interrelationship of microorganism and plant species is also involved in the formation of different humus types, thus Kendrick and Burges (1962) have shown the different sequence of fungal colonization in a suspected podzol-forming species (*Pinus sylvestris*—scots pine) compared with mull humus-formers and Stringer (1974) identified a similar fungal succession on litter of *Ulex europaeus* (gorse), another soil acidifier (Grubb and Suter, 1971).

Many other organic matter forms have been identified and related to environment, fauna and microorganisms, for example Kubiena (1953) named an intermediate between mull and mor as *moder*, characteristic of the pH 4–5 range and containing large quantities of microarthropod faecal pellets, hence the alternative name of insect mull (p. 99). Various partially decomposed materials have been identified as semi-terrestrial humus forms (Kubiena, 1953), occur in wet or waterlogged habitats and are generally referred to as *peats*. Discussion of production and decomposition in peat-forming sites is given in Heal and Perkins (1978) and, because these are the soils which have provided, in their macro-fossil remains and pollen spectra, most of our recent palaeo-ecological knowledge, have been very well studied and described (Moore and Bellamy, 1973; Moore and Webb, 1978).

Pedogenesis is a function of parent material, climate and topography but it is also strongly affected by plant, animal and microorganism species composition and is sensitive to human interference. Leaching and, to some extent podzolization, is promoted by a number of plant characteristics such as low nutrient content (particularly calcium), high fibre (lignin) content and the presence of organic materials which decrease palatability and/or inhibit microorganisms: these may include polyphenols and derivatives such as tannins. Kubiena (1953) cites such plants as raw-humus plants: they include many of the Ericaceae and various conifers. Grubb *et al.* (1969) showed fairly conclusively that the two common heath plants *Calluna vulgaris* (ling) and *Erica cinerea* (bell heather) could reduce a soil surface from pH 5–6 to pH 4–5 within a decade. The leguminous *Ulex europaeus* (gorse) was equally active in acidification.

Most of the data concerning acidification of podzolization by particular species is derived from forestry: Ovington (1953) compared the soil pH in the depth range 0–15 cm with that below 30 cm and found, on unafforested sites, that the surface was 0·7 to 1·6 pH units more acid than the deep soil. On sites afforested for 20 years or more this trend was reversed by some broad-leaved species including *Quercus robur* (oak) and *Castanea sativa* (chestnut) which increased the surface pH by 0·3 units. Some conifers on some sites also gave surface increments, for example 0·2 units by *Pseudotsuga menziesii* (douglas fir). Conifers in general either caused, or permitted, surface acidification comparable with that of the unafforested sites, examples being *Pinus nigra* (Corsican pine) 1·9 units reduction; *P. contorta* (lodgepole pine), 1·6; *P. sylvestris*, 0·2; and *Larix decidua* (larch), 1·1. Some species interacted very variably with soil type, for example the *P. menziesii*, cited above, which increased the pH of an acid soil, caused a reduction of 1·4 pH units on an originally alkaline soil. Smith (1975) found similar changes following afforestation in an upland area of South Wales, for example *Picea sitchensis* (sitka spruce) caused (or permitted) an acidification ranging from 0·1 to 0·45 pH units within 20–30 years. The mineral soil horizons studied were the A_1 layers of four soil types and the greatest reduction in pH was associated with the originally least acid soil.

Zinke (1962) has suggested that an acidification mechanism may be associated with the effects of specific trees and showed, for example, that the soil directly below the crown of a *Pinus contorta* tree was 1·3 pH units more acid than that outside the canopy. Many conifers may be active in acidification, compared with hardwood species, because of their enhanced capacity for cycling aluminium and consequently causing base desaturation of the surface soil (Messenger, 1975; Messenger *et al.*, 1978). Jarvis and Duncan (1976), investigating the relationship between *Calluna vulgaris* and *Pteridium aquilinum* (bracken) on acid heathland soils have shown that *P. aquilinum* may reverse the trend to acidification under *C. vulgaris*. A picture thus emerges of pedogenesis in which the long-term vegetation cover may well be as significant as any other factor and biotic effects such as grazing and human interference may alter the course of soil formation.

Many archaeological investigations of 'fossil' soils buried under tumuli or other artefacts have revealed considerable changes of soil type and vegetation since the time of construction (Dimbleby, 1962, 1965, 1975, 1976). Buried soils dating from the late Neolithic or Bronze Ages in northern Europe are usually less leached than those of the overlying younger surfaces or of the modern surrounding soils. In extreme cases the buried soil may be a Brown Earth in which pollen analysis reveals a former deciduous forest cover, but the modern soil is a podzol bearing Ericaceous heathland. It is tempting to suggest that removal of forest by burning, felling or grazing-prevention of regeneration, has promoted this change by enhancing the trend to leaching which deciduous tree cover, in some cases, retards or prevents. To place this suggestion in perspective it must be appreciated that soil acidification and podzolization, in northern Europe, have featured in previous interglacial periods, without the interference of man (Godwin, 1975).

The nature of soil organic matter

Soil organic matter represents the equilibrium between input, originating from primary photosynthetic production, and the degradative and resynthetic processes associated with soil-dwelling organisms of all kinds. Most naturally occurring organic compounds may thus be found, at some time, in soil.

The input is largely of carbohydrate, lignin compounds, fats and proteins with smaller quantities of, for example, free amino acids, alkanes, terpenoids, carotenoids, flavonoids, alkaloids, polyphenols, resins and others. These compounds are easily decomposed and have a short residence time in the soil; for example, carbohydrates are rapidly metabolized by both animals and microorganisms and even the more stable cellulose component is prone to fungal attack and to digestion by lumbricids and other animals which secrete cellulase or have cellulose-attacking gut flora (McLaren and Peterson 1967).

Minor constituents

Carbohydrates thus occur in low concentrations in the soil, much of it being polysaccharide, secondarily derived from microbial products such as bacterial cell walls. Soils rarely contain more than 0·25–0·3% of polysaccharide, though exceptions may be found in peat soils, where it may form 10% or more of the organic matter. Soluble sugars do occur but only in minute quantities, probably as equilibria with enzymatic conversion processes. Lipids are generally broken down quickly but the cuticular waxes may persist even to the extent of forming fossil deposits. Pollen grains and spores may have a very long life in some soils, their sporopollenin coats being very stable (Brooks *et al.* 1971). For this reason, soil pollen analysis has been used in palaeo-ecological studies (Dimbleby, 1961) though it is susceptible to problems of differential decay and contamination which do not arise in peat pollen analysis (Haavinga, 1974).

Proteins and amino acids are also rapidly attacked by microorganisms though chromatographic studies show most amino acids to be present at very low concentrations. The majority of soil organic-N is difficult to characterize: some 3–4% (about 25 g m^{-2} in a fertile soil) is present in living biomass, predominately microorganisms (Parsons and Tinsley, 1975) but the remainder is intimately bound to either humic substances or to clay minerals. About 40–50% of this nitrogen is hydrolysed to α-amino-N or amide-N by hot 6 M hydrochloric acid (Parsons and Tinsley, 1975) but there is no evidence that proteins as such can be identified in the soil organic matter except perhaps for some extracellular enzymes (Ladd and Butler, 1975). The former concept of a soil 'ligno-protein complex' should perhaps be abandoned.

It is of pedogenetic significance and relevant to plant nutrition that soil organic nitrogen is essentially insoluble in water and thus protected from leaching as it is only very slowly 'leaked' to the inorganic form by microbial metabolism. The same may be said of the other main anionic nutrient elements for which there is no substantial ion-exchange capacity. These are phosphorus and sulphur of which the main soil content is organic (Cosgrove, 1967; Anderson, 1975a, b). Half to two-thirds of the soil phosphorus may be inositol hexaphosphate derived from phytin (*myo*-inositol polyphosphate), most other sources such as phospholipids, nucleic acids and sugar phosphates being rapidly dephosphorylated in soil. One of the factors limiting phosphorus availability is consequently the rate of dephosphorylation of inositol phosphate by microorganism phosphatases.

Sulphur is present in a wide range of organic compounds, many of which are not yet characterized (Freney, 1967). The quantities involved, as sulphur-amino acids associated with organic macromolecules, are appreciable and another bulk constituent, which is hydriodic acid reducible, seems to contain mainly sulphate esters, possibly of bacterial polysaccharides. The carbon-bonded sulphur of the amino acids and these reducible esters may make up 95% of the soil organic reserve of sulphur (Anderson, 1975b) and, again, the limiting factor in availability may be the rate of microbial mineralization.

Organic acids may accumulate in soils, produced both by plant roots and by microorganisms. The commonest are those involved in the tricarboxylic acid cycle but other aliphatic acids, sugar acids and aromatic acids do occur. The plant growth substances, such as indole acetic acid, also contain carboxylic acid groups while specialized aromatic acids are synthesized by the lichens. Some workers have attributed considerable pedogenetic significance to such acids as they may mobilize inorganic materials by acidification–solubility effects or by chelation. Certain other minor organic constituents of soil may also be implicated in soil-forming processes, for example by mobilization of sesquioxides. Bloomfield's work, cited above, suggests that polyphenols may be important in this role (Stevenson, 1967).

Most of the remaining minor constituents are found in trace quantities and the soil enzymes are of considerable interest in this group. In the past, various workers have treated the soil as an 'organism' and described processes of 'soil metabolism'. Generally this is a reflection of overall microbiological activity in the soil but the presence of free enzymes does suggest that some organic transformations may be considered as whole-soil processes. The most difficult problems in this field are the investigation of the origin of enzymes and the localization of their functional sites in the soil. They may originate as microbial extracellular enzymes, by autolysis of bacterial cells and from soil animals, plant roots or residues. Many different enzymes have been isolated from soils and it seems likely that they are adsorbed on the surface of clay particles (Skujins, 1967) or associated with humic macromolecules (Ladd and Butler, 1975).

Major constituents—humic complexes

Most of the organic matter in soils is in the form of humic complexes comprising a mixture of diverse phenolic polymers (Hurst and Burges, 1967; Schnitzer, 1976) which are more stable than the starting materials. These compounds are tightly bound to the clay colloids and have been described as the clay–humus complex. They are difficult to characterize because the necessarily severe chemical extraction techniques may cause breakdown or alteration during processing. The classical extraction is depicted opposite (Black, 1965; Flaig *et al.* 1975):

Important characteristics of the humic substances are their abilities to form complexes with metal ions or other organic compounds and to react with clay minerals. Many complexes of fulvic acids are water soluble and may be the principle agents in transporting otherwise sparingly soluble metals and organic compounds in soils and into natural waters (Schnitzer, 1976).

Humic acid is generally dark coloured, its greys and browns contrasting with the yellow-brown of fulvic acid. The chemical composition is similar as characterized by oxidative breakdown products, and Hurst and Burges (1967) suggested that the main difference was the lower molecular weight of the fulvic acid. Felbeck (1971) has similarly noted that humic and fulvic acids are by no means homogenous and distinct fractions. Stout *et al.* (1980 in press) also claim that the classical extraction does not differentiate between materials of varying stability but different periods or intensities of acid hydrolysis do separate fractions which differ in stability in the soil as measured by ^{14}C age.

Oxidative degradation of humic acid produces a high proportion of phenolic products while reduction produces a range of units similar to those encountered

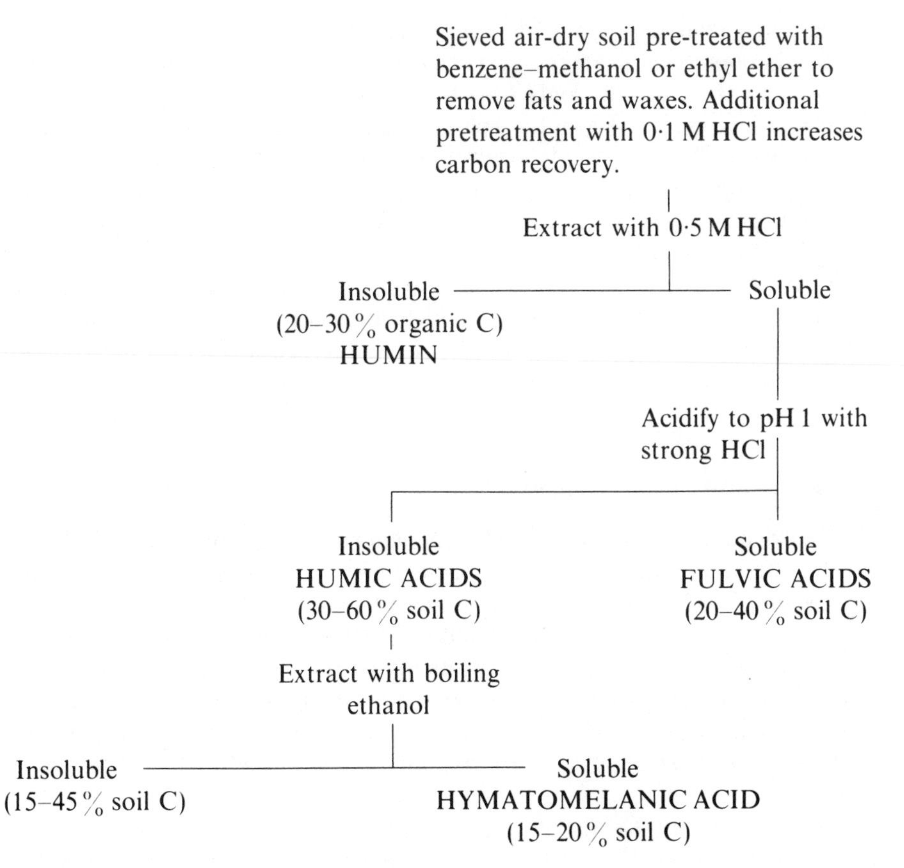

in microbial breakdown of lignin. They include various phenylpropane-derived units such as hydroxy–cinnamic and ferulic acids, substances also found as precursors in the biosynthesis of lignin (Isherwood, 1965). Reductive cleavage also produces 1.3.5-substituted rings based on the parent molecule, phloroglucinol. These are most likely to be derived from flavonoids but whether these are of direct plant origin or produced by microbial biosynthesis is not known.

Hurst and Burges (1967) summarize the ideas concerning humic acid synthesis, suggesting that the molecules grow by stepwise addition of phenolic units liberated by microorganism metabolism and that cross-linkages gradually develop. At any time the molecule contains monomers in all states of binding, ranging from loosely attached new additions to tightly condensed core units. The wide variety of degradation products suggests that there is no great rigidity of configuration; 'They may best be regarded as polycondensates of those phenolic units immediately available in a particular micro-area of the soil. These units are principally derived from lignin and plant flavonoids of the overlying vegetation'. In a more recent review, Schnitzer (1976) suggests, as a result of X-ray diffraction studies and electron microscopy, that the structure of the low molecular weight fulvic acid is relatively 'open'. The carbon skeleton consists of a broken network of poorly condensed aromatic rings bearing appreciable numbers of aliphatic chains and the rings are only loosely joined by hydrogen bonding between hydroxyl and carbonyl groups. The consequent flexible 'sponge-like' surface, with voids of differing size, is ideally suited to the surface-trapping of ions or organic compounds. Humic acids are larger but otherwise similar except that more energetic linkages between rings via C–C or C–O bonds would give greater stability and less surface activity.

Figure 4.9 indicates some of the unit structures and bondings which may be involved in the composition of the humic complexes. Several reviews of humus biochemistry have appeared recently (Haider *et al.* 1975; Schnitzer, 1976; Gieseking, 1975) but despite the volume of publication the problem as yet eludes critical solution.

Properties of the humic complexes

Soil organic matter is the critical component responsible for converting a dust-like or mud-like mixture of compacted mineral matter to a structurally aggregated, easily deformable, material with considerable pore space which provides aeration under most conditions. Unstructured soils may become anaerobic even with fairly low water contents as the capillary spaces of the matrix are very small. The organic macromolecules presumably bind clay particles with bondings such as R—COO—Ca—clay which can bridge adjacent clay particles, forming a network of clay particles and macromolecules: the clay–humus complex.

Different soils contain varying amounts of organic matter ranging from $<1\%$ in raw soils such as lithosols and regosols to 80–90% in organic histosols. The content is not uniform with depth for, as stressed above, nutrient-rich soils of high biological activity show strong mixing of organic matter into both *A* and *B* horizons while oligotrophic, leached soils have a marked stratification of organic matter, most of it lying on the surface and a little penetrating to the *B* horizon. Table 4.1 shows organic profiles of two contrasting soil types and the associated variation of cation exchange activity.

It is not surprising that the humic complex is active in cation exchange, since humic and fulvic acids contain 35–47% oxygen of which 50–100% may be involved in functional groups such as carboxyl and hydroxyl (Ladd and Butler, 1975).

The data of Table 4.2 show that cation exchange capacity of clay-rich soils is not strongly influenced by the organic component but in sandy soils it may be the main source of exchange. Carboxylic acid, hydroxyl and imide groups may all lose a hydrogen ion in exchange for a metal cation. The participation of the various groups varies with pH and with the cations involved. Most soil organic matter has a cation exchange capacity of 100–400 me/100 g and thus contributes about 1–20 me/100 g of exchange capacity to normal soils with an organic content of 1–5%. In some ombrogenous peats the organic exchange activity is of

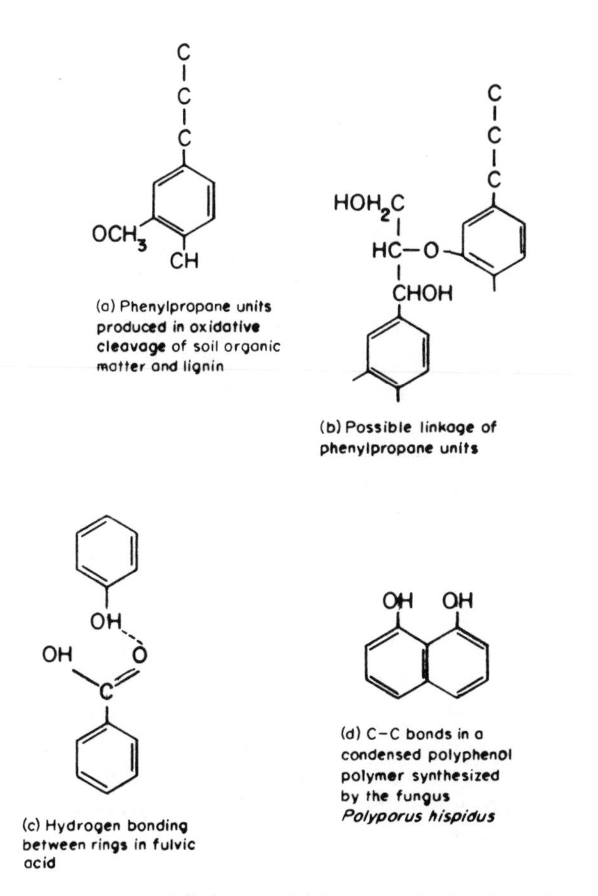

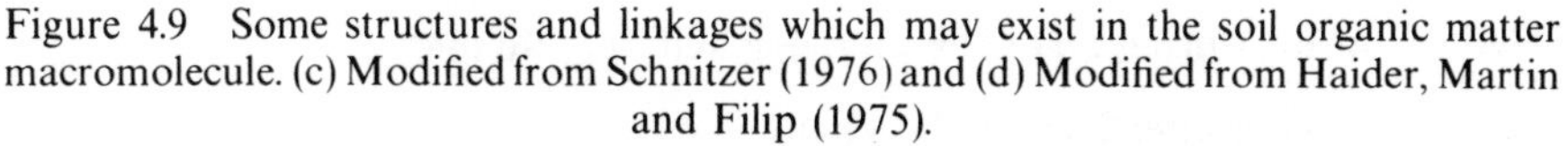

Figure 4.9 Some structures and linkages which may exist in the soil organic matter macromolecule. (c) Modified from Schnitzer (1976) and (d) Modified from Haider, Martin and Filip (1975).

extreme importance as it permits the removal of low concentrations of nutrient cations from percolating rainwater; the only source of plant nutrients in such soils.

The foregoing discussion has been related to temperate zone soils which have featured most prominently in the pedological and ecological literature. Processes of organic matter accumulation and decomposition appear to be similar under tropical conditions, being related to the continuum of earth's surface temperature and wetness conditions. Generally speaking, there is a tendency for the degradation of organic matter to outstrip its production as temperature increases, hence, from poles to equator, there is a gradient of decreasing accumulation of soil organic matter. Superimposed upon this is a wetness effect: high water content slows the diffusive access of oxygen to soils and reduces the rate of organic decomposition. Figure 3.1 outlines these temperature and water relationships. Soil nutrient content and pH also influence this relationship, increased acidity and oligotrophy reducing the bacterial population and encouraging organic accumulation. The general consequence is that well-drained tropical soils are often of low organic content and show little litter accumulation but, under wetter or nutrient-deficient conditions, tropical peats or podsols may form.

Many workers have noted that the ratio relationships between soil organic carbon and some other elements appear to have pedogenetic significance. Carbon/nitrogen ratios in particular have been widely examined and endowed with varying degrees of significance. The C/N ratio is variable from soil to soil but a number of generalizations may be made. When plant litter first enters the soil it is of high C/N ratio, often exceeding 50:1, though the nitrogen-rich litter of some Leguminosae may be as low as 20:1 (Burges, 1967). The early stages of decomposition rapidly remove carbohydrate and fats so that the ratio closes, approaching a value of about 10 in calcium-rich mull humus. By contrast, the mor humus of podzols has a much wider ratio which is usually 20 and may reach 50. The highest values are often found in the surface horizons of podzols as they may contain elemental carbon, the charcoal remnants of intentional or accidental burning. Inflated values of this type are not pedologically significant as charcoal carbon is virtually inert in soil.

The relative constancy of C/N ratios in many different soil types is also reflected in similarities of elemental constitution, functional group content and cation exchange behaviour of the humic material (Kononova, 1961). It is not suggested that the humic complex has the same composition in all soils but it is likely that there is a basic similarity of constitution and properties which is the consequence of microorganism metabolism and plant composition rather than of soil type and chemistry.

Methods of measuring and characterizing soil organic matter

Soil organic matter content is usually expressed as a percentage of total dry weight and often as organic carbon content, a more meaningful expression when the composition of the organic matter is not known. Organic carbon may be measured by wet oxidation with a dichromate–sulphuric acid mixture ($Cr_2O_7^{2-}$) or by combustion of dry samples. Most wet oxidation techniques rely on the reduction of the chromic ion by the organic matter and titration of the excess oxidant. Allison (1965) gives an account of two variants of this technique and also describes the combustion method in which carbon is determined gravimetrically. Both wet oxidation and combustion techniques may also be used in conjunction with a carbon dioxide absorbing and measuring arrangement. The normal $Cr_2O_7^{2-}$ oxidation measures only organic forms of soil C while dry combustion measures, in addition, elemental C. Soils which contain free carbonates require special precautions when gravimetric or CO_2 absorption analysis is used.

Techniques for the extraction and fractionation of organic matter may be found in Black (1965) and Hesse (1971) with more general reviews in Felbeck (1971), Schnitzer (1971) and Scharpenseel (1971). The detailed techniques presented by Black (1965) also include those for characterising the soil fauna and microflora.

SOIL PARTICLE SIZE DISTRIBUTION

The International Society of Soil Science classification of soil particle size distribution (texture) was outlined in Chapter 3. Stones and gravel above 2·0 mm diameter are excluded from the textural classes, of which there are four: coarse sand (2·0–0·2 mm), fine sand (0·2–0·02 mm) silt (0·02–0·002 mm) and clay (0·002 mm). Separation of the size fractions is preceded by oxidation of organic matter and removal of inorganic cementation to break down structural aggregates. The fine fractions, silt and clay, are removed from the coarser material by a sedimentation technique and the coarse fraction dried and graded by sieving. Stones and gravel are retained by a 2·0 mm perforated plate sieve, the coarse sand is retained on a 0·2 mm wire sieve, while the fine sand passes through. The silt and clay fraction is graded according to sedimentation velocity in a liquid column; it is assumed that clay and silt both have the same specific density and hence their sedimentation rates will be influenced only by relative diameter. Sedimentation may be followed gravimetrically either by taking pipette samples from the column or by using a suitable hydrometer to follow the liquid density change as the particles settle. Details of the procedures may be found in Day (1965) and Piper (1944).

Soil texture directly influences soil–water relationships, aeration and penetrability through its relationship with interparticle pore space. These factors are, of course, also influenced by the degree of structural aggregation discussed below. Indirectly, there is a further relationship with soil nutrient status as the clay fraction is the main source of many plant nutrients and of cation exchange activity. Sandy soils thus tend to be inherently nutrient-deficient and, because of their high porosity, to lose nutrients by leaching in humid climates.

Figure 4.10 shows an arbitrary textural classification of natural soils which may be used with laboratory particle size analyses or, more crudely, with 'finger tests' in the field. It also serves as a framework to relate particle size to water availability: the highest values are associated with a uniformly moderate particle size (silt) or with a balanced size range (silty and sandy loams). Low values occur in predominantly sandy soils in which the inter-particle spaces are so large that they are mainly air-filled at −30 kPa (approximately field capacity; see Chapter 5). Soils with a high clay content generally have small pore sizes from which much of the water is not extracted at a water potential of −1500 kPa (permanent wilting percentage). Salter and Williams (1967) have used this approach to estimate available water in the field. In addition to influencing water availability, particle size also influences water infiltration rate during rainfall. Coarse, open- textured soils manifest high and maintained infiltration rates but soils with a high proportion of silt- and clay-sized particles may produce a surface 'pan' when exposed to raindrop beating or surface run-off. Panning is also related to the degree of aggregation as it is caused by the blockage of soil pores with

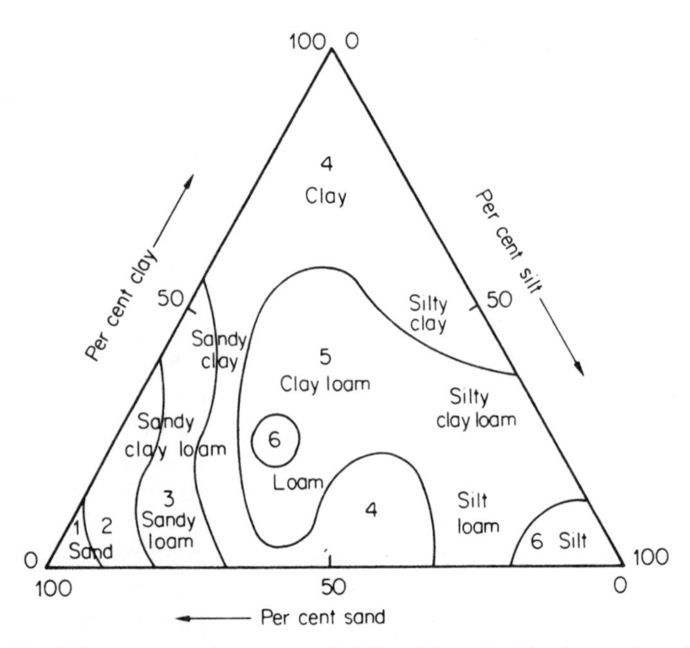

Figure 4.10 Soil texture and water availability. The triangle shows the relationship between soil-survey textural types and the co-variation of sand, silt and clay content. The triangle is contoured in cm water available per cm depth of soil in the water potential range −0.03 to −1.5 MPa (field capacity to permanent wilting point). 1 = 0.00–0.05; 2 = 0.05–0.10; 3 = 0.10–0.15; 4 = 0.15–0.20; 5 = 0.20–0.25; 6 = 0.25–0.30 cm water. Data from Black (1968).

suspended clay/silt particles which are liberated from the surface only if the soil is poorly structured. Protection by a vegetation canopy and good soil structure plays a major part in stabilizing soil surfaces against panning, run-off and soil erosion. The relationship between water matrix potential, water retention and pore size is more fully discussed in Chapter 5.

The relationship between cation exchange capacity and particle size distribution is shown in Figure 4.11, which indicates a positive correlation between CEC and clay content. The wide scatter of points derives from the variable organic matter content of the samples and the unspecified nature of the clay mineral. The samples are all from the lower parts of profiles and have comparatively small organic contents. Clay content also correlates well with the supply of many plant nutrients; for examples, Figure 4.11 also shows exchangeable potassium plotted against clay content for the same samples.

SOIL STRUCTURE: AGGREGATION AND POROSITY

Individual mineral particles in soil may be free from each other or, more commonly, bound together by chemical and biological effects to form *aggregates* (peds) ranging from sub-mm to several cm in equivalent diameter, each

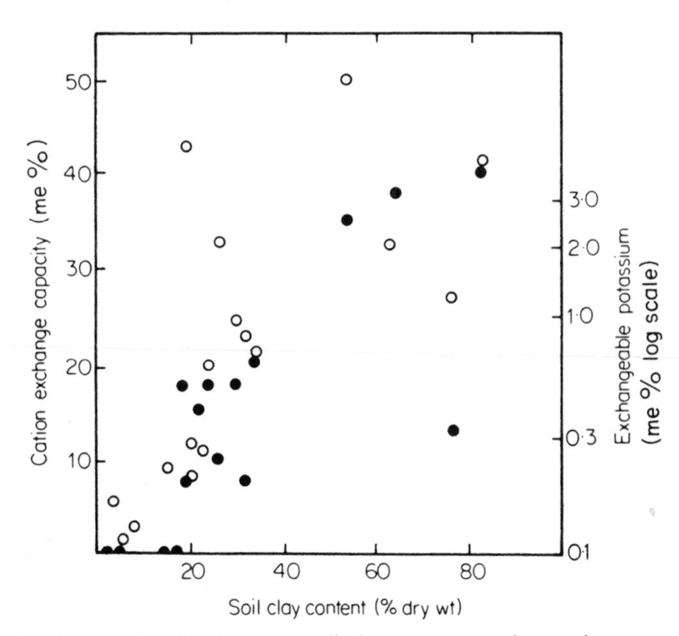

Figure 4.11 The relationship between soil clay content, cation exchange capacity (○) and exchangeable potassium content (●). Data from Soil Surv. Staff (1960).

containing a very large number of textural particles. Between these peds are *structural pores* created by the effects of root penetration, animal burrowing and shrinkage of clays during dry weather. The combined shape and size characteristics of the ped–pore complex is termed soil *structure*. Structural pores reduce the *bulk density* (weight per unit volume) of the soil: close-packed mineral particles with a mean density of 2·65 kg dm^{-3} (quartz) would have a bulk density of 2·0 kg dm^{-3} if the pore space was 25 % by volume. Soil bulk densities commonly lie between 1·0 and 2·0 kg dm^{-3} and substantially lower values are usually associated with very high organic matter content or sometimes the presence of low density minerals such as volcanic ash in Andepts (Andosols). Bulk density is most easily measured as the dry weight of a known volume core sample (Blake, 1965).

Large structural pores remain air-filled at water potentials very close to zero (saturation) thus improving aeration in wet soils. They also provide pathways for rapid infiltration of rainwater and permit easy deformation of the soil by penetrating roots or burrowing animals both of which are unable to enter consolidated layers of low porosity soil. Deformation and crushing of structural pores may however be a problem in agricultural soils or heavily trampled sites.

Structure is difficult to quantify as it may be easily damaged by the measuring process and for this reason, many different techniques or modes of expression have been described (e.g. Bryan, 1971; Emerson *et al.* 1978). It may be defined by the size, proportion and geometry of aggregates, or by measurement of pore space, either total or air-filled at a specified water content. Agriculture, conservation and erosion control may require measurement of the stability of the aggregates to raindrop impact, sheetwash or deformation under pressure.

Definition and measurement

Aggregation is usually reported as the percentage of textural particles bound together into peds of larger than a specified size (or size-range). It is commonly measured by a mechanical wet sieving technique in which a gently crushed sample is placed on a sieve or sieve-stack which is repeatedly agitated in water. After a specified number of agitation strokes, the dry weights of the retained material, corrected for gravel (and sand if less than 2 mm), are expressed as a percentage of the total gravel-free weight of the sample. Large numbers of replicates must be processed as aggregation is spatially very variable within a soil type and the structure may be more or less damaged during sampling (Kemper and Chepil, 1965).

There is little agreement concerning size-grades and various values, usually between 0·5 and 5·0 mm, have been chosen for various investigations. Aggregation is often expressed as a single value coefficient for statistical convenience despite the inevitable loss of information. Examples are: geometric mean diameter of all aggregates; mean weight-diameter calculated as the sum of mean diameter times the weight proportion in each fraction and; coefficient of aggregation, a value which is proportional to the reciprocal of the total surface area of the aggregates. Bryan (1971) gives references to sources of calculation.

The stability of aggregates to wetting, raindrop impact and pressure is a further important characteristic of soil structure. Wet-sieving for size distribution is, to some extent, a measure of water stability but more vigorous methods have been used, for example exposure to simulated raindrops of known weight and kinetic energy (Pereira, 1956; Ghadiri and Payne, 1977). The simplest technique for measuring resistance to pressure deformation is the hand penetrometer which indicates the resistance to penetration of the soil by an impact driven, or spring-loaded, probe (Davidson, 1965). The instrument is simple, rapid to operate and its readings may be interpreted as a measure of soil shear-strength bearing capacity or, within similar soil types, may be related to soil bulk density (e.g. Liddle and Moore, 1974).

Soil pore space may be measured as the difference between the bulk density of the dry soil and the mean density of the mineral particles; manometrically by measuring pressure–volume relationships in a pycnometer or by using a suction plate (Chapter 6) to measure the volume of water removed from the soil by known reductions of water potential (Vomocil, 1965). The first two techniques are most suited to measurement of total pore-space while the last gives air-filled porosity at defined water potentials. The shape and continuity of pores may be measured by pouring a thin slurry of plaster of paris or of silicone rubber monomer on the soil surface and then excavating the resultant casts. For special purposes, in particular the relation of root or animal distribution to pore size, the measurement of pore cross-sectional area in thin sections may be used to define porosity.

Formation of pores

Russel (1971) notes that, in addition to shrinkage cracks, root decomposition channels and faunal burrows, plant roots may also propagate cracks by localized drying of the soil and also by forcing their way into pores of smaller diameter than the growing point. Pores formed by all of these processes may be stable and long-lived or may collapse by crumbling of the pore wall or slumping of whole aggregates under wet conditions. Russel discusses the stability of pores in relation to organic matter and microorganism activity, concluding that roots have a direct influence in nourishing rhizosphere bacteria whose capsular slimes and gums stabilize the crumbs. Allison (1968) made this point very strongly, suggesting that the rhizosphere zone provides, simultaneously, nearly ideal conditions for both aggregate formation and aggregate stabilization by incorporation of bacterially synthesized macromolecules. Foster (1978) found carbohydrate gels up to 50 μm thick around grass roots in Australian soils. The efficacy of grassland cover in promoting aggregation is almost certainly due to its rapid and prolific root production and the associated dense microflora. Allison also suggested that the coating of aggregates with clay skins (cutans) in argillic horizons and inorganic cementation by sesquioxides, the latter in tropical soils, also aids aggregate stabilization. The hydraulic conductivity of clay cutans with a particle-spacing of about 3 nm may, however, limit water-flow to roots (Greenland, 1979).

Satchell (1967) discusses the role of earthworms in creating pore space and aggregate stability. After casts are voided, fungal hyphae develop in them, giving an initial stability which is short-lived but probably replaced by cementation with bacterially produced polysaccharides. Kubiena's (1953) statement that nearly all aggregates in forest mull soils 'are earthworm casts or residues of them' is frequently true. As the annual turnover of *A* horizon soil through the earthworm gut is fairly small, this would suggest that the stabilization must endure for fairly long periods. Jacks (1965) commented: 'it may be justified to make the rather sweeping generalisation that in the two great plant ecologically distinguished worlds of the soil, the movement of plant roots is the major structure-forming factor in grassland soils, and the movement of animals in forest soils'.

The effect of worms in aerating soils is probably not immediately great, since their annual production of burrows, as measured by surface voiding of worm casts, is not more than 5–10 % of the soil in the main rooting zone, this soil often already having a pore space of 40–60 %. Though the burrows are not a large annual contribution to total pore space they must also be regarded in the light of the potentially long residence time of aggregates—there is not much evidence relating long-term build up of pore space to earthworm activity but the spongy earthworm mull soils referred to by Kubiena (1953) suggest that they have considerable influence. The worm burrows are also important because they open to the surface and provide large-diameter channels for water infiltration.

Low (1972) attributed the maintenance of porosity in an otherwise structurally unstable soil to earthworm burrowing and Ehlers (1975) showed that the earthworm channels, though representing only 0·2% of the soil volume, contributed substantially to the infiltrability of the soil.

Structure and plant roots

The structuring of soils is of considerable importance in relation to root penetration. Wiersum (1957) showed that roots would not pass through rigid pores of less than about 0·15–0·20 mm, roughly equivalent to the mean root diameter. The upper value corresponds to the pore size of an unaggregated sandy soil with a uniform particle diameter of 0·8 mm. Growth can only occur freely in such soils if the particles can be forced apart by the extending roots: this requires plenty of available pore spaces to absorb the deformation. Taylor and Ratliff (1969) have shown that both root growth and plant yield are reduced as soil strength (resistance to deformation) increases. Russel and Goss (1974) grew plants in beds of glass ballotini to which pressure could be applied and found 10 kPa (about one-tenth of atmospheric pressure) sufficient to reduce root extension rate by 50%.

Many soils have a total porosity of 40 to 50% but most of this is less than 1·0 μm in equivalent diameter and is contained within aggregates. Roots are thus confined to inter-ped fissures, though root hair penetration of aggregates may assist in ion-uptake as their cores may be less oxygenated and chemically or microbiologically different from the outer layers (Greenwood and Goodman, 1967; Hattori, 1973; Smith, 1976). In this context it is interesting that root hairs are better developed in well aggregated soils (Greenland, 1979) and often absent in flooded conditions when ped aeration gradients would be less marked.

Many of the pores penetrated by roots are relatively large and this, coupled with shrinkage of roots when they lose water, may cause loss of contact with the soil. This may be overcome by mucigel production (Jenny and Grossenbacher, 1963; Oades, 1978) which, by providing a bridge between root surface and soil, forms a path for nutrient and water movement as well as sustenance for the rhizosphere microflora. Weatherley (1976) explains the resistance to water absorption shown by rapidly transpiring soil-rooted plants in terms of such gaps, though Cowan and Milthorpe (1968) claim that their calculations show a significant reduction of this resistance by water vapour transport to the root surface.

Newman (1976) considered that the water conductivity of the root hair walls is considerably less than that of the parallel radial pathway in the soil. This supports the interpretation of root hairs as nutrient absorbing surfaces and as binders which help to maintain root–soil contact. A great deal more observation and experimentation is needed in the study of the interaction between roots, root hairs and soil pores.

Structure of natural soils and modification by agriculture

A few studies of natural soils have been made, for example Bryan (1971) compared 16 British and Canadian soils of which some examples are given in Figure 4.12. The high pH, nutrient-rich soils generally have a higher proportion

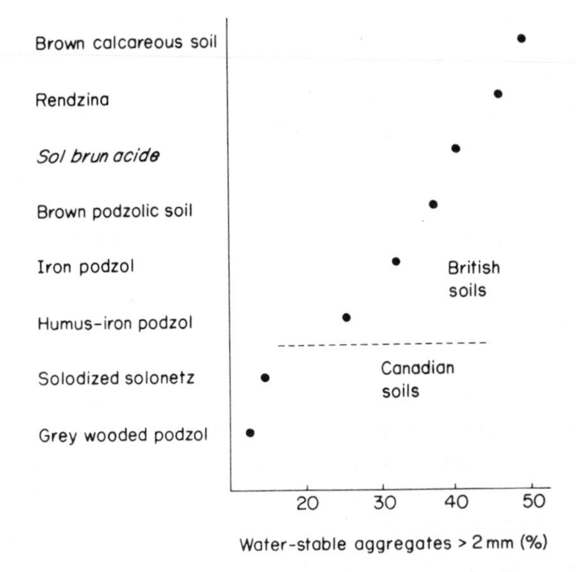

Figure 4.12 The relationship between soil type and degree of soil aggregation. Data of Bryan (1971).

of large aggregates and the Canadian soils tended to be less well aggregated than British counterparts, probably because of their higher montmorillonite and sodium contents, differences which may originate both from parent material and climatic P/E regime. Organic carbon content is often correlated with degree of structuring (Wooldridge, 1965) but not all workers have found this relationship. Clay soils are usually better aggregated than sandy types as their surface-active particles permit bonding by organic and ionic bridges while clay cutans may stabilize peds by coating their surfaces.

Many agricultural experimenters have attempted to relate crop growth to aggregate size, stability or soil porosity but much of the work is difficult to assess as the aggregation-promoting treatments may affect crops as much as the changes in structure. One possible solution is to use a well aggregated soil as a control, the treatments being varying degrees of crushing and compaction of the same soil. Taylor and Gardner (1963) used this technique to measure the penetration of a compacted soil layer by cotton (*Gossypium hirsutum*) roots. At a metric potential (p. 150) of −20 kPa (field capacity), root penetration was reduced by over 70 % as soil bulk density increased from 1·55 to 1·85 g ml^{-1}. The main cause was apparently increased resistance to mechanical deformation, and reducing the water potential to only −77 kPa totally inhibited penetration even at a bulk density of 1·75, probably because the drier soil has greater mechanical strength.

Ploughing of grassland or other soils rapidly decreases aggregate size, stability, pore space and population sizes of soil fauna and microorganisms (E. W. Russel, 1973; R. S. Russel, 1977). The use of farmyard manure and other organic materials in agriculture is, in part, motivated by the need to maintain structure which not only influences plant growth but also the trafficability of the soil and its resistance to water and wind erosion. The use of synthetic aggregating agents has been investigated experimentally but their cost often makes their use prohibitive in agriculture. They constitute a range of long-chain polymers such as polyvinyl alcohol, polyacrylic acid and polyacrylamide which function by hydrogen bonding or ionic bonding to clay particles. Their use has incidentally provided another experimental means of investigating soil structure (Emerson *et al.* 1978).

SOIL AERATION

As noted above, aeration is closely related to water content and, in the context of anaerobic soils, will be discussed in Chapter 10. Various workers have noted, for aerobic soils, that the nitrogen + argon content remains close to that of the external atmosphere (79 % v/v) while oxygen and carbon dioxide vary in complementary proportions to make up the remaining 21 % (Van Bavel, 1965).

Experimental investigation of root growth in relation to aeration suggests that quite low levels of oxygen may be tolerated, but the accompanying increase in carbon dioxide may inhibit growth, water uptake and nutrient absorption in some species when it exceeds 4–5 % in the soil atmosphere. Some studies also show that the CO_2 level must exceed the O_2 level for toxic effects to occur (Sheikh, 1970). The duration of exposure is also important, the toxicity of CO_2 often remaining hidden in short experiments. Most experimental work has been undertaken either by flushing nutrient solutions with the requisite gas mixtures or by forcing the gas through the soil or some other rooting medium; neither of these techniques fully simulates the natural root environment.

Van Bavel (1965) describes both suction and diffusion techniques for sampling the soil atmosphere in the fields. In the former case a sample is taken by inserting a hypodermic needle to the required depth and withdrawing a sample while the latter method permits diffusive equilibration of a buried sample chamber before the gas is extracted. The suction technique has been criticized as it sometimes appears to cause changes in composition, presumably due to differential gas density/transport effects. Using methods of this sort, most workers have found appreciable O_2 contents in the majority of soils unless they were almost entirely water-saturated. As a water table is approached, from above, the O_2 content gradually falls and the CO_2 content rises proportionately. Normal, dry soil rarely has a CO_2 content of more than 0·5–1·0% or an oxygen content of less than c. 20%. Between about 15 to 30 cm above the water table, according to soil texture and structure, the CO_2 content reaches c. 5%. This corresponds with a very wet soil condition: a water potential of −3 kPa, much wetter than field capacity. With even wetter conditions the CO_2 and O_2 concentrations may become equal at about 10%.

These effects are closely dependent on soil texture and structure. Coarse-textured or well-structured soils have higher gaseous diffusive transfer rates than fine-textured or poorly structured soils under wet conditions. This is because they contain many large pores which remain gas-filled. The water-filled pores of the fine-textured soils form a potent barrier to gaseous diffusion; oxygen, for example, diffuses ten thousand times more slowly in water than in gas. The relationship is reversed, quite often, in dry soils as the finer textured soils often have a greater total pore space and provide a larger gas-filled cross-sectional area for diffusion. The major part of the gaseous transfer in the soil is diffusive: Kimball and Lemon (1971), for example, showed that the effects of wind turbulence caused only a slight change in the rates of surface gas exchange even in coarse-textured soils. Changes in atmospheric pressure and temperature must result in some gas exchange but probably not fast enough to influence the equilibrium between diffusive movement and microorganism metabolism.

INVESTIGATION OF ROOT GROWTH AND DISTRIBUTION

Despite the inherent difficulties of technique, studies of root distribution have been made since the early years of plant ecology. Amongst them is a remarkable example: Weaver's (1926) classic monograph on the root systems of North American crop plants and native indicator species. Weaver made his observations by dissecting out and mapping roots in the walls of soil pits but he cites earlier workers who used washing-out techniques after stabilizing the roots by thrusting closely spaced pins through an isolated soil monolith. This 'pinboard' technique and other profile-face mapping methods have been widely used to investigate root-distribution in wild and cultivated ecosystems (Kutschera, 1960; de Roo, 1969; Bohm, 1979).

Extraction of the whole root system gives some impression of its condition at the time of sampling but it is not a suitable technique for displaying the dynamic aspects of growth: the time course of root initiation, the extension and ageing of roots and the exploitation and re-exploitation of rooting volume. The most satisfactory approach to this problem has either been the use of glass-sided containers or the installation of glass panels in trench walls in the field. Though the rooting conditions are not entirely natural, a great deal of information has been gained by measurement of root extension and time lapse photographic studies of the distribution of roots and soil organisms. One of the most extensive investigations reviewed by Rogers and Head (1969), with an underground laboratory installation (Rogers, 1969), has given information on the seasonality of root growth in perennial woody species, the length of life of corticated and decorticated roots and the influence of various environmental factors. The quantity of organic matter returned to the soil by sloughing of the cortex approximates to half the dry matter of the young root and as it occurs at a root age of only 2–3 weeks, represents a very large input. Disruption of the decaying cortical tissues is largely due to the activity of soil animals: small worms and arthropods.

Bohm (1979) has given a review of other underground laboratory installations and also describes a cheaper technique using embedded glass tubes, the inner walls of which are inspected using a periscope.

The measurement of root weight has been discussed by Lieth (1968b) and Newbould (1968). Samples may be removed by cutting out soil blocks of known volume, or with coring tools, and the roots removed by gentle washing and sieving procedures. These techniques are fraught with difficulties: the sampling is laborious and, because the variability is usually high, demands extensive replication while the washing procedure causes the loss of some fine rootlets. To reduce the first problem, various investigators have designed powered core-sampling devices which are reviewed by Bohm (1979), who also describes a range of washing and sieving procedures.

The high replicate variability is usually a reflection of the heterogeneity of the soil environment which is discussed in Chapter 10 in relation to the niche concept in competition. The biogenic nature of soil formation must lead to spatial variability even when parent materials are very uniform. Consider, for example, a young dune soil: when the first plant colonists appear they will cause a localized accumulation of organic matter, a localized specialization of the microflora and fauna and, on death, the localized release of mineral nutrients which were gathered from a large soil volume. Some of these nutrients, such as nitrogen, phosphorus and sulphur, are not rapidly mobilized from organic material and, thus, a three-dimensionally variable mosaic of chemical conditions is imposed on the soil and influences root growth on both a macroscopic and a microscopic scale. The existence of 'pattern' in vegetation and the 'clumping' of some species due to the convergence of rhizomes of different individuals, discussed by Kershaw (1963), the mutual exclusion of rooting volumes noted by Rogers and Head (1969) and the localized, long-term accumulation of nutrients by plants (Goodman and Perkins, 1959) all have relationship to this small-scale heterogeneity which makes root study, in the field, so difficult.

In addition to the horizontal and vertical variation in root exploitation of soil volume, there is also considerable variation in the size and age of roots found in different parts of the soil. The distribution of the finest, youngest, absorptive roots may differ from that of older roots, and also changes rapidly with time. Meyer and Göttsche (1971) investigated root biomass distribution in a *Fagus sylvatica* (beech) stand and found that fine roots (2–0·5 mm diameter) and active root tips were most abundant in the superficial organic layers of an acid brown soil but roots of 5–2 mm diameter reached maximum development in the *A* and upper *B* horizons of the mineral soil (Figure 4.13). They also noted that the comparative content of active root tips was a sensitive index of soil condition; for example, the

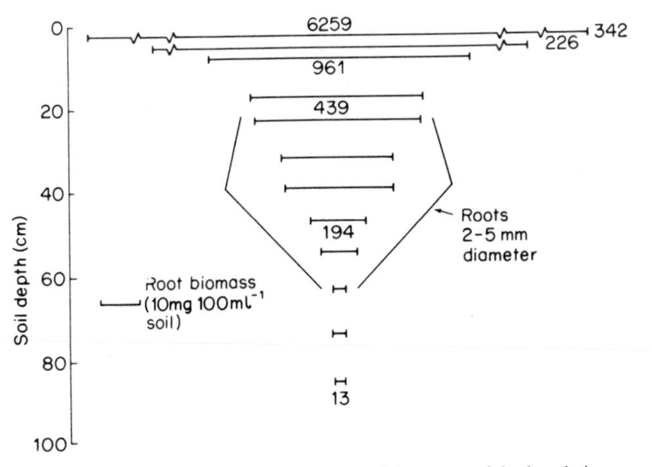

Figure 4.13 Distribution of *Fagus sylvatica* root biomass with depth in an acid brown earth. Data of Meyer and Gotsche (1971). Horizontal bars represent fine-roots (0.5–2 mm) and coarser roots (2–5 mm) are shown by the outer line. Central numbers are root tips per 100 ml soil.

maximum number of root tips per 100 ml of soil, in the *A* horizons of *F. sylvatica* stands, varied between 556 in a eutrophic brown earth and 46,600 in a podzol, presumably reflecting the need for additional absorptive area in the nutrient-deficient soil.

Root distribution may also be studied autoradiographically by injecting the stem with, for example, $H_2^{32}PO_4^-$ and, after equilibration, binding a slab of soil in contact with the photographic film. This technique has the advantage of distinguishing between living and dead roots but gives only circumstantial evidence of physiological function in the soil. Bohm (1979) has described a technique in which the plant leaves are exposed to $^{14}CO_2$ and isotopic dilution of $^{14}C/^{12}C$ used to assess the rate of root carbon turnover.

More critical studies of ion-uptake in the vicinity of the soil–root interface have been made by measuring the depletion of specific labels such as $H_2^{33}PO_4^-$, $^{35}SO_4^{2-}$, $^{36}Cl^-$, $^{65}Zn^{2+}$, $^{45}Ca^{2+}$. Potassium uptake has been simulated in such studies using $^{86}Rb^+$ (Nye and Tinker, 1977). The depth at which nutrient absorption is proceeding has been investigated by injecting radioisotopes into the soil at known depths (Bohm, 1979) and tritiated water has similarly been used in water-uptake studies (McWilliams and Kramer, 1968).

Plant Responses

Chapter 5

Plants and radiant energy

INTRODUCTION

Most physiological studies of photosynthetic carbon assimilation have been undertaken in completely artificial conditions. The quest for experimental repeatability has demanded the exposure of known small areas of leaf tissue to relatively low, controllable light intensities in conditions which permit variation of any required environmental factors. Often excised leaf tissues—whole leaves or leaf discs—have been used, sometimes with the added insult of suspension in bicarbonate buffer as a CO_2 supply.

The foregoing is not written in criticism: indeed much of our present understanding of leaf photosynthesis stems from such studies. It does, however, draw attention to the difficulty of measuring and realistically interpreting the photosynthesis of an entire leaf canopy or of individual plants within a canopy. There may be times when the uppermost leaves of the canopy are above light saturation and CO_2 limited while the basal leaves of the same canopy are near or below the light compensation point. At the same time, substantial differences of light quality, leaf temperature, leaf water balance and stomatal aperture may be found through the depth of the canopy. In some situations, adjacent leaves may even be fixing carbon by different biochemical pathways!

From this chaos of variation must be drawn sufficient information to answer questions concerning the competitive partitioning of energy use between individual leaves or between species, to extrapolate photosynthetic measurements to meet the needs of a production ecological study (p. 333) and to interpret or predict the effects of changing environmental factors on the day to day or seasonal photosynthetic functioning of an ecosystem and its component plants.

This chapter is devoted to the photosynthetic behaviour of the individual leaf or plant and the whole canopy when unlimited by factors other than radiant energy supply and atmospheric CO_2. It is thus concerned with the different strategies which have been developed by plants to trap radiant energy, to concentrate carbon from a very dilute atmospheric background and to cope with the self-created shade environment of the canopy by intraspecific morphological and physiological adjustment or interspecies photosynthetic niche exploitation. Teleologically a plant may be considered as a mechanism which has optimized its photosynthetic function within the constraints of a particular environment to give, at the very least, a small annual carbon gain—a survival strategy—and, at the best, a large carbon gain representing a substantial proportion of the total ecosystem carbon budget. This would be the strategy of a successful dominant species.

STRATEGIES OF CARBON GAIN

The present-day atmosphere contains CO_2 at 0·03% v/v, a very low concentration which is photosynthetically limiting for many plants at radiant flux densities exceeding 25–50% full sun. The problem of harvesting this sparse crop of carbon without an active air-pumping system has been solved by a remarkably uniform feat of 'biological engineering' amongst all mesophytic terrestrial vascular plants. The solution is the diffusion-exchanger consisting of a very thin leaf lamina comprising an external waterproof cover (epidermis) enclosing the photosynthetic mesophyll which is permeated by an extensive gas-filled pore space communicating with the exterior through closable stomatal pores. Stomata and gas-filled pore space are found in all vascular plant sporophytes with the exception of a very few much-reduced aquatic thalloid species (Raven, 1977).

More than two-thirds of the mesophyll cell surface abuts on intercellular space, the chloroplasts are never more than 10–20 μm from water-saturated air and the maximum length of diffusion path from exterior to the centre of the mesophyte leaf rarely exceeds 500 μm. With fully open stomata the CO_2 scavenging efficiency of this system is quite remarkable: an agricultural grass sward in a temperate climate may frequently show gross canopy photosynthetic rates of 20–30 g CO_2 m^{-2} h^{-1}, equivalent to the total CO_2 content of 30–46 m^3 of air. The efficiency of the carbon dioxide trapping mechanism depends on the porosity of the leaf lamina and the exposure of a CO_2 dissolving water surface at the mesophyll–intercellular space boundary. The photosynthetic leaf is consequently a remarkably good evaporator of water: while assimilating 30 g CO_2 h^{-1}, a square metre of grass sward, may lose 0·5–1·0 kg H_2O h^{-1}, an amount which could exceed the remaining available water in a moderately dry soil. The physiological dilemma is imposed of necessity: the stomata must close and, in their closing, the ingress of CO_2 will be limited.

It is useful to look at the carbon gainwater loss dilemma through a number of hypothetical strategies and to compare these with the real systems which have arisen by evolutionary adaptation. Some obvious possibilities are:

(i) Produce a CO_2-permeable, water-impermeable coating for evaporating surfaces: a natural anti-transpirant.

(ii) Modify the physical nature and function of the leaf so that stomatal closure most effects water loss and least effects carbon gain.

(iii) Decrease the shoot:root ratio and accept the inevitable respiratory cost.

(iv) Store water internally so that CO_2 may be assimilated in dry conditions.

(v) Limit radiant energy absorption by leaf to reduce evaporative latent heat flux.

(vi) Increase the affinity of the biochemical CO_2 trapping mechanism for CO_2: this will increase the ratio of CO_2 gain to water loss.

(vii) Transfer the period of CO_2 assimilation to a time when atmospheric humidity is highest and radiant energy supply is lowest (night?).

Taking these possibilities in sequence, the first appears to be a physical impossibility for living systems. 'Neither man nor nature has been able to produce a material which has a high permeability to CO_2 but a low permeability to H_2O' (Raven, 1977). The second possibility has been developed and it has long been shown that some degree of stomatal closure limits water loss before it reduces CO_2 uptake (Gaastra, 1959; Meidner and Mansfield, 1968). This is because the major resistance to CO_2 assimilation lies in the aqueous phase of its diffusion pathway and in the biochemical reactions of fixation. The resistance of the gaseous diffusion pathway may represent as little as 10% of the total barrier to CO_2 uptake, and stomatal closure must be almost complete before this resistance much exceeds the mesophyll-biochemical component. By contrast there is little resistance to evaporation from wet cell surfaces and the main limitations to leaf water loss are stomatal and boundary layer resistance of which the latter is much reduced by wind. Meidner (1975) gives a convincing argument that the CO_2 uptake and water loss systems are actually separated from each other, CO_2 being absorbed through wet mesophyll cell walls but most water evaporation taking place from the internal face of the leaf epidermis, including the guard cells, adjacent to the sub-stomatal cavity: this shortening of the internal water diffusion pathway would make the stomatal mechanism an even more efficient differential controller of water loss compared with CO_2 gain.

Decrease of shoot:root ratio is a common consequence of exposure to drought and also appears as a genotypic adaptation to arid conditions. It increases the ability to absorb soil water while limiting transpiration loss. Similarly, some mesophytes respond to drought by increasing succulence and the development of massive water storage tissues is a common genotypic means of escaping drought damage. Extensive discussion appears in chapter 6 (p. 162 ff). Both of these adaptations have a serious respiratory cost which will be discussed below. The modification of leaves to limit radiant energy absorption (p. 20) again reduces water loss but must involve a photosynthetic sacrifice unless the system is already light saturated.

The remaining two possibilities are different in nature, each requiring biochemical modifications of normal photosynthetic behaviour to produce either a more efficient CO_2 trapping mechanism or to divorce CO_2 trapping from a need for light. The early years of biochemical photosynthetic research were dominated by pathway studies on green unicellular algae, notably *Chlorella* spp or on *Spinacia oleracea* (spinach) chloroplasts. It was amongst these organisms that the 'normal' photosynthetic pathway of CO_2 fixation to produce 3-phosphoglycerate as a first product was described (Calvin and Benson, 1948). This is the three-carbon commencing compound of the Calvin cycle and these plants now usually referred to as C-3 plants. It was not until 1966 that Hatch and Slack reported the discovery of an entirely new carbon fixing path in which the first products are four-carbon acids (malic, oxaloacetic and aspartic, for example). Plants with this pathway are now called C-4 plants. Similarly, though 'dark fixation' of CO_2 by plants to form organic acids had been known since the early 1800s it was not until the 1960s that this Crassulacean acid metabolism (CAM) was seen as a major contribution to the carbon balance of such plants (Neales, 1975). These two alternative pathways, C-4 and CAM are interpretable as an evolutionary answer to the need for strategies (vi) and (vii) above.

To explain the ecological implications of the C-3, C-4 and CAM photosynthetic pathways it is necessary to outline the path of carbon in each (Figure 5.1). In the C-3 system which is common to the majority of vascular plants and to all non-vascular plants, CO_2 is accepted by a five-carbon compound, ribulose bisphosphate (RUBP), mediated by the enzyme RUBP carboxylase. Unfortunately for the plant, RUBP carboxylase also acts as an oxgenase which, in competition with CO_2 trapping, converts a substantial amount of RUBP to glycollic acid which is either *photorespired* to CO_2 or leaked to the environment by some algae (Figure 5.1a). No function for photorespiration has been described and the carbon loss *via* glycollate has been interpreted as a mark of the evolutionary conservatism of the RUBP carboxylase which first evolved before photosynthesis oxygenated earth's atmosphere (B. N. Smith, 1976). The present day oxygen concentration of 21 % v/v is strongly inhibitory to C-3 photosynthesis because of this effect and photosynthetic rates increase with decreasing oxygen concentration down to 2 %, the Warburg effect, first described in 1920.

50 % or more of the assimilated carbon may be lost again to photorespiration (Zelitch, 1975), thus RUBP effectively has a fairly low affinity for CO_2 unless the CO_2:O_2 ratio is higher than the normal atmospheric value. Strategy (vi), an increase in the CO_2 affinity of the trapping mechanism, is thus a possibility, and the C-4 pathway is the evolutionary development which has provided it. CO_2 is accepted by phospho-enol-pyruvate (PEP) to produce the C-4 compound, oxaloacetic acid, by a process which is not oxygen inhibited (Figure 5.1b). The C-4 product is exported as a malate from the mesophyll cells to specialized bundle sheath cells where it is decarboxylated and the CO_2 fed into conventional C-3 photosynthesis at a concentration sufficiency high to outweigh the effects of oxygen competition. PEP thus acts both as a CO_2 trap and concentrating mechanism. The relative trapping efficiencies of RUBP and PEP are reflected in the CO_2 compensation points of the two systems: usually 50–100 v.p.m. for C-3 and 0–5 v.p.m. for C-4. The compensation point is the minimal CO_2 concentration to which photosynthesis can scavenge a closed system and also represents the minimal intercellular CO_2 concentration which will be established during photosynthesis.

The CAM system (Figure 5.1c) is similar to the C-4 pathway in trapping CO_2 with PEP but the compartmentation of malate decarboxylation is temporal and not spatial. The CAM plant opens its stomata at night and traps CO_2, which accumulates as malate to substantial concentrations (100–200 $\mu M\ g^{-1}$ fresh

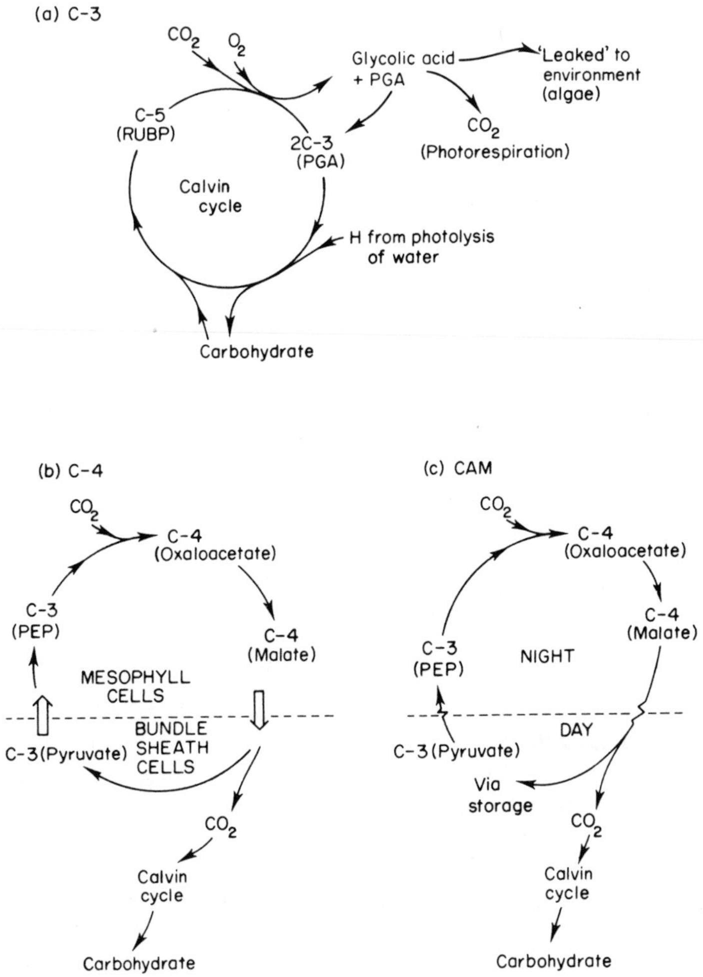

Figure 5.1 Biochemical strategies of carbon gain. RUBP = ribulose diphosphate. PGA = phosphoglyceric acid. CAM = Crassulacean acid metabolism. Arrows represent biochemical pathways and block-arrows indicate physical transport.

weight). At dawn the stomata close, hermetically sealing the intercellular space system, and CO_2 from malate decarboxylation is then processed by the C-3 pathway with no accompanying water loss. The non-autrophic trapping of CO_2 at night, followed by daytime stomatal closure, is entirely compatible with strategy (vii) and is almost entirely confined to plants of arid habitats. Further discussion of the water conserving and other ecological aspects of these photosynthetic options is given on p. 182. Table 5.1 is a synopsis of the main ecologically relevant comparisons between the three photosynthetic pathways.

Table 5.1 Ecologically relevant characteristics of plants with C-3, C-4 or CAM carbon assimilation*

	C-3	C-4	CAM
a. Usual morphology	Meso-	Meso-	Succulent (at least in mesophyll)
b. Stomata open	Day	Day	Night
c. Maximum photosynthetic rate ($mg\ CO_2\ dm^{-2}\ h^{-1}$)	20–40	40–60	10–15 (dark fixation)
d. Photorespiration (× dark respiration)	3–5	<0·1	3–5
e. Light saturation (× full sunlight)	Commonly 0·3–0·5	Often >1	Not relevant
f. CO_2 compensation point (v.p.m.)	50–100	0–10	Not relevant
g. Maximum CO_2 gradient (external–internal v.p.m.)	*c.* 280	320 +	Not relevant
h. Temperature optimum (°C)	10–30	30–45	<30?
i. Water use efficiency ($g\ H_2O/g\ C(H_2O)$)	600	300	50–150 or less
j. Ecological tolerance range	Wide	Moderate	Narrow (specific habitat)

* Information from Ludlow in Lange *et al.* (1976) and Kluge and Ting (1978).

CARBON BALANCE

The photosynthetic plant meets first the dilemma of CO_2 gain versus water loss but once carbon has entered the autrophic metabolic pathway there is a further problem: some proportion of the assimilated carbon is immediately required to fuel the respiratory needs of maintenance, growth and reproduction. It is known that some plants inherently respire more slowly than others and that shading may induce lower rates of respiration and so it may be postulated that some plant-respiratory loss of carbon is unnecessary. If this loss could be reduced, the net photosynthetic efficiency of the plant would be greater.

Dark respiration usually represents about 10% of the CO_2 fixation rate in bright light (Zelitch, 1975) but, because many of the plant parts are either shaded in the lower canopy, or non-photosynthetic, the quantity of carbon lost to dark respiration may range between 30 and 70% of gross photosynthesis (Zelitch, 1971). Photorespiration, which is 'switched-on' by glycollate production in the light (Figure 5.1a) may represent at least 50% of net photosynthesis. Photorespiration can be supressed by reducing the oxygen concentration (Bjorkman, 1971) or by blocking the biochemical pathway of glycollate production (Zelitch, 1975). In either case, the reduction of photorespiration increases net photosynthesis and the plant suffers no apparent harm. It seems then that almost half of the gross photosynthesis of normal mesophytes is lost to plant respiration and that some of this loss is unnecessary.

The green plant ranges from the unicell through low-growing herbaceous types to the giant forest tree with a large investment in structural tissue. These various morphologies are accompanied by an equally diverse range of photosynthesis–respiration patterns. The very high primary production:biomass ratio of many algal communities, discussed in chapter 11 (p. 336) is partly a consequence of their limited development of structural and storage tissue. Similarly, many aquatic macrophytes, supported by their own buoyancy in water, are rather efficient in terms of net photosynthesis. Terrestrial plants, however, are faced with problems of mechanical support, development of root systems for both anchorage and water absorption, water storage and conservation and the need for metabolite storage tissue to cope with environmental instability. These problems are met by production of non-photosynthetic tissues which carry a penalty of respiratory loss and there comes a time in plant development when additional respiratory tissue becomes a burden to the plant.

Went (1957) cites the example of strawberry in which non-photosynthetic tissue can build up to such an extent that the plant barely reaches compensation point at 50 $W\,m^{-2}$ (1500 foot candles), an irradiance at which younger plants grow satisfactorily. Evans (1972) discusses this relationship and also shows the effect of temperature in the much larger proportion of gross photosynthesis which may be dissipated in respiration of tropical forest trees compared with temperate zone trees. Any adaptation which decreases shoot:root ratio, for example drought and cold tolerance, or endurance of high grazing intensities, automatically incurs an increased respiration: photosynthesis ratio. This cost may be recouped in the sense that the period of carbon-gain may be extended during drought or it may be accepted as a payment for survival, the stressed habitat offering a niche to the adapted, but slower growing species in which it escapes from the competitive pressures of stress-unadapted plants. Mooney (1975) presents an interesting cost/benefit analysis of the carbon cost of various physiological and morphological strategies for survival of environmental stress.

The relatively enormous photorespiratory loss of C-3 plants is avoided by the C-4 group: the trapping and concentrating efficiencies of the PEP system is so great that any photorespiratory CO_2 is recycled almost as soon as it is produced, furthermore the favourable CO_2:O_2 ratio in the bundle sheath probably limits glycollate production (Bishop and Reed, 1976). The apparent lack of photorespiration permits C-4 plants, when photosynthesizing under optimum conditions, to have very high rates of CO_2 assimilation (Table 5.1) compared with most C-3 plants, a fact which has attracted plant breeding research into the feasibility of transferring the C-4 pathway into C-3 plants (Bishop and Reed, 1976).

PHOTOSYNTHESIS AND LIGHT

Plants show a great diversity of adaptation to light environment with a several hundred-fold difference in saturation value between sun plants and the extreme shade plants of the rain forest floor (Bjorkman, 1975; Boardman, 1977). These differences stem from differences in morphology and anatomy or biochemistry of photosynthesis and respiration. The morphological and anatomical relationships with light intensity have long been studied. Haberlandt (1884) wrote: 'A sun-leaf may be thrice as thick as a shade-leaf on account of the more abundant development of its palisade tissue.' High light intensity and water shortage promoted the formation of thicker, more sclerified leaves (Shields, 1950) and the truly xeromorphic leaf is usually characteristic of open, high light-intensity habitats. The thin shade-leaf, often with a single palisade mesophyll layer, has a much larger area to weight ratio than a sun leaf, carries a lesser proportion of respiring tissue and reaches photosynthetic saturation at a relatively low light intensity. Larcher (1969) elegantly showed these differences for the Mediterranean sun-plant *Quercus ilex* with thick sclerophyllous leaves and *Fagus sylvatica*, a tree with a cool-temperate distribution. *F. sylvatica* produces two leaf types: fairly thick sun-leaves in the upper crown and thin, horizontally orientated shade leaves in the lower canopy. Figure 5.2 shows the light saturation curves for

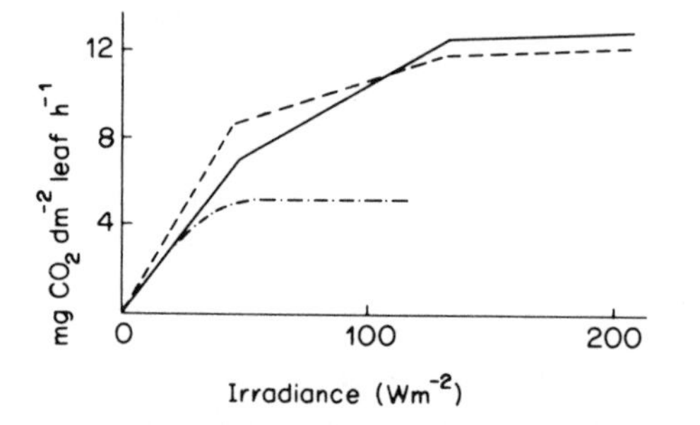

Figure 5.2 Light saturation curves for *Fagus sylvatica* sun leaves (– – – – –), shade leaves (–·–·–·–) and *Quercus ilex* sun leaves (———). The values of irradiance have been approximately converted from the author's illuminance measurements using the nomogram of Figure 2.13. Adapted from Larcher (1969).

these two leaf types and for the more extreme sun leaves of *Q. ilex*. The shade leaves saturate at a very much lower intensity that the sun-leaves of beech and the sclerophyllous leaves of the oak saturate at the highest intensity.

The effect of reduced respiration in shade species is well known (Figure 5.3). Bjorkman (1968) found shade leaves which respired five times more slowly per unit area than sun-leaves. The light compensation point of such shade plants is consequently much lower than that of sun plants. In reviewing the work of Bjorkman and his co-workers, Boardman (1977) notes that the compensation point of *Alocasia macrorhiza*, a Queensland rain forest floor species, is 50 times lower (0·05–0·2 nE $cm^{-2} s^{-1}$) than for the sun-species. Tropical rain forest floor species also illustrate the fact that shade-leaves do not always have the large area to weight characteristic which is so common in temperate forests: many, in fact, have relatively thick leaves (Bjorkman, 1975). *A. macrorrhiza* also shows photosynthetic saturation at a very low quantum flux density and, in its forest habitat, receives an average of only 22 μE $cm^{-2} d^{-1}$ compared with 300 times this

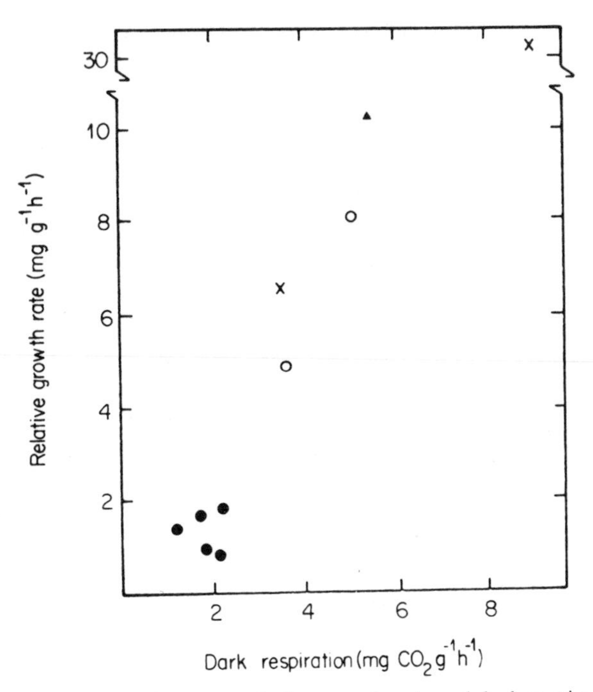

Figure 5.3 The relationship between relative growth rate and dark respiration rate of leaf discs taken from plants of differing light intensity habitats. The relative growth rates were measured over 15 days of sunny summer weather. ●—shade tolerant species; ○—shade intolerant species; +—crop plants; ▲—arable weeds of high light habitats. Data of Grime (1965).

amount received by the C-4 desert plant *Tidestromia oblongifolia* when it is at peak growth rate (Bjorkman, 1975). As might be expected, the light-saturated photosynthetic rate of *T. oblongifolia* is about 15 times greater than that of *A. macrorrhiza*. The desert plant has the phenomenal relative growth rate of 30 % d^{-1} at 45 °C: a biomass doubling in less than 3 days (Bjorkman, 1975)! These two plants are extreme examples of, respectively, opportunist and survival photosynthetic strategies.

The most commonly noted biochemical difference between shade- and sun-species is reduced leaf content of the primary carboxylating enzyme RUBP carboxylase (carboxydismutase) in shade plants (Bjorkman, 1966). Further work has shown that a whole range of enzymes and intermediaries in the photosynthetic system are also depressed in shade-species (Bjorkman, 1975). The largest difference appears in total soluble protein and RUBP carboxylase protein: this has been interpreted in terms of the high energy cost of protein synthesis: the sun-plant gains its potentially very high light saturation value and photosynthetic rate at the expense of a substantial respiratory cost which prevents it from surviving in a low-light habitat. Similarly the shade-plant with limited investment in protein has a low respiratory rate and tolerates long periods of low light intensity, but its CO_2 trapping mechanism and electron transporting ability is limited, causing it to light-saturate at a low intensity and photosynthesize at a maximum rate much lower than that of a sun-plant (Raven and Glidewell, 1975). This behaviour is not confined to vascular plants, for example Raven and Smith (1977) described similar relationships for sun- and shade-species of giant-celled green algae.

The sun- and shade-characteristics may, in a limited degree, be produced by pre-adaptation of plants or even individual leaves at the requisite light intensities. *Atriplex patula* (orache), a sun-species, when grown at three irradiances (20, 6·3 and 2 $mW\,cm^{-2}$ PAR) in pretreatment, shows photosynthetic saturation at, respectively 4·3, 3·0 and 0·8 $mW\,cm^{-2}$ (Bjorkman *et al.* 1972). This lowest value is still 20 times greater than the light saturation value for a true shade plant such as *Alocasia macrorrhiza* and, irrespective of pretreatment, no sun plant can maintain positive carbon balance at very low light intensities; thus shade tolerance is first and foremost genetically determined despite having some phenotypic plasticity. Many workers have demonstrated this genetic control, ecological races from open and shaded habitats differing in their capacity for shade tolerance or high-light photosynthesis (Boardman, 1977).

The *A. patula* plants pretreated with low light intensity also showed reduced respiration rates: dark respiration was approximately proportional to light saturated gross photosynthesis. Similar results have been reported for other species (Boardman, 1977) and, as discussed in the next section, this may have a marked influence on net photosynthesis of a deep canopy if lower leaves may acquire the lowered respiration characteristic.

The topic of sun- and shade-plant photosynthesis has been well reviewed by Boardman (1977) and Table 5.2 presents a synthesis of the main differences which have emerged from these studies. Most C-4 species are sun-plants and, because of

Table 5.2 Characteristics of sun and shade leaves (Boardman, 1977)

	Sun	Shade
1. Leaf morphology and anatomy		
a. Thickness	Greater	Lesser
b. Mesophyll layers	More	Less
c. Stomata/area	More	Less
d. Petiole vasculation	Increased	Reduced
2. General physiology		
a. Light compensation (ne $cm^{-2}\,s^{-1}$)	*c.* 10	Down to 0·05
b. Light saturation (% full sunlight)	20–30	1–3
c. Quantum efficiency	Low	High
d. Dark respiration rate	High	Low
e. Photo-inhibition	No	Yes
f. Stomatal and mesophyll conductance	High	Low
g. Chloroplasts: granal stacks	Small	Large
3. Biochemistry		
a. Chlorophyll/dry weight	Large	Small
b. Soluble protein	High	Low
c. Chlorophyll/soluble protein	7–14	2–5
d. RUDP carboxylase	High	Low

their different metabolism and shade intolerance, are not included in this tabulation.

LEAF AND CANOPY

In Chapter 2 the penetration of canopies by short-wave radiation was discussed in relation to the Monsi and Saeki (1953) model:

$$R_i = R_o e^{-K_{ext} L} \tag{2.6}$$

If the uppermost leaf layer of a canopy is just light saturated and if photosynthesis is linearly related to light intensity, the model would predict a progressive exponential reduction in photosynthetic rate of each sequential leaf layer. With sufficient canopy depth, the lowest layers will be below light compensation point and represent a respiratory drain on canopy net photosynthesis. Stern and Donald (1962) discussed the implications of this situation for a clover-grass sward: an optimal leaf area (L) index would exist when the lowest leaf layer is just at light compensation point. In this condition the light-harvesting by the canopy will be most efficient and the crop growth rate maximal but further increase of L will reduce crop growth rate.

Real canopies do not behave in accordance with this model for a number of reasons. In mixed-species canopies the subordinate species are adapted by selection to a shade environment and, in both mixed- and single-species canopies, the lower leaves have been pre-conditioned, by a period of exposure, to shade. McCree and Troughton (1966) showed for *Trifolium repens* (white clover) that photosynthetic yield continues to increase with leaf area index to its maximum of 10–11 (Figure 5.4b). This implies that an optimum leaf area index does not exist

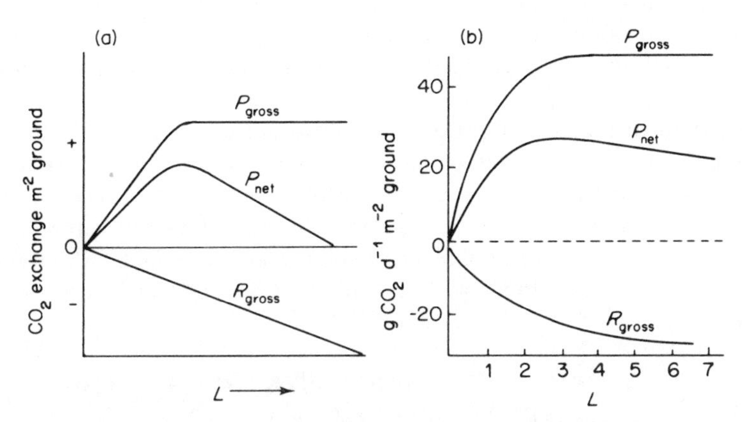

Figure 5.4 (a) The relationship between gross photosynthesis (P_{gross}), net photosynthesis (P_{net}) and respiration (R_{gross}) assuming respiration to increase in proportion to leaf area index (L). (b) Measured values of P_{gross}, P_{net} and R_{gross} for canopies of *Trifolium repens* of differing leaf area index in direct illumination in controlled environmental conditions. Adapted from McCree and Troughton (1966).

even for a leafy canopy such as that of *T. repens.* This unexpected finding is explained in the terms of the relationship between photosynthetic and respiratory activity of each additional leaf layer. The lower leaves of the canopy become less active in respiration than the younger leaves above and consequently the respiration of the whole canopy is proportional not to leaf area index (Figure 5.4a) but approximately to gross canopy photosynthesis (Figure 5.4b). This is the same relationship shown for shade preconditioning in the previous section.

The relationship between total plant respiration and gross photosynthesis is often difficult to establish, with the result that field investigations of canopy photosynthesis and crop growth rate have often given more ambiguous results than the controlled environment experiments of McCree and Troughton (1966). Difficulty of measuring root respiration was suggested by Seehy and Peacock (1975) as one reason for their failing to find a significant correlation between daily crop growth and canopy net photosynthesis in various stands of forage grasses. Despite the problems of shade adaptation and the change in spectral quality as light penetrates the canopy, Seehy and Peacock did show a highly significant linear correlation between extinction coefficient (K_{ext}) and crop growth rate for seven of the eight species which they studied, suggesting that the Monsi and Saeki equation (2.6) for canopy penetration does give some general estimate of potential canopy production.

The larger the range of light-saturated photosynthetic rates amongst the leaves and species of the canopy, the greater the number of photosynthetic niches available thus permitting the development of deep, species-rich and efficient light-harvesting canopies such as those of rain-forests. The adaptations needed are tolerance of reduced PAR levels and of spectral enrichment in the far-red, red and green wavebands (p. 23).

Forest floor plants show pigment modifications adaptive to these conditions, for example increased chlorophyll b compared with chlorophyll a content, so improving absorption in the mid-spectral range (Boardman, 1977). The colour of deep-water red algae is given by the pigment phycoerythrin which is similarly adaptive.

In addition to these modifications, the low-light adaptations of deep shade species make them more quantum efficient. *A. macrorrhiza*, on overcast days, has a quantum efficiency of 0·07 mole CO_2 Einstein^{-1} for most of the day. This is very close to the maximum possible efficiency. On clear days the value drops to 0·05 mole E^{-1} as sunflecks of almost full solar intensity make up about 60% of the forest floor photon flux and overload the shade adapted photosynthetic systems.

MEASUREMENT OF PHOTOSYNTHESIS IN THE FIELD: GROWTH ANALYSIS

Gasometric methods

Characterization of individual leaf or plant photosynthesis, in the field, is now most often undertaken by using an enclosure through which air is passed and an infrared gas analyser (IRGA) to monitor changes in CO_2 caused by photosynthesis–respiration balance: the technique thus measures net photosynthetic CO_2 exchange (Sestak *et al.* 1971). Care must be taken in chamber design to prevent overheating, to control humidity and to avoid so much CO_2 depletion that stomatal opening is induced. The walls of the chamber must be transparent to short-wave and long-wave radiation, thus maintaining a normal net radiation balance. These problems are discussed in Sestak *et al* (1971). The most satisfactory chamber is usually the 'open' type in which a large air-flow is maintained over the plant and then passed, in total or part, to the IRGA before venting to waste. Controlled leakage may be used as an 'air-seal' against ingress of external air and to make enclosure of the plant quick and easy (e.g. Wolf *et al.* 1969).

Some alternative methods are measurement of oxygen evolution or of ^{14}C incorporation from labelled CO_2 (p. 340). The former is difficult with a high atmospheric oxygen background but is commonly used in aquatic studies where low solubility of oxygen reduces this problem. Fixation of ^{14}C has long been used in aquatic ecology and is now becoming more popular as a conveniently portable terrestrial technique (e.g. Fernandez, 1978).

Whole canopy studies are less easily undertaken gasometrically as even the best air-conditioned chambers disrupt normal air-flows and diffusion patterns. For example, Ordway (1969) made serious aerodynamic criticism of the Odum 'giant cylinder' technique which Odum and Jordan (1970) used to measure CO_2 exchange in tropical forest (p. 344). To avoid such problems attempts have been made to use micrometeorological techniques to calculate CO_2 fluxes through canopies (p. 30) but so far have only proved satisfactory in very uniform, large stands of crop or forest (p. 32).

Photosynthetic bioassay and growth analysis

Attempts to bioassay canopy photosynthesis have been made, for example Leach and Watson (1968) used solution-cultured *Beta vulgaris* (sugar beet) plants exposed at different depths in crop canopies and measured their weight gain with time. This technique may be very satisfactory if the phytometer has a similar photosynthetic response to the crop or other vegetation in which it is used, and if it is pre-conditioned to the canopy shade environment to which it will be exposed. Leach and Watson found that photosynthetic rates were reduced to 4% of the upper canopy maximum under a crop 60 cm deep: net photosynthesis was always positive at the base of the canopy but pre-conditioning the phytometers to the shade environment might have revealed rates more in excess of light compensation.

The long-term consequence of the photosynthetic process is weight gain by the whole plant. Blackman (1919) showed that plant growth is a phenomenon in which the rate of increase of size or weight is often governed by existing size and as such may be expressed in the form of Kelvin's compound interest law:

$$W_1 = W_0 \, e^{R(T_1 - T_0)} \tag{5.1}$$

where W_1 and W_0 are the dry weights at the beginning and end of the growth period; e is the base of natural logarithms (2·718); $(T_1 - T_0)$ is the growth period and R, the *Relative Growth Rate*, is defined as the weight gain per unit of plant weight per unit time. The instantaneous value of R is:

$$R = \frac{1}{W} \cdot \frac{dW}{dT} \tag{5.2}$$

In classical *Growth Analysis*, R may be partitioned to show the relative contributions to weight increment made by the inherent efficiency of the photosynthetic process and the efficiency with which tissue is deployed as photosynthetic surface:

Relative Growth Rate = *Unit Leaf Rate* × *Leaf Area Ratio*

$$\frac{1}{W} \cdot \frac{dW}{dT} = \frac{1}{L_A} \frac{dW}{dT} \times \frac{L_A}{W} \tag{5.3}$$

$$R = E_A \times F$$

Units are a source of difficulty: formal SI expression of $g\,m^{-2}\,s^{-1}$ leads to difficulty with sizes of numbers or time periods. It is often convenient to express analyses as $g\,dm^{-2}$ for a convenient non-SI time unit e.g. day, week or month.

Unit Leaf Rate (E_A), sometimes called Net Assimilation Rate, is the weight gain per unit of photosynthetic area per unit of time and is a direct function of net photosynthesis. *Leaf Area Ratio* (F) is the ratio of photosynthetic area (L_A) to total plant weight: broadly it represents the ratio of photosynthesizing to respiring tissue within the plant.

During periods of rapid growth, plant weight increases exponentially and Ashby (1937) showed that the mean of such a weight gain is:

$$\frac{W_2 - W_1}{\log_e W_2 - \log_e W_1} \tag{5.4}$$

If weight data are collected at two harvests the value of R may thus be calculated as below (Williams, 1946):

$$R = \frac{W_2 - W_1}{T_2 - T_1} \times \frac{\log_e W_2 - \log_e W_1}{W_2 - W_1} = \frac{\log_e W_2 - \log_e W_1}{T_2 - T_1} \tag{5.5}$$

likewise:

$$E_A = \frac{W_2 - W_1}{T_2 - T_1} \times \frac{\log_e L_{A_2} - \log_e L_{A_1}}{L_{A_2} - L_{A_1}} \tag{5.6}$$

Leaf area ratio (F) may be calculated from the exponential means of L_A and W but many workers have assumed F to be linearly related to time (Hunt, 1978) in which case:

$$F = \frac{(L_{A_1}/W_1) + (L_{A_2}/W_2)}{2} = \frac{F_1 + F_2}{2} \tag{5.7}$$

Leaf area ratio may usefully be subdivided into two components:

Leaf Area Ratio = *Specific leaf area* × *Leaf weight ratio*

$$F = \frac{L_A}{L_W} \times \frac{L_W}{W} \tag{5.8}$$

where L_W = leaf weight. Specific leaf area gives some impression of the efficiency with which leaf tissue is displayed to trap solar radiation and the leaf weight ratio is a measure of the leafiness of the plant.

These classic methods of growth analysis, first developed from the ideas of Blackman (1919) by West *et al* (1920) and by Gregory and his co-workers (Gregory, 1918, 1926), have found extensive use in analysing the growth of crop plants gradually extending into various aspects of ecology. For example, Grime and Hunt (1975) present numerous relative growth rates for wild plant species grown in standardized conditions which may be used in the interpretation of ecological strategies. Effects of environmental factors may similarly be studied, for example Ashenden *et al.* (1975) used the technique to investigate the effects of water stress on growth of *Dactylis glomerata* (cocksfoot grass) populations. Further applications will appear through the book. Much more detailed treatment of growth analytical techniques is given by Evans (1972), Hunt (1978) and Causton and Venus (1981).

The use of growth analysis in crop-stands and natural ecosystems has been further refined by the addition of two other concepts: that of *Leaf Area Index* (L), defined as the ratio of leaf area to ground area, and *Leaf Area Duration* (D), the integrated value of the relationship between leaf area index and time (Watson, 1947).

$$L = L_A/P_A \tag{5.9}$$

where P_A is the projection area of the ground over which L_A is deployed.

The usefulness of the leaf area index concept should have become apparent in chapter 2 where the Monsi and Saeki (1953) model of light penetration was discussed, and from the preceding section in which canopy photosynthesis was related to their model.

Leaf area duration (D) represents what Watson (1947) called the 'Whole opportunity for assimilation' and represents the 'leafiness' of the vegetation cover throughout the growing period:

$$D = \frac{(L_1 + L_2)(T_1 + T_2)}{2} \tag{5.10}$$

where L_1 and L_2 are leaf area indices at T_1 and T_2.

In the analysis of crop growth the concept is easily understood if the growing 'leafiness' of the canopy more or less coincides with the seasonal increase of solar radiation, thus giving a relatively simple relationship between D and absorption of PAR. Evans (1972), however, shows marked disparity between the times of development of 'leafiness' in different crops. Interpretation of the relationship between *D* and photosynthetic behaviour will be much more difficult in such cases.

LIMITATION OF PHOTOSYNTHESIS

Net photosynthesis of whole canopies is limited by a complex of environmental interactions with plant genotype-phenotype many of which are either neglected or do not arise in conventional laboratory studies of single-leaf or whole plant photosynthesis. Table 5.3 is based on the summary given by Sestak *et al.* (1971) and presents the most important of the individual effects.

Despite the complexity of these controls (the table contains over 30 individual headings and sub-headings), various attempts have been made during the past 20 years to model the process of canopy photosynthesis, with the main aim of optimizing crop yield by providing the plant breeder and the agronomist with varietal and cropping management 'design data'.

One of the earliest attempts was that of de Wit (1959) who produced a very simple mathematical model based on empirical response curves of individual leaves to irradiance level and penetration of the closed canopy by diffuse light and direct sunlight from any solar angle. The model permits calculation of instantaneous photosynthetic rate of canopy: integration, with suitable variation of parameters for time of day and season, and for respiratory loss, gives the potential photosynthetic production (p. 346) of the canopy over any required growth period. Under favourable conditions the model gave good correspondence with experimental measurements but longer-term estimates deviated markedly from the experimental data because the model did not include compensation for summer drought or winter cold effects.

More recent models have included far more environmental information: a very readable account is given by Lemon *et al.* (1971) who produced a comprehensive model of the soil–plant–atmosphere system which included prediction of photosynthesis. Equations have been developed which deal with energy and matter fluxes at an upper boundary above the crop canopy and a lower boundary at the soil-air interface (Figure 5.5). Between these boundaries, equations define crop-submodels of canopy architecture (leaf angle and area), the distribution of light, wind and vertical diffusivity (p. 20 ff) with height. A set of leaf-submodel equations relate stomatal resistance and photosynthesis to light intensity, respiration to temperature and boundary layer resistance to windspeed.

Inputs at the upper boundary are solar angle, direct solar radiation, diffuse radiation, air temperature, windspeed, carbon dioxide and water vapour concentration. At the lower boundary the soil surface water potential is used to predict evaporation, and soil CO_2 transfer to the atmosphere is included.

The computation flowchart of the model (Figure 5.6) shows the input of crop and boundary conditions followed by derivation of the light distribution pattern. This is then used to establish a leaf energy balance via stomatal, net radiation and wind exchange equations. Sufficient data is then available for the photosynthetic computation and the remaining flowchart steps check the fit of the computed

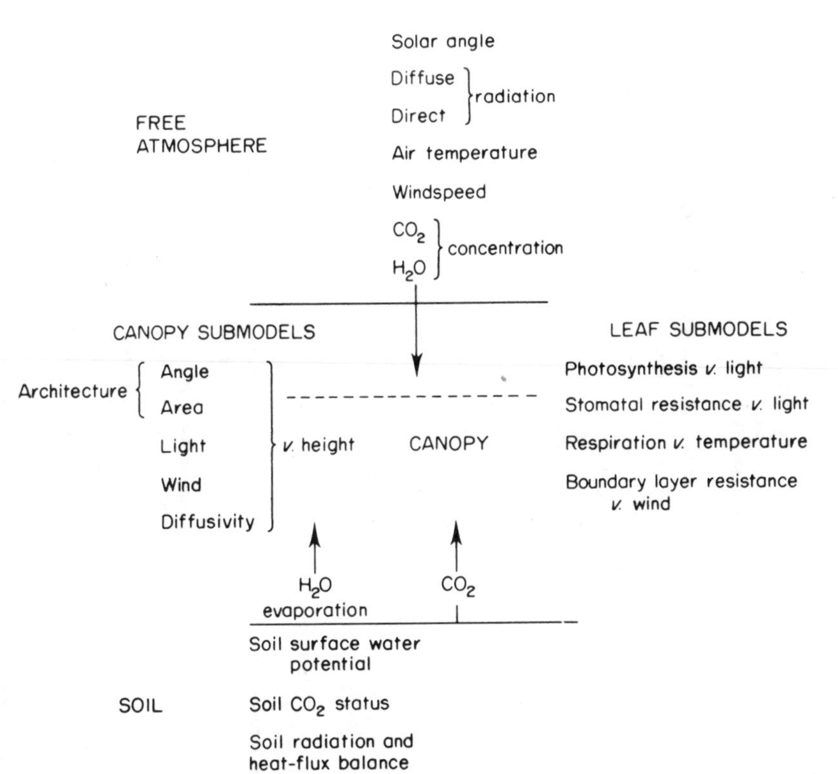

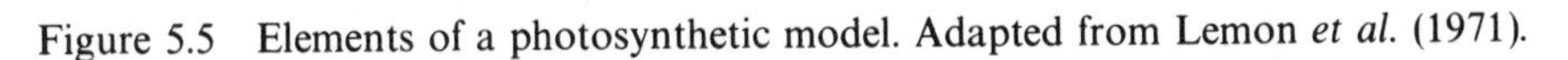
Figure 5.5 Elements of a photosynthetic model. Adapted from Lemon *et al.* (1971).

model with the requirement that all energy and mass sources and sinks should equate to zero.

Weaknesses in the model were sensitivity to soil surface wetness, which is difficult to measure and reduces accuracy of latent and sensible heat flux predictions; difficulty of predicting stomatal status, which is overriding to water and CO_2 exchange, and, finally, inadequate modelling of turbulent exchange within the canopy. The stomatal problem derived from its complexity of control, including at least four environmental feedbacks, ageing and hormonal effects. Despite these criticisms the authors claimed that the model gives fairly accurate prediction of net photosynthesis, which they used to investigate the effects of canopy architecture and of CO_2 concentration.

Leaf angle had a great effect on net photosynthesis in the simulations, a high leaf angle permitting better light penetration but the authors warn against the uncritical acceptance of this finding as a breeding criterion. No one structure is likely to prove 'ideal' in the real world of changing solar angle during the day and with latitude. Their investigation of CO_2 concentration was intended to assess the feasibility of CO_2 fertilization in the open air comparable with that which is used in commercial glasshouses. The model showed that the best increase of daily net photosynthesis would be no more than 10–20% even if 450 kg CO_2 ha^{-1} h^{-1}

Table 5.3 Limiting factors in photosynthesis

Factor	Field of influence	Control by e.g.	Chapter reference
	Physical environmental variables directly affecting photosynthesis		
1. 0·4–0·7 PAR			
a. Instantaneous quantity	Photosynthetic rate; stomatal opening	Solar angle: cloud; canopy or water depth	5
b. Quality	Photosynthetic rate	Canopy or water depth	5
c. Duration	Total photosynthetic production	Latitude-season; aspect	11
2. CO_2 concentration and diffusion gradient to chloroplast	Stomatal opening; CO_2 exchange, photosynthetic rate at light saturation	World-wide not very variable but gradient may be steeper for C-4 plants. Water-plants have very different CO_2 problem	5
	Physical environmental variables indirectly affecting photosynthesis		
3. All-wave radiation			
a. Instantaneous quantity	Net radiation balance; leaf temperature and transpiration	Solar angle; cloud; canopy or water depth	2
b. Duration	Period of supra- or sub-optimal leaf temperature or water balance	Latitude-season; aspect	2
4. Specific wavelength radiation	Photoperiod and photomorphogenetic control of e.g. leaf expansion, leaf morphology, bio-chemical status etc.	Latitude-season (photoperiod); canopy or water depth	5,8
5. CO_2 concentration	Stomatal opening effects on water loss and leaf cooling	See 2	
6. Windspeed	Influence on leaf and canopy boundary layer	Regional and local meteorological conditions	2
7. Air temperature	Plant and leaf temperature; water balance	Latitude-season; solar-angle; cloud; transpiration rate etc.	2,8
8. Air humidity	Transpiration rate; leaf temperature	Latitude-season; solar-angle; cloud; transpiration rate	
9. Soil water			
a. Potential	Plant water potential; stomatal opening etc.	Precipitation-evaporation balance	6,7
b. Mobility	Water movement to root; plant water potential	Soil water potential; soil particle-size distribution	6,7
c. Quantity available	Time until limitation is encountered	Soil and rooting depth; soil texture	6,7
	a, b, c above all operate through effects on plant water potential which then have a multifactorial influence on stomatal aperture, gas exchange, photosynthetic pathways, leaf temperature control, cell expansion and tissue		

d. Waterlogging	Adverse chemical environment for roots; toxicity	Precipitation-evaporation; Topography; drainage	3,4,9,12
10. Mineral nutrition			
a. Deficiencies	Diverse effects on root systems and photosynthetic organs	Geological parent material; Pedogenesis	
b. Toxicities and imbalance	As above	As above and pollutants	
11. Seasonal cycle: interaction of many of the above listed single factors	Plant development and growth	Latitude-season	8
Biological environmental variables indirectly affecting photosynthesis			
12. Competition	Shading; induced deficiencies etc.	Ecosystem structure and development	13
13. Herbivory and pathogenesis	Loss of photosynthetic tissue or assimilates; modification of growth pattern	Ecosystem type	13
	Genotype-phenotype characteristics affecting photosynthesis		
14. Leaf area (LA) manifested as LA ratio; specific LA; LA index and LA duration	Efficiency of light-harvesting; metabolic balance	Genetic and environmental	5
15. Leaf display by both phyllotaxic relationships and leaf angle	Effect on light harvesting and self-shading	Mainly genetic	5
16. Leaf structure; anatomical and optical properties	Use of available light; net radiant balance	Genetic and environmental	
17. Leaf diffusion resistance			
a. Variable stomatal	Homeostatic to water; influence CO_2 exchange	Plant water balance	
b. More or less invariant; cuticular and internal	Cuticular limits water-loss; air-water interface of mesophyll often limits CO_2	Mainly genetic; some environmental developmental influence	6,7
18. CO_2 capture mechanism (e.g. C-3, C-4 and CAM)	Controls efficiency of light-use, CO_2 trapping and water use	Basically genetic; CAM may be environmentally facultative	5
19. Composition of photosystems (e.g. Sun-shade leaves)	Efficiency of light and CO_2 use	Genetic and environmentally inducible	5
20. Chloroplast shape, structure and distribution	Efficiency of light trapping; CO_2 diffusion resistance in mesophyll	As above	
21. Concentration of photosynthesate and translocation	May limit photosynthesis by assimilate accumulation.	Combined effect of phenotype and environment	
22. Developmental stage and leaf age	Photosynthesis; respiration balance	As above	
23. Endogenous rhythms	Stomatal control of gas-exchange; metabolic patterns	Genetic and as above	
24. Adaptation and pretreatment	Capacity of photosynthetic system to cope with limits	Phenotype and environment	

were released from a pipe grid at ground level. CO_2 enrichment using very heavy applications of animal manure to the soil surface would scarcely alter canopy CO_2 concentration sufficiently to alter photosynthetic rate. The model did suggest that the present rate of increase of fossil fuel CO_2 in the atmosphere ($0{\cdot}7$ v.p.m. year^{-1}) could, within a century, increase net photosynthesis by

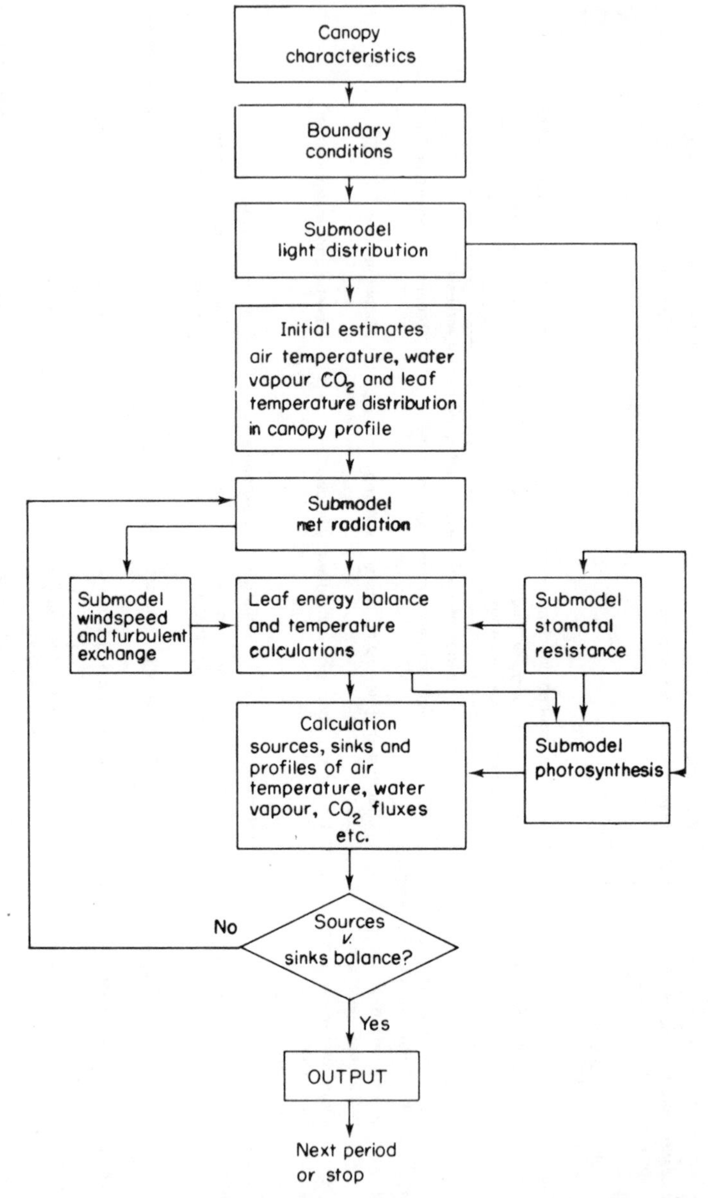

Figure 5.6 Flowchart of the SPAM canopy photosynthesis model. Adapted from Lemon *et al.* (1971).

10–20 %. A further useful observation was the importance of wind turbulence in mixing atmospheric CO_2 into the crop canopy: breeding for enhanced canopy air-flow may be worthwhile.

Models such as that of Lemon *et al.* (1971) operate predictively by depending on a fundamental knowledge of physiological responses, for example of the relationship between leaf photosynthesis and irradiance. Similar models have been developed by other workers, for example the ELCROS model of de Wit *et al.* (1971), which is a sophisticated product of de Wit's original attempts (1959) to model canopy photosynthesis mentioned above.

An alternative approach is empirical modelling in which existing field data relating whole canopy response to environmental factors is used to produce a model, without the need to resort to theoretical prediction of physiological function of any single plant organ. An example may be found in the ELM model developed by the U.S. IBP Grassland Biome Study (Sauer, 1978) in which a producer submodel predicts gross photosynthesis of the canopy in relation to more than six environmental factors and vegetation phenological stage. The ELM model also contains submodels for the various environmental factors and for consumer and decomposer function. The IBP Tundra Biome Study similarly produced the Swedish ABISKO model described by Bunnel and Scoular (1975) which also operates by identifying and quantifying the transfers between biomass pools based on empirical values.

PHOTOSYNTHETIC RATES OF DIFFERENT TYPES OF PLANT

The realized photosynthetic rates of plants are very often limited by the environmental factors which have been discussed above, but in the absence of such limitation, photosynthesis becomes a function of photosynthetically available irradiance and sometimes CO_2 concentration. Thus in eutrophic waters, algal photosynthesis is often a linear function of light intensity (Wassink, 1975). Records of maximal photosynthetic rates are consequently useful in comparing plants of different taxonomic status or physiological ecological type as they generally reflect variations in the inherent capability of the plants to harness solar PAR.

Table 5.4 summarizes the available data on maximal photosynthetic rates and, as might be predicted, in the absence of limitation by water stress the C-4 pathway with its high CO_2 trapping efficiency gives very large photosynthetic yields in climates which satisfy the generally high temperature optimum of the process. Herbaceous vascular plants have a very wide range of photosynthetic maxima as might be expected from their diversity of shade and sun habitat adaptation. Deciduous tree leaves, for reasons which are not clear, have substantially lower rates and, presumably because of their xeromorphic structure, conifers and other evergreen temperate zone species have very low rates. Raven (1977) by calculation suggested that absence of intercellular space increases CO_2 diffusive resistance sufficiently to reduce the photosynthetic rate of an otherwise

Table 5.4 Maximal photosynthetic rates for different types of plant

	mg CO_2 g dry wt^{-1} h^{-1}
Herbaceous shade plants (C-3)	10–30
Herbaceous sun plants (C-3)	30–60
Herbaceous C-4 plants	60–140
Conifers and evergreen temperate spp.	3–18
Mosses	2–4
Lichens	0.3–2
Algae (plankton)	*c.* 3
Pteridophyta	2–3
Submerged Tracheophyta	*c.* 7
CAM plants (dark fixation rate)	1–1.5

Data from Larcher (1975), Cooper (1975) and others. Values are recorded in terms of dry weight so that very dissimilar organs may be compared.

comparable organ by a factor of about ten. This suggestion is supported by the low photosynthetic rates of mosses and lichens. These plants have very low maximal rates, probably again because of the high resistance to diffusive CO_2 exchange in solution. The last category, the CAM plants are not really comparable, the dark CO_2 fixation not being an energy trapping process and the illuminated rate being achieved with closed stomata using stored CO_2. These photosynthetic rates may also be compared with the long-term productivity of the different vegetation types and Figure 11.13 on p. 355 shows that maximum productivity, without limitation, is a function of maximum photosynthetic rate despite its being confounded with seasonal canopy duration, day-length and many other factors.

Chapter 6

Plants and water deficit: soil, plant and atmosphere

INTRODUCTION

Man's early attempts at agriculture, cradled in the arid Near-East, were probably more often doomed to failure by drought than by any other hazard. The first physiological and ecological problem to be attacked by the human race was that of water shortage, and irrigation one of the first solutions. Oates and Oates (1976) document the use of irrigation channels by 6000 B.C. in Mesopotamia, the relatively sophisticated *shaduf* (water lift) was used after 2000 B.C. and the first labour-intensive underground water channels (*quanats*) of Iran were built by 1000 B.C. (Sherratt, 1980).

The struggle to understand and alleviate the effects of water deficit continues today at both theoretical and applied levels and attracts worldwide interest in response to the serious depletion of water resources in many regions.

The need of plants for water and the consequence of its shortage pose one of the most complex physiological problems in which experimental isolation of the plant from its normal environment is of doubtful value. Plant–water relationships must generally be considered ecologically as a field problem of microenvironmental physics or submitted to laboratory attempts at environmental simulation.

SOIL–PLANT–ATMOSPHERE CONTINUUM (SPAC)

The rooting zone of the soil, the plant body and the lower layer of the atmosphere behave as a continuum in relation to water transfer and must be considered as a whole in any complete analysis (Phillip, 1966). Soil water movement takes place in the liquid phase by capillary flow and in the gaseous phase by molecular diffusion or mass flow. In moist soil capillary transfer predominates but it is possible that vapour transfer may be important in rather dry soils (Cowan and Milthorpe, 1968). Water transport in the plant occurs by liquid-phase viscous flow in channels or by diffusive transfer through membranes. Final escape to the atmosphere is by vapour-phase molecular diffusion across the intercellular spaces of the mesophyll to the stomatal pores. From this point turbulent diffusion in the leaf and canopy boundary layers completes the transfer to the atmospheric sink, but molecular diffusion may become important under protected or very calm conditions. The main components of the SPAC are shown in Figure 6.1.

Solar radiation is the primary energy source for the water transport process in the SPAC, the sink being the latent heat change in the evaporation of water at the mesophyll cell surface and at the soil surface. Some of the energy for evaporation is drawn from environmental long-wave radiation, sensible heat transfer in the SPAC and water-vapour deficit in the free atmosphere, but these all depend secondarily on the effects of absorption of solar radiant income (Chapter 2).

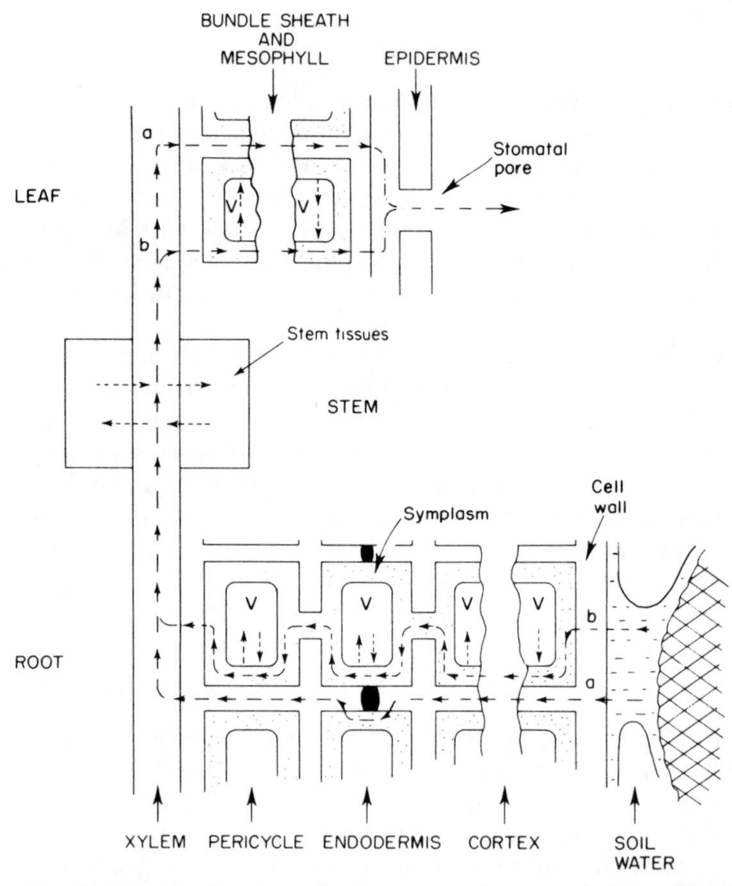

Figure 6.1 Flow paths in the soil–plant–atmosphere continuum (SPAC). Two alternative pathways are shown in parenchymatous tissue, either (a) the cell wall pathway or (b) the symplast pathway. Vacuoles are not involved in the direct mass-flow pathway (→—→—→) but they do slowly equilibrate with local ψ_{wall} or $\psi_{symplast}$ as water status changes (– – – →). Diffusion in the leaf air space is shown by (– · – · →). After Newman (1976) and Weatherley (1969).

SOIL AND PLANT WATER: TERMINOLOGICAL DEVELOPMENT

During the growth of any science the concepts and terminology which are developed may have to be changed to accommodate new advances. The terminology of water deficit in soils and plants has suffered repeated modification and alteration of usage, and now appears confusing to students and occasional readers.

Various suggestions have been made for a unified thermodynamic terminology, amongst the earliest being those of Schofield (1935) and Edlefsen (1941). From 1960 onwards, following the publications of Taylor and Slatyer (1960) and Slatyer and Taylor (1960), the term 'water potential' (ψ) has gradually been adopted. Water potential is a function of the difference between the chemical potential of water in the specified system and that of pure, free water at the same temperature, elevation and at atmospheric pressure. The difference between the chemical potential of a substance in two parts of a liquid or gaseous system determines the direction in which the substance will spontaneously move.

$$\psi = \frac{(U_w - U_w^0)}{\overline{V}_w} \tag{6.1}$$

where ψ = the water potential (J m^{-3}); U_w = the chemical potential of water in the system (J mole^{-1}); U_w^0 = the chemical potential of pure, free water at the same temperature and elevation as the water in the system (J mole^{-1}); and $\overline{V}_w$ = the partial molal volume (m^3 mole^{-1}); $\overline{V}_w$ is approximately equal to the volume of one mole of water.

It is an intensive property of the system (analagous to temperature), transfer taking place from regions of high to regions of low water potential. The units of water potential, J m^{-3}, are dimensionally the same as those of pressure and it is recommended that water potential should be described by multiplets of the pascal (Pa), the SI name for N m^{-2} (Incoll *et al.* 1977).

Pure, free water at atmospheric pressure is arbitrarily defined as having zero water potential while water in any other condition has a potential which is less than or greater than zero according to:

$$\psi_{\text{net}} = \psi_{\text{pressure}} + \psi_{\text{solute}} + \psi_{\text{matric}} \tag{6.2}$$

ψ_{pressure} (ψ_p) may be positive or negative in value. Positive ψ_p may be identified as a hydrostatic pressure, thus a 10 m water column exerts a pressure of approximately 0·1 MPa. A negative ψ_p is typical of water under tension, for example in a barometer column or in xylem lumina. A column with a height of 10 m (approximately a one atmosphere column) would have $\psi_p = -0{\cdot}1$ MPa at the top of the column.

ψ_{solute} (ψ_s) is related to the osmotic effect of dissolved solids and is almost invariably negative. A molar solution of a molecular or ionic species will have $\psi_s = -2{\cdot}27$ MPa at 0 °C if the solute shows ideal osmotic behaviour. Ions in electrolytes behave as independent osmotically active particles, thus seawater which is approximately 0·5 M NaCl, because it is ionized, has $\psi_s = -2{\cdot}0$ to $-2{\cdot}5$ MPa.

ψ_{matric} (ψ_m) is a function of the porous matrix in which water is held by capillary effects, its interfaces with air being bounded by curved meniscus surfaces. ψ_m for water and a wettable matrix is always negative, the water being under tension created by the concave meniscus. Occasionally, soils and some organic matrices may be non-wettable and the pore water will be under convex menisci which exert a positive pressure. Figure 6.2 shows the relationship between capillary rise and the effects of the mineral matrix on soil water.

To complete the consideration of water in biological systems it is necessary to relate water potential in the vapour phase to the components of water potential in the liquid phase which are responsible for water movement in the soil and within plant tissues. This is relatively easy, as Equation (6.1) is related to vapour pressure by:

$$\psi = \frac{(U_w - U_w^0)}{\bar{V}_w} = \frac{\mathbf{R}T \log_e(e_p/e_s)}{\bar{V}_w} \tag{6.3}$$

where e_p = the vapour pressure of water in the system (Pa); e_s = the vapour pressure of pure, free water at the same temperature and elevation as water in the

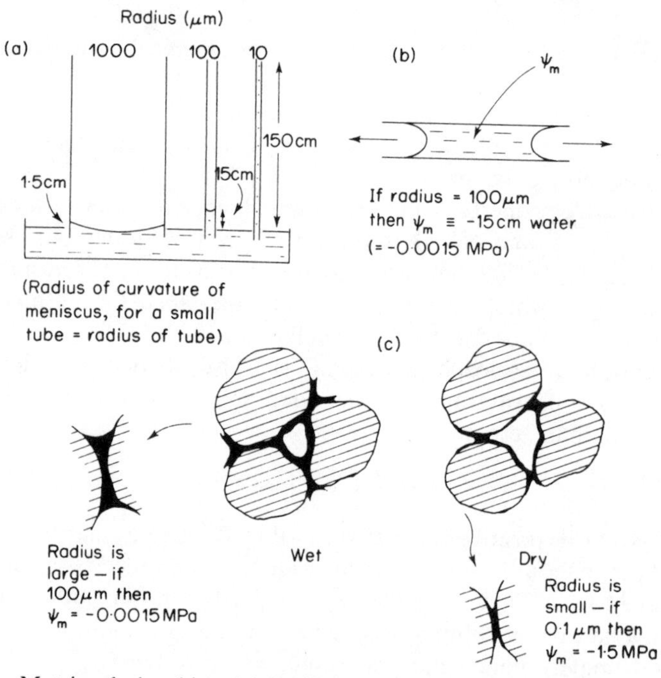

Figure 6.2 Matric relationships. (a) Capillary rise in a tube is a function of bore-radius which, for small tubes, equals the meniscus radius. (b) Capillary activity in an isolated portion of tube reduces the water potential in exactly the same way that it causes capillary rise in (a). (c) The activity of menisci between soil particles is directly comparable with the situation in (b). At a given equilibrium water content the radii of curvature of the menisci will all be equal and correspond to the matric potential of the soil.

system (Pa); **R** = the Universal Gas Constant ($J\,mole^{-1}\,K^{-1}$); and T = the temperature (K).

It should be noted that diffusion of water molecules in the vapour phase is a function of concentration gradient ($kg\,H_2O\,m^{-3}$ air) and that this is *logarithmically* related to equivalent water potential, consequently quite small leaf to air vapour pressure gradients are able to generate very large gradients of water potential (frequently 10 to 100 MPa).

This relationship makes possible the measurement of the water potential gradient between the leaf mesophyll surface and the external atmosphere providing that the external water vapour pressure is known and it is assumed that the saturation vapour pressure is manifested at the mesophyll cell surface. This simplistic assumption neglects the fact that leaf and atmosphere may differ in temperature.

The relationship between water potential and vapour pressure also provides a convenient laboratory tool for measuring tissue or soil water potential. Pressure, dissolved solutes or matric capillarity not only alter water potential but produce a proportional variation in saturated vapour pressure. A tissue or soil sample may be equilibrated with an enclosed air system and a psychrometric (wet-bulb temperature depression) technique used to measure the vapour pressure (Figure 6.3).

The magnitude of the water potential changes associated with naturally occurring gradients of water vapour pressure are very great, for example air at 20 °C and 50% relative humidity ($e_p/e_s \times 100$) has a water potential of −93·4 MPa (Slavik, 1974): much smaller departures from saturation create appreciable gradients, thus ψ_{vapour} (ψ_v) for air at 20 °C and 99% r.h. is −1·36 MPa. Temperature change has a very strong influence on ψ_v, thus an increase of 1·0 °C, from 19 °C and saturated air ($\psi_v = 0{\cdot}0$), creates a saturation deficit equivalent to −8·0 MPa. This will be of particular importance in the leaf–atmosphere vapour diffusive transfer system, since leaf temperature and air temperature may differ by several degrees. It also poses severe problems of temperature control in laboratory psychrometric systems involving vapour equilibration.

The water potential terminology has only been accepted universally during the past decade and there is possibility of confusion concerning units, which the contents of Table 6.1 may help to clarify. Milburn (1979) gives an excellent introductory account of plant water relationships.

Table 6.1 Unit conversions in the varying usage of plant and soil water

Pressure				Energy mass^{-1}	Head	
(bar)	(atmosphere)	(dyne cm^{-2})	(Pa) (N m^{-2})	(J kg^{-1})	(cm H_2O)	(cm Hg)
1·0	0·987	1 × 10^6	1 × 10^5	1 × 10^5	1019·7	75·0

The range of ψ_{soil} and ψ_{plant} is most conveniently expressed in MPa (e.g. 0 to −100 bar is equivalent to 0–10 MPa). Water vapour partial pressures are more convenient in kPa (e.g. 0 to 100 mbar is equivalent to 0 to 10 kPa). Expression in these forms avoids awkward powers of ten.

TERMS USED TO SPECIFY SOIL AND PLANT WATER STATUS

Soil

Soil water potential

This is an expression of the total reduction of water potential in the soil which is attributable to matric and solute effects. External pressure makes no measurable contribution to soil water potential unless the profile is flooded. A column of 1 m of water would impose a positive ψ_p of only 0·1 MPa and, consequently, soil water potential is usually defined as

$$\psi_{soil} = \psi_{solute} + \psi_{matric} \tag{6.4}$$

Its value may range from zero in saturated soil to less than −10 MPa in dry soils. Extremely low values (−30 to −40 MPa) may be found in soils which are not

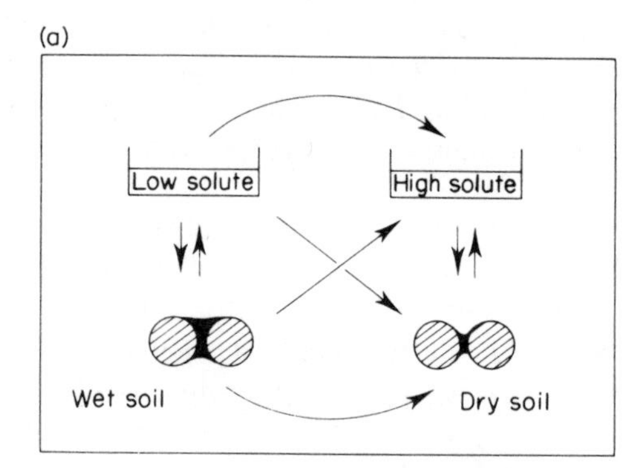

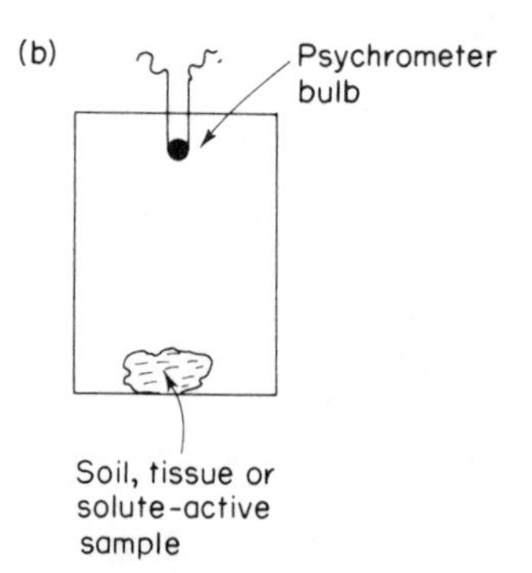

Figure 6.3 (a) The equivalence of solute potentials (ψ_s), matric potentials (ψ_m) and vapour pressure potentials (ψ_v). In a closed system vapour transfers would occur in the directions indicated: the equilibrium arrows show the wet soil and low concentration solution and the dry soil and high concentration solution to be in vapour equilibrium. (b) Vapour equilibrium of a sample (soil or tissue) in a closed chamber permits psychrometric measurement of water potential (see Table 6.2 and Barrs 1968).

only dry but contain high concentrations of salts. ψ_s values are appreciable only in arid-zone salt-rich soils and in salt-marshes.

Field capacity

This is the water content of an undisturbed soil (% oven-dry weight) after saturation by rainfall and followed by the cessation of gravitational drainage. In practice it involves sampling about 48 hours after rain. The technique is valid only for soils which are out of capillary contact with a water table. For various reasons, to be discussed later, the field capacity percentage may be considered as a soil 'constant' and is equivalent to values of ψ_{matric} ranging from −0·01 to −0·03 MPa according to soil type. As its determination is rather awkward, attempts have been made to simulate F.C. values by equilibrating soil samples with suction-plate water potentials of −0·01 or −0·03 MPa or by centrifuge-draining samples at 1000 g for 30 minutes. The water content in the latter case has been called the *moisture equivalent* (M.E.). There is, however, no universal relationship between suction-plate or M.E. values and field capacity.

Permanent wilting percentage

Where field capacity is used to specify the upper limit of soil water storage, permanent wilting point may similarly express the lowest level of easily available soil water. The determination is made by growing test plants in watertight containers of soil until several leaves are formed. At this stage the soil surface is sealed and the plants left until wilting occurs. If recovery cannot be induced by placing the plants in a water-saturated atmosphere, the soil water percentage is then measured. For many years P.W.P. was accepted as a soil 'constant' which differed little with the species of test plant, but the development of techniques for measuring *in situ* soil water potentials has revealed that the water potential corresponding to the permanent wilting percentage may range from (approx.) −1·0 to −2·0 MPa. The apparent discrepancy arose because a comparatively small change of volumetric water content may induce a very large change in water potential in this region of the characteristic curve (see Figure 6.5). The mean water potential value of −1·5 MPa has, however, been widely identified with the P.W.P.

Slatyer (1957) criticized the concept of P.W.P. as a single-value constant on the grounds that loss of turgor will be a reflection of the water potential of the leaf tissue which may, with species and environment, range between −0·5 and −20 MPa according to the solute potential of the leaf tissue sap at the time when wilting occurs. The limited changes of soil water potential around the P.W.P. appear almost insignificant compared with the possible range of leaf values.

The water which may be extracted from soil between the field capacity percentage and the permanent wilting percentage has been described as the *available water*; this is, volumetrically, a function of the pore size distribution and hence of soil texture and structure. The term is used with the implication that water beyond the P.W.P. is not easily available for plant growth and also suggests the concept of 'equal availability' which was first put forward by Veihmeyer and Hendrickson (1927) and generated a controversy which has only recently approached a settlement (p. 172).

The concept of equal availability between F.C. and permanent wilting point is now discredited but it satisfied the practical need to control application of irrigation water using the knowledge that a particular soil at P.W.P. would require a specified volume of water per unit area to restore it to field capacity. This both permitted economy and protected against waterlogging. The calculation is facilitated if soil water content is expressed volumetrically (volume per soil volume) rather than gravimetrically but the two values are easily interconvertable using soil bulk density ($kg\,dm^{-3}$) as a conversion factor.

Now that it is possible to monitor *in situ* soil matric potential (Table 6.2), irrigation amounts may be calculated accurately without the soil having to dry to the P.W.P., a practice which is known to limit photosynthetic production (Slatyer, 1967; Boyer, 1976a). The characteristic curve of volumetric water content against soil matric potential is used as the basis for the calculation and water may be applied at any predetermined potential. In both agricultural and natural ecosystems these characteristic curves, coupled with measurements of soil depth or rooting volume, give a useful estimate of total water available for transpiration.

Plant

Plant water potential

Plant water potential is a measure of the mean water potential of the cells in all tissues of the plant; the cell water potentials deriving from the balance between solute potential and pressure potential (turgor pressure) in living, vacuolated cells, and from the negative hydrostatic pressure in the non-living cells of the xylem, and in the microporous structure of cell walls. Whole-plant water potential is a less meaningful concept than the water potential of individual tissues of organs; for example, leaf water potential, which is a controlling factor in water loss to the atmosphere via the stomatal pores. Leaf water potential is one of the characteristics governing stomatal aperture and is also likely to establish an equivalent hydrostatic potential throughout the adjacent xylem, the resistance to flow in tracheids and vessels being comparatively low.

The total water potential of a plant tissue made up of a homogeneous collection of living cells, e.g. parenchyma, comprises $\psi_p + \psi_s + \psi_m$ where p, s and m are subscripts representing mean values, for all cells in the system, of p: hydrostatic (turgor) pressure within the cell; s: the solute potential of the vacuolar contents; and m: the matric potential attributable to the interaction between cell water and the microporous structure of the cell, mainly in the free space.

Table 6.2 Some methods of measuring soil water status. See Black (1965), Slavik (1974) and Rawlins (1976)

Technique	Principle	Application*
(a) Calibration techniques		
Suction plate	Equilibration of sample soil matric potential with imposed suction through a water-saturated capillary plate (limited 0 to −0·1 MPa)	LS. Construction of soil matric potential *v.* water content characteristic curves
Pressure plate or pressure membrane	Equilibration of sample soil matric potential with imposed gas pressure through a water saturated capillary plate or membrane	LS. Construction of soil matric potential *v.* water content characteristic curves 0 to −10 MPa
Vapour equilibration	Soil samples are vapour equilibrated with solutions of known vapour pressure (water potential). Gravimetric determination of water content. Time-consuming	LS. Construction of soil water potential *v.* water content characteristic curves
Tensiometer	Equilibration of soil matric potential with hydrostatic pressure in porous pot linked to a pressure gauge (limited 0 to −0·1 MPa)	FSC. Useful in monitoring matric potential of undisturbed moist soils or experimental containers. Direct water potential reading
Psychrometer	Wet-bulb thermometer (thermocouple or thermistor) equilibrated with water vapour in soil atmosphere	FLSC. Reading is a measure of total soil water potential
Pressure bomb	Applied pressure squeezes water from capillary matrix	FLC. Direct but not accurate reading of matric potential
(b) Measuring techniques		
Neutron scatter	Detection of backscatter of slow neutrons by hydrogen nuclei. Fast neutron source and slow neutron detector lowered into borehole. Changes in water content are mainly responsible for changes of hydrogen content. Difficult in swelling soil.	FS or C. Integrated value for a volume of soil. Needs calibration against water potential or content
Gamma ray absorption	Density of soil changes with water content. Source and detector lowered into adjacent boreholes. Difficult in swelling soil	FCC or S. Needs calibration
Resistance blocks	Buried porous block containing electrodes. Electrical resistance of unit is calibrated against water content. Not very accurate in moist soils	FCS or C. Needs calibration against soil water potential or content
Gravimetric	Soil samples are weighed, dried at 105 °C, reweighed. Damage to site by repeated sampling	FLS. Water content

* F, Field and controlled environment; L, laboratory; S, spot reading; C, continuous reading.

ψ_m is a fairly small component of the total water potential and does not change rapidly with time in intact tissues, consequently the total water potential is usually shown as:

$$\psi_{\text{tissue}} = \psi_p + \psi_s \tag{6.5}$$

ψ_p is usually positive in living cells and ψ_s negative. If the two are numerically equal then

$$\psi_{\text{tissue}} = \psi_p + \psi_s = 0$$

and the individual cells of the tissue are fully turgid. If the cells lose so much water that they become completely flaccid then $\psi_p = 0$ and $\psi_{\text{tissue}} = 0 + \psi_s$. In this latter case the driving pressure of water uptake will be directly related to the solute potential of the vacuolar contents. These relationships are shown graphically in Figure 6.4.

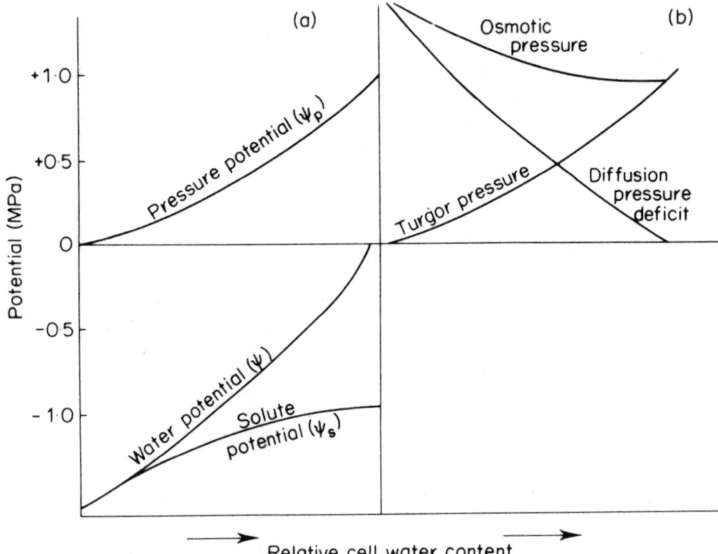

Figure 6.4 (a) The relationship between cell (or tissue) water content and the components of cell (or tissue) water potential (ψ). ψ_p = pressure potential of vacuolar contents; ψ_s = solute potential of vacuolar contents. The same data are plotted in (b) as a Hofler diagram using the former diffusion pressure terminology.

The normal situation during the diurnal changes of water balance is that the relative water content (see below) is less than 100% and ψ_p, consequently, less than its maximum value but more than zero. The driving pressure for water uptake will thus be

$$\psi_{\text{tissue}} = \psi_p + \psi_s = <0.$$

A tissue of which $\psi_{\text{tissue}} = 0$ has no tendency to take up water but if ψ_p falls then the increasingly negative value of ψ_{tissue} is reflected by an increasing tendency for the tissue to take up water. For water to move into a cell or tissue, ψ_{cell} or ψ_{tissue} must be less than the external water potential which could be ψ_{soil}, ψ_{xylem} or $\psi_{other\,tissue}$. Rapid changes of ψ_{tissue} reflect changes of ψ_p; osmotic adjustment of ψ_s does occur in response to water deficits or external solutes but relatively slowly (hours or days rather than seconds).

Below ground, living tissues are in contact with soil to the exterior and the non-living xylem elements to the interior. Above ground the external contact is with the water-sink of the dry atmosphere. For water to move upward through this continuum the following conditions must be satisfied:

$$\underset{(=\psi_m+\psi_s)}{\psi_{soil}} > \underset{(=\psi_p+\psi_s)}{\psi_{cortex}} > \underset{(=\psi_p)}{\psi_{xylem}} > \underset{(=\psi_p+\psi_s)}{\psi_{mesophyll}} > \underset{(=\psi_{vapour})}{\psi_{atmosphere}}$$

where the subscripts refer only to terms within a pair of brackets.

Relative water content

When an isolated plant organ or tissue is placed in water it will absorb water until it reaches 'full turgidity'. The relationship between the initial and the turgid water content has been expressed in various forms; water deficit (Stocker, 1929), saturation deficit (Oppenheimer and Mendel, 1939) and relative turgidity (Weatherley, 1950) are examples. Weatherley's relative turgidity has recently and more accurately been described as *relative water content* (Ehlig and Gardner, 1964; Slatyer and Barrs, 1965) and is defined as

$$\textit{Relative water content} = \frac{\text{Fresh wt.} - \text{Dry wt.}}{\text{Turgid wt.} - \text{Dry wt.}} \times 100. \tag{6.6}$$

Relative water content is a plant water characteristic which is much simpler to measure under field conditions than water potential and it is also possible to make specific calibrations so that relative water content may be converted to water potential (Slatyer, 1961c) though there are some difficulties arising from the change in the relationship with plant age (Millar *et al.* 1968).

BEHAVIOUR OF WATER IN SOILS

The physical description of the status and movement of water in the non-homogeneous, three-phase matrix of a water-unsaturated soil is a problem which has exercised soil physicists for many years. Briggs (1897) visualized the retention of water in terms of the force provided by the curved capillary menisci between soil particles and proposed the classification of soil water as hygroscopic, capillary and gravitational. Hygroscopic water is held by the surface forces of the soil particles, capillary water by the interparticle meniscus effects and gravitational water is surplus to this holding capacity and drains from the soil. The range of water potential corresponding to the draining potentials of the various capillary pores in the soil includes the optimum region for plant growth.

The capillary retention concept was further developed by Buckingham's (1907) treatment of water flow in terms of *capillary potential* in which the flow of water through soil is considered analogous to the flow of heat or electricity through a conductor. Buckingham was one of the first workers to treat soil water in terms of its relative energy status and the term capillary potential is synonymous with matric potential.

The relationship between soil water content and matric potential is of hyperbolic form (Figure 6.5) and a function of soil texture.

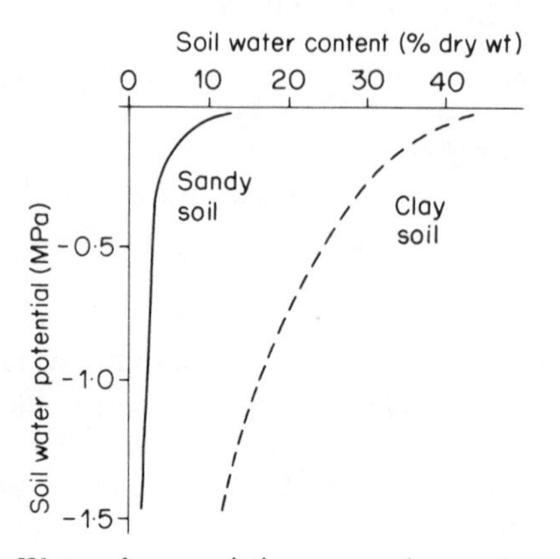

Figure 6.5 Water characteristic curves of a sandy and a clay soil.

A coarse textured soil contains mostly large pores (200 to 2 μm equivalent diameter) which empty very easily as the soil dries (Figure 6.2). Because it lacks much small pore space a very small further water loss causes a very large decrease in matric potential (Figure 6.5). By contrast, in a fine textured soil, there is a great deal of pore space in the smaller size range (2 μm to 0·2 μm equivalent diameter). Water loss from these pores causes a much less abrupt decline in matric potential because a substantial proportion of the total water content is stored in these pores. A plant rooted in a sandy soil thus has very free access to water stored in large pores but then has no further reserve. Once the soil has dried to about $-0{\cdot}2$ MPa, water deficit and wilting may rapidly follow whereas a clay soil, if well structured, has a substantial water store between 0 and $-0{\cdot}2$ MPa but can continue to supply a considerable amount to the permanent wilting point ($-1{\cdot}0$ to $-2{\cdot}0$ MPa).

Buckingham's (1907) analogy between capillary potential (ψ_m), as a driving force of water flow, and electrical potential as a source of electrical current may be used for infinitesimal gradients of ψ_m:

$$J = \frac{\psi_m^1 - \psi_m^2}{r} \tag{6.7}$$

where J = the water flux ($m^3 m^{-2} s^{-1} = m s^{-1}$); ψ_m = the matric potential superscripted for two points (1 and 2) in the soil which differ only slightly in ψ_m

(Pa); and r = the resistance to water flux between points 1 and 2 ($s\,m^{-1}\,Pa^{-1}$).

It cannot, however be used for large differences of ψ_m as, unlike electrical resistance in the Ohm's law analogue, the resistance of the soil to water-flux (r) is dependent on the prevailing water potential because water-filled pores empty as ψ_m decreases, so reducing the cross-sectional area of the flow path. Thus a soil which may have a fairly high water conductivity when wet poses a very strong barrier to water movement when it is dry (Figure 6.6).

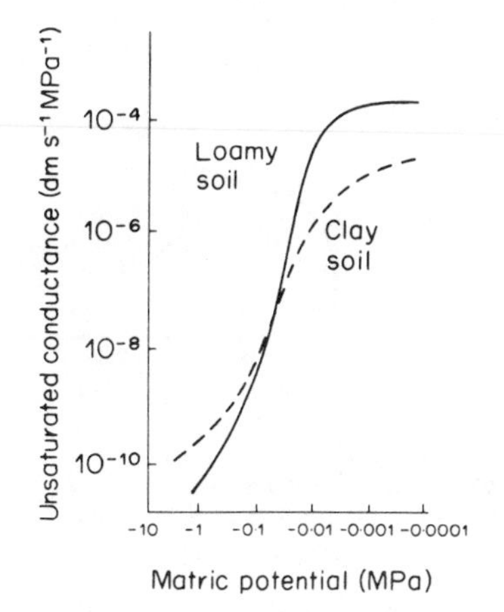

Figure 6.6 The relationship between unsaturated hydraulic conductance and soil water potential in a medium textured loam and in a clay soil.

Water flow through saturated soil under positive water pressure gradients (ψ_p) may be treated as if resistance remains constant as, by definition, all pores in the flow-path remain water-filled. The relationship is expressed by Darcy's law, a variant of the Hagen–Poseuille equation for viscous flow through smooth capillaries:

$$J = -H_y \frac{d\psi_p}{dx} \tag{6.8}$$

where H_y = hydraulic conductance = $1/r$ for a water-filled system ($m\,s^{-1}\,Pa^{-1}$); $d\psi_p/dx$ = the gradient of ψ_p over distance x (Pa).

The sharp inflection of the conductivity:water potential curve explains the relative constancy of the field capacity percentage and the abrupt cessation of gravitational drainage after rain. Early work of Veihmeyer and Hendrickson (1927) illustrates the lack of mobility of water in unsaturated soil.

Figure 6.7 shows the results of an experiment in which a slab of wet soil was sandwiched between two slabs of dry soil for 144 days. The two water distribution curves show the extreme slowness of capillary movement under these conditions. For the same reason water often penetrates soils non-uniformly, the downward movement of 'wetting-fronts' interacting with the localized removal of water by

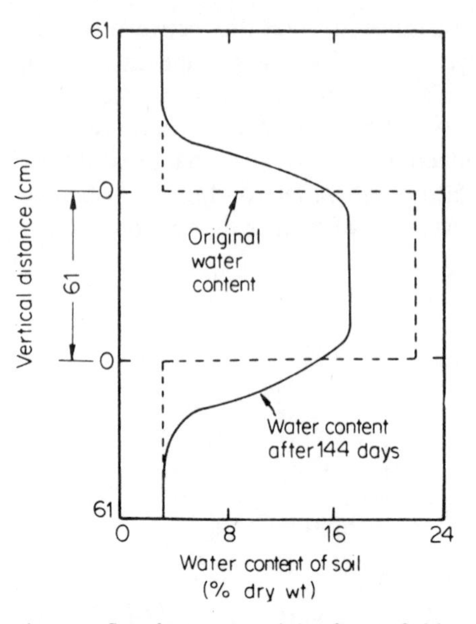

Figure 6.7 Unsaturated water flow between a slab of wet (field capacity) soil and two adjacent slabs of dry soil during a period of 144 days. After F. J. Viehmeyer, *Hilgardia*, **2**, 125–291 (1927).

roots to establish complex moisture profiles which may be detected by direct gravimetric sampling as in Specht's (1957) study of an Australian sclerophyll heath (Figure 6.8a), or by arrays of measuring instruments as in Etherington's (1962) work in a natural stand of *Agrostis tenuis* (bent grass) (Figure 6.8b). Youngs (1965) gives an excellent theoretical account of these aspects of water movement in soils.

The low conductivity of soils at water potential less than −0·1 to −0·2 MPa has prompted the suggestion that very steep gradients of water potential may develop around rapidly absorbing roots, with the consequence that water uptake may become self-limiting even in fairly moist soils (Richards and Loomis, 1942; Weatherley, 1951; Macklon and Weatherley, 1965; Etherington, 1967). There is no unambiguous evidence for this view, which has been criticized on both theoretical and experimental grounds by Newman (1969a, b). For example, in Macklon and Weatherley's experiments *Ricinus communis* (castor oil) plants, growing in soil or in water, were exposed to different relative humidities; the plants in soil showed a marked decline in leaf water potential with high transpiration rate, whereas the leaf water potential of the water-rooted plants was unaffected. The authors explained the difference by postulating the formation of drying zones around individual roots, but Newman suggests that the usually accepted values of hydraulic conductivity for soils are too high to give such steep gradients of ψ_m around individual roots. Weatherley (1976) has contributed the further suggestion that the resistance is located at the soil–root interface and depends on loss of contact between root and soil caused by shrinkage of roots on

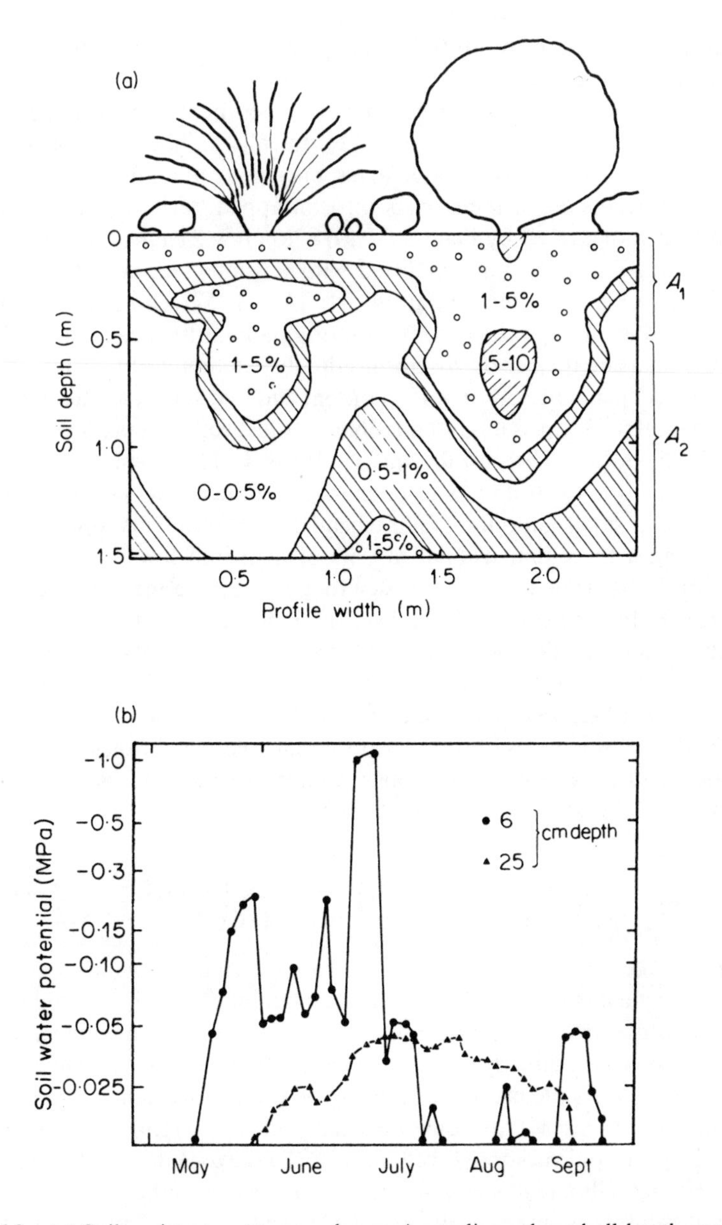

Figure 6.8 (a) Soil moisture pattern under an Australian sclerophyll heath vegetation subsequent to the fall of 24 mm rain when the whole profile was near wilting point. Stemflow and canopy drip have caused marked irregularity of rewetting. Reproduced with permission from R. L. Specht, Dark Island Heath, IV. Soil moisture, *Aust. J. Bot.*, **5**, 137–150, Figure 2 (1957). (b) Changes in soil water potential profiles during the summer months in S.E. England. Records from two depths in a sandy soil under grass. Etherington (1962).

drying. He cites values of up to 40% reduction in root diameter on drying to support this contention.

The settlement of the controversy awaits experimental measurement of water potentials in the soil adjacent to roots; technically a difficult problem. Plants in the field, under conditions of water stress, may also obtain further water supplies during the extension of their roots into untapped parts of the soil. This was elegantly demonstrated by Davies (1940) who grew *Zea mays* (sweet corn) in a long, rectangular soil container fitted with tensiometers at different distances from the plant, and by Majmudar and Hudson (1957) who showed that lettuce plants could grow to maturity on a soil originally at field capacity without any further irrigation, their root systems gradually extending further into moist soil.

The interdependence of soil water content and hydraulic conductivity poses an experimental problem as it is not possible to adjust soils to a predetermined water potential by adding water to the surface. If this is done a small portion wets to field capacity and the rest remains unchanged. Some workers have injected water in localized, small amounts (Whitehead, 1965) but this only succeeds in establishing a three-dimensional mosaic of high and low potentials. For this reason most experimenters have chosen to use 'water regimes' in which the whole soil mass is returned to field capacity after drying to a predetermined water potential. Using small volumes of soil, a few recent attempts have been made to control water potential using suction plates of various types, for example, Etherington (1962) used plates of porous polyvinyl chloride, and Babalola *et al.* (1968) encased thin slabs of soil in semi-permeable membranes so that external osmotica could be used to establish equilibrium potentials. None of these methods is suitable for the growth of large plants.

TRANSPORT OF WATER IN THE SOIL–PLANT–ATMOSPHERE CONTINUUM

Water makes most of the journey from root to leaf in the water-filled lumina of non-living xylem elements, lifted against the pull of gravity by the reduced water potential caused by transpiration of water from leaf to atmosphere, a process which is dependent, directly or indirectly, on solar radiation as its energy source.

Figure 6.1 shows the probable pathway of water movement and it is apparent that only for the short distance between soil and xylem (100–200 μm) and between xylem and mesophyll cell surface (10–20 μm) does the water have to traverse living cells and it is very likely, even in this part of the pathway, that much of the movement is in cell walls external to cytoplasm (Tanton and Crowdy, 1972a, b; Milburn 1979) or, at least, external to the vacuoles (Newman, 1976). It is only at the endodermal barrier that the radial cell-wall pathway is sealed by the Casparian strips and the water is constrained to passage through the cells.

Current thinking is that this water transport system is a physically driven mass-flow of water through an effectively non-living microporous system and that equilibration of component water potentials with the vacuoles of living cells is lateral to the main pathway and not a component part of it.

The physical demands of such a system are (i) a lowering of water potential at the leaf surface, of sufficient magnitude to account for the rise of water in the tallest of trees, and (ii) the existence of continuous root to leaf water columns showing sufficient tensile strength not to rupture under these large hydraulic potential gradients. There is no problem in explaining the lowering of leaf water potential caused by transpiration; from a thermodynamic point of view, evaporation of water into air at the temperatures and humidities which prevail in the leaf canopy provides more than sufficient energy. The second requirement, the maintenance of uninterrupted water columns, was first considered by Dixon and Joly (1894) and Dixon (1914), and led to the development of the cohesion/tension theory of water rise which is widely accepted. Work by Milburn and Johnson (1966), who used a sensitive microphone and amplification equipment to detect the sound pulse from cavitation in xylem elements, has shown that the water columns break down quite freely even in herbaceous plants. This poses a problem in the acceptance of the cohesion/tension theory concerning the mechanism by which the water columns might be regenerated, but it does remain the best overall explanation of the water transport process.

The flux rate of the transpiration stream will be directly proportional to the magnitude of the water potential gradient between soil and external atmosphere and inversely proportional to the total flux-resistance of the pathway. This is most simply modelled by a modification of equation (6.7):

$$J = \frac{\psi_{\text{soil}} + \psi_{\text{air}}}{r_{\text{total}}} \tag{6.9}$$

This relationship was used by Van den Honert (1948) to treat water transport as a catenary function which may be rewritten in water potential terminology:

$$J = \frac{\psi_1 + \psi_2}{r_{1-2}} = \frac{\psi_2 + \psi_3}{r_{2-3}} = \frac{\psi_3 + \psi_4}{r_{3-4}} \quad \text{etc.} \tag{6.10}$$

Where subscripts 1, 2, 3, 4, etc. represent different sequential water potentials and resistances of the transfer pathway. If these component equations are solved for realistic values of water potential for soil, root, leaf and atmosphere, for example $-0{\cdot}1$, $-1{\cdot}0$, $-1{\cdot}5$ and -100 MPa being substituted for ψ_1, ψ_2, ψ_3 and ψ_4, and if r_{1-2} is arbitrarily taken as unity, then $r_{2-3} = 1{\cdot}7$ and $r_{3-4} = 66$. Thus the major resistance to water transport may be seen to be the leaf to atmosphere vapour transfer.

Under steady-state flow conditions through the system illustrated in Figure 6.1 the gradient of water potential will not be uniform throughout the SPAC but will show relatively abrupt changes in passing from one part of the pathway to another according to the component resistances. At any one time the highest resistance in the pathway will be a limiting factor to transport and as Heath (1967) has noted, resistances on the atmospheric side of the mesophyll evaporating surfaces will be 'protective' to the plant while any resistance below this level is potentially 'harmful'. The relatively high value of the leaf to atmosphere resistance and the fact that it may be homeosatically controlled by the stomata is thus of crucial importance to the plant. Figure 6.9 shows the range of water potential values which may be encountered within the SPAC under conditions of adequate supply and of increasing deficit. The great decrease in water potential at the leaf–atmosphere junction is a reflection of the high resistance of this portion of the pathway.

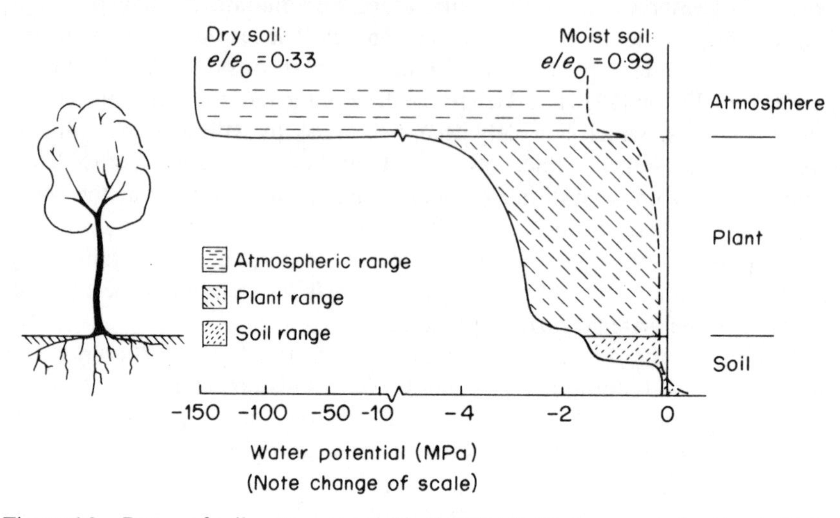

Figure 6.9 Range of soil water potential in the SPAC. Water vapour fraction in air in the range $e_p/e_s = 0.33$–0.99 and temperature above 5 °C.

Transpiration

The total resistance to diffusive loss of water vapour by a leaf (r_l) comprises an external boundary-layer and eddy-diffusive resistance which is manifested over any type of evaporating surface, and three component resistances of the leaf system which are related to leaf size and anatomy. These are the resistances to transfer of water from the mesophyll cell surface into the intercellular space, the resistance to transfer through the stomatal pores to the exterior and the resistance of the parallel pathway through the cuticle to the exterior:

$$r_l = r_m + \frac{r_s r_c}{r_s + r_c} + r_a \qquad (6.11)$$

where r = the resistance to gaseous diffusion of water vapour in air ($s\,m^{-1}$), and the subscripts l, m, s, c and a, respectively, denote leaf, mesophyll, stomata, cuticle and external.

The values of r_m and r_c, in the short term, remain constant and the value of r_c is often very high compared with the minimum value of r_s, in particular for xeromorphic leaves (p. 180ff). The consequence is that resistance to transpiration is mainly a function of stomatal aperture, which is homeostatically controllable by the plant, or a function of r_a when this is substantially larger than r_s. The external resistance is, in part, a function of leaf size and external morphology but it is also very sensitive to the micrometeorological conditions within the canopy, particularly windspeed.

Typical values of r_c, r_s and r_a, respectively, might be 3000, 100 and 40 s m^{-1} for a mesophyte (Cowan and Milthorpe, 1968), thus

$$\frac{r_s r_c}{r_s + r_c} = \frac{100 \times 3000}{3100} = 97\,\text{s m}^{-1}$$

with fully open stomata but, as the stomata close, r_s will approach infinity and the value will become 3000 s m^{-1}. If no other resistance in the pathway were limiting, the effect of stomatal closure would be dramatically reflected in a reduction of transpiration to 3·2% of its previous value. This is because the linear approximation of Fick's law for diffusion may be written in a form analagous with Ohm's law in which diffusive flux varies inversely with resistance.

If r_a is large by comparison with r_s then changes in stomatal aperture will have much less influence on the total resistance of the pathway and transpiration would be insensitive to stomatal movement. This will be the case in still air, as a large value of r_a depends on the presence of a deep boundary layer of unstirred air in which the water vapour concentration is near the maximum. Milburn (1979) gives calculated values for the depth of the boundary layer ranging from 4 mm in still air to 0·4 mm with a wind velocity of 10 m s^{-1}. This would give a tenfold decrease in the resistance of the pathway and thus the value of r_a = 40 s m^{-1}, cited above, might easily increase to 200 s m^{-1} and would then be considerably larger than the minimum stomatal resistance. Transpiration is thus strongly affected by air movement (Figure 6.10) but it should be noted that still air conditions are rarely encountered in the field, consequently those factors which influence guard cell opening also markedly affect ecosystem transpiration. Monteith (1973, pp. 195–9) discusses the relationship between individual leaf stomatal resistance (r_s) and the 'bulk stomatal resistance' of a canopy (r_{st}) which is used in his variant of the Penman equation for potential evapo-transpiration (p. 32, Equation (2.14)). With a closed canopy the value of r_{st} bears the same relationship to whole canopy water loss as does r_s to water loss from a single amphistomatous leaf.

Stomatal opening and environmental factors

Increased light intensity and reduced concentration of CO_2 in the leaf intercellular space both usually cause stomatal opening. The effect of light is probably mediated by photosynthesis CO_2 uptake. Leaf water stress causes closure and usually overrides all other controls as might be expected of a homeostatic mechanism. High CO_2 concentration, low light intensity and a large gradient of water vapour pressure, from leaf to exterior, all promote closure. The effects of temperature are in dispute, probably because it is so difficult to control it independently of the vapour pressure gradient from leaf to air. Endogenous

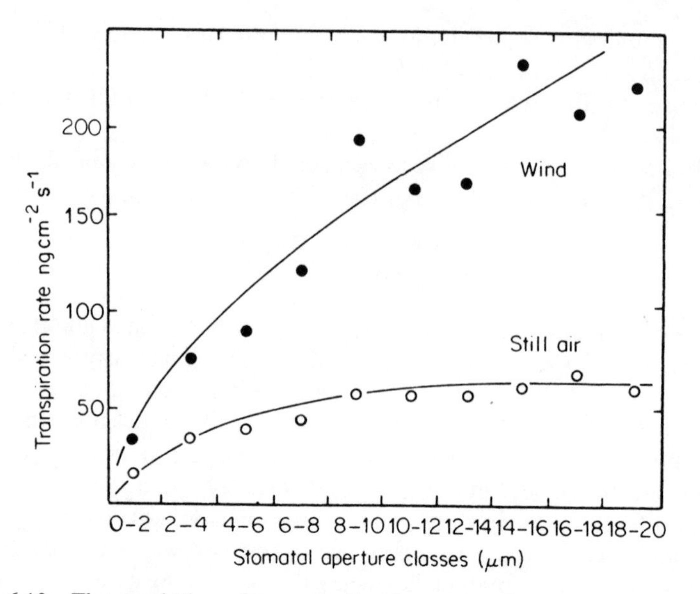

Figure 6.10 The regulation of transpiration by stomatal aperture in *Zebrina pendula* leaves under moving and still air conditions. The dashed lines show predicted behaviour based on calculated diffusive resistances. Data of Bange (1953).

rhythms generally reinforce the normal pattern of diurnal opening which is established by environmental factors in the absence of water stress. Recent reviews are given by Raschke (1976) and Hall and Schulze (1976).

The general pattern of stomatal opening during the day and closure at night is not universal, for example it is reversed in those succulents and a few non-succulent plants which show Crassulacean Acid Metabolism (CAM) which is discussed extensively on pp. 127–130. CAM is interpretable as a survival strategy in which the need to conserve water is overriding to the extent of a relatively inefficient CO_2 trapping process being substituted for the normal daytime C-3 or C-4 processes (Neales 1975).

A further departure from a simple diurnal opening and closing pattern occurs in high radiation environments, the stomata closing for a period near mid-day when the solar income is greatest. This may be caused by transient water deficit or by leaf overheating stimulating respiratory CO_2 production and causing closure by raising the CO_2 concentration in the intercellular space (p. 19).

Water stress ultimately overrides all other controls and causes stomatal closure. Many workers (Hsiao *et al.* 1976) have suggested that this occurs when leaf water potential falls to a threshold value. An alternative explanation is that the epidermis, which is in poor hydraulic contact with the mesophyll, becomes water stressed in response to cuticular transpiration and it is this mechanism which is the controller of stomatal aperture (Jarvis, cited in Landsberg and Butler, 1980). This would explain recent observations that leaf to air vapour pressure gradient strongly influences stomatal opening (Hall and Schulze, 1976).

TISSUE WATER DEFICITS

The flow resistance of the plant tissues is not negligible and tissue water deficits will build up during periods of heavy transpiration; these decline slowly during the following night. Such deficits have been attributed to root resistance, and Kramer (1938, 1940) showed that de-rooted plants or plants with roots killed in boiling water had much lower internal resistances and accrued much smaller transpirational deficits. It is possible, however, in soil, that the drying zones previously suggested (p. 160) may cause much larger resistances than those of the roots.

Such changes in plant water deficit (and consequently water potential) are a reflection of the long-term lack of equilibrium between plant and environment. Figure 6.11 shows the diurnal course of leaf water potential in *Capsicum*

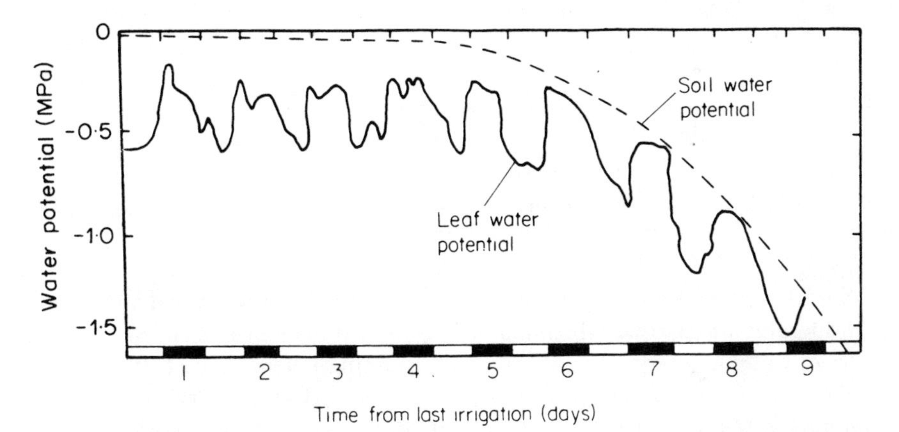

Figure 6.11 Diurnal changes of leaf and soil water potential of a pepper plant rooted in a clay loam soil. Reproduced with permission from W. R. Gardner and R. H. Nieman, *Science*, **143**, 1460–1462, Figure 1 (1964). Copyright 1964 by the American Association for the Advancement of Science.

frutescens (pepper) during a period when soil water was becoming depleted (Gardner and Nieman, 1964). The irregularity of the ψ leaf curve shows the daily reduction followed by slow nocturnal recovery when transpiration demand is low. The ability of a plant to adjust to very sudden changes of light intensity and evaporating conditions may be as important as its ability to cope with long periods of intense, unchanging radiation. Under transient state conditions it is difficult to predict flow rates from the known water potential gradients as many of the resistances in the SPAC are potential-dependent.

Large changes of water potential cause variations in the turgidity of unthickened cells and, in lignified tissues, the reduction of water potential may cause a shrinkage as the water columns, which are under tensile stress, adhere to the walls of the lumina and cause contraction of individual elements. Diurnal fluctuations in the diameter of various plant organs have been measured, for example recording dendrometer traces for tree trunks show daytime contractions which provide indirect evidence for the existence of very large negative water potentials in the xylem (Figure 6.12).

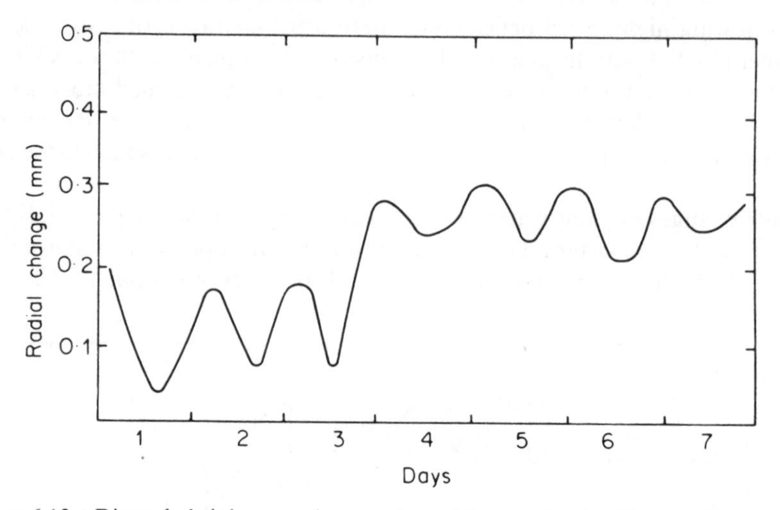

Figure 6.12 Diurnal shrinkage and expansion of the trunk of a *Pinus resinosa* tree in summer conditions. Data of Kozlowski (1964).

Shrinkages of such magnitude suggest that the sapwood must represent a considerable water-store which may be calculated to release between 10 and 50 litres of water in a large tree. Waring and Running (1976) suggest that this, and other tissue-stored water, may represent at least a 10-day reserve, even in mid-summer. Kozlowski (1972b) gives an extensive review of such work.

Klepper (1968), following the diurnal pattern of leaf and fruit water potential and fruit diameter in *Pyrus communis* (pear), showed that leaf and fruit water potential could fall to −1·5 to −2·0 MPa during the midday period. This fall was accompanied by a withdrawal of water from the fruit, causing a considerable reduction in diameter. At midday there was a slight, transient rise in leaf water potential which was attributed to stomatal closure; this was also observed in *Vitis vinifera* (grape). Her experiments were conducted in conditions of intense atmospheric water deficit (30 °C, r.h. 20 %) which caused water loss to exceed supply during the day but recovery was rapid as the evening advanced.

WATER DEFICITS AND PHYSIOLOGICAL RESPONSES

Plant water deficits originating from low soil water potential or high transpiration demand have consequences which involve all of the physiological functions ranging from primary biochemical processes to overall effects on whole plant physiology (Crafts 1968; Kozlowski 1968a, b, 1972a, 1976, 1978).

Figure 6.13 indicates the range of sensitivity of relevant physiological processes to tissue water potential but the ecological interpretation of variations in such

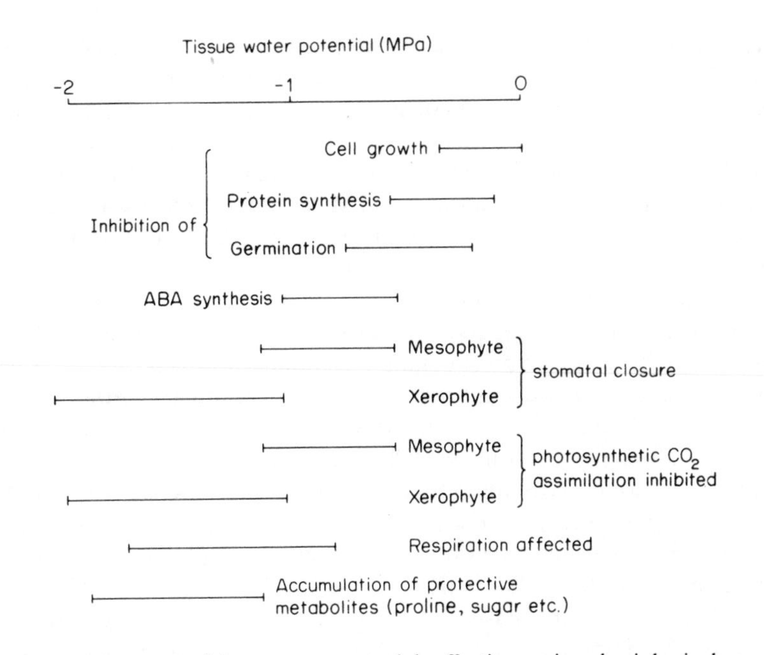

Figure 6.13 The range of tissue water potentials affecting major physiological processes. ABA = abscisic acid. Modified from Hsiao *et al.* (1976) and other authors.

responses is complex and dependent on stage of life-cycle and interaction with other environmental factors. Thus, at germination sensitivity to water deficit might confine a species to a wet habitat though the adult plant may be relatively tolerant of drought. Extreme examples are found amongst the Pteridophyta which include some of the most truly drought-tolerant vascular plants (Oppenheimer, 1960; Parker, 1968) and yet many are confined to wet habitats or microsites because the gametophyte generation requires free water for antherozoid transfer prior to production of the adult sporophyte. This is well marked in the fern flora of Britain; over 70% of the widespread species show a strong western bias to their distribution which is almost certainly a reflection of the wetter Atlantic coast climate (Jermy *et al.*, 1978).

Seed germination

Germination may be slowed or prevented by water deficit and the degree of sensitivity is specifically variable. Harper and Benton (1966) germinated seeds of various species on porous glass plates (filter discs) through which water was supplied under suction, ψ_m ranging from 0 to $-0{\cdot}02$ MPa. Spiny and reticulate seeds were very sensitive to reduced ψ_m, while smooth seeds showed increasing sensitivity with greater size. Mucilaginous seeds were least sensitive, all of these results suggesting that contact between seed and soil interstitial water is a critical factor.

Harper and Benton found considerable germination of some species at −0·02 MPa while Owen (1952) recorded germination of *Triticum vulgare* (Wheat) in soil drier than the permanent wilting point (*c.* −1·5 MPa). Kauffman and Ross (1970) germinated *Triticum vulgare* and *Lactuca sativa* (lettuce) in soils which had been equilibrated to specified matric potentials. *T. vulgare* germinated well at −0·8 mPa but not at all at −1·5 MPa, but *L. sativa* was much more sensitive, being inhibited by more than 50% at −0·23 MPa and failing to germinate at −0·4 MPa.

Solute potentials reduce germination less than equivalent matric effects (Hillel, 1972) which might be expected if seed-surface contact is a critical factor. The seed coat is likely to be 'leaky' to most common osmotica, in which case water will be imbibed relatively easily from an external solution. Natural soil osmotica such as sodium chloride, as might be expected, show some relationship with the nature of the habitat, halophytes generally germinating in relatively high sodium chloride concentrations which are inhibitory to glycophytes (Koller, 1972). Some responses are more complicated, for example saline pre-treatment enhancing subsequent germination in non-saline water, perhaps interpretable as a requirement for rainwater flooding.

Some desert plants are sensitive to rain and its duration depending on the leaching of water-soluble inhibitors; thus the Californian *Filago californica* (cudweed) will not germinate in moist soil, but any rainfall up to 20 cm increases germination: beyond this amount it declines again to a minimum at 89 cm and is then unaffected by further rain (Lang, 1965).

Water uptake and transpiration: inorganic nutrient uptake

Unless water deficit imposes permanent stomatal closure, water uptake is closely related to the absorption of solar radiant energy which drives the transpiration stream, the affect of the storage time-lags being relatively slight. Under almost any conditions the plant will transpire sufficient water to satisfy its needs for movement of inorganic solutes in the xylem. There is some evidence that variation in the transpiration rate may modify nutrient uptake by controlling mass flow of solutes to the root surface. This will be most marked in the case of those elements which are not entirely supplied by diffusion through soil solution (Oliver and Barber, 1966). There is strong evidence that some dissolved materials are carried into the plant passively with the transpiration stream, for example Hutton and Norrish (1974) found the silicon content of *Triticum vulgare* lemmas and glumes to be very strongly correlated with the total water transpired during growth.

In soils it is difficult to separate the role of root extension in tapping fresh nutrient supplies from that of diffusion and mass-flow towards and within the root. All of these processes may be differentially affected by water stress and for this reason there has been controversy concerning the relationship between transpiration rate and ion accumulation. These and other factors are discussed in a review by Viets (1972) and by Nye and Tinker (1977).

Serious drought may indirectly influence nutrition by inhibiting nitrogen fixing and decomposer microorganisms, for example Reichert *et al.* (cited in West and Skujins 1978) found a one-hundredfold decrease of free-living nitrogen fixation during drying of a desert soil-crust from 0 to −0·4 MPa. Symbiotic fixation may be similarly affected (Sprent, 1976) by the impact of water deficit on the host plants. Kovda *et al.* (1979) give further information concerning desert soils and suggest very rapid resumption of nitrogen mineralization following rain, with most biochemical activity confined to the surface 3 cm of soil.

Photosynthesis and growth

Evidence has gradually accumulated that photosynthesis is reduced even by relatively small water deficits and certainly by water potentials higher than the −1·5 MPa permanent wilting point. Schneider and Childers (1941), for example, compared the photosynthetic rates of watered and unwatered apple trees and found a steady decline during the 5 days prior to the first observed wilting. At this time both photosynthesis and transpiration had fallen to *c.* 40 % of the control values. More recently Ashton (1956) conducted similar experiments with sugar cane during five drying and re-irrigation cycles but in this case the decline occurred more abruptly after the build-up of a considerable soil water deficit.

Clark (1961) compared the photosynthetic response of *Picea glauca* and *Abies balsaminae* to soil water deficit. During the early drying cycles in his experiments, the photosynthesis of *A. balsaminae* was reduced to zero by drying to the P.W.P., but that of *P. glauca* only fell to about one-third of the maximum value. In this case the decline in photosynthesis again occurred gradually. Similar results were obtained by Brix (1962) with *Lycopersicon esculentum* and *Pinus taeda*, the reduction commencing after the leaf water potential had fallen to −0·7 MPa in the former and −0·45 in the latter plant.

The major response of photosynthesis to water deficit is through the decrease in stomatal conductance of CO_2 consequent on water stress-induced closure. Transpiration and photosynthetic CO_2 exchange may be described by the following two diffusion equations (Bierhuizen, 1976):

$$J = \frac{\Delta H_2O}{r_a + r_s} \tag{6.12}$$

$$P = \frac{\Delta CO_2}{r'_a + r'_s + r'_m} \tag{6.13}$$

where J is the transpiration rate (kg m^{-2} s^{-1}); ΔH_2O is the concentration gradient of water vapour (kg m^{-3}); r is the diffusive resistance to water vapour, subscripted a and s for boundary layer and stomata respectively (s m^{-1}); P is the net rate of CO_2 uptake (kg m^{-2} s^{-1}); ΔCO_2 is the concentration gradient of CO_2 (kg m^{-3}); and r' is the diffusive resistance to CO_2, subscripted as above and also m for mesophyll (s m^{-1}).

Species with a low mesophyll resistance for CO_2 are likely to have a relatively high net photosynthesis and, consequently a low transpiration ratio (H_2O/CO_2) but if the value of r'_m is high then stomatal closure will reduce transpiration more than net photosynthesis, for example Stanhill (1972) observed a decline in dry matter production of 10 % and water-use of 30 % when citrus orchard soils dried from −0·01 to −0·3 MPa. Further discussion appears in chapter 5 (p. 127 ff) where the water-use efficiencies of C-3 and C-4 photosynthetic pathway species are discussed.

Relatively small reductions of soil water potential may affect both photosynthesis and overall growth (Jarvis, 1963; Etherington, 1967), observations which conflict with the early views of Viehmeyer and Hendrickson (1927) who advanced the theory that growth and transpiration were not reduced until the soil reached the permanent wilting point. Their papers initiated a long-standing controversy concerning the concept of 'equal availability' of water between the field capacity and permanent wilting points. They upheld this view for almost 25 years (Viehmeyer and Hendrickson, 1950) but in an extensive review Richards and Wadleigh (1952) produced a great deal of evidence against the equal availability hypothesis and Stanhill (1957) analysed data from 80 papers describing water regime experiments and reported 66 cases in which growth was reduced before the P.W.P. was reached. Experiments with osmotica also support the concept of gradual reduction of growth by decreasing potential (e.g. Slatyer, 1961a).

Growth response to water deficit varies with the stage of the life-cycle and also with the physiological mechanism through which it is mediated. The growth-analytical approach shows that both net assimilation rate and leaf area ratio components of relative growth rate may be reduced by water deficit. This was shown, for example, by Etherington (1967) for *Alopecurus pratensis* (meadow foxtail grass) (Table 6.3). The reduction of net assimilation rate conforms with the lowered photosynthesis described above and the reduced leaf area ratio suggests that water stress has limited full tissue expansion. This is confirmed by the partitioning of the *A. pratensis* leaf area ratio into its two components, leaf

Table 6.3 The influence of soil water deficit on growth of *Alopecurus pratensis* (Etherington, 1967)

	Soil water regime	
	Maintained field capacity	Dried to −0·5 MPa before watering to field capacity
Net assimilation rate (g dm² week⁻¹)	0·22	0·19
Leaf area ratio (dm² g⁻¹)	0·86	0·74
	Leaf area ratio data partitioned into its two components	
Leaf area/leaf weight (dm² g⁻¹)	2·98	2·57
Leaf weight/plant weight (g g⁻¹)	0·29	0·29

area/leaf weight and leaf area/plant weight; the former shows a reduction in the dry treatment; but the latter is unaffected.

Slatyer (1967) cites many workers who have found reduced cell enlargement under water stress. For example, Ordin (1958, 1960) found a marked relationship between turgor pressure and cell enlargement in *Avena* coleoptiles by using osmotica of different permeating characteristics; mannitol, which was relatively non-permeating, caused the greatest reduction in extension. Impaired leaf expansion caused by soil water stress has been shown by Wadleigh and Gauch (1948) for cotton. Their measurements of leaf length show a decreasing rate of elongation which first became apparent at a soil water potential of −0·5 to −0·6 MPa and reached zero at −1·3 to −1·4 MPa. Jarvis (1963) found complete cessation of leaf expansion in *Saxifraga hypnoides* (mossy saxifage) at −0·05 MPa and in *Filipendula vulgaris* (meadow sweet) at −0·2 MPa. The above experiments were carried out in pot culture and the overall measurements of soil matric potential may be considered reliable but it must be realized that the water potentials at the root surfaces are only a matter of assumption and may differ from the bulk soil values. Boyer (1976a, b) suggests that mid-day leaf water deficits are sufficient to inhibit leaf cell expansion and that, if such deficits are repeated for several days they will cause an irreversible reduction in leaf area compared with well-watered controls. Drought experiences which limit leaf area ratio may thus secondarily inhibit the photosynthetic production of a whole season.

Reduction of leaf area by failure of tissue expansion or suppression of leaf initiation usually lowers the ratio of shoot to root dry weight despite the reduced availability of photosynthesate for root production. Etherington (1962), for example, found this effect with swards of two grass species: *Alopecurus pratensis* and *Agrostis tenuis* (bent grass). An irrigation regime simulating a very dry temperate zone summer (10 cm water) was compared with a control field capacity regime (45 cm water). The dry treatment reduced mean shoot/root ratio to 1:9 from 4·5 in the control.

Reduced shoot/root ratio may be interpreted as a diminished ratio of transpiring to absorbing surface, and might well be advantageous during the period of stress. In the next chapter (p. 179) it is noted that low shoot/root ratio is of adaptive significance in arid-zone plants and appears to be selected for in any habitats which experience predictable drought.

Respiration

Most workers who have measured respiration in relation to water deficit have observed that it is not so rapidly suppressed as the synthetic processes of metabolism, and in some cases moderate water stress has increased respiratory rates. Brix (1962), in his work with loblolly pine and tomato *Pinus taeda* and *Lycopersicon esculentum*, found that photosynthesis was reduced to zero by leaf water potentials of −1·0 to −1·5 MPa, but respiration was not reduced below *c.* 60 % of the maximum rate. In *P. taeda* increasing the deficit beyond this point caused a rise in respiratory rate reaching 140% of the original value at −3·4 MPa. In most cases increasing deficit reduces respiration: Jarvis and Jarvis (1965) showed a nearly linear relationship between increasing solute potential (0 to −1·4 MPa) of bathing solutions and decreasing respiration of root sections of several tree species. In photosynthetic plants water stress may induce a net weight loss due to the more rapid suppression of photosynthesis than of respiration, but in storage organs such as seeds, the net rate of loss will be reduced as respiration declines.

Translocation

Experiments with ^{14}C-labelled metabolites and various herbicides show that the phloem translocation rate is reduced by water deficit but not to the same extent as the reduction of photosynthesis caused by equivalent deficits. The phloem operates under a positive hydrostatic pressure which, if the Munch hypothesis is accepted, derives from the high solute content of the sieve tubes and consequent osmotic withdrawal of water from adjacent xylem elements. Any reduction of xylem ψ_p might, thus, be expected to correspondingly reduce phloem ψ_p, interfering with its transport function. Such an effect is dramatically shown by Hall and Milburn (1973).

Biochemical changes

Some attempt has been made to interpret observations of the physiological consequences of water deficit at the biochemical level, but great experimental problems exist as most techniques require the disruption of cell or tissue systems in aqueous media with consequent changes of localized water potentials. The greatest success has been achieved in following long-term gross changes of plant composition, for example, the early work of Petrie and Wood (1938a, b) showed reduced synthesis of proteins from amino acids under water stress conditions.

Concentrations of soluble carbohydrates and their derivitives are usually increased by water deficit (Parker, 1968; Levitt, 1972). Such effects are possibly related to osmoregulation and it is significant that osmotically induced water stress causes similar effects in algal cells which produce glycerol, mannitol and galactose-glycerides by mobilizing insoluble polysaccharides (Kluge, 1976). Soluble nitrogen compounds are also formed under water stress in both algae and higher plants: these include proline, other amino acids and quaternary ammonium compounds (Kluge, 1976; Binet, 1978). Ability to form proline has been correlated with drought tolerance (Hurd, 1976) and it is also formed in response to saline conditions (Stewart and Lee, 1974), a topic further discussed in relation to halophytism on p. 287.

Heat, chilling and freezing stress may all cause accumulation of soluble materials which have variously been described as osmoregulators and cryoprotectants. This is significant as such stresses may secondarily induce water deficit and it seems likely that osmotic adjustment of cytoplasmic versus vacuolar fluids may be the key role of these materials.

Endurance of cytoplasmic desiccation is of obvious physiological interest but the biophysics and biochemistry of the mechanisms are, as yet, poorly understood. Drought damage seems to accrue from the deformation of cells and cytoplasm with consequent damage to membrane systems. Protective mechanisms may be gelation of cytoplasm or vacuolar contents, increased elasticity of cytoplasmic structures and the production of protective agents which influence these characteristics: these include soluble sugars, polyols, peptides and other materials (Parker, 1972). Levitt (1972) discusses these problems extensively, including the suggestion that aggregation of proteins by intramolecular S–S bonding during dehydration shrinkage might be the source of desiccation damage. Bewley (1979) provides a detailed review of desiccation tolerance in plants.

Hormonal changes are induced by water stress, and attention has recently focused on abscisic acid (ABA) as a natural antitranspirant which functions by causing stomatal closure without interfering with CO_2 absorption at the mesophyll surface (Meidner, 1975). A more complex, hypothetical model of cytokinin and ABA relationships to other plant processes has been developed by Itai and Benzioni (1976).

A few workers have attempted more sophisticated biochemical investigations of processes such as photosynthesis, respiration and nucleic acid replication in relation to water stress but serious difficulties arise from the need for aqueous media, and Boyer (1976a) has warned against the uncritical use of osmotica in experiments with isolated chloroplasts as the effects may differ from those of drought stress in the intact leaf.

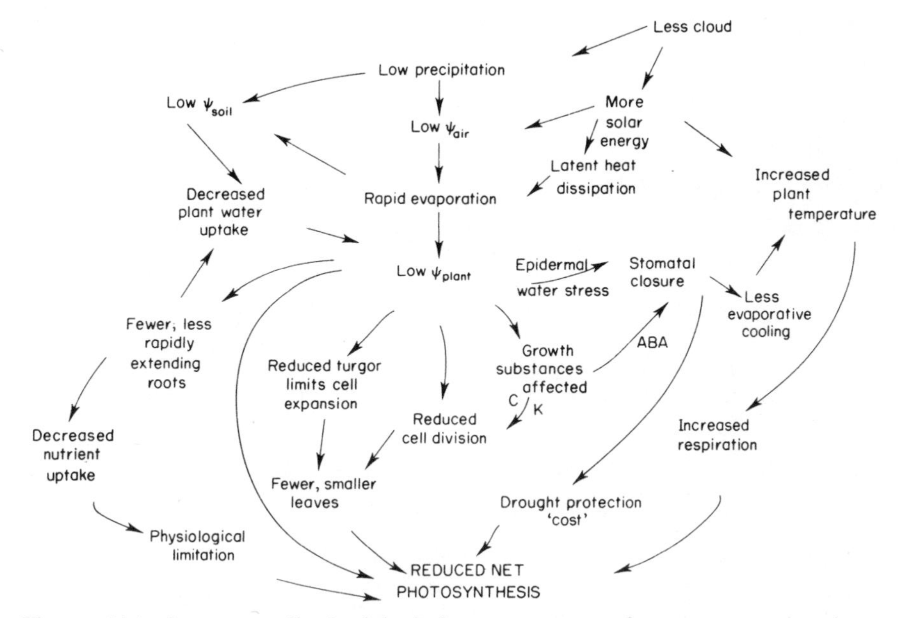

Figure 6.14 Some overall physiological consequences of water stress in plants. ABA = abscisic acid. C = cytokinin. K = kinetin. After an idea of Fritts (1976).

SUMMARY

Turgor is probably the cell characteristic which is most quickly affected by the onset of water stress, transient or permanent, and it is also likely that its sensing is the mechanism which leads to transduction of the internal pressure change into the branching chain of physiological and biochemical responses which result in reduced photosynthetic carbon fixation, failure of tissue expansion and the longer-term effects of reduced leaf area and absorptive root surface. Fritts (1976) has produced an elegant synopsis of these interacting events to explain the impact of water deficit on tree ring width (Figure 6.14).

Chapter 7

Plants and water deficit: ecological aspects

XEROMORPHY AND DROUGHT TOLERANCE

The existence of the terms xerophyte, hydrophyte and mesophyte implies that the distribution of plants may be influenced by their physiological response to water. This is an uncritical viewpoint which neglects the fact that water availability may show a very wide range at any site, but the concept of a xeric habitat as one experiencing a high frequency of 'deficit-days' is useful.

It is perhaps unfortunate that the term xerophyte is so widely used as it has such a breadth of connotation as to be misleading. Maximov (1929), in his early classification of the xerophytes, was aware that the limits of the group are ill-defined. The term is more useful when expressed as relative xeromorphy but even so the criteria used for its evaluation are too numerous for convenience. Oppenheimer (1960), in a literature survey, favours a very broad interpretation embracing all of the anatomical, morphological and physiological modifications which may assist the plant to cope with environmental water deficit.

Levitt (1972) has produced a classification of water-stress tolerance (overleaf) in which all categories except the ephemeral, escaper strategy may be considered broadly as xerophytic modifications:

Stress escape—ephemerals

The ephemeral, drought-escaping, strategy is epitomized by many desert summer-annuals which appear at irregular intervals of 5 to 20 years at any one spot, following thunderstorm-showers (Went, 1957). Some examples are *Euphorbia setiloba*, *E. micromera* (spurges) and *Pectis paposa* (a composite) which sense rainfall of adequate duration by loss of a soluble inhibitor. Others, for example *Baeria chrysostoma* (goldfields), are very sensitive to osmotic stress and germinate only when the soil solution becomes sufficiently dilute (Went, 1957).

Many desert shrubs utilize a partial escape-strategy: they also germinate following summer rain of sufficient duration and the drought-susceptible seedlings are thus guaranteed a full winter season to develop deep roots or

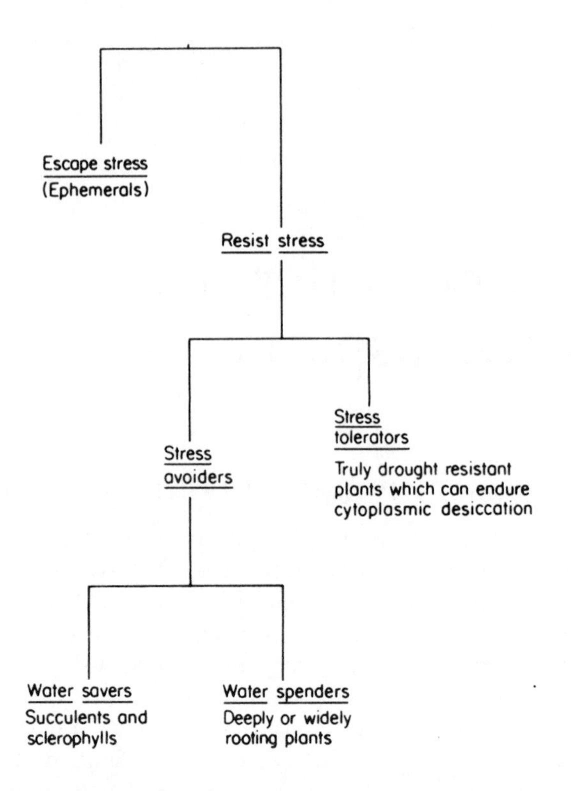

otherwise become drought tolerant before the onset of the next summer's stress. Further information may be found in Went (1979) and Koller (1972).

Even the moister environment of the temperate zone may impose serious water stress and an escape-strategy is shown by those winter-annuals which germinate in autumn to flower and set seed before the onset of summer drought. Seed of such species as *Stellaria media* (chickweed) and *Phacelia purshii* (scorpion weed) may be dormant when shed, autumn germination being guaranteed by a requirement for a high after-ripening temperature. Any seeds which fail to germinate are stratified into new dormancy by winter-chilling and thus persist as a seed-bank until the following autumn (Baskin and Baskin, 1976; Newman, 1963; Ratcliffe, 1961).

Stress tolerance

Amongst the plants which resist stress, Levitt's stress-tolerators are not well represented amongst the angiosperms, the majority of which are killed by relatively little cytoplasmic desiccation. One group of vascular plants, the Pteridophyta, does contain many truly tolerant species, mainly ferns, some of which can desiccate to air-dryness and recover on rewetting. Rouschal (1937–8) showed that *Ceterach officinarum* (rusty-black fern) would tolerate drying to equilibrium with zero relative humidity for 5 days, experiencing massive cytoplasmic water loss which amounted to 98% of the plant dry weight! This 'poikilohydric' behaviour is distinguished from the more normal 'homoihydry' of most angiosperms which die at relatively small tissue water deficits but also have some control over water loss. Poikilohydry is fairly common amongst non-vascular plants such as bryophytes, some algae and some lichens. Parker (1968) cites examples which include survival of *Nostoc commune*, a blue-green alga, on a herbarium sheet for 87 years. Mechanisms of cytoplasmic desiccation tolerance are discussed in the previous chapter (p. 175).

Stress avoiders

The majority of angiosperms have solved the problem of water stress successfully by mechanisms other than tolerance and inhabit all but the most extreme of desert environments. Their stress avoidance is usually by a combination of water conservation or storage with either an extensive root system or the rapid production of roots under favourable circumstances. The importance of a deep and profuse root system was emphasized by Weaver's (1926) classic study of North American sandhill species.

Desert habitats produce extremes of root development, for example roots of *Tamarix aphylla* (tamarisk) have been traced horizontally as far as 30 m from the shrub and *Prosopis flexuosa* (mesquite) roots have been found at a depth of 53 m in North America (Drew, 1979)! These deep-rooted plants which tap stored ground-water (phraeatophytes) belong to Levitt's 'spender' class and may transpire enormous quantities of water: *Tamarix gallica* has been estimated to lose 2000 $m^3 ha^{-1} y^{-1}$ which may exceed the annual precipitation to its habitat (Drew, 1979 in Goodall and Perry 1979).

Gulmon and Mooney (1977) have recently described the strategy of the desert shrub *Tidestromia oblongata* (honeysweet) which, unlike many such plants, grows and transpires freely during the hottest part of summer in Death Valley, California (p. 204). The plant establishes from seed on temporary alluvial fans with a supply of stored subsoil water. It has a restricted life of 5 to 7 years during which a large amount of seed is set and dies when the water reserve is depleted. This is in striking contrast to similar shrubs such as *Atriplex hymenelytra* (Desert holly saltbush) which grow in the same area but become dormant during the summer and are able to survive a water potential of −6 MPa in their rooting zone. *A. hymenelytra* is thus very water-conservative and much longer lived.

Many water-storing succulents form ephemeral shallow root systems in response to rain: thus, within 8 hours of rainfall, *Opuntia puberula* produces new adventitious roots despite having endured several months drought. Rather than dying during drought, the roots of some plants become dormant, for example those of *Pinus halepensis* (Aleppo pine), in the European Mediterranean, produce a suberized layer protecting the dormant meristem from desiccation when the soil water potential falls to about −0·7 MPa, and extension is resumed within a few days of rain (Drew, 1979).

In addition to these mechanisms for utilizing soil water, some desert plants may directly absorb water vapour (Levitt, 1972). Lange *et al.* (1975a) suggest that the annual photosynthetic production of some desert lichens is mainly related to dew absorption. Because of the strong sensitivity of saturated vapour pressure to temperature it is difficult to be certain whether such absorption is direct from the vapour form, or caused by water condensation on radiation-cooled surfaces, but the effect on the plant will be the same. Went (1975) described the anomalous occurrence of a 'forest' vegetation of *Prosopis tamarugo* (mesquite) in the Atacama Desert of Chile. The dead, spongy root cortical tissues remain permanently water saturated and Went suggested that this water is supplied from the atmosphere by vapour absorption. Tritiated water supplied to the leaves appears in the wet root cortical tissues, thus supporting the suggestion.

The enormous root systems of many drought-tolerant plants result in a low shoot/root ratio which may be interpreted either as a part of the spender strategy of maintaining sufficient absorptive area for full transpiration, or part of a saver strategy when coupled with other above-ground modifications which conserve water. Abrupt reductions of shoot/root ratio may be found in some desert plants which shed either leaves or whole branches at the onset of the dry season. Many examples are discussed by Kozlowski (1973), one of the most extreme being *Zygophyllum dumosum* (bean caper) in the Negev Desert which may lose nearly 90% of its transpiring surface.

Orians and Solbrig (1977b) have suggested a cost–income model for root–leaf balance in which the thin, low cost mesophytic leaf gives the best return for investment when water is freely available and is expendable in times of adversity. The thick, lignified leaf which is necessary to maintain photosynthetic production at low soil water potential represents greater investment and is not expendable thus the evergreen habit of sclerophyllous plants is a consequence of this relationship and has the built-in disadvantage of a slower growth rate when water is not limiting (Figure 7.1).

Sclerophylly and leaf modification

The saver strategy of sclerophylly is best developed in Mediterranean and desert ecosystems but it may be encountered in lesser form in most other climatic zones where water deficit or other stress tolerance is at a premium. It is characterized by evergreen, small leaves, thick cuticles, sunken stomata at high density, leaf hairiness, thickened or sclerified cell walls and narrow xylem elements which, presumably, are less subject to cavitation at low water potentials. Despite all of these modifications which may be interpreted as limiting water loss and giving protection against the consequences of low cell turgidity and overheating, very low water potentials may be encountered in sclerophyll plants. Orians and Solbrig (1977) cite values as low as −4·8 MPa for *Acacia gregii* (cat's claw) in Arizona, never rising above −1·6 MPa at any time of year. Even lower values have been encountered, for example −16·3 MPa for *Artemesia herba-alba* (a sagebrush) in the Negev (Richter, 1976).

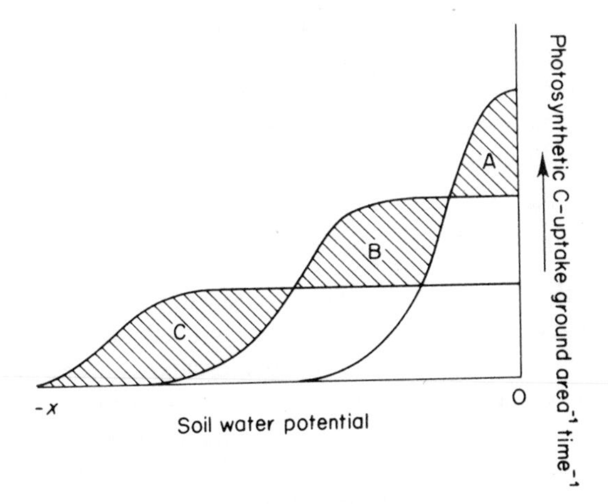

Figure 7.1 A cost–income model of plant morphological strategies which optimizes carbon-gain *v.* tolerance of water deficit. Modified from Orians and Solbrig (1977b). A: Mesophyte with thin leaved (low cost), large area canopy. B: Intermediate. C: Xerophyte with thick lignified leaves, limited leaf area and a deep and/or extensive root system. In the hatched areas the indicated strategy will outcompete the others but at high water potential C will be slow growing because of its large investment in protective and underground respiring tissue.

When the cell contents have lost turgidity, cell water potential (ψ_s) is equal to the vacuolar solute potential (Equation 6.5) and it is not surprising to find that values of ψ_s are correlated with minimal leaf water potentials in many sclerophyll species. Walter (1979) shows, for the Sonoran Desert shrub, *Encelia farinosa* (brittlebush), a wet season ψ_s of $-2{\cdot}2$ MPa but as drought becomes more severe this falls to $-3{\cdot}2$ MPa, smaller, more hairy leaves are formed and finally at $-4{\cdot}0$ MPa the leaves are abcissed.

The low solute potentials of such leaves may be interpreted as extending the operating range of tissue water potential in which some turgidity may be maintained while the strength given by thickened cell walls and sclerenchyma prevents structural deformation during wilting, protecting the cell contents from mechanical damage.

Another mechanism which effectively modifies shoot/root ratio, but reversibly, is the leaf-rolling mechanisms of grasses which depends on specialized hinge-cells, the bulliform cells. Some controversy has surrounded leaf-rolling (Parker, 1968), as it does not occur in some species until lethal water deficits have accrued. Recent experimental evidence for *Oryza sativa* (rice) showed that artificial rolling of leaves could reduce transpiration by as much as 50% (O'Toole and Cruz, 1979) and this would certainly be predictable from the effect of rolling on the boundary layer thickness (p. 165) of a hyperstomatous leaf.

Leaf rolling also influences interception of solar radiation and increases forced-convective cooling, thus reducing leaf temperature and available energy for evaporation. The very hairy leaves of many desert species and the presence of salt-crystal coats on some *Atriplex* spp. (salt bush) may also control leaf net radiation by reflection of both visible and infra-red radiation.

Photosynthetic pathways and succulence

Because the paths of water and carbon dioxide movement are common, there is no strategy of water conservation which does not reduce carbon fixation. Some evolutionary progress has been made towards escape from the dilemma, the C-4 and CAM photosynthetic pathways being more water-use efficient than C-3. Neither of these processes are without disadvantage compared with C-3, thus the daily photosynthesis of a full CAM plant is limited by the amount of CO_2 which can be trapped and stored at night and there is also a requirement for low night temperature otherwise respiratory CO_2 output becomes a substantial fraction of dark fixation. Similarly C-4 plants, with a few exceptions, have a very high temperature optimum and their photosynthesis is inhibited by low day temperatures. Further discussion appears on pp. 127–30.

Normal C-3 photosynthesis remains the most universally successful strategy when combined with water conserving or seeking characteristics. Competitive limitation confines both CAM and C-4 to relatively specialized niches despite their undoubted water-use advantages. CAM plants are commonest in the predictable regime of summer-dry deserts but the irregular precipitation of hot deserts, such as the Negev, excludes them from all but limited habitats such as shaded crevices. C-3 plants with dormancy or leaf-shedding mechanisms are better fitted to survive continuous high temperature (Lange *et al.* 1975b). Some species, for example several members of the genus *Dudleya* (Crassulaceae) in California, have been shown to switch from normal C-3 photosynthesis to CAM with the onset of drought and thus combine the survival strategy of CAM with more productive photosynthesis in the wet season (Troughton *et al.* 1977). This ability is characteristic of leaf succulents rather than stem succulents which are permanent CAM plants.

Succulence is associated with CAM as a presumed mechanism to provide vacuolar space for storage of malic acid without interfering with cytoplasmic pH. It does, however, provide a large water reserve which may reach several tonnes in a giant species such as the cactus *Carnegiea gigantea* (Saguaro). Such plants close their stomata quickly in drought and have thick cuticles: their transpiration rates are sometimes many thousand times less, weight for weight, than those of leafy mesophytes. *Echinocactus wislizienii* is recorded as losing only 0·05% of its fresh weight daily, falling to 0·015% as it dries. This cactus is able to survive 6 years without water. The effect of the cuticle is shown by the enormous 87% daily fresh weight loss of some cacti when peeled (Oppenheimer, 1960).

Stem and leaf succulents of arid habitats which are often also CAM plants should not be confused with the outwardly similar succulent halophytes of salt marshes and salt deserts. These have a high vacuolar concentration of electrolytes, usually sodium and chloride, and solute potentials below $-0{\cdot}1$ to $-0{\cdot}5$ MPa. They are generally intolerant of tissue water deficits (Orians and Solbrig, 1977a) and their root systems must be lost or at least become dormant in dry soil otherwise tissue water would be lost to the soil. The CAM mechanism permits long-term stomatal closure and recycling of internally generated CO_2 perhaps even in plants which do not use the mechanism for trapping exogenous CO_2 (Daniels unpublished data).

Very few CAM succulents are halophytes but, despite this, desert plants, particularly those of the most arid and unpredictable climates, do encounter problems of excess electrolytes in soils where concentration by evaporation has occurred. In some species, for example those of the sclerophyllous genus *Tamarix* (tamarisk or salt cedar), the plants are equipped with active salt-glands which effectively desalinate the soil of the rooting zone by returning salt to the soil surface. Many *Atriplex* (saltbush) species have the same effect, loading salt into the epidermal bladder hairs to concentrations over three times those in the leaf lamina (Osmond, 1979).

COMPARATIVE STUDIES: DIFFERENTIAL SENSITIVITY TO WATER STRESS

The modifications which permit xerophytes to survive drought were identified in some detail by the end of the last century (Schimper, 1898, citing in particular the work of Wiesner, 1875–94). Characteristics which were associated with arid climates, dry soils and 'physiologically droughted' habitats were largely identified by teleological assumption though Schimper does mention the occasional experimental device, for example the production of more mesomorphic shoots by subjecting a xerophyte to a water-saturated atmosphere.

By the time of Maximov's (1929) classic account of plant water relationships a substantial amount of comparative experimentation had already been published, in particular comparisons of xerophyte and mesophyte transpiration rates, transpiration efficiencies and drought survival characteristics. A long-standing controversy has surrounded Maximov's assertion that many xerophytes transpire as fast or faster, on an area basis, than mesophytes. Some of his cited data are certainly puzzling but it is significant that much of the information was collected from plants which were not water-stressed. The need to make measurements with water-deficient plants was, perhaps, not fully realized as there was controversy concerning the degree to which transpiration could be controlled by stomatal closure.

More recent measurements of the stomatal and cuticular resistance to water vapour loss in mesophytes and xerophytes support the intuitive conclusion that sclerophyllous modifications, in particular a thickened cuticle, are associated with high resistance to water-loss when stomata are fully closed under drought conditions (Table 7.1). A xerophyte and a mesophyte might both have $r_s = 500\,s\,m^{-1}$ with fully open stomata, and given similar conditions, including

Table 7.1 Resistances to water vapour loss in mesophytes and xerophytes (after Cowan and Milthorpe, 1968; and Nobel, 1974)

	Resistance ($s\ m^{-1}$)	
	Stomatal minimum (r_s)	Cuticular (r_c)
Mesophytic crops and deciduous trees	40–1000	2000–5000
Conifers	1000–3000	—
Xerophytes	500–2000	10,000–40,000

boundary layer resistance, would be transpiring at similar rates per unit area. At moderate to high windspeed the transpiration rate would be very sensitive to stomatal aperture and full closure would restrict water loss almost entirely to cuticular transpiration. Even assuming the highest cuticular resistance cited for mesophytes, the ratio of xerophyte r_c/mesophyte r_c would be 40,000/5000, in other words the minimal transpiration rate per unit area of the xerophyte might be one-eighth that of the mesophyte.

Comparisons of species, or even of population within species, in which structural and physiological differences are not so extreme, has thrown considerable light on the selective forces which operate on plants in regularly water-stressed habitats. Experimental approaches have extended from simple water regimes in which dry weight gain has been measured after exposure to a series of drying cycles, to more sophisticated procedures in which water potentials have been instrumentally monitored or controlled with osmotica and measurements have been made not only of growth but also of constituent processes such as photosynthesis and respiration. Monitoring of tissue water balance, stomatal opening, transpiration, metabolic and control function biochemical changes have all contributed to such studies.

A good example of a population study is the work of Ashenden *et al.* (1975) with *Dactylis glomerata* (cocksfoot grass). Seed was collected from six sand-dune sites chosen for their differing height above an arbitrary datum: these heights were shown to be correlated with the rate of soil drainage and thus with relative wetness. Seed populations were grown in a controlled environment under three water regimes which imposed minimum soil water potentials of −0·03, −0·17 and −1·18 MPa. Unfortunately these regimes were maintained by adding less than the amount of water necessary to return the compost to full water capacity and there must be some doubt about the uniformity of water distribution through the depth of the soil (see p. 160).

Figure 7.2 shows the effect of the driest treatment on net assimilation rate and relative growth rate. The dry treatment reduced relative growth rate, mainly through its effect on net assimilation rate, but the reduction was much greater in the wet site than in the dry site populations. Compared with control plants of *D. glomerata* (Aberystwyth S37) most of the dune populations are to some extent tolerant of water deficit.

Leaf area ratio was not strongly affected by water stress and showed no correlation with site of origin. This is rather unexpected as other workers have

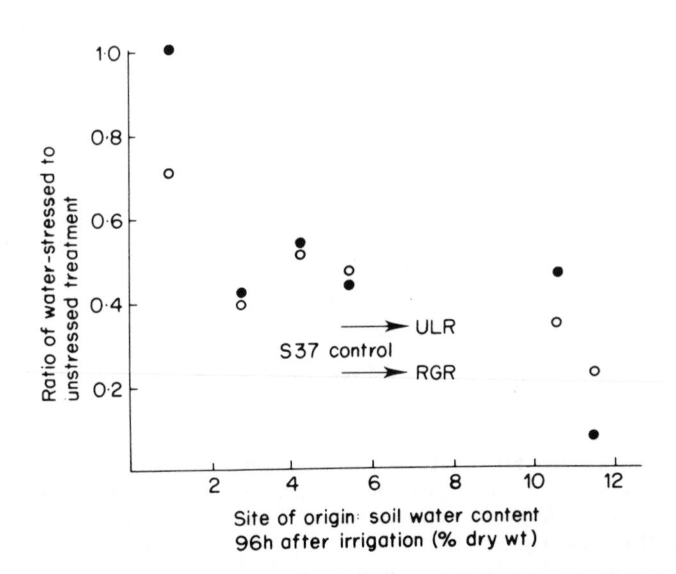

Figure 7.2 The effect of water stress on unit leaf rate ULR (○) and relative growth rate RGR (●) of *Dactylis glomerata* populations originating from sand dune and slack sites differing in their drainage characteristics. The values are plotted as ratios of stressed to unstressed treatments against soil wetness of the site of origin. The ULR and RGR ratios for control plants of *D. glomerata* cv. 537 are shown by the arrows but cannot be related to the water content scale. Data from Ashenden *et al.* (1977).

shown leaf expansion to be inhibited by slight soil water stress. Etherington (1967), for example, found both leaf area ratio and net assimilation of the grass *Alopecurus pratensis* (meadow foxtail grass) to be equally reduced by a minimum soil water potential of −0·5 MPa (Table 6.3) and Jarvis (1963) recorded cessation of leaf extension at water potentials between −0·05 and −0·3 MPa. This discrepancy may be accounted for by the low irradiance (about $\frac{1}{10}$ full sunlight), used by Ashenden *et al.*, which would have limited evaporative demand and prevented the development of very low leaf water potentials. This is a problem which is commonly encountered in all controlled environment investigation of water deficit effects.

Ashenden (1978) has also reported a number of other adaptive differences between these *D. glomerata* populations. In a spaced plant trial the area of the youngest fully expanded leaf was much smaller in the two driest site populations compared with either the control or the wettest site population and, as might be expected, the total transpiration loss of the two dry site populations was less than that of the control or of the wettest site population whether water-stressed or not (Figure 7.3).

Comparative experiments also confirm the suggestion that rooting depth and extent is related to drought tolerance. Cook (1943) compared eight populations of the grass *Bromus inermis* (awnless brome) and found total root length and maximum depth of rooting to be positively correlated with drought resistance.

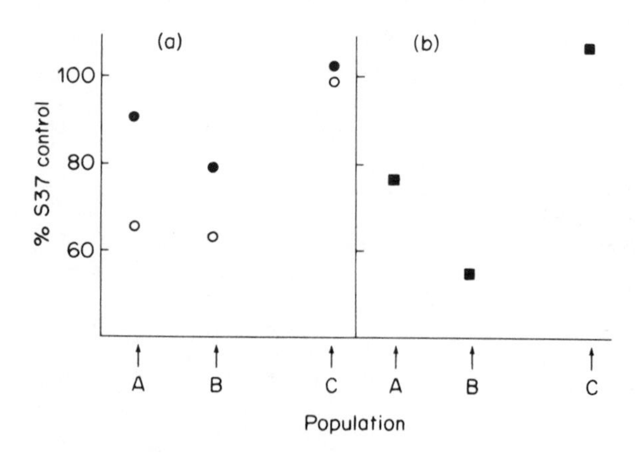

Figure 7.3 (a) Transpiration rate (wt area^{-1} time^{-1}) and (b) leaf area of largest expanded leaf in a spaced plant trial of the two driest (A and B) and the wettest (C) sand dune population of *Dactylis glomerata* shown in Figure 7.2. Both results are plotted as percentage of an S37 cultivar control. Data of Ashenden (1978). ○—stressed. ●—unstressed.

Extreme soil water deficit reduces root weight and, in many crops (Hurd, 1976), the greatest yield is given by varieties which maintain the most root in drought conditions.

Plants of arid climates often depend on stored water to survive the dry season, and differences in drought tolerance may be related to the ability to reach deep supplies. McWilliams and Kramer (1968) used tritium labelled water to show uptake from at least 1·2 m depth by the Australian pasture grass *Phalaris tuberosa*, which is much more drought tolerant than the shallow-rooted *P. minor*. Goode (1956) found a similar relationship between *Lolium perenne* (rye grass) and *Poa annua* (annual meadow grass) in English orchard pasture swards. In this case the deeper rooted *L. perenne* is more drought resistant but overgrazing reduces its root penetration so that both species are susceptible to drought and in competition for available water.

The timing of root activity is important in stored water exploitation: the successful alien weed grass *Bromus tectorum* (cheatgrass) has become a problem in North American rangeland and its ability to continue root extension at low soil temperature allows it to pre-empt the use of winter-stored water and outcompete the desirable dominant *Agropyron spicatum* (bluebunch wheatgrass) (Harris and Wilson, 1970). Fernandez and Caldwell (1975) followed the root phenology of three semi-desert shrubs and showed a clear progression of growth from the upper to the lower part of the soil profile as summer drought advanced. *Atriplex confertifolia* (shadscale sagebrush) continued to produce new roots at the greatest depth in the driest part of the year and this is probably the principle factor enabling it to maintain a positive carbon balance at this time.

The detection of differential drought tolerance between populations of a species or between species within a single habitat suggests that fairly delicate adaptations of morphological and physiological characteristics may exist, even in response to local variations of soil depth, topography and other factors which modify water supply. Many experimental studies of both wild plants and agricultural species or varieties confirm this suspicion. Bannister's (1964a, b, c) work with *Erica cinerea* (bell heather) and *Erica tetralix* (cross-leaved heath) is an example of such a comparison. *E. cinerea* is a species of dry, heath soils while *E. tetralix* occupies much wetter, usually peaty soils. Bannister's observations showed that *E. tetralix* is very sensitive to soil water deficit compared with *E. cinerea*, the death of the plants being preceded by strong depression of relative water content and transpiration rate which did not occur so markedly in *E. cinerea* (ling) (Figure 7.4). Hence, at a whole-plant physiological level, there was

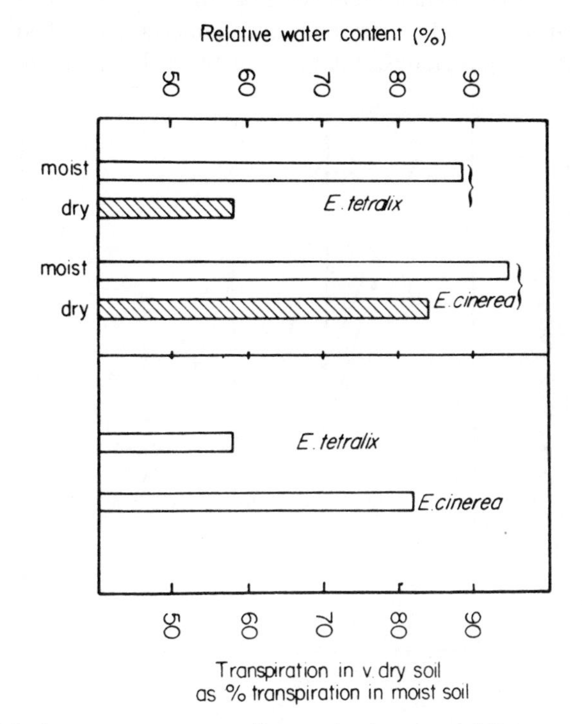

Figure 7.4 Relative water content and transpiration rate of *Erica tetralix* and *Erica cinerea* after 2 h at 45 % relative humidity; 20 °C in a moist soil and in a very dry soil (below PWP). Data of Bannister (1964c).

pronounced species difference which could account for the respective habitat tolerances. This difference was further reinforced at the wet end of the water availability scale by a differential response to waterlogging (p. 310). In the same study *Calluna vulgaris* was compared with the two *Erica* species and, though showing a reduction of relative water content and transpiration rate on droughting, remained unharmed under the same conditions as those lethal to *E. tetralix*. This suggests that *C. vulgaris* may have an inherent desiccation tolerance and account for the observation that its habitat range overlaps that of *E. cinerea* and *E. tetralix*.

Agricultural experimentation to isolate drought-resistant varieties or species has stimulated a great deal of comparative work. Even very simple screening methods may play a part; Carrol (1943) for example, exposed sod-samples of various pasture grasses to a slow drying process and assessed drought resistance on the basis of numbers of plants surviving after 21 days. Among the intolerant species were *Poa trivialis* and *P. nemoralis*, while *P. pratensis* (meadowgrasses) and *Festuca rubra* (red fescue) were the most tolerant.

Though much can be done with simple techniques, increased sophistication of instruments has led to the introduction of physiological methods of some complexity. Comparative measurement of photosynthesis in relation to drought resistance of various cereal varieties has been undertaken by Todd and Webster (1965) using infrared gas analysis to monitor CO_2 exchange. Figure 7.5 shows the

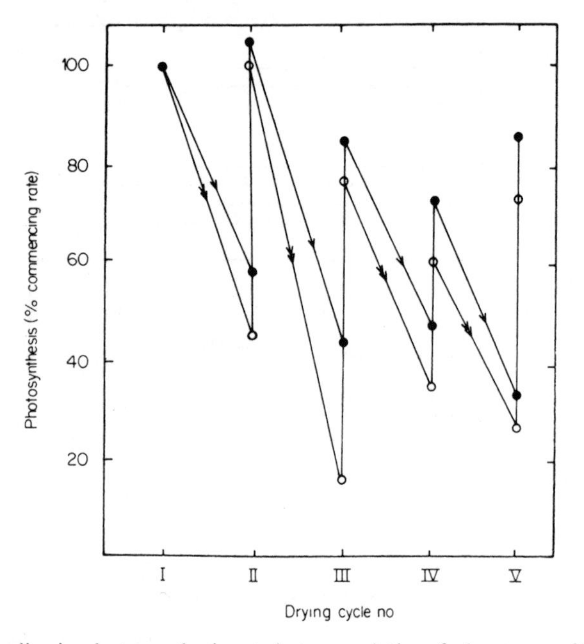

Figure 7.5 Decline in photosynthetic rate in two varieties of wheat caused by exposure to five drying–rewetting cycles. Var. Kan-King ●, Var. Ponca ○. Data of Todd and Webster (1965).

course of photosynthesis during five drought cycles in a tolerant variety (Kan-King) and an intolerant variety (Pinca) of wheat; the degree of drought tolerance was assessed by a conventional drought cycle/survival test. The photosynthetic measurements showed the sensitivity of the two varieties to drought very much more quickly than the conventional test, and the authors proposed that the technique could be used in screening populations for drought resistance.

Klikoff (1965) found a strong correlation between the response of photosynthesis to water deficit and the habitat distribution of some dominant species of the Sierra Nevada timberline. The sedge *Carex exserta* inhabits dry montane meadows, *Potentilla breweri* (cinquefoil) most meadows and *Calamagrostis breweri* (Brewer's reedgrass) wet meadows in which the water potential never falls below −0·03 MPa. By contrast the dry meadows develop water potentials of less than −1·5 MPa by late summer. Figure 7.6 shows the

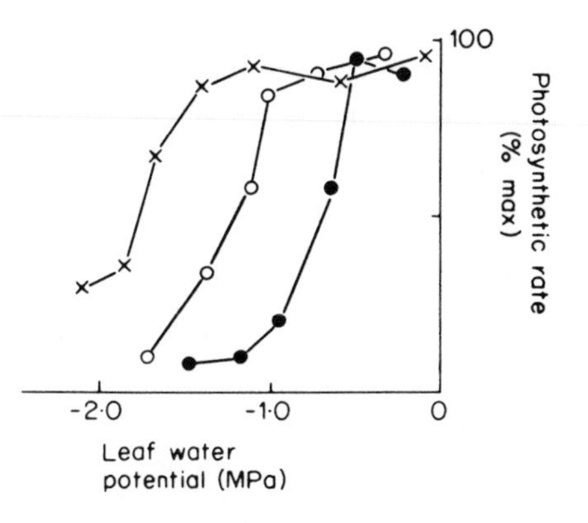

Figure 7.6 Relationship between photosynthesis and leaf water potential of *Calamagrostis breweri* ●; *Potentilla breweri* ○; and *Carex exserta* ×. Redrawn with permission from L. G. Klikoff, Photosynthetic responses..., *Ecology*, **46**, 516–17, Figure 1 (1965).

relationship of photosynthesis to leaf water potential determined by making measurements on pot-grown plants at various times after watering; the response to water stress shows a remarkable parallel with the distribution of the plants.

Comparative experiments are ecologically useful but it must be remembered that some water deficiency situations are difficult to simulate under experimental conditions. Water availability varies widely in both space and time but extreme deficits are rare in the temperate zone, occurring possibly only once or twice each century. Nonetheless, they will cause far-reaching effects in the ecosystem, ranging from a temporal reduction in the population of plants with a short life-cycle to the total exclusion of other, longer life-cycle plants. Today's ecosystem is a reflection of yesterday's environment and its study requires a knowledge of the range of climatic variables; this is nowhere more apparent than in the distribution of precipitation and potential evaporation in which identical annual means may be accompanied by very different bio-climates of water balance.

Odening *et al.* (1974) made a similar comparison of three North American desert shrubs which differ in their drought tolerating strategies. *Larrea divaricata* (creosote bush) is a truly tolerant evergreen; *Encelia farinosa* (brittle bush) produces a new crop of smaller more sclerophyllous leaves as the dry season progresses, then sheds all of its leaves, while *Chilopsis linearis* (desert willow) is winter-deciduous and copes with the dry summer by deep-rooting coupled with leaf xeromorphy. Figure 7.7 shows the relationship between leaf water potential and maximum photosynthetic rate established from field, experimental garden and phytotron measurements on intact plants. *L. divaricata* is capable of net carbon gain down to −8·0 MPa or less but the other two species are more sensitive to leaf water stress. Despite this, *E. farinosa* is successful because it can extend its photosynthetic season by developing a new crop of more tolerant leaves and finally escapes the most extreme drought by leaf abcission. The combination of a phraeatophytic habit with the protection afforded by xeromorphic leaves permits *C. linearis* to photosynthesize throughout the summer but it should also be noted that its carbon uptake is not reduced to zero until leaf water potential falls to almost −4·0 MPa: this may be compared with the −0·5 to −1·0 MPa which will totally inhibit most mesophytes.

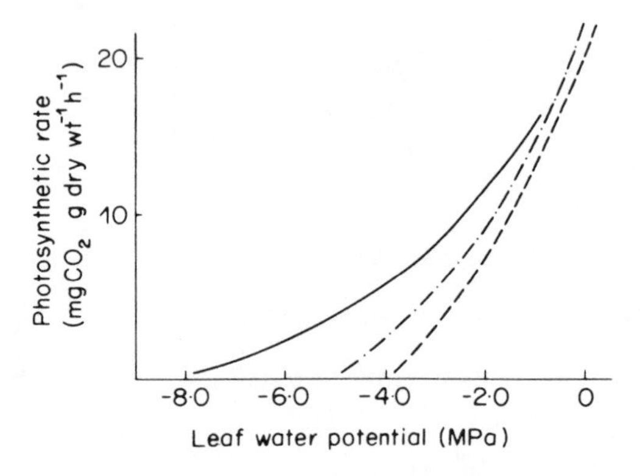

Figure 7.7 Relationship between leaf water potential and photosynthesis of three desert shrubs; *Larrea divaricata* (evergreen); –·–·– *Encelia farinosa* (drought-deciduous and leaf-variable) and ----- *Chilopsis linearis* (winter-deciduous, deep rooting hraeatophyte). Modified from Odening *et al.* (1974).

Plants of much less extreme habitats differ in drought susceptibility of the photosynthetic process. Bazzaz (1979) gives data for a range of North American tree species and relates their sensitivity to successional status, with the conclusion that sensitivity increases as succession proceeds (Figure 7.8). Bazzaz suggests that, as succession proceeds, so the habitat becomes less physiologically demanding, pioneer species more often having to cope with environmental extremes such as water stress. The higher absolute photosynthetic rates of early succession species may again be a reflection of the greater insolation of their unshaded habitats.

The effects of occasional drought will vary not only with duration and intensity but also with the stage of the plant life-cycle. Milthorpe (1950) found three distinctly different degrees of susceptibility to desiccation injury during the early stages of germination and seedling growth in wheat. For example, until the emergence of the first leaf the loss of 98 % of the tissue water did not cause serious damage. A single period of deficit may have a long-term effect; C. T. Gates

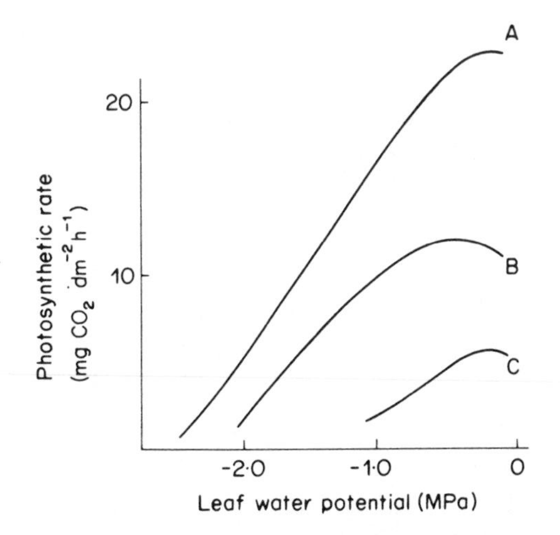

Figure 7.8 Relationship of photosynthetic rate to leaf water potential in N. American trees of differing successional status. A: *Acer theophrasti* (early). B: *Sassfras albidum* (mid). C: *A. saccharum* (late). Derived from Bazzaz (1979).

(1955a, b; 1957), for example, showed that one wilting treatment of young tomato plants caused a persistent alteration of net assimilation rate and nutrient status (particularly phosphorus). It seems likely that a single drought at an early stage in the growth of a plant may alter its competitive ability and affect the ecosystem balance for a season or more (Figure 7.9).

FIELD INVESTIGATION

The distribution of wild plants in relation to climatic wetness, their behaviour in response to short-term climatic fluctuations, the measurement of *in situ* soil water status and the elucidation of ecosystem water balance all provide a core of observational knowledge around which to build the experimental investigations suggested in the last section.

For many years botanists have attempted the subjective interpretation of plant distribution in relation to water; the habitat notes of most flora provide many examples. Critical distribution maps give a more definite picture of such relationships, for example, Figure 7.10 shows the distribution of *Nardus stricta* (matgrass) in the British Isles in relation to precipitation. The apparently high positive correlation cannot, of course, be taken as proof of a causal relationship as the high rainfall of the northern and western regions is associated with areas of greater altitude and these in turn are correlated with older, weathering-resistant, nutrient-deficient rocks. However, a distribution of this sort suggests a relationship which could be investigated by experimental means.

The work of Woodell *et al.* (1969) is an example of a more critical attempt to correlate plant distribution with rainfall: Figure 7.11 shows the density of *Larrea divaricata* stands in California plotted against mean rainfall. Except for one

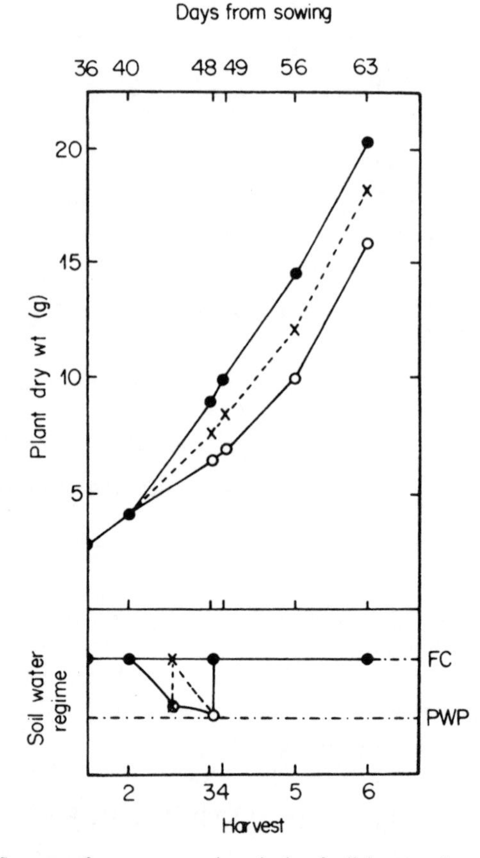

Figure 7.9 The influence of two repeated periods of wilting on the subsequent growth of tomato plants. The lower diagram indicates the soil moisture regime. FC, field capacity; PWP, permanent wilting point; ●, FC control; ○, allowed to dry to PWP without rewatering; ×, dried to PWP, watered and redried. Reproduced with permission from C. T. Gates, *Water Deficits and Plant Growth*, Vol. 2, © Academic Press, New York, 1968, p. 159, Figure 3.

deviant stand which the authors explain by its differing degree of exposure, there is a markedly linear relationship of density to rainfall. Again causality cannot be proved but the authors suggest that the increased spacing in the drier sites is caused by competition for water. *L. divaricata* is a drought-tolerant plant found in some of the most arid regions of North and South America and for this reason rather extensive studies have been made of its ecological relationships and drought tolerance. Chew and Chew (1965) considered that the plant showed a true physiological tolerance of drought rather than avoidance, while Yong (1967) isolated at least two ecotypically differing populations of which the Californian type had a lower water requirement than that of more southerly States. This latter point must be taken as a warning against placing too much reliance on distributional studies alone without any check on the genetic constitution of the populations.

A similar use of climatic data may be found in Felgan and Low (1967) who studied the surface/volume ratio of a columnar cactus (*Lophocereus schotii*) at various sites in Northwest Mexico; the ratio was found to fall with reduced precipitation and this was taken as circumstantial evidence of a trend to increased xeromorphy. Again, experimental work is needed to prove the assumption.

Using shorter-term rainfall data, Zahner and Donelly (1967) correlated the growth-ring width in *Pinus resinosa* (red pine) with rainfall and calculated water deficits. Their multiple regressions of previous year's rain and deficit and current year's rain and deficit gave a correlation coefficient of 0·91. Relationships such as this, with short-term variables, may be taken generally as better evidence for causality than the long-term correlations described above.

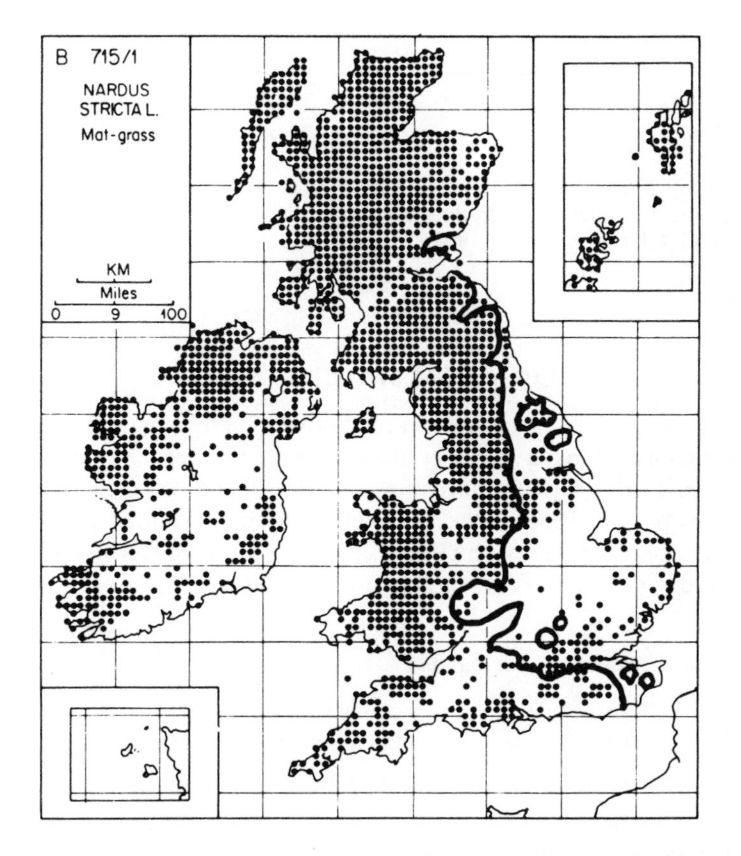

Figure 7.10 Distribution of *Nardus stricta* in Britain in relation to rainfall. Areas enclosed by or to the left of the heavy line have a rainfall exceeding 100 cm y^{-1}. Reproduced with permission from F. H. Perring and S. M. Walters, *Atlas of the British Flora*, Botanical Soc. of the British Isles, 1962, Map 715/1.

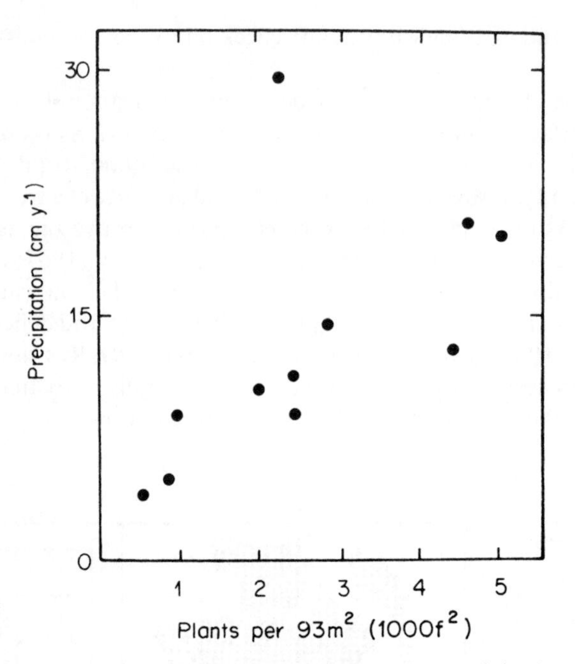

Figure 7.11 The density of *Larrea divaricata* in relation to precipitation at a number of desert sites in N. America. Reproduced with permission from S. R. J. Woodell, H. A. Mooney and A. J. Hill, Behaviour of *Larrea divaricata*, *J. Ecol.*, **57**, 37–44, Figure 2 (1969).

ECOSYSTEM WATER BALANCE

The increased study of plant–water relationships at the physiological level has been accompanied by a growing interest in the water economy of whole ecosystems. In the agricultural context studies of the water requirements of irrigated crops extend back to the late nineteenth century (Jensen, 1968) but more critical analysis of the ecosystem water budget has awaited recent developments in technique, instrumentation and recording facility.

Methods of measuring water balance

Rutter (1968), in a review of water consumption by forests, describes six basic methods which may be used in the measurement of water balance.

(1) Micrometeorological estimation of the vapour flux from the canopy.
(2) Transpiration measurement.
(3) Measurement of interception.
(4) Measurement of soil water depletion.
(5) Lysimetry.
(6) Stream gauging on experimental watersheds.

Micrometeorological estimation of vapour flux

Vapour flux away from the upper surface of the canopy may be calculated by the aerodynamic or Bowen Ratio techniques discussed on pp. 30–32, or estimated as potential evapotranspiration (p. 32) from available meteorological data.

Transpiration

Most early field determinations of transpiration were made by the quick weighing, cut leaf or branch method, the assumption being that water loss continues unchanged for a short time after excision. With some plants, particularly branches of trees and shrubs, this may be valid, but in others there are rapid changes (Ivanov, 1928; Franco and Magelhaes, 1965). With suitable calibration the technique can be used, but it is better avoided.

A more recent development is the heat-pulse method, in which a localized area of the stem is warmed for a short period; the arrival of the heat pulse in the xylem contents is then noted at a higher level, using a thermocouple or thermistor sensor. This gives a direct measurement of the transpiration stream velocity and, if the cross-sectional area of the functional xylem is known, it may be converted to volumetric loss. Unfortunately, the functional extent of the xylem varies seasonally by gas embolism of vessels or blocking with tyloses, thus introducing some error into the estimate.

Various attempts have been made to measure transpiration loss by using 'transpiration tents', made of different types of plastic sheeting and supplied with air by large-capacity fans. The changing water content of the air passed through the tent may be measured with various psychometric sensors, either by condensation or absorption or by infrared gas analysis. The greatest problems with this method are the controlling of the enclosure microclimate so that it remains comparable with the exterior, and the expense of high volume-air-conditioning equipment.

Interception

Precipitation intercepted by the upper surface of the canopy is either re-evaporated directly, contributes to stem-flow, or may be absorbed. This re-evaporation is termed interception loss and is estimated by difference using standard rain gauges outside or above the canopy, ring funnels to measure stem-flow to the ground, and further gauges below the canopy to assess throughfall and canopy drip. Massive replication of such gauges is needed to cope with the horizontal variation of canopy distribution and the exposure of the gauges in clearings or above the canopy is fraught with difficulties caused by turbulence which may seriously influence the catch (Leyton *et al.* (1968)). A typical study is described by Ford and Deans (1978).

Soil water depletion

This may be measured using any of the standard techniques, the rooting zone either being permanently instrumented with tensiometers and resistance units or periodic gravimetric samples or neutron scatter readings being taken. The neutron scatter technique has the advantage of giving a direct volumetric estimate of soil water, which is more accurate than the conversion, using a bulk density factor, needed for tensiometers or resistance units. Slatyer (1961b) gives a detailed account of the methodology of a water balance study in which all of these techniques were used in an arid zone *Acacia aneura* (mulga) woodland.

Lysimeters

The simplest, though most expensive, approach to the problem of crop water use is lysimetry. The lysimeter is a large soil container which is hydrologically isolated from the surrounding soil and permits the measurement of water income and loss. The derivation of the term implies the collection and measurement of drainage water but it has since been applied to large soil containers in which the water budget may be elucidated either by weighing or by soil water measurement. Hudson, in the discussion appended to a paper by Van Bavel and Reginato (1965), suggested the generic term 'Evapotranspiration gauge' and the specific epithets, lysimeter, weighable lysimeter and weighable container. It is also of considerable importance to specify whether the gauge contains repacked soil or encloses an undisturbed monolith. Hudson (1965) discussed the construction and specification of evapotranspiration gauges.

Penman (1963) reviewed the subject, commenting on the need for the crop cover on the gauge to be continuous with the canopy of the surrounding crop. Failure to ensure this leads to excessive exposure and overestimation of water loss by the 'oasis' or 'clothes-line' effect which is discussed by Stanhill (1965). The largest contributions to the oasis effect are the advection of sensible heat from the surroundings and evaporation from the 'sides' of a canopy arising from insufficient wind 'fetch' for typical profile development. Penman's comparison of results of Mather (1954) and McCloud and Dunavin (1954) provided an excellent example: Mather's gauges were exposed so that the contiguous canopies were level, but in McCloud and Donavin's work the crop projected about 10 feet above its surroundings, thus giving an estimate of water loss which apparently exceeded the available radiant energy income by a large amount.

Etherington and Rutter (1964) used an array of 28 lysimeters to compare the response of two grasses, *Alopecurus pratensis* and *Agrostis tenuis* (meadow foxtail and bent grass) to water deficit and waterlogging. In this instance, water use was followed with tensiometers and resistance units and showed that water potential profiles similar to those occurring in the field could be simulated. The highest irrigation level was intended to maintain field capacity and was controlled by checking for drainage after each weekly irrigation: the subsequent irrigation could then be increased or decreased as necessary. This technique gave an almost exact correspondence between the applied water volume and the calculated weekly potential transpiration. Simulation of a natural environment, in addition to avoidance of the oasis effect, requires the establishment of soil, root and water profiles similar to those of the surroundings. Van Bavel and Reginato (1965) discussed these problems and also suggested that water should be removed from the base of the soil block to avoid impeded drainage at the soil/air interface. This effect has been investigated by Coleman (1946), who showed that it might have a considerable effect on the behaviour of water in lysimeters.

Experimental watersheds

Watersheds which are geologically sealed by an impermeable layer may be considered as gigantic natural lysimeters, the drainage of which may be observed in stream-flow. Reinhart (1967) has described techniques for calibrating such areas: streamflow may be correlated with climatic records, before and after a treatment such as clear-felling, or the streamflow from the treatment watershed may be compared with an untreated control. The 'watertightness' of the catchment may be checked, for example by comparing chlorine input in precipitation with output in streamwater: as there is no appreciable soil storage of this element the two values should be about equal (Likens *et al.* 1977).

Most studies show increase in run-off following disafforestation, averaging about 20 cm worldwide (Hibbert, 1967). Likens *et al.* (1970) measured an increase of almost 40 % in streamflow (from 87 to 120 cm) 2 years after the clearance of an experimental watershed at Hubbard Brook, New England. Changes of vegetation type may also influence streamflow, for example Swank and Douglass (1974) found that replacement of a mixed oak-hickory deciduous forest cover (*Quercus–Carya*) by *Pinus strobus* (white pine), reduced predicted runoff by 20 % after 10 years (Figure 7.12). They explained this reduction in terms of the greater interception and re-evaporation capacity of the conifer canopy.

Forest cover, in general, has a lower albedo (p. 10) than herbaceous vegetation, since a considerable amount of reflected radiation is absorbed by the undersurfaces of leaves in the upper parts of deep canopies. Viewed from the air it is very obvious that this effect is more marked in coniferous canopies compared with deciduous and it may be that the lower albedo coupled with greater surface roughness of coniferous canopies will increase their capacity for transpiration.

The general consensus now is that interception and re-evaporation are the most important factors in reducing run-off from afforested catchments (Ford and Deans, 1978) and this effect is greater for forest than for grassland, becoming as great as 38 % for coniferous cover in upland Britain (Institute of Hydrology, 1976).

Disafforestation and replacement of forest cover by a shallow canopy increases soil wetness and has been suggested as an explanation for the initiation of blanket-bog formation and other soil deterioration in Britain and other northern temperate areas following tree removal in late Neolithic and Bronze Age times (Dimbleby, 1962; Moore, 1975). Soil hydrological changes approaching 10 to

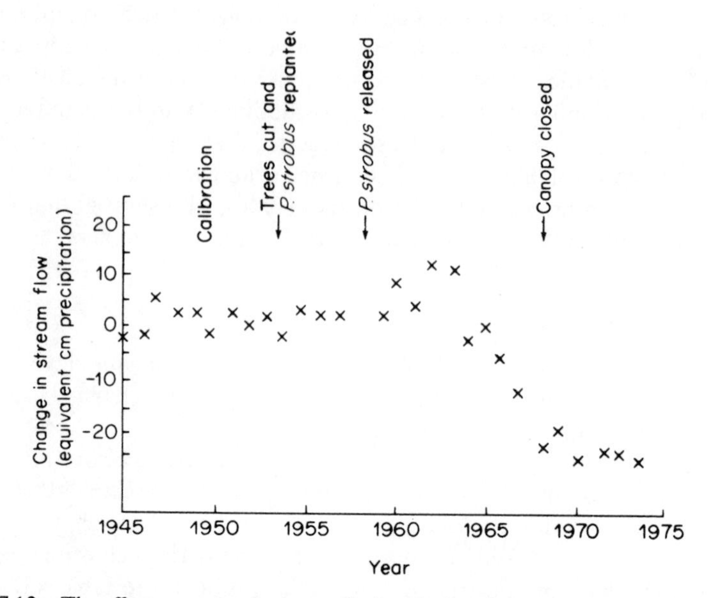

Figure 7.12 The effect on watershed runoff, of replacing *Quercus–Carya* deciduous forest with *Pinus strobus*. Data of Swank and Douglass (1974).

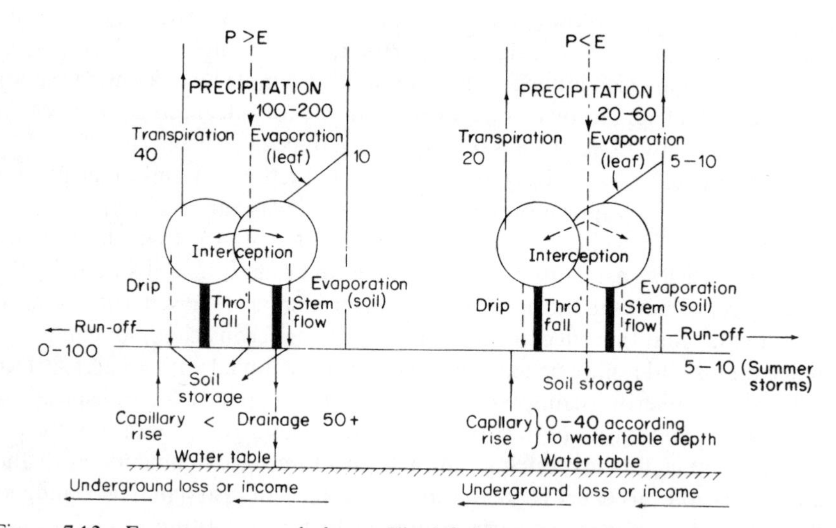

Figure 7.13 Ecosystem water balance. The left-hand diagram shows the path of water through the system when precipitation exceeds transpiration and evaporation. Rain may pass the canopy as throughfall or be intercepted by foliage from which it may either evaporate directly or be channelled into stemflow or leaf-drip and increase the input to the soil surface. Infiltration replenishes the soil storage reservoir or leads to drainage. If the infiltration capacity is exceeded, lateral run-off may occur, reducing the input to some parts of the soil and increasing it to others. Lateral movement below the water table may also be a source of income or loss. Water uptake by plants and its loss, by transpiration,

represents the main return pathway to the atmosphere, reinforced by some evaporation from the soil surface. Between periods of rain the soil storage capacity buffers these changes and may be replenished by capillary rise from the water table.

The right-hand diagram shows the inversion of the dominant direction of water movement when evaporation exceeds precipitation: much less water enters the soil and capillary rise may be all important both in supplying the plant with water and in determining pedogenetic processes. Surface enrichment with solutes may occur compared with the strong leaching processes which are encountered when downward drainage of water occurs.

Appended figures indicate water balance in temperate–oceanic and temperate–continental climates (cm y^{-1}).

20% of total run-off might easily increase soil waterlogging by raising a watertable or exceeding the infiltration capacity of an impermeable soil. Rogerson (1976), for a forest soil in northern Arkansas, found that disafforestation maintained the soil water deficit near zero during the whole year compared with the uncut control which developed a deficit of more than 200 mm by late summer.

Figure 7.13 represents the pathways of water movement within an ecosystem and gives some indication of the relative quantities involved under different climatic conditions and vegetation cover.

Chapter 8

Climate and plant response

CLIMATE

Long-term climatic records are an essential background to the local measurements so far discussed. The seasonal cycles of temperature, day-length, rainfall, humidity and wind exert a strong control over physiological and reproductive processes which is reflected in ecosystem structure and function. The incidence of short-term oscillations such as droughts, floods, gales and extreme frosts may also be determinate of species distribution of competitive behaviour, sometimes even when their incidence is on a time scale of many years. The patterns of climatic and altitudinal species distribution which characterize the world's ecosystems are basically determined by these interactions and, of course, by their relationship to soil formation.

Climatic effects of this type are excessively complex as they interact differently at various stages in the plant life-cycle. At all stages, population phenotypic plasticity is a reflection of genetic constitution, hence the enormous variety of species which are confined to different ecosystems throughout the world. The following scheme indicates some of the possible interactions:

Between the extremes of extravagant success shown by forest dominants and the clinging survival of stress tolerant desert and tundra plants, lie the innumerable species and their genetic sub-populations which fill, so far as their phenotypes allow, virtually all of the earth's available niches. Even the snows of the high latitudes, arid desert surfaces and all but the hottest of igneous springs, support a population of chlorophyllous organisms. If the distribution of these plants is investigated the most obvious world-wide patterns, especially of the specialist terrestrial plants and marine littoral algae, are those of climatic zonation.

The same cannot be said for the distribution of substratum types: igneous rocks occur randomly with reference to climatic zones whilst the palaeoclimatically determined nature of some sedimentary rocks has lost any relationship to modern climate as a result of plate-tectonic continental drift over the great time-spans of geology. So strongly does climate override the finer patterns of parent materials that earlier workers (e.g. Clements, 1916) suggested that succession proceeded toward a climatically determined monoclimax. Whilst this view is no longer tenable, it is very obvious that the present day climatypes, which have

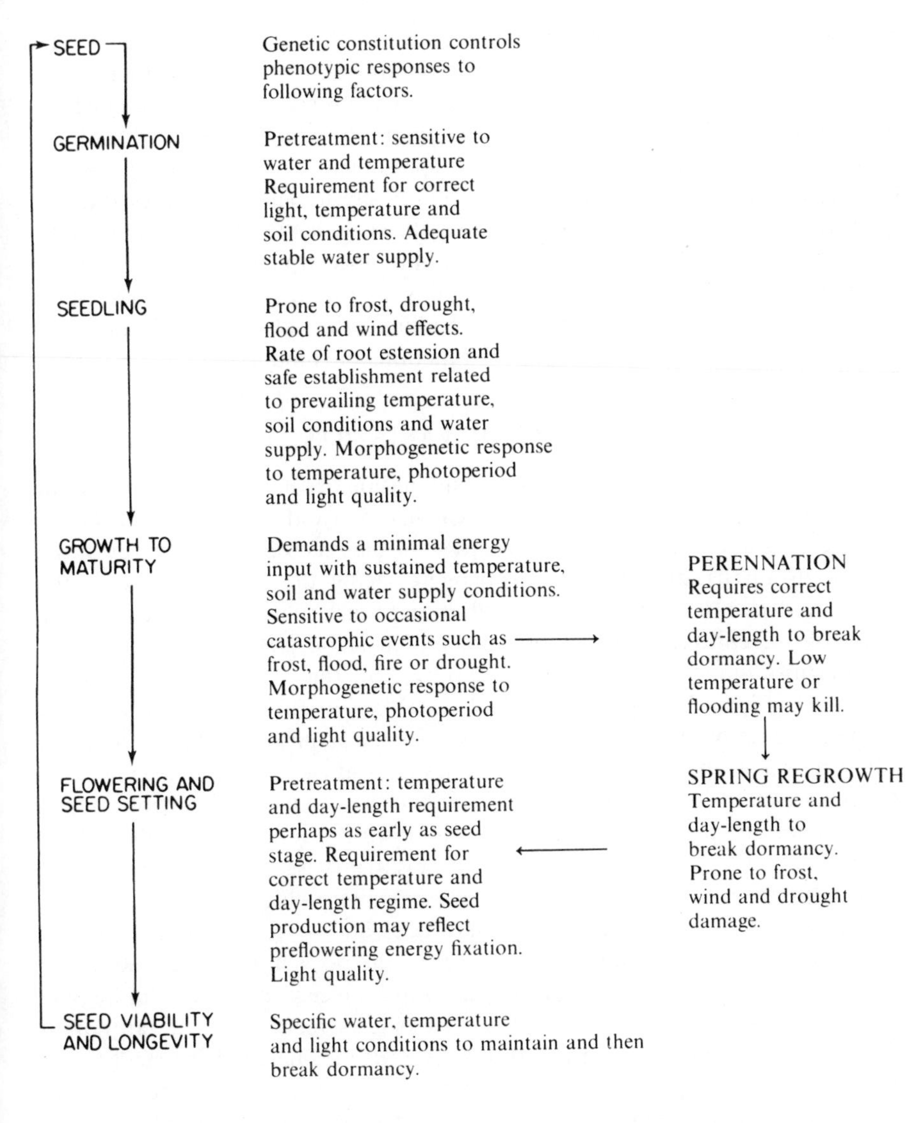

prevailed for almost 10,000 years since the Devensian glaciation of the northern hemisphere, and with some changes of rainfall for much longer periods in the equatorial belt, have permitted the co-evolution of plant, animal and microorganism species together with their soil environments.

Many of the world's vegetation types show remarkable convergance of physiological ecological characteristics amongst taxonomically unrelated groups, confirming the suspicion that if a niche exists, selection and evolution will cause it to be filled. The Mediterranean climatype is perhaps the most startling case of a discontinuous vegetation type, occurring in southern Europe, northern and southern Africa, South Australia, and South America, in many cases with few or no species in common but an almost uncanny physiognomic similarity. The underlying mechanisms of these resemblances are now attracting the attention of eco-physiologists (Dunn *et al.* 1976). The early plant geographers described the obvious similarities of morphology and anatomy which appear as unifying characteristics of the various vegetation types but it is now becoming obvious that there are also distribution patterns of many physiological attributes, thus the geography of C-3, C-4 and CAM photosynthesis is related to climatic wetness and temperature, the temperature optima of tundra, alpine and lowland plants are very different while responses to day-length and thermoperiod have important ecological effects. These, and many other physiological responses which control species distribution, are functions of climate.

SOURCES OF VARIATION

The 'weather machine' is driven by a combination of solar radiant energy input and the motions and relative motions of earth, moon and sun. If the earth were a flat disc constantly facing the sun its surface would receive radiant energy at the solar constant irradiance. As a hemisphere facing the sun it has twice the illuminated area of a disc and the distribution of surface irradiance varies as the cosine of the zenith angle (sine of the angle of incidence). Because of this, about half the instantaneous income falls on 14·6% of the earth's surface and the remainder is spread unevenly over 35·4% (50% is in darkness).

The presence of the atmosphere exaggerates the geometrical effect. The earth spins on its axis once in 24 hours and circles the sun every 365·25 days: because the axis is 'set' in space and is tilted at *c.* 66° to the orbital plane (ecliptic), each pole apparently tilts toward and then away from the sun during the year.

The result of this axial tilt and orbital effect is to impose a seasonal variation of day-length and radiant income. It represents a winter reduction of mean radiant income by 40% of the maximum in mid-latitudes, and 10% in the equatorial belt. The seasonal impact becomes even greater as the poles are approached.

Because of the zenith cosine effect, low latitudes receive more radiant income than high latitudes, but the long-wave loss shows relatively little latitudinal variation, thus from the pole to *c.* 40° loss exceeds income and between 40° and the equator income exceeds loss. These differences of radiation balance result in the world pattern of mean annual temperature ranging from about 27 to 28 °C on the equator to −16 to −20 °C in tundra environments, with even lower values on polar ice-caps. Mid latitudes in continental areas have annual means as low as −4 °C rising to 10 to 11 °C in oceanic conditions while, between these extremes, warmer Mediterranean climates are near 17 °C. These differences of temperature would be much greater if it were not for the transport of heat by forced convection and by the latent heat bank of the atmospheric wind system.

If the earth's surface was entirely ocean the climatic patterns of temperature, wind circulation and precipitation would be remarkably symmetrical, but the asymmetric distribution of continents and their interactions with the prevailing directions of the wind systems generates an equivalent asymmetry of continental climatypes and vegetations (Walter, 1979). The effect of continental land masses is an interaction between the advection of warm moist air to the windward coasts and the scavenging of water from rising air as it enters the land-mass with consequent formation of inland areas of reduced precipitation. Thus, as a function of prevailing wind direction, distance from the ocean, and distribution of high ground, there is generated a rather variable pattern of climatic wetness superimposed on the latitudinal temperature gradients, seasonal and day-length cycles.

In addition to these relatively predictable patterns there are also more localized changes of temperature with altitude which may, in extreme conditions such as the west coast of Mexico, create a series of vegetation zones which range from tropical forest to arctic-alpine within a few tens of kilometres. The temperature lapse-rate for dry air is 10 °C km^{-1} and 6 °C km^{-1} for water-saturated air, which is warmed by the recovery of latent heat as water condenses. In the tropics even a modest 2000 m mountain will have an annual mean temperature, at the summit, of only 7 to 8 °C; this is equivalent to a temperate climate. With similarity of temperature, resemblance between altitudinal and latitudinal zonation ceases, at least in the tropics: compare the seasonless tropical montane environment with the continuous day of the short tundra summer.

Seasonal changes of day-length provide organisms with a calendar to which developmental changes are physiologically locked, particularly in high latitudes where 'prediction' of winter cold or summer drought may save the plant from untimely development and death. These programming mechanisms are controlled either by direct sensing of day-length or by response to the thermoperiod which may be a function of the 24 h cycle or longer, seasonal cycling.

PLANTS, TEMPERATURE AND LIGHT

Adult plants

Temperature change within an ecosystem is complex, cycling slowly up and down with season, diurnally with solar radiant heating and randomly with meteorological fortuity so that the precise value of any daily temperature, or even seasonal mean, is unpredictable. It is the element of unpredictability which poses the greatest problem for plants as the upward or downward 'spikes' of the temperature–time curve may cross boundaries of physiological tolerance or thresholds of stimulus for such responses as flowering and dormancy breakage.

Despite the fact that temperature is one of the most obvious environmental differentials, remarkable little information exists concerning its effects as a physiological control mechanism in natural ecosystems. A great deal of our information, as in so much eco-physiology, comes from agricultural comparisons of varieties, populations or species, but the results so generally point to provenance as a strong controlling factor in physiological response to environmental temperature and its changes, that it is not unreasonable to suspect that similar differentials will be found in wild plants with an equivalent range of origin.

Plant responses to temperature may conveniently be subdivided as: (*a*) sensitivity of primary metabolism at normal temperatures; (*b*) stress effects at abnormal temperatures; and (*c*) control responses.

Metabolism and consequent activity

Metabolic processes, cell division and elongation growth respond to rising temperature by increasing to an optimum value beyond which the rates decline, finally to zero at temperatures which may be biochemically damaging. Within this apparently uniform response pattern plants show a wide range of environmentally adaptive variation in temperature optima, upper and lower temperature limits and temperature rate gradients. This is particularly marked where major alternative biochemical pathways may be selected, for example C-3, C-4 and CAM photosynthesis.

The C-4 process generally has a high temperature optimum compared with C-3 (Figure 8.1) and is also much more sensitive to temperature change, having a

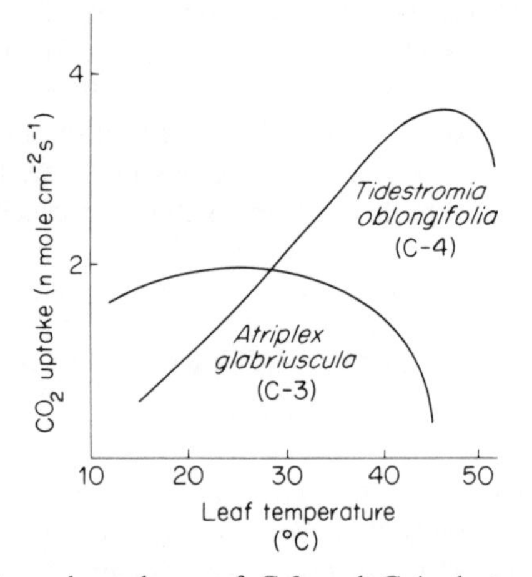

Figure 8.1 Temperature dependence of C-3 and C-4 photosynthetic pathways. *T. oblongifolia* is a hot desert species and *A. glabriuscula* from cool coastal California. O_2 2 %; CO_2 320 vpm; PAR 160 nE cm^{-2} s^{-1} (Modified from Bjorkman 1975).

response curve which is flattened in the temperature range below the optimum, a relationship which indicates that CO_2 fixation is primarily limited by carboxylation resistance *sensu* Monteith (1963). The ranges of optima are 20–30 °C for C-3 and 30–40 °C for C-4 plants. The drought survival strategy of CAM plants fits them best for a desert environment, though Kluge and Ting (1978) note that the CAM strategy reduces evaporative cooling both through stomatal closure and through the succulent morphology with reduced surface–volume ratio. Though very tolerant of high temperature, for example *Opuntia* spp. (prickly pear) shows growth between 0 and 50 °C, there is some suggestion that CAM may not be entirely well adapted to very hot daytime conditions, for example in the Negev desert the dark fixation of CO_2 by the Asclepiad *Caralluma negevensis* was inhibited by high night temperatures, and high day temperatures seriously increased respiratory carbon-loss (Lange *et al.* 1975b). In these conditions the plant is confined to shaded rock crevices and is not so well adapted to the hot desert environment as arido-active C-3 shrubs.

Within particular photosynthetic processes the temperature optima may be specifically variable and may also be altered by pre-treatment, temperature conditioning or cyclically by photo- and thermoperiod control effects. Bauer *et al.* (1975) cite upper limits of 35–39 °C for arctic species and 40–48 °C for temperate herbs and deciduous trees. Most of these are C-3 plants: the upper limits for C-4 plants are between 55 and 61 °C as might be expected from the strong representation of arid-zone and dry-temperate grassland and desert sun-plants amongst their number. Both optima and upper limits may be increased 3–4 °C by preconditioning at high temperatures.

Photosynthesis is similarly variable in response to cold; temperatures well above freezing causing strong inhibition in some tropical species. Taylor and Rowley (1971) exposed the tropical cereal, *Sorghum bicolor* cv hybrid NK145 (millet) to a temperature of 10 °C for 3 days. On returning to 25 °C the photosynthetic rate returned to only 3·2 % of the pre-stress value compared with a return to 51 % by *Lolium multiflorum* (Italian ryegrass) a temperate Mediterranean grass.

Contrasting with the sensitivity of hot climate species, arctic and alpine plants have remarkably low temperature minima for maintenance of positive net photosynthesis. Bauer, *et al.* (1975) cite a summer minimum of – 6 °C for the two circumpolar species, *Ranunculus glacialis* (glacial buttercup) and *Oxyria digyna* (mountain sorrel). All species for which they cite data have winter minima which are 3–4 °C lower than the summer values, probably a combined conditioning and photo-/thermoperiodic effect. Lichens and mosses show even more extreme values. *Xanthoria parietina* (Yellowscales' lichen) continuing to photosynthesize down to – 18 °C (Lange, 1965) and *Racomitrium lanuginosum* (alpine hair moss) to – 10 °C (Kallio and Karenlampi, 1975). Billings (1974) gives an extended review of low temperature effects including photosynthesis.

In the normal temperature range, respiration is much more temperature sensitive than photosynthesis, having a Q_{10} of 2·0–2·5 compared with 1·2–1·5 for C-3 net photosynthesis (e.g. Figure 8.1). Larcher (1969) suggested that 'dark respiration appears to be primarily adapted to the temperature environment of the plant habitat... by adaption to higher temperature the respiratory rate will be reduced'. An excellent example was given by Eagles (1967) who measured respiration of a Danish (cool) and Algerian (hot) population of *Lolium perenne* (ryegrass) over a range of temperatures. The Algerian strain had a respiratory rate three times greater than the Danish strain at 5 °C but one-fifth less at 30 °C (Figure 8.2). Similarly Larcher (1969) showed the cool climate species *Picea abies*

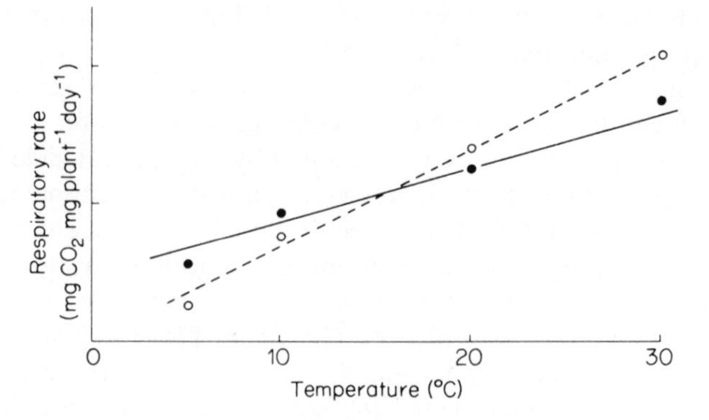

Figure 8.2 Temperature dependence of respiration rate in strains of *Lolium perenne* from a hot Algerian climate (●) and from a cool Danish climate (○). Data of Eagles (1967).

(Norway spruce) to have a respiratory rate well in excess of photosynthesis at 30 °C but, at the same light intensity and temperature, the desert shrub *Artemesia tridentata* (sagebrush) still retained a positive photosynthetic balance.

The strong temperature sensitivity of respiration has a number of consequences for the field physiological function of the plant. One is related to the suggestion that mid-day stomatal closure may be caused by respiratory CO_2 evolution if leaves become overheated by radiant loading (p. 19). Heath and Orchard (1957) showed that high leaf temperature in *Allium cepa* (onion) increased intercellular CO_2 concentration and they suggested that this caused stomatal closure. Similarly, Bjorkman *et al.* (1970) found that the CO_2 compensation point increased by 59 v.p.m. as leaf temperature rose from 5 °C to 36 °C in the C-3 *Atriplex patula* (common orache). Zelitch (1971) suggested that dark respiration doubles and photorespiration more than doubles for a 10 ° increment in the 20–35 °C range though this is controversial, as Holmgren and Jarvis (1967) found photorespiration to be slightly less temperature sensitive than dark respiration. However, the net result will always be a CO_2 compensation increase with temperature in C-3 plants.

The work of Bjorkman *et al.* (1970) suggests that a C-3 plant will show stomatal closure in bright mid-day sunlight but a C-4 plant is likely to keep its stomata open until hydro-active closure overrules the CO_2 effect. At this stage the greater CO_2 trapping efficiency of the C-4 plant will allow it to continue photosynthesis, despite much reduced stomatal aperture (p. 127 ff). The stomatal limitation of water loss will result in a warmer leaf but the higher temperature optimum of the C-4 process is suited to this, hence the eminently suitable adaption of the C-4 plant to the arid, sunlit environment.

Beyond the effects of temperature induced changes of the stomatal aperture, plant water relationships are also very sensitive to rapid temperature change because the water vapour saturation pressure in air is so strongly temperature dependent (p. 27); saturated air at 20 °C has a water potential of 0·0 Pa but an increase of 0·5 °C takes it to less than −4 MPa! Such effects are particularly marked when passage of cloud-shadows or flickering of sunflecks cause changes of air temperature or of the differential between air and leaf temperature.

The carbon balance of the whole ecosystem likewise responds to the differential temperature sensitivity of the two processes, thus tropical rain forest may have a high gross productivity compared with a temperate deciduous forest but, because of the discrepancy in respiratory consumption, a directly comparable net production. Evans (1972) cites a high forest in Thailand with gross primary production of 52·5 t ha^{-1} y^{-1} compared with 23·5 t ha^{-1} y^{-1} for Danish *Fagus sylvatica* (beech) woodland but similar net primary production figures of 9·0 and 9·6 t ha^{-1} y^{-1}, respectively.

The decomposer organisms which terminate the ecosystem food-chains and close mineral cycles also show the same respiratory temperature sensitivity differential by comparison with net photosynthetic carbon fixation. In both terrestrial and aquatic systems, most decomposers are mesophilic fungi and bacteria with temperature optima between 20 and 37 °C and minima of 5–10 °C, below which temperature very little decomposition proceeds (Dickinson and Pugh, 1974) though a few exceptional low temperature optima have been recorded for microorganisms from high latitude tundra soils. This relationship accounts for the relatively larger accumulation of soil organic matter in well-drained soils of cooler but otherwise comparable climates (p. 49).

It is well known that cell division is very temperature sensitive, for example Clowes (1961) cites information for *Pisum sativum* (Pea) which shows the mitotic cycle time to extend from 14 h to 25 h as the root-tip temperature falls from 30 °C to 15 °C and to 260 h for *Vicia faba* (broad bean) root tips at 3 °C. Little or nothing is known of the plant-geographical correlations of such sensitivity.

The extension growth consequent on vacuolation of cells produced from meristems is similarly temperature sensitive and often shows substantial temperature differential responses related to provenance or micro-habitat. Harris and Wilson (1970) attributed the success of *Bromus tectorum* (cheatgrass) as an annual weed competitor of *Agropyron spicatum* (bluebunch wheatgrass) to the much greater elongation rate of *B. tectorum* roots at temperatures as low as 3 °C. *A. spicatum* roots ceased to grow at 8–10 °C and, as a result, winter and spring growth of *B. tectorum* roots permits it to pre-empt the use of stored winter precipitation by *A. spicatum* during the dry summer months of the north American rangeland.

Cooper (1975) showed that provenance exerted strong control over the temperature sensitivity of leaf elongation in *Lolium perenne* (perennial ryegrass) and *Phalaris tuberosa* (tuberous Canary grass): most work with agriculturally useful grasses has shown similar results for both leaf expansion and tiller production (e.g. Eagles and Østgard, 1971) but it is difficult to separate the morphogenetic effects of temperature and thermoperiod from direct effects on vegetative growth. Peacock (1975), for example, noted that the temperature of the stem apex in *Lolium perenne* rather than the overall air-temperature to which the leaves were exposed governed the rate of leaf expansion. Shoot growth at low temperature permits plants to utilize environmental resources which might otherwise be lost to them. In a stressed situation this might make the difference between survival or extinction rather than influencing competitive balance as in the case of *Bromus tectorum* root extension. Tiezen and Wieland (1975) noted, in tundra conditions, the very rapid exsertion of grass leaves from their sheaths at low spring temperatures. These leaves had formed at the end of the previous growing-season and were protected from frost damage by enclosure in the sheath tissue. Early exsertion permits rapid development of photosynthetic capacity at the beginning of a very short growing season and provides an alternative strategy to winter survival of chlorophyllous tissues as in mosses and lichens.

Little information is available, but observation suggests that the rather low heat conductivity of some soils must delay the warming of deeper horizons during the spring. In extreme cases this effect preserves a permafrost horizon (p. 54) and disturbance of the organic surface layer or any other interference with surface thermal balance may permit increased depth of thawing with resultant soil erosion and thermokarst formation (Ives, 1974; Pruitt, 1978). Ecosystems with stratified root systems may in consequence show changes of competitive balance according to the amount of winter soil-cooling which has occurred and the rate of air-temperature increase in different years. Good examples are furnished by the geophyte *Pteridium aquilinum* (bracken fern). The size of the *P. aquilinum* clonal patches is known to be affected by winter frost (Watt, 1954, and p. 393) but it is also noticeable that the emergence of the frond croziers from the soil may be delayed by up to one month by a preceding, colder than average, winter such as occurred in Britain in 1962–3 and 1978–9 (Etherington: unpublished data). This is almost certainly related to slow soil warming at depth. *Calluna vulgaris* (ling), with which cyclic competition often occurs (Watt, 1955), is more cold-tolerant than *P. aquilinum*, shallower rooted and consequently less sensitive to deep-soil temperature. The grass *Holcus mollis* (creeping soft grass) which often occurs in *P. aquilinum* stands is also shallow rooted and again may be favoured in these conditions.

Temperature stress effects

Low temperature stress is usually met in the form of frost damage, though some tropical plants show ultrastructural deterioration of tissues at temperatures as much as 10 °C above freezing (Taylor and Rowley, 1971). Plants vary widely in tolerance of freezing and this specificity is reflected in the ecology of species distribution and phenology. Heber and Santarius (1973) list causes of injury as: (1) low temperature damage *per se* (to cell components); (2) water expansion on freezing and subsequent crystal growth due to water and vapour migration down water potential gradients causing tissue disruption; (3) shrinkage of cells and cytoplasmic damage caused by such water loss; (4) increased concentration of solutes by 'freezing-out' causing osmotic, solubility and pH-change problems.

Plants may be subdivided into four categories of freezing tolerance: (1) tender; (2) limited hardiness; (3) tolerant, woody; (4) extremely tolerant, woody. Tender plants are killed by the first autumn frost and do not tolerate temperatures below −1 to −3 °C. A marked example in Britain is *Hydrocotyle vulgaris* (pennywort) which, in dune-slacks and other wet habitats, is reduced to a gelatinous pulp by the first frost while the adjacent evergreen leaves of *Juncus* spp. (rushes) survive to tolerate complete winter freezing (Etherington: unpublished data). Burke *et al.* (1976) include *Zea mays* (maize) and the cultivated Cucurbitaceae in this category. Most tropical herbs which never normally encounter frost behave in the same manner. Plants of limited hardiness may tolerate temperatures down to −25 °C by forming ice 'glaciers' in the intercellular space of unimportant tissues such as the cortex. The glaciers open frost-cracks which reclose on thawing (Idle, 1966), doing relatively little tissue damage. Below this temperature either dehydration of the cytoplasm or intracellular ice-formation is lethal. Freezing of cell contents does not occur before this critical temperature is reached and many hardy and semi-hardy plants contain natural 'antifreeze' compounds amongst which are sugars, their derivatives such as glycerol, organic acids and amino acids (Heber and Santarius, 1973). These presumably function by lowering the freezing point of the cell contents of indispensible tissues.

Tolerant woody plants show deep supercooling of the xylem and of critical tissues such as apical meristems, cambium and flower primordia. At a temperature of about −40 to −41 °C homogeneous ice nucleation occurs and intracellular ice formation destroys the tissues so that this temperature is a minimum for survival (Burke *et al.* 1976; Rajashekar and Burke, 1978). The most tolerant of all woody species show no deep supercooling and survive by formation of intercellular ice, mainly in the cortical tissues, with accompanying dehydration of other tissues. At the completion of acclimation and freezing, these plants will then tolerate −196 °C, the temperature of liquid nitrogen (Levitt, 1972).

Burke *et al.* (1976) show distribution maps for the mid-North American *Quercus rubra* (red oak) which deep supercools to −40 °C and for the more northerly *Populus tremuloides* (quaking aspen) which does not supercool and has no lower limit of temperature survival and naturally encounters minimal temperatures below −40 °C. Kaku and Iwaya (1978) found deep supercooling limits of Japanese trees to occur in three ranges related to provenance. Evergreen trees such as *Camellia japonica* (camelia) and less hardy deciduous trees (e.g. *Rhus succedanea*—wax tree) nucleated at −16 to −20 °C, hardy deciduous trees (e.g. *Castanea crenata*—chestnut) at −23 to −25 °C and very hardy deciduous trees (e.g. *Alnus hirsuta*—hairy alder) at −25 to −34 °C. These four representative species will, respectively, tolerate temperatures of −18, −15, −27 and −40 °C, and come from Southern Japan, mid-Japan, Northern coastal Japan and Northern inland Japan. Further examples of plant geographical range of freezing tolerance are given in Table 8.1.

The tolerance of liquid nitrogen temperatures by fully frozen, very hardy woody plants in which tissues are dehydrated is of interest, since poikilohydric plants (p. 175) which are cytoplasmically resistant to dehydration and may become air-dry are also resistant, in the dry condition, to −196 °C (Larcher,

Table 8.1 Limits of low temperature injury (From data compiled by Larcher, 1973)

Type of plant		Minimum temperature experienced for 2 hours or more without injury (°C)	
Poikilohydric Plants			
Lichens	High latitude	−80	All to −196 when dry
Forest mosses	High latitude	−15 to −25	All to −196 when dry
Saxicolous mosses	High latitude	−30	All to −196 when dry
Tropical mosses		−5 to −15	All to −196 when dry
Specialized angiosperms, e.g. *Ramonda myconi*		−9	All to −196 when dry
Homoiohydric Plants			
Tropical evergreen rain forest leaves		+5 to −2	
Mediterranean sclerophylls		−6 to −13	
Temperate trees	Leaves	−25 to −35	
Temperate trees	Buds	−25 to −40	
Alpine dwarf shrubs		−20 to −70	
Boreal conifers		−40 and below	

1973). Such plants are mosses, lichens, some pteridophytes and a very few angiosperms (Table 8.1). Pretreatment hardening is also an important factor in frost tolerance and may be induced artificially or occurs naturally as a result of differing microclimatic conditions within the plant canopy: underground and near-the-ground organs tend to be less hardy than exposed parts of the plant (Table 8.2). Hardiness is also strongly correlated with season and with the age and type of tissue involved. Table 8.2 gives some examples of artificial hardening and also shows the variation of frost hardiness at different heights within a forest canopy and below the soil. Similar variation is found between marine littoral algae which will tolerate frost (−15 to −40 °C) compared with sublittoral species which are never normally frozen and die below −2 °C (Larcher, 1973). Thus

Table 8.2 Artificial and naturally induced frost hardening in plants. (Data from Larcher, 1973)

Species	Hardening treatment during autumn (days at −10 °C)		Post-hardening lethal temperature (°C)
Picea (spruce)	0		−25
	6		−34
	24		−39
Betula (birch)	0		−30
	6		−50
	24		−60
			Lethal temperature range (°C)
Temperate forest:	Roots and rhizomes		−6·0 to −13·5
Plant organ and position	Evergreen leaves (cm above litter layer)	3–5	−11·5 to −14·5
		5–10	−11·5 to −18·0
		10–20	−13·0 to −20·0

insulation, by water, snow, soil or insensitive plant tissue such as bud scales may also play a substantial part in climatic cold tolerance by providing an avoidance mechanism.

Seasonal variation of hardiness is very marked in woody species, for example Mair (1968) shows the apical bud of *Fraxinus ornus* (ash) to tolerate −27 °C in mid-winter rising to −3 °C as the buds begin to swell in spring. During the same period the tolerance of the vascular cambium in 2-year-old wood rises from −39 to −5 °C. Figure 8.3 shows the seasonal variation of hardiness for different tissues

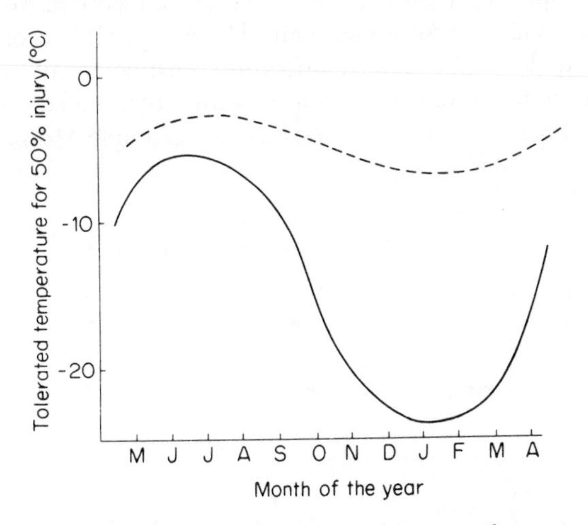

Figure 8.3 Seasonal variation in frost hardiness of *Quercus ilex* stem cambia (———) and root cambia (-----). Injury assessment by tetrazolium staining. Data of Larcher (1973).

of the Mediterranean *Quercus ilex* (evergreen oak): the shoot cambium is most tolerant and most seasonally variable whereas the root, protected by soil, is least tolerant and shows little annual variation. The leaves have an intermediate tolerance, reaching −15 °C in the winter: this is of interest as the tree is frequently planted as an ornamental in oceanic northern Europe and occasionally shows severe frost damage to leaves. Leaf age substantially effect tolerance: expanding leaves are most sensitive, and very often the most likely to experience frost, while mature leaves are much more tolerant: leaves of the temperate evergreens *Hedera helix* (ivy) and *Ilex aquifolium* (holly) leaves are damaged by −1.5 °C while expanding but become tolerant of −4·5 to 5·0 °C at maturity (Larcher, 1973). These plants are only semi-hardy and are frequently damaged by frost in Britain, where both species reach the northernmost part of their range. *I. aquifolium* also has a strong Oceanic component to its distribution and both distribution characteristics are probably related to frost incidence.

Intraspecific heritable variations of hardiness are also found and, as might be expected, hardy populations have been selected by extremes of climate; thus Robson and Jewiss (1968) showed *Festuca arundinacea* (tall fescue) ecotypes from

North Africa to be much more susceptible to freezing damage than north-temperate forms. The mediterranean populations continue growth in winter and it is probably this lack of dormancy, coupled with the production of soluble carbohydrate cryoprotectants by the northern forms, that accounts for the differences (Figure 8.4).

Heat and cold tolerance are similar in being related to provenance, pretreatment and age. Plants of hot climates may encounter very high air temperatures and also suffer extreme radiant loading (chapter 2, p. 144 ff). A remarkable example is the 56·7 °C air temperature recorded by Mooney *et al.* (1975) in Death Valley, California, while Huber's (1935) record of succulent tissues more than 20 °C above air temperature must be one of the highest over-temperatures ever measured. High temperature environments support plants which may not be able to survive elsewhere, for example Mooney *et al.* (1975)

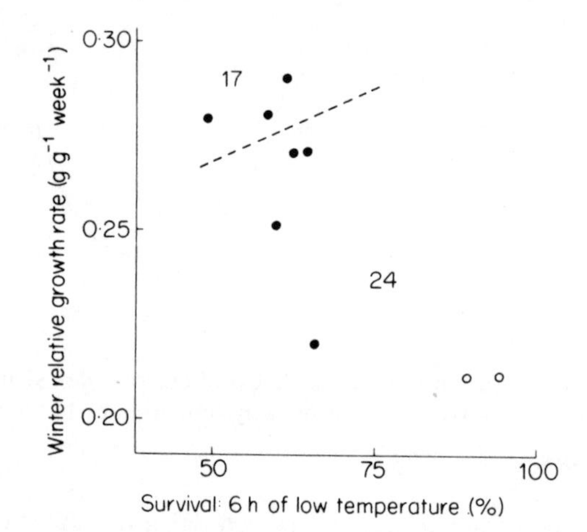

Figure 8.4 Relationship between low temperature tolerance and winter relative growth rate in nine varieties of *Festuca arundinacea*. Temperate varieties ○; Mediterranean varieties ●. Appended figures are mean values of soluble carbohydrate (%) from varieties above and below the broken line. Low temperature survival was the overall mean of six treatments between −5 and −15 °C. Data of Robson and Jewiss (1968).

transplanted the California cool-coastal endemics *Atriplex hastata*, *A. glabriuscula* and *A. sabulosa* (oraches) to experimental plots in Death Valley. Despite irrigation these cool-adapted species died by mid-summer, contrasting with the desert endemic *Tidestromia oblongifolia* (honeysweet) which shows maximum photosynthesis at 45 °C and grows during the summer months at a daily maximum temperature of 46 °C. This C-4 sun-plant may double its biomass in less than 3 days at 45 °C and at even 50 °C grows faster than *A. glabriuscula* does under any conditions (Bjorkman, 1975)!

Subtropical succulents and C-4 plants head the list of temperature tolerance (Table 8.3) with limits between 50 and 60 °C but the range is much less marked

Table 8.3 Heat tolerance of plants (summer condition) (from data compiled by Larcher, 1973)

Type of plant		Temperature endured for 30 minutes without damage (°C)
Subtropical succulents and C-4 plant leaves		+50 to +60
Mediterranean sclerophylls		+50 to +55
Temperate tree leaves		c. +50
Alpine dwarf shrubs*		+47 to +59
Boreal conifers		+44 to +50
Lichens		+35 to +46 (+70 to +100 dry)
Marine algae (cold ocean)		+25 to +32
Freshwater algae		+40 to +45
Hot spring	Cyanophyta	+70 to +75
	Bacteria	+70 to +90

* Experience high radiant loading.

than it is for cold, thus even northern evergreen species will survive 40 to 50 °C. Poikilohydric organisms, such as lichens, may be more tolerant in the dry condition but, when hydrated, they have similar heat limits to vascular plants. The only organisms exceeding these survival temperatures are the thermophilic prokaryota of hot springs which may tolerate 80 to 90 °C (Tansey and Brock, 1978). Few eukaryotic algae tolerate more than 40 °C and cold ocean algae have unusually low temperature maxima (Table 8.3).

The generalized ecology of plants is a reflection of the interaction of the selective forces to which they have been exposed for many generations and, because the range of temperature oscillation is relatively predictable in most habitats, the ecological differentiation of high and low temperature tolerance is manifested as a plant geographical characteristic, hence the possibility of classifying into 'temperate' or 'tropical' temperature response groups. Wild plants may encounter temperatures which are injurious but rarely lethal to whole populations: it is only in transplant experiments and amongst horticultural and agricultural plants that such fatalities are frequently encountered with plants which are outside their normal climatic range. Figure 8.5 is a summary of the ecological groupings and possible physiological mechanisms by which plants survive temperature stress.

Control responses

Day-length and temperature provide reference signals by which plant function may be programmed. Breakage of winter dormancy, onset of spring growth, induction of flowering and the timing of autumnal leaf-fall are all examples in which the failure of the programme would expose the plant to harmful or lethal circumstances. In high latitudes the need for winter dormancy and the synchronization of reproductive growth with the most favourable season is a prerequisite of survival. This might seem less so in the tropics but climatic data reveal the fact that only a quarter of the total area has a continuously humid climate in which seasonal growth might be unnecessary: the remaining three-quarters of the tropical area is substantially seasonal with well marked dry

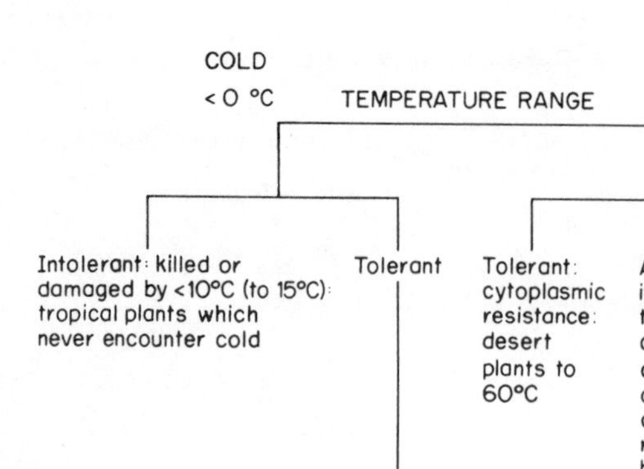
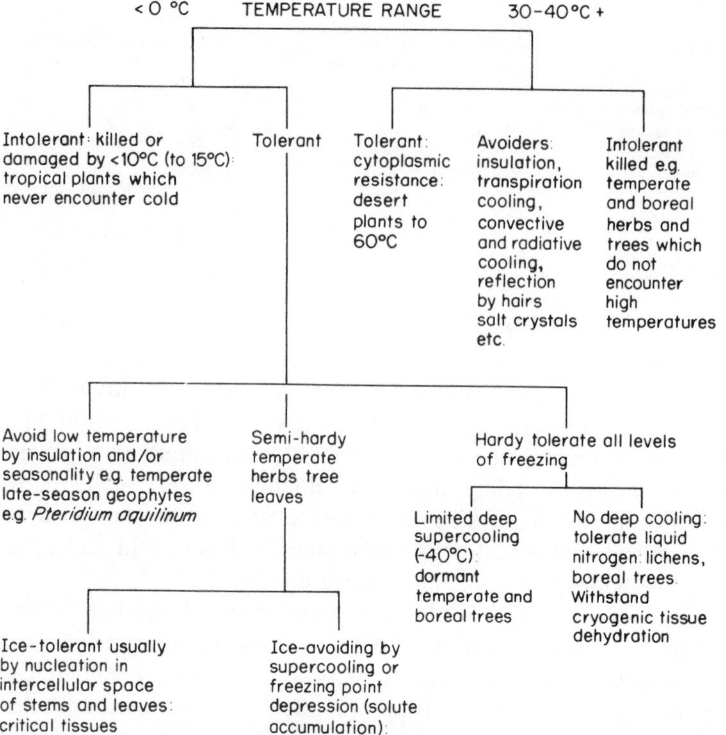

Figure 8.5 A synopsis of cold and heat tolerance mechanisms in plants.

periods, in some cases so extended as to give a desert environment (Sanchez, 1976).

Plant growth, even in the moist tropics, is not necessarily continuous and most plants show one or more 'flushes' of growth each year. These may be under photo- and thermoperiodic control or related to endogenous growth regulators. All tropical plants experience a uniform length 'short-day' photoperiod and most respond physiologically as short-day plants (see below) but many are sensitive to even the slight seasonal changes in the tropical day length, for example, flowering in some varieties of rice is inhibited by a 10 minute change (Sanchez, 1976).

There is then a world-wide need for control mechanisms which adjust vegetative and reproductive growth to periods of optimum temperature and water supply and impose dormancy when cold or drought and high temperature are beyond the limits for survival. The mechanisms which have evolved are the sensing of short-days or long-days, sensing of the period and magnitude of diurnal temperature change and seasonal 'accumulation' of a temperature experience (often expressed as day-degrees beyond a threshold temperature). In some cases, most often seed germination, presence or absence of light has a control function and in all cases the sensing mechanism is affected by the spectral quality of the light.

Adult plants pass unfavourable periods in a dormant state which may range from a shallow *pre-dormant* condition which may be imposed by the gradual onset of cold or drought, through deep *innate-dormancy* which is related to the growth regulator status of the plant, to *post-* or *imposed-dormancy* in which innate dormancy has been broken by an environmental signal but the conditions have not yet improved sufficiently for growth to re-start. These stages may be ecologically interpreted as permitting the plant to re-emerge from dormancy if favourable late-season conditions occur, for example, an unusually warm autumn, but the onset of innate dormancy prevents growth from being initiated by an untimely midwinter warm period or mid-drought rainfall. Finally imposed-dormancy, which only commences following a suitable timing stimulus, permits immediate growth when the environmental conditions improve.

It is probably simplistic to make this interpretation, because different species of plant show a variety of multifactorial responses to the complex of photo- and thermoperiodic signals but, because dormancy and its breakage are so strongly related to species, climate and even to the provenance of the individual, it is reasonable to assume that the system best adjusts vegetative growth and reproduction to the average of many seasons' climatic oscillation.

Herbaceous species of high latitudes which survive winter as dormant underground structures or freezing-tolerant shoots generally enter dormancy as a response to decreasing autumn day length or, in the case of spring and early summer-flowering geophytes, to increasing day-length and temperature (Vegis, 1963). The northern temperate grasses *Molinia caerulea* (purple moor grass) and *Nardus stricta* (mat grass) overwinter respectively either as a buried root system with surface tiller buds or as an entire, but dormant, plant. In both cases spring growth can only start after innate dormancy has been broken by a few weeks of winter cold below a threshold temperature (Cooper, 1965). Similarly the bulb-forming and rhizomatous geophytes of temperate climates (e.g. *Narcissus pseudonarcissus*—daffodil; *Convalaria majalis*—lily of the valley) require 12–16 weeks below 10 °C to break dormancy (Went, 1957; Rees, 1972) and contrast strongly with tropical bulbs such as *Hippeastrum* spp. which may commence growth whenever the *prevailing* temperature permits. The need for several weeks below threshold temperature has considerable ecological value, preventing the 'fine-tuning' of dormancy breakage to a single low temperature exposure which could have disastrous consequences of precocious development.

Plants which encounter summer drought may have a high-temperature imposed dormancy, for example, *Phalaris tuberosa* (tuberous Canary grass) and *Trifolium subterraneum* (subterranean clover) of Mediterranean climates, enter a temperature-controlled dormancy in early summer. The reason why many hot-climate plants lack winter hardiness is that they have the wrong temperature requirements to experience winter dormancy if planted in temperate climates. The lack of dormancy in the frost-sensitive *Festuca arundinacea* (tall Fescue) North African populations (Robson and Jewiss, 1968) is a good example.

The control of dormancy in cold climate woody plants is essentially similar to that of herbaceous species, shortening autumn day-length being the usual trigger for formation of overwintering buds, winter cold experience breaking winter dormancy, and increasing temperature and day-length finally initiating spring growth. The majority of north temperate tree genera such as *Salix*, *Populus*, *Larix* and *Betula* show this pattern of short day dormancy with winter cold breakage and it is also found in temperate-zone fruit trees such as *Malus pumila* (apple), *Prunus domestica* (plum) and *P. persica* (peach). Typically the winter experience must be between 10 days and 2–3 months below 0–7 °C (Kramer and Kozlowski, 1960) and failure to satisfy this requirement, coupled with short day induction of dormancy, is a problem often encountered when such trees are grown in low latitudes.

In some cases increasing daylength of the spring may replace the cold requirement serving as a 'fail-safe' dormancy breaker in unusually warm winters but there is a variety of response, thus in *Fagus sylvatica* (beech) short days induce dormancy and this may then be broken by increasing day-length without an intervening cold period but the innate dormancy of *Pinus sylvestris* (Scots pine) can only be broken by a period of winter cold (Wareing, 1969). This is likely to be related to the more northerly distribution in Europe of *P. sylvestris* than *F. sylvatica.* Woody plants, even intraspecifically, are rather closely adapted to natural day-length conditions and both *P. sylvestris* and *Picea abies* (Norway spruce) show marked ecotypic differentiation of photoperiodic dormancy induction in populations from different latitudes and altitudes (Woolhouse, 1969b).

Photoperiod and temperature also have morphogenetic effects related to vegetative growth and the onset of flowering. Some of these seem to have a rather finely defined relationship with plant distribution, for example Woodward (1979a, b) compared the British upland grasses *Phleum alpinum* (alpine timothy) and *Sesleria albicans* (blue moor grass) with two lowland species, *Phleum bertelonii* (timothy) and *Dactylis glomerata* (cocksfoot). The lowland species were sensitive to temperature between 0 and 20 °C but the upland species were not: the differential sensitivity was by a morphogenetic effect of temperature on leaf area ratio, rate of leaf initiation and expansion. Temperature sensing in the *Phleum* spp. seedlings resides in the coleoptile and first leaf and operates by controlling specific leaf area. Woodward's results in part explain the distribution of these grasses, the temperature sensitive lowland species being unable to perform satisfactorily in the colder upland environment. The morphogenetic effect of temperature on leaf area ratio occurs in other types of plant, for example Woodward and Pigott (1975) showed a similar temperature differential effect between the two semi-succulant species *Sedum telephium* (orpine) of lowland habitats, and *Sedum rosea* (roseroot), a montane plant.

Many plants are sensitive to diurnal fluctuation of temperature as well as the mean temperature. Went (1944) termed the phenomenon thermoperiodicity and also drew attention to the remarkable effect of night temperature (nyctotemperature) on some species, for example the Californian annual *Baeria chrysostoma* (goldfields) grows well at a day temperature of 26·5 °C but the same temperature is lethal at night (Lewis and Went, 1945). Went (1957) explained the sharp reduction of species diversity at the timberline of the Sierra Nevada in California by the critical change of nyctotemperature within a few hundred metres of 3300 m, the timberline altitude.

The classic study by Clausen *et al.* (1948) of the *Achillea millefolium* complex (yarrow) in North America showed very clearly that climatic ecotype differentiation was substantial. They sampled at different altitudes on a latitudinal transect across central California at 38 °N and established the photoperiodic and temperature responses of the different populations. The *A. millefolium* complex contains the hexaploid *A. borealis* from the Pacific coast to the Sierra Nevada foothills and from there, eastward, the tetraploid *A. lanulosa*. The different populations are exposed naturally to about the same photoperiod but experience very different diurnal and seasonal temperature regimes. A race of *A. borealis* from the cool Pacific coastal plain showed a 36 % reduction in height growth in response to increasing night temperature from 7 to 17 °C while a race from the inland, warm San Joaquin valley produced the opposite response, the 7 °C night depressing the growth by 63 %. Along the remainder of the transect similar correlations of temperature response to climate of origin were found and, as might be expected, no substantial differences in photoperiodic response. Collections of northern and southern races showed similar differentiation but with more involvement of day-length. The plants responded to different controlled environments in flower initiation as well as growth.

The experiment discussed above showed that, in later stages of development, temperature experience and photoperiod have a strongly controlling influence on the induction of flowering. As a generalization it has been observed that low latitude species have a short-day requirement for flowering and, as with dormancy induction, this may be inhibited by brief illumination during the night and is really a long-night phenomenon (Hamner, 1940). By contrast many higher latitude plants have a long-day requirement and this is often compounded by the need for a winter cold experience termed *vernalization.* The ecological value of these differences is again obvious, flowering of temperate and boreal species being calendared to the summer period, the vernalization requirement preventing untimely autumn flowering. Tropical plants may have a dry season to evade and this would usually be achieved by a sensitive short-day strategy preventing flowering in the slightly longer days of the dry season.

Vernalization refers to the accelerated induction of 'ripeness to flower' by a period of cold and it may be experienced at any age from seed to adult plant in the pre-flowering condition. Ripeness to flower is attained when all conditions except one have been satisfied for flowering to occur: the final condition is usually photoperiod. The term vernalization is used only when the response is inductive (delayed) and is not applied to immediate responses to low temperature. The phenomenon was observed in wheat, but not named, as long ago as the 1830s: the Russian equivalent of the term vernalization was contributed in 1929 by T. D. Lysenko (Salisbury, 1963).

Because of its agricultural and horticultural implications a great deal of work in vernalization has been with cultivated plants: in the case of winter- and spring-cereals which behave as winter- and spring-annuals the response is open to ecological interpretation as also is the conversion of the first year rosette of many biennials to a ripe-to-flower, erect second-year morphology. Examples are (Salisbury, 1963) *Hyoscyamus niger* (henbane), *Oenothera biennis* (evening primrose) and *Beta vulgaris* (sea beet and cultivated forms). They are generally mid- and high latitude species, and the vernalization requirement permits 1 year of vegetative growth before the relatively massive seed production of the second year. Harper (1977) suggests that the biennial strategy is interpretable as an extension of the winter annual habitat in which growth is made both in the autumn of one and the spring of the next calendar year. Because they have the potential of surviving beyond 2 years if prevented from flowering and seeding, biennials may best be described as monocarpic perennials which are usually able to complete the life cycle in 2 years. The winter annual habit of the ancestors of cultivated cereals, and of many other plants, is suited to an annually unstable habitat which provides bare soil and freedom from competition. The extension of this behaviour to bienniality, particularly coupled with seed longevity, exploits the less frequently disturbed but still unstable 'wasteland' habitat which was originally provided by dunes, cliffs and sporadic drought or fire damage. In winter annual, summer annual and biennial habit, vernalization provides a most suitable signal for locking reproduction to the seasonal cycle, modulated by an endogenous 'ripe to flower' effect and to a final requirement for increasing day length.

Prince *et al.* (1978) and Marks and Prince (1979) have described an excellent example of vernalization interacting with other temperature and light effects to synchronize the life history of *Lactuca serriola* (prickly lettuce) with the seasonal cycle. Clapham *et al.* (1962) describe *L. serriola* as an overwintering or biennial herb but Prince *et al.* refute this with the suggestion that it is either a winter or a summer annual according to circumstances: in a sense intermediate between annual and biennial in behaviour. Figure 8.6 shows that the plant sheds seed in late summer and these, having no dormancy, germinate immediately unless buried. The plants grow throughout the winter and are vernalized as rosettes. In spring they elongate and flower in response to increasing day-length and die after seed production: a winter annual behaviour pattern. Seed which overwinters in the soil is vernalized by 28 days at 2–4 °C, germinates in spring and flowers the same year as a summer annual. Buried seed which is vernalized in winter does not germinate unless exposed to light and it becomes devernalized by a short exposure to 25 °C. If this seed is uncovered during the autumn it germinates but untimely flowering does not occur, because of this devernalization.

It is interesting to note that the sand-dune and grassland biennial *Hyoscyamus niger* (henbane) which has been classic material for the study of vernalization (Salisbury, 1963) shows the same division of behaviour, some plants acting as summer annuals, but in this case the difference is genetically controlled by a single gene: the annual race has no cold requirement and flowers in the late summer of

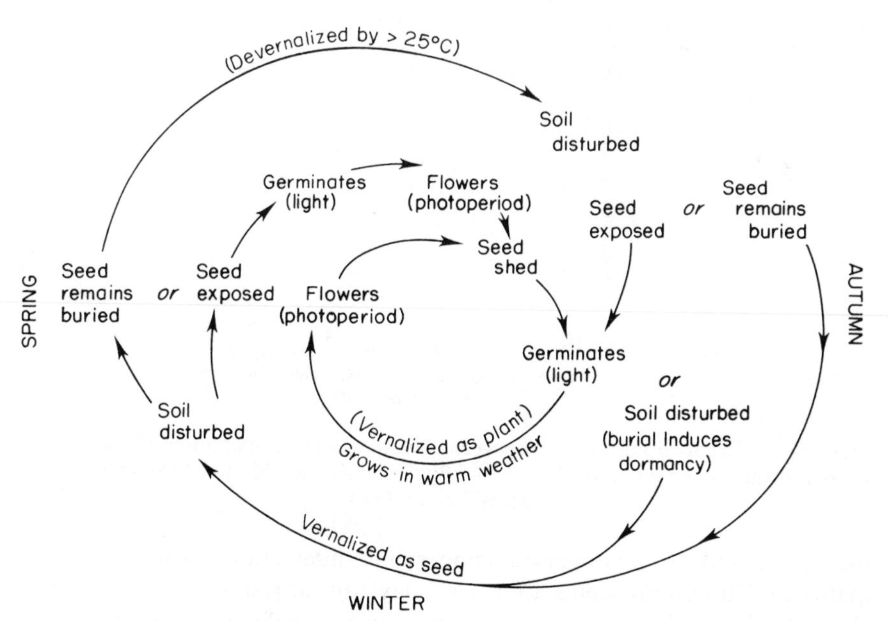

Figure 8.6 The annual behaviour (outer cycle) and winter annual behaviour (inner cycle) of *Lactuca serriola* in response to varying environmental conditions. Controlling factors are indicated in brackets. Adult plants which fail to flower are killed by frost during the next winter. Data of Prince *et al.* (1978) and Marks and Prince (1979).

its germination year but the biennial race remains permanently vegetative unless the rosette receives winter cold. This vernalization requirement is absolute as also is the need for long days to permit flowering in both races.

The temperature requirement and duration for vernalization varies with provenance and is closely correlated with the prevailing winter temperature and the need to programme flowering to the optimum summer period. The Mediterranean grass *Phalaris tuberosa* (tuberous Canary grass) shows an inverse relationship between necessary vernalization time and the lowest winter temperatures (Figure 8.7). Thus plants from more extremely Mediterranean climates with a warmer winter and drier summer are ready to enter spring growth and flower sufficiently early to avoid summer drought.

In summary, it may be said that flowering is locked to the seasonal cycle by a sophisticated interaction of vernalization, photoperiodic and thermal responses, all of which show continuous variation within and between species.

By selection over many generations this has produced an enormous diversity of flowering behaviour which is suited not only to the pattern of the average season but also has co-evolved with the life-cycle timing of pollinating insects. Salisbury (1963) has tried to represent the diversity of control by factorial interaction of 11 photoperiodic response types, 7 temperature regimes including diurnal

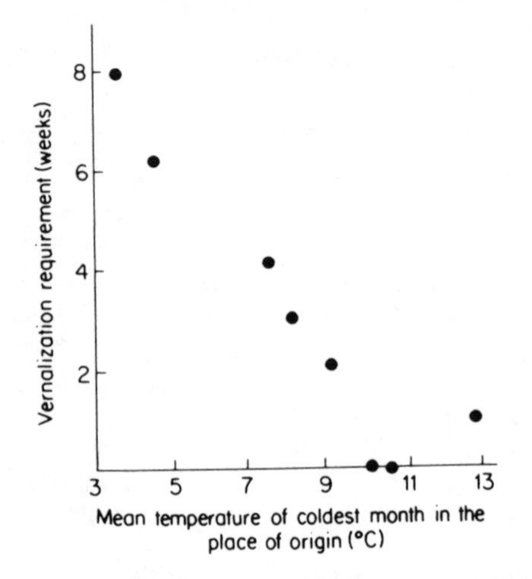

Figure 8.7 Negative correlation between lowest winter temperatures of habitat and vernalization time required by varieties of *Phalaris tuberosa*. Modified from Cooper and McWilliam (1966).

oscillation and 10 photoperiod–temperature interactions. This produces a matrix of 770 responses plus a further 7 day-neutral responses to temperature alone! Little wonder that all plants seem to behave differently and that there is scope even for a wide range of intra-specific variation. A recent review with many examples of photoperiodic and temperature responses is given by Vince-Prue (1975).

Seeds: germination

The plant seed is a survival capsule for genetic material which gives protection from environmental hazards and also uncouples some species from the effects of time by providing a soil seed-bank which has a half-life ranging commonly from months to decades. Despite the necessarily very low metabolic rate of such deeply dormant seeds they must be capable of sensing the environmental changes of temperature, light and wetness which provide not only the conditions for germination but satisfy the longer-term demands of establishment and growth.

Control mechanisms; temperature, light and dormancy

Seeds are very often innately or inducibly dormant when first discharged and may become buried in soil or plant litter. In the case of most summer-annual and perennial species of seasonal climates, the dormancy is discharged during the winter months by experience of cold or by endogenous biochemical changes. As the temperature rises during the spring the seed needs to 'know' whether it is sufficiently near the surface for successful germination: this signal may be provided by light or perhaps by the sensing of diurnal temperature variation which will be greater near the surface.

Huiskes (1979) presents germination data for *Ammophila arenaria* (marram grass) exposed to constant and diurnally fluctuating temperatures following different periods of cold treatment. Constant temperature is unfavourable (Figure 8.8) but a low temperature oscillating regime gives substantial germination after a long cold treatment. A higher temperature oscillating regime gives good germination almost irrespective of cold pretreatment. These results are entirely consistent with the ecology of the plant which germinates freely in the spring in the surface sand of mobile dunes where temperature amplitude is large. *A. arenaria* is a temperate species and most winters would provide 35–50 days at 5 °C. Burial by only 1 cm of sand inhibits germination by 50 % and it is likely that

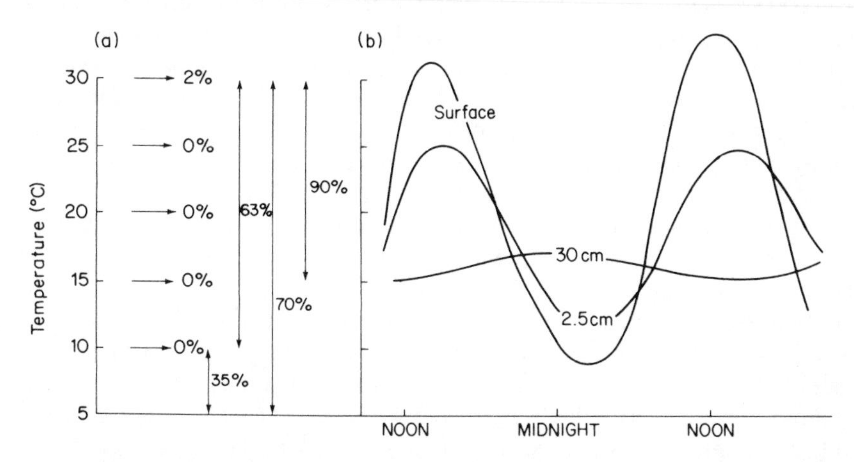

Figure 8.8 (a) Percentage germination of *Ammophila arenaria* seed in different temperature regimes. Horizontal arrows signify constant temperature and vertical arrows are diurnal oscillations between two temperatures. Data of Huiskes (1979). (b) Diurnal changes of temperature at the surface and at 2.5 cm and 30 cm depth in bare dune sand; S. Britain, June (adapted from Willis *et al.* 1959a). Comparison with (a) suggests that germination is likely to be maximal if seed is buried between 0 and 2.5 cm deep.

sensitivity to temperature fluctuation serves as a depth gauge. Dormancy breakage by burial and low temperature is often termed *stratification* consequent on the horticultural practice of burying dormant seeds between layers of moist sand during the first winter after harvesting: effective temperatures range from 1–10 °C.

Dormancy may, in some seeds, be imposed by high temperature: a typical pattern in winter annuals of the temperate zone and of many Mediterranean plants which need to escape summer drought. Winter annuals are germination-inhibited by the soil temperatures which occur during summer in their normal habitats. These temperatures also promote after-ripening so that, as temperatures fall in late summer, a flush of germination begins. Any seeds which do not germinate are returned to dormancy by stratification (1–10 °C) in winter and consequently cannot germinate until the following autumn (Baskin and

Baskin, 1976). In some cases there is a marked interaction between soil temperature, after-ripening and soil moisture, thus species such as *Teesdalia nudicaulis* (shepherds cress), which after-ripen quickly, then become very sensitive to soil water stress (Newman, 1963).

Harper (1977) has suggested that a reserve of dormant seeds in the soil may be a measure of the unpredictability of the habitat; thus the fixed dune system provides a high level of predictability from year to year and loses most of its soil seed reserve annually. By contrast the disturbed soils which give a home to weedy annuals were, before the advent of agriculture, highly unpredictable and contain enormous seed banks of great longevity. Roberts and Stokes (1966) reported up to 86,000 seeds m^{-2} in soil under vegetable crops, but the flush of germination following ploughing is now rarely observed as weedkillers are routinely used in agriculture. The long persistence of such seed banks may often be seen when old grassland sites are disturbed, for example in road construction, to be followed by several years' spectacular growths of weeds such as *Papaver rhoeas*, *P. dubium* (poppy) and *Sinapis arvensis* (charlock) despite previous cultivation having been several decades previously. Wesson and Wareing (1967) describe such a flush of growth on ploughing of 6-year-old grassland, the species including not only *S. arvensis* but *Stellaria media* (chickweed), *Polygonum aviculare* (knotgrass) and *Veronica persica* (grey speedwell).

Buried seeds are in a state of deep dormancy which is broken after soil disturbance, usually by exposure to light. Wesson and Wareing (1967) showed that only two of 22 common weed species germinated in the dark and even these germinated more freely in light. Many of the species which they studied have no light requirement when the seeds are first shed, but burial induces dormancy: opportunist weedy species may thus germinate immediately if the conditions are suitable or accumulate as a seed bank which is drawn upon after disturbance. The induction of dormancy by burial may be caused by a gaseous inhibitor, possibly ethylene, which has been shown to act as both dormancy inducer and germination promoter in different species (Olatoye and Hall, 1973).

The presence of a seed bank may buffer plant communities against species loss following environmental catastrophes such as fire and drought, though its species composition is frequently not well correlated with that of the existing vegetation (Harper, 1977). In many cases the seed bank is of earlier successional species, for example the seeds of *Juncus tenuis* (rush) and *Danthonia spicata* (oat-grass) were present in all stages of North American old-field succession from 1 year after abandonment to 80-year-old *Pinus strobus* (whitepine) forest (Livingston and Allessio, 1968). Tree seeds are not so common in soil seed banks and, when present, usually represent successional species such as *Betula verrucosa* and *B. pubescens* (birches) in northern Europe. Farnsworth and Golley (1974) comment that successional species are also the usual tree seeds in tropical forest systems.

The influence of a soil seed bank may be compounded by the existence of a bud bank thus, in northern European heathland systems, recovery from fire is by germination of the copiously produced and preserved seed of various ericaceae (*Calluna vulgaris*: ling, and *Erica cinerea*: bell heather) and by seedlings of *Ulex* spp. (gorses) but also by regeneration of all of these from the burned stem bases (Fritsch and Parker 1913; Fritsch, 1927). *Pteridium aquilinum* (bracken), probably one of the world's most successful invasive plants (Harper, 1977) also survives fire by buried buds.

Seed dormancy and germination is a complex topic involving the interaction of endogenous biochemical controls with a complex of environmental factors. Reviews may be found in Kozlowski (1972a, b, c) Heydecker (1973), Kahn (1977) and Taylorson and Hendricks (1977).

Control mechanisms; canopy and light quality

Control of germination, growth and flowering by light is strongly wavelength dependent because it relies on specific absorption characteristics of light receptor pigments. The *phytochrome* pigment consists of a protein molecule with an attached chromophore light sensor which absorbs red light (600–700 nm) with consequent conversion to a far-red (700–760 nm) absorbing form. This change is reversible by exposure to far-red or occurs spontaneously and slowly in the dark:

$$P_r \underset{730\,\text{nm}}{\overset{660\,\text{nm}}{\rightleftharpoons}} P_{fr}$$

The phytochrome 'low-energy' response is extremely sensitive and may be triggered by low light intensities or very short exposure to higher intensities. It is usually saturated by irradiances of less than $100\,\text{W}\,\text{m}^{-2}$ and often less than $10\,\text{W}\,\text{m}^{-2}$ (Smith, 1975).

Plants are commonly exposed to much higher irradiances than this and it is to be expected that other light sensitive systems may operate in these conditions. Smith (1975) describes these as 'high-energy' responses which are involved in a wide range of control processes including germination, photoperiodism, leaf and stem extension. Their specific sensitivity is in the blue (400–500 nm) part of the spectrum and also in the far red: the blue system may involve a caretenoid or flavin receptor while phytochrome is probably the high-energy, far-red sensor.

Over 40 different physiological processes in the Angiosperms are cited as showing response to the low-energy phytochrome system (Smith, 1975). In some cases the same process shows either inhibition or promotion, by exposure to red light, according to species and it must be assumed that adaptive selection has resulted in a variety of biochemical control responses to the P_r:P_{fr} balance according to environmental need. Even within species there is considerable variation, for example the polymorphism of seed germination response to light (and other factors) which spread the period of germination and permit the species to take best advantage of short-term environmental fluctuation (Harper, 1977).

The ecological significance of the low and high-energy sensing mechanisms is that they allow the plant not only to register exposure to light but also to measure the duration of exposure and to sense the change in quality caused by shading under a leaf canopy. The enrichment of sub-canopy light with far-red has already been discussed in chapter 2 (p. 23): it may be expressed as the ratio of quantum flux density at 660 nm to that at 730 nm. Monteith (1976) cites this as 1·1 for full sunlight and only 0·1 after passage through a single leaf. Most work in this field has been physiological but recently the ecological implications have begun to be studied. Changes in light quality under canopies have been experimentally related to plant morphogenesis, for example shade-intolerant plants are more sensitive to far-red promotion of stem and petiole elongation (incipient etiolation) than shade-adapted plants (Holmes and Smith, 1977a, b; Morgan and Smith, 1979) and seeds of many open habitat species will not germinate if exposed to far-red light, presumably preventing suicidal germination beneath a canopy (Smith, 1973; King, 1975; Frankland, 1976). Heterophylly in some submerged aquatics is controlled by the red:far-red ratio (Bodkin *et al.* 1980) and sub-canopy light modifies the growth of some Bryophytes (Hoddinot and Bain, 1979). Experimental work is difficult and open to criticism, since spectral simulation of sub-canopy light is difficult, but it is reasonable to conclude that one function of phytochrome is 'to tell the plant where it is'.

The high energy far-red response may serve as the timing sensor in photoperiodism: P_r and P_{fr} cycle at a rate which is related to irradiance and the cycle ceases in the dark (Vince-Prue, 1975) but more physiological work is needed in this field. Reviews of these topics may be found in H. Smith (1975, 1976).

PRECIPITATION AND EVAPORATION: CLIMATIC WETNESS

Climatic wetness is, in a sense, secondary to, and controlled by, radiant income and temperature regimes which would be manifested even in the absence of water. However, the abundance of water and its physically anomalous behaviour makes earth's surface very different from that of the adjacent planets and provides an essential basis for all known forms of terrestrial life.

Water is anomalous, amongst related hydrides, in its elevated melting and boiling points caused by hydrogen bonding between molecules, which is substantial even in the liquid condition. At terrestrial temperatures liquid water can thus exist in quantity, and a further anomaly, the temperature of maximum density being 4 °C above the melting point, is crucially important, since ice would otherwise sink and thermal turnover could not prevent substantial, probably total, freezing of the oceans.

With the exception of aquatic and sub-aquatic systems the availability of water for plant growth is governed by the precipitation: evaporation ratio and by low (sub-zero) temperature. The consequences of water deficit or excess are discussed in chapters 6, 7 and 10: this section is limited to a brief discussion of the terrestrial distribution of precipitation:evaporation patterns and factors which affect them.

According to the size and latitude of land-masses, the precipitation:evaporation regime varies substantially as a function of the proximity of oceans and the effect of wind systems in water vapour transport (Figure 8.9). This generalized

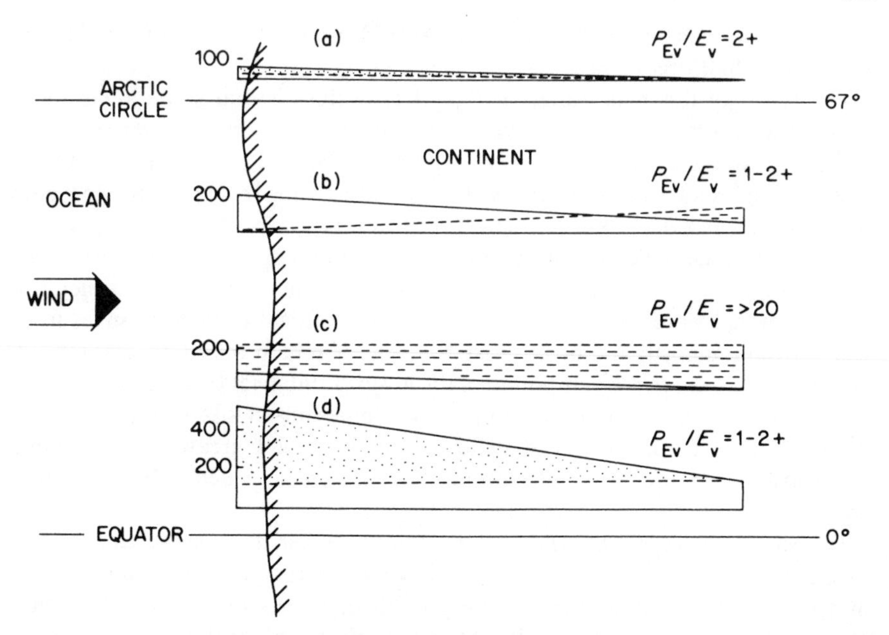

Figure 8.9 Geographical distribution of precipitation, potential and actual evaporation in tundra (a), temperate (b), arid sub-tropical (c), and humid tropical (d) ecosystems. Precipitation ———. Potential evaporation - - - - -. Water surplus indicated by stipple and deficit by broken hatching. Data of Budyko (1974) and Lockwood (1974).

pattern is reflected in the vegetation zones of the earth, but within land masses, there are micro-variations related to both altitude and the interaction of high ground with wind-systems. Thus, in a country such as Britain which is about 1000 km in length and not more than 500 km wide, but situated in a steep gradient of climatic change from oceanicity to continentality, the highest annual precipitation is 5000 mm and the lowest 500 mm (Atkinson and Smithson, 1976). This is as large a ratio as any between major world climatypes with the exception of extreme deserts.

Evaporation (E_v), shown in Figure 8.9, represents actual water loss when limited by water availability. Potential evaporation is the maximum possible loss from a free water surface and potential evapotranspiration (P_{E_v}) that from a closed vegetation canopy with unlimited water supply (p. 32). The value of P_{E_v} is overridingly governed by solar radiation and temperature and consequently has a quasi-sinusoidal diurnal and seasonal fluctuation (Ward, 1976). Over the oceans evaporation reaches nearly its theoretical value in relation to latitudinal solar radiation distribution (see map in Lieth and Whittaker, 1975). On land, E_v may be much less than P_{E_v} if it is limited by aridity, thus in tropical areas

$$P_{E_v} \equiv E_v = 1500\text{–}2000 + \text{mm y}^{-1}$$

but in the arid sub-tropics though evaporation from the oceans may be of this same magnitude, the adjacent land-mass E_v may be less than 100 mm y^{-1} and even in temperate climates with $P_{E_v} = 1000–1500\,mm\,y^{-1}$, aridity may reduce E_v to less than half this value.

The result of this wide range of P_r and P_{E_v} values is that some areas suffer substantial soil moisture deficits which impede plant growth and cause surface enrichment of soils with solutes from capillary rising soil solution (p. 92). Other areas with an excess of P_r over E_v show strong soil leaching (p. 48) and may develop bog or other hydromorphic soil and vegetation types (p. 75 ff).

Britain may again be used as an example of the possible range of variation in a limited geographical area. The mean annual soil moisture deficit (difference between actual water content and water content at field capacity) ranges from near zero in the north and north-west, to over 150 mm in the south-east (Ward, 1976), This last figure is a compound of the lowest rainfall ($500–600\,mm\,y^{-1}$) and the highest P_{E_v} of more than $550\,mm\,y^{-1}$ with some run-off loss during the winter months. In arid zone climates deficits may be much greater, for example $P_r = 100\,mm\,y^{-1}$ and $P_{E_v} = 1000\,mm\,y^{-1}$ would create a deficit of more than $900\,mm\,y^{-1}$.

The ecological impact of varying P_r/E_v ratios is seen in its most extreme form in the differentiation of various world vegetation types ranging from totally water-limited desert systems to the profuse growth of forests in the humid tropics. Because P_r is, on a very local scale, of low variability, minor differences between vegetation types are often related to the value of P_{E_v} and consequently to topography, because steep-gradient sunward slopes receive a great deal more solar radiation than north-facing slopes which, in Northern temperate and higher latitudes experience no direct sunlight during most of the year if steeper than *c.* 30° (Geiger, 1965). These differences have a direct influence on soil temperature which is reflected in soil moisture status, for example Geiger (1965) cites data for Northern Europe which show soil water contents of only 10 % w/w during August and September on south-facing slopes but values of *c.* 20 % on equivalent north slopes. The relationship is not entirely simple, since long-wave radiant exchange, wind effects and their interaction with receipt of precipitation all play a part in controlling the effects of topography and aspect on soil water (Geiger (1965) presents an extensive discussion).

Tansley (1939) described the deeper, moist soils which occur on the shaded northern slopes of the chalk downs in South-east England and Perring (1959; 1960) was able to group plant species according to their aspect occurrence on chalk as well as demonstrating the effects of aspect and slope on soil depth, calcium carbonate content and other characteristics. The occurrence of arctic-alpine species in Britain, at the warmer limits of their ranges, is often confined to former glacial corries which owe their existence and present day microclimate to their north or north-easterly aspect. Gleason and Cronquist (1964) show a dramatic photographic example in the southern Corderillas (U.S.A.) of south-facing, sunward slopes which carry a rather sparse grassland of *Agropyron spicatum* (blue-bunch wheatgrass) while adjacent northern slopes are heavily forested with conifers (probably *Pinus ponderosa*—ponderosa pine). They also describe the incursion of *Populus tremuloides* (aspen) onto northern slopes at the

Forest boundary of the prairy grassland province. Both of these examples are attributable to aspect control of soil water availability.

Goodman and Bray (1975) describe the failure of revegetation of mining-spoil related to high soil temperature which may reach 60 °C with an air temperature of 29 °C. North-facing slopes show better establishment and east–west ploughing may be used to give cooler, moister ridges.

WIND

Transfer of mass, momentum and heat in the canopy and its relationship to wind has already been discussed in chapter 2 (p. 30) but wind has other ecological impacts which include direct modification of physiology and morphology, windthrow and resultant modification of soils and ecosystem successional status, control of altitudinal tree limits by exposure, secondary effects through snow-drifting and transport of pollen, seed, dust, seaspray, pathogens, predators and air-pollutants.

Grace (1977) reviews the experimental evidence for wind modification of plants which seems to be an anatomical trend to xeromorphy with increasing development of sclerenchyma, reduced specific leaf area and consequent increase in stomatal frequency (stomata mm^{-2}). Anatomical responses are particularly pronounced in the woody tissues of trees. Extensive zones of reaction wood form, their adaptive significance being to confer strength, often by increasing diameter in the direction of the prevailing wind. Extreme wind-exposure causes remarkable changes in morphology of trees, ranging from minor deformation of the canopy in the direction of the prevailing wind to complete 'flagging' of the crown in which no branches develop on the windward side. This may be due to branch death through desiccation of immature tissues, or windward branches may be so deformed by the production of reaction wood that they grow downwind around the trunk. The effect has been used in mapping the micro-distribution of prevailing wind direction (Oliver, 1960; Holroyd, 1970; some useful photographs are reproduced in Tranquillini 1979).

Windthrow, the overturning of a tree and the upper part of its root system, is a usual consequence of senescence but may also affect young or vigorously mature trees if they root in situations which limit rooting depth, such as waterlogged soils, shallow soils over solid rock or an impermeable 'pan'. It is a common problem in afforestation of difficult soil types: in 1953 for example, a single storm caused windthrow of about one million m^3 of timber in Northern Scotland with a standing value (1953) of £4 $\times$ 10^6 (Steven, 1953). A similar gale in 1968 caused an even larger loss of 1·6 million m^3, the worst affected areas having experienced 6–7 hours of wind in excess of 100 km h^{-1} (Holtam, 1971). Trees of between 6 and 12 m in height, on upper windward slopes with surface-water gleys or peat soils were at worst risk.

The soil movement caused by windthrow is substantial, a single event completely overturning 5–10 m^2 of complete soil profile. Pedogenesis is deflected, a new microrelief created and germination sites for successional species provided.

Denny and Goodlet (1956) attributed the natural regeneration of *Pinus strobus* (white pine) in parts of presettlement Pennsylvania, U.S.A. to such effects.

Tree limits and the altitudinal limits of other species are controlled by wind-exposure though lower temperatures at high altitude also play a part. The formation of wind-flagged and cushion-form trees (krummholz) at the timberline is a characteristic effect of wind- and frost-induced desiccation damage: typical examples are *Picea englemannii* (Englemanns spruce) in North America and *Pinus cembra* (Siberian pine) in the European alps. The short growing season at timberline altitudes prevents the maturation of some newly formed tissues which enter the winter in a frost- and desiccation-sensitive condition, very often succumbing on the windward side to produce the flagged krummholz condition (Wardle, 1974; Tranquillini, 1979). Wind exposure may limit other species, for example *Pteridium aquilinum* (bracken) in Northern Europe shows very characteristic upper limits which are related to local topography and wind. The plants also show changes of morphology and anatomy caused by wind exposure (Bright, 1928).

Persistent snowdrifting in arctic and arctic–alpine habitats is a function of wind direction and topography and produces well-defined patterns of plant distribution. Snow cover provides winter insulation and a constant supply of melt-water during summer but also shortens the available growing season. It also modifies the quality of light available just prior to complete thawing. The gradual recession of snow-beds during the early summer produces a characteristic concentric pattern of plants. In Northern Europe, the innermost area is bare soil surrounded by a liverwort and moss community (typically *Anthelia juratzkana* and *Polytrichum sexangulare*) followed by *Salix herbacea* (dwarf willow) and then *Vaccinium myrtillus* (bilberry). The most windswept areas which may even be free of snow in winter carry freezing-tolerant species such as *Empetrum hermaphroditum* (crowberry) and *Loiseleuria procumbens* (dwarf azalea).

Wind-transport plays an important part in the distribution of some pollen and seed, for example pollen has been collected over the mid-Atlantic (Erdtman, 1937) and pollen counts from the Bishop Rock lighthouse south-west of the Scilly Isles have occasionally been quite large (Hyde, 1956). The nature of the dissemination of particulates from a point source has been extensively studied in relation to pollen, seeds, dust and aerosols. Generally the concentration declines with distance according to a negative power-law (Chamberlain, 1975), for example grass pollen had fallen to 40 % of the source concentration at 60 m, and would be less than 1 % at 1 km downwind (Raynor *et al.* 1972). Despite this depletion, pollen of anemophilous species is produced in such quantity that substantial amounts remain airborne for long periods. During high pollen production periods the count may reach $1000\,m^{-3}$ corresponding to about $14\,\mu g\,m^{-3}$ which may be compared with the usual $20–60\,\mu g\,m^{-3}$ of particulates and aerosols encountered in urban air (Chamberlain, 1975).

The long-range transport of pollen and dust-sized or 'parachuted' seeds has considerable ecological significance for plant breeding and dispersal. It is perhaps no coincidence that some of the highly competitive species, which colonize disturbed habitats, such as *Chamaenerion angustifolium* (narrow-leaved willowherb) and *Tussilago farfara* (coltsfoot) have such plumed or pappus-bearing seeds (Grime, 1979). The interpretation of palaeo-ecological pollen spectra must be undertaken with circumspection for the same reason.

The spread of spore-borne diseases is akin to pollen dispersal and, in some cases, is closely related to prevailing wind direction. The long-distance dissemination of *Puccinia graminis* (wheat stem rust) in North America is a classic example, uredospores sometimes spreading suddenly for a thousand or more km from the south, autumn sown crops in Mexico and Texas infecting spring-sown wheat in the northern U.S.A. and Canada. The return journey is made later in the year (Gregory, 1973). Insect predators are similarly wind dispersed, for example the winged generations of aphids are weak fliers and depend on wind for dispersal and, incidentally, often carry with them plant viral diseases.

Transport of seaspray aerosols is substantial, about $10^9\,t\,y^{-1}$ of sodium chloride falling on land surfaces (Dyrssen, 1972). This may make a contribution to cycling, particularly of trace nutrient elements, but also in conjunction with wind, may injure coastal vegetation: salt spray damage may extend 25–55 km inland (Grace, 1977). A single gale in the spring of 1972 occurred just after the opening of *Fagus sylvatica* (beech) buds in South Wales and caused almost total loss of exposed parts of canopies for over 10 km inland (Etherington, unpublished data). There is some difficulty in separating the effects of salt and wind (Boodle, 1934) but close to the coast the characteristic blackening of young leaves may be seen in even relatively wind-sheltered spots which receive seaspray, suggesting that the salt effect is considerable.

The transport of atmospheric pollutants is a strongly wind-dependent process and may involve long distances, thus sulphur dioxide from midland England and the German Ruhr has been implicated in the acidification of rain and surface waters in Scandinavia (Engstrom, 1971; Hutchinson and Havas, 1980). Gaseous pollutants will travel further downwind than particulates and aerosols, though SO_2 will gradually oxidize and, as the sulphate, gathers atmospheric moisture and ends its travels as a more easily impactable or precipitable aerosol.

Chapter 9

Mineral nutrition

INTRODUCTION

Plant physiologists have collected an enormous amount of data concerning the uptake and influence of the various mineral nutrient elements but, because of the complexity of the soil–plant system, the majority of these studies have been made in solution culture. Many investigations of nutrient uptake have, indeed, been made with excised roots. Despite these artificial approaches, much of the accumulated information has had implications for the ecologist, as well as defining the physiological problems of nutritional relationships. Of greater interest for the ecologist are the early field experiments which established, for many crop plants, the gross effects of nutrient elements on plant growth and physiological function. Work of this type was originally stimulated by Liebig's ideas concerning the mineral nutrition of plants and is directly responsible for the development of the modern artificial fertilizer industry. Russel (1973) provides a useful historical account of the agricultural development of plant nutritional studies.

The major advances during the second half of the nineteenth century were the realizations that plants require phosphates and salts of the alkali metals, that non-leguminous plants have a high requirement for nitrogen compounds and that artificial fertilizers incorporating these materials could maintain crop yield, for many years, in the absence of organic additions. Further developments, with even stronger ecological implications, were the discoveries that different crops had different inorganic nutrient requirements and that nitrogen might be 'fixed' from the atmosphere by the organisms in leguminous plant root nodules and by free living organisms in some soils.

The ecologist's interests in plant nutrition are threefold, comprising: (i) behavioural studies of the soil–plant system in the context of naturally occurring nutrient ranges; (ii) investigation of differential responses between and within species; (iii) investigation of competitive relationships between and within species (Clapham, 1969). Most early nutritional physiology has not helped in this sphere as crop plants or convenient 'laboratory plants' were used. However, the techniques of whole-plant physiology and the methods of soil analysis, originally developed with agriculture in mind, have provided an extensive background of experimental method with which to tackle these new questions.

Bradshaw (1969) stresses the importance of nutrition as an ecological factor by noting that there are virtually no British grasslands which do not respond to the addition of one or more of the plant macronutrient elements. This observation may be extended to the majority of the soils and vegetation types of the world; if water is not a limiting factor, nutrient additions produce a response, either in yield or in species composition. In each case this suggests that nutrient deficiency is a limiting factor in either photosynthetic productivity or competitive ability.

Recent ecological investigations suggest that variation in soil nutrient levels, at all scales, influences species composition and growth of vegetation. At the smallest scale, vertical variation in the soil profile and horizontal variations, within a few centimetres, occur, and are of great theoretical importance in plant competition and the maintenance of species diversity by niche differentiation (see Chapter 13). On a larger scale the differing plant associations of various soil types partially owe their existence to differences in nutrient status.

Responses to local or regional nutrient variation may span the range from deficiency to toxicity and result in differential behaviour of plant species or ecotypes. The consequent relationships of nutritional physiology and competitive efficiency, cause complex interactions between cohabiting species.

NUTRIENT REQUIREMENT

Plants require 15 to 20 essential elements and the absence of any one of these will cause failure. Three of the elements, C, H and O, which are the constituents of the structural and primary energy storage compounds, are not usually described as nutrients. The remaining elements are subdivided into the macronutrients, N, P, K, S, Ca and Mg which are required in comparatively large amounts, and the micronutrients such as Cu, Zn, B, Cl, Mo, Mn and Fe which are required in smaller amounts.

All of the nutrient elements, with the exception of N, are usually derived from the weathering products of the soil minerals, though it should be noted that rainfall may contain significant amounts of nutrients, particularly S and Cl and, under extreme conditions, may be the sole source of plant nutrients in ombrogenous soils. Nitrogen differs in its ubiquitous presence as atmospheric N_2, from which soil-dwelling, or root nodule bacteria and other microorganisms, may fix it as organic N.

Most, if not all, plant nutrient elements are absorbed from soil-solution in the ionic form: the common metals are taken up as cations and the macronutrients N, P and S as oxyanions. This distinction is ecologically important as it is related to the nutrient conservative characteristics of soils, thus one convenient classification of nutrient and other elements absorbed by plants is 'cationic' or 'anionic'. Another possibility is classification related to physiological function, thus the covalently bonded elements N, P and S which have an essentially structural role may be separated from those which exist as salts and complexes (Table 9.1).

Table 9.1 A functional classification of some nutrient elements absorbed by vascular plants (modified from Clarkson and Hanson, 1980)

Covalently bonded with C, H and O		Salts and complexes			
			Conformation control		
Reduced	Oxidized	Osmotic and ion-balance	Enzymes	Other molecules	Redox function
N	P	K	Mn	Mg	Fe
S	S	Ca	Zn	Ca	Cu
		Mg	Cu	Cu	Mo
		(Na)	Mo		Mn
		(Cl)			

The three elements of the former group confer special properties by their covalent bonding into organic molecules. Nitrogen may be considered to 'substitute' for carbon but its five valency electrons introduce an asymmetry which underlies the highly reactive properties of many biological molecules. Phosphorus occurs in the fully oxidized tetroxyanion form and serves as a strong covalent linkage or binding site through the resonating phosphoryl bond (Clarkson and Hanson, 1980). Most functional sulphur is in the reduced sulphydryl form and acts as an oxygen 'substitute' which contributes to metal complexing, redox properties and protein stabilizing activity under redox rather than hydrolytic control

Those elements which enter ionic and reversible complexing reactions with biological materials may be sub-grouped into those which have simple osmotic or ion-balancing roles, those which influence molecular conformation including that of enzyme proteins and those which facilitate redox reactions. (Table 9.1).

Ecologists have rarely discussed the physiological role of essential elements, a surprising omission as the selective pressures which have resulted in a specific element filling a particular niche in organism and ecosystem function are of fundamental ecological interest and may help in elucidating the geological history of biological evolution. Many major ecosystem processes would either not exist or would be very localized without the involvement of trace amounts of specific elements with specialized functions.

A striking example is molybdenum, a group-5 transition metal and the heaviest element demonstrably essential to plants. It is involved in plant nitrate reductase enzymes and also in the nitrogenase enzyme of N-fixing bacteria and actinomycetes. During succession, the increase of nitrogen fixing and absorbing capacity is keyed to these two enzymes and consequently is under the control of minute amounts of molybdenum. Similarly, the first appearance of these enzymes in the geological past opened the door to massive increases of biological carbon fixation over that greater part of earth's surface which is inherently deficient in combined nitrogen.

Another recent biochemical hypothesis has similar implications for the ecological history of land-plants. Lewis (1980) has suggested that boron has an essential role in the biosynthetic pathway of lignin without which the 'superapoplast' of vascular plants could not have evolved (Raven, 1977). Without xylem, terrestrial plants would be limited to bryophytes and algae and, if the hypothesis is correct, boron is the key to an increase in potential primary production of ten times or more in many habitats (see p. 355).

The changing of the primeval reducing atmosphere to our present-day oxygen-rich system raises further questions concerning the biochemical and ecological options for selection of specific elements in metabolic redox reactions. Clarkson and Hanson (1980), for example, note that copper has been exploited in enzymes for which no other element has the properties to substitute. The same could be said of the molybdenum flavoprotein nitrate reductase, of which the redox potential is in the correct range for nitrate reduction. Would the same evolutionary 'choices' have been made in a different redox environment? an enigmatic ecological question.

MINERAL NUTRIENTS IN THE SOIL

Nitrogen

Most natural soils contain between 0·01 and 0·25 % N in the surface layers (Bear, 1964) but the content is less in the deeper horizons. Some peat soils contain larger amounts: up to 2–3 %. It is usually assumed that most of the soil-N is organic though recent work has shown that ammonium-N may become bound between the lattice layers of expanding clays such as montmorillonite (Bremner, 1967). In some subsoils up to 50 % of the nitrogen may be held in this way.

It is safe to generalize, for surface soils, that most of the N is organic and, on hydrolysis, up to half of it may be recovered as amino-N. The soil-N exists as protein or amino acids, the binding sites being on the clay fraction or amino units may be incorporated in the humic complex molecules; earlier workers often described the humic complex as ligno-protein (Quastel, 1963). Attempts to isolate proteins from soil organic matter have, however, always failed, except for showing the ratios of hydrolytically produced amino acids to be compatible with the constitution of an 'average' protein (Felbeck, 1971, p. 105).

The ramainder of the organic-N is of varied composition, comprising amino sugars, purines and pyrimidines at very low levels, traces of amino acids and numerous other N-compounds. As all of these constituents occur at low concentrations, there remains considerable doubt concerning the composition of at least half of the soil organic-N which does not hydrolyse to amino-N (Parsons and Tinsley, 1975).

Nitrogen is absorbed by roots as nitrate in aerobic soils and, consequently the nitrogen-supplying status of the soil depends more on the rate of mineralization of organic-N than on the total N content. Most of the nitrate-N is in soil solution, very little being adsorbed by any form of anion-binding (Fried and Broeshart, 1967).

Nitrogen fixation; free-living organisms

Most nitrogen in soils originates from fixation by microorganisms because very few parent materials are nitrogen-rich, having originated either from igneous activity at temperatures above which N-compounds are volatile or from sedimentation in oxygen-poor environments which favour denitrifying organisms.

All ecosystems, including extremes ranging from desert crust-soils to antarctic lakes, show some nitrogen-fixing ability (McGregor and Johnson, 1971; Horne, 1971). This was once demonstrable only by analysis for total organic nitrogen but, about 30 years ago, more critical measurement became possible using the mass-isotope ^{15}N as a label (Stewart, 1966). The mass-spectrometry makes this a cumbersome technique and since 1966 it has been supplemented by the introduction of the acetylene-reduction method. The nitrogenase enzyme which is responsible for nitrogen-reduction will also reduce acetylene to ethylene which is easily measured by gas chromatography. Samples are exposed, in field or laboratory, to a 10% acetylene mixture with argon and oxygen. Large numbers of samples may be handled and, coupled with the accuracy of the method, this has resulted in the screening of many organisms and ecosystems for even low intensity fixing ability (Burris, 1975; Paul, 1975).

Nitrogen fixing ability has been demonstrated only in prokaryotic organisms of which the free-living representatives are either bacteria or heterocystous blue-green algae. The bacterial N-fixers are drawn from diverse physiological groups which include aerobes, anaerobes, photosynthetic bacteria and specialists such as some of the anaerobic sulphate reducing bacteria (Stewart, 1973; 1975). The most widespread free-living N-fixers are members of the aerobic Azotobacteriaceae, mainly *Azotobacter* spp. in temperate soils and *Beijerinkia* spp. in the tropics. Nitrogen fixation by these bacteria is favoured by a relatively low oxygen supply as the nitrogenase enzyme is strongly sensitive to free oxygen. For this reason fixation is favoured by soil wetness, similar conditions to those which promote denitrification. *Azotobacter* has a pH optimum near neutrality and is incapable of fixing below about pH 6·0 but *Beijerinkia* has a broader pH spectrum from about pH 3·0 to 7·0 and also a lower calcium requirement (Virtanen and Meittinen, 1963).

The ecological significance of free-living nitrogen fixation is most apparent in pioneer and successional situations. Photosynthetic organisms such as blue-green algae have a substantial colonizing advantage and are widespread as soil-crusts (Mulder *et al.* 1969). In some desert ecosystems such crusts form the only source of fixed nitrogen (McGregor and Johnson, 1971). The most common blue-greens found in soil are species of *Anabaena*, *Nostoc* and *Cylindrospermum*.

Blue green algae are rare below pH 4·5 but this is much too acid for most N-fixing bacteria and consequently acidic ecosystems may depend on algae for much of their nitrogen. This was the case in a Swedish tundra described by Granhall and Selander (1973) in which the blue-green algae were associated with a *Sphagnum* moss carpet in the acidity range pH 4·5 to 6·0.

The high temperature, wetness and neutral soils of lowland-rice culture strongly favours blue-green algae and their economic importance has been shown by soil inocculation experiments in which the induced algal N-fixation has increased rice yield up to 20% (Venkataraman, 1975).

It has been suggested that bacterial nitrogen fixation makes a relatively unimportant contribution to the nitrogen economy of soils as it will be restricted by energy supply. *Azotobacter*, for example, requires about 50 g of carbohydrate to fix 1 g of nitrogen (Campbell and Lees, 1967). This may be the case in agricultural soils but most natural ecosystems are strongly nitrogen deficient and in soils above pH 6·0, genera such as *Azotobacter* may provide some species of plant with survival levels of nitrogen in intensively competitive circumstances. This would be more likely if the bacteria were associated with the rhizosphere zone of particular species. Hassouna and Wareing (1964) suggested that *Azotobacter* might inhabit the rhizosphere of the dune pioneer *Ammophila arenaria* (marram grass) thus providing nitrogen in exchange for organic metabolites diffusing from the root surface.

The tropical counterpart of *Azotobacter*, *Beijerinkia*, is present in greater numbers in the rhizosphere zone and there is evidence that the plants with which it is associated receive appreciable benefit from the nitrogen supply (Mulder *et al.* 1969; Dobereiner and Day, 1975). *Beijerinkia* incidentally has a much wider pH spectrum (pH 3·0–7·0) and lower calcium requirement than *Azotobacter* (Virtanen and Meittinen, 1963).

Nitrogen fixing is favoured by soil wetness and high temperature, thus a temperate zone clay soil showed a hundredfold increase of fixing capacity as its water content increase from 13 to 40%, a change which would also enhance its denitrifying activity (Day *et al.* 1975). The highest free-living fixation values are associated with the wet tropical conditions of rice culture, values between 14 and 69 kg N ha^{-1} being recorded during the growing season. Mature ecosystems are less effective than successional or disturbed habitats, thus values of 0·5 to 5 kg N ha^{-1} y^{-1} have been recorded for a range of natural tundra, grassland and forest systems compared with 39 kg N ha^{-1} y^{-1} for a cleared temperate woodland (these data are presented in Stewart, 1975).

Nitrogen fixation; symbiotic organisms

The largest contribution to biological N-fixation is from bacteria of the genus *Rhizobium* which inhabit root nodules of leguminous plants. These bacteria are more efficient than free-living forms, fixing about 0·15 to 0·25 g N per g C compared with 0·01 to 0·02 g N for *Azotobacter* (Pate and Herridge, 1978; Mulder, 1975). Not only is the C/N efficiency of *Rhizobium* greater but also its N-fixing life is much longer: about 1 month compared with 2–4 hours for *Azotobacter* (Mulder, 1975).

Nitrogen-fixing root nodules are also found amongst about 13 genera of non-leguminous plants. They have been called *Alnus*-type nodules and contain a non-bacterial symbiont believed to be an Actinomycete of the family

Dermatophylaceae (Nutman, 1976). Some of the commonly infected genera are *Alnus* (alder), *Myrica* (bog myrtle), *Hippophae* (seabuckthorn), *Casuarina* (she oak), *Ceonothus* (blue blossom and other names) and *Arctostaphylos* (bearberry or manzanila), most of which are associated with soils in which nitrogen is deficient because of wetness, acidity, aridity or juvenility.

Bacterial symbionts which do not form nodules have also been described, the most important probably being *Spirillum lipoferum* in association with various tropical grasses. With *Zea mays* (sweetcorn) it fixes up to 2 kg N ha^{-1} d^{-1}, a rate quite comparable with that of nodule rhizobia (Von Bulow and Dobereiner, 1975).

Blue-green algae are symbiotic with both vascular plants and other organisms. The most well known is the lichen symbiosis of fungi and algae of which about 10% involve N-fixing blue-greens. Lichens are essentially pioneer and stressed habitat organisms and it seems a little puzzling that so small a proportion should carry the advantage of N-fixation.

The association with vascular plants varies in intimacy, for example *Anabaena azollae* inhabits intercellular spaces in the fronds of water ferns, (*Azolla* spp.) but *Nostoc* filaments occur inside the glandular cells of the petiole bases in *Gunnera* spp. Many of the Cycadaceae have coralloid masses on the roots: these contain either *Anabaena* or *Nostoc*, in this case intercellularly. Further information is given in Carr and Whitton (1973) and Nutman (1976).

More is known of the *Rhizobium* association than any other N-fixing symbiosis. An infected leguminous root forms a parenchymatous nodule by proliferation of pericyclic tissue. When mature this is often surrounded by a vascular sheath and its inner parenchyma cells are packed with branching Y- or X-shaped *Rhizobium* bacteroids. The tissue is coloured pink by the unique pigment, leghaemoglobin which is present in the cytosol surrounding the bacteroids and carries oxygen inward to them. The nitrogenase enzyme is thus placed in an optimum oxygen environment with sufficient to satisfy the large respiratory energy demand but not sufficient to compete for reducing power, a problem which often limits free-living nitrogen fixation in well aerated soils.

The plant vascular system provides metabolic substrate, placing the leguminous plant or other symbient in a delicate cost-benefit relationship with its nitrogen-fixing guest. The host receives a large proportion of the fixed nitrogen, sometimes in excess of 90% as it is exported from the bacteroids, presumably in $-NH_2$ form, and is directly assimilated by the plant (Shanmugam *et al.* 1978).

The overall ecological consequence of symbiotic N-fixation is that the host plants are freed from dependence on combined nitrogen in the soil and are consequently able to colonize, or compete in, very nitrogen-deficient environments. The benefit of the system must be delicately balanced against cost otherwise all plants would be associated with N-fixers or would have developed inherent fixing ability. Agricultural research is currently pursuing the latter possibility, perhaps by 'genetically engineering' bacterial N-fixing ability into the crop-plant genome. The fact that evolution-selection has not arrived at this goal suggests that it may have energetic or other disadvantages.

Though less than that of free-living fixation, the energy cost of symbiotic N-fixation is fairly high at 4·0 to 6·5 g carbon per g nitrogen (Pate and Herridge, 1978). Nodulated roots respire at almost twice the rate of nitrate-fed non-nodulated roots and require 11 % to 13 % increase in daily photosynthetic carbon fixation (Ryle *et al.* 1979). However, this carbon need may be outweighed by the N-mediated gain in photosynthetic production as a survival advantage in a nitrogen-deficient or competitively limited environment.

The nitrogenase enzyme consists of two components, one an iron-protein, the other an iron–molybdenum protein, thus molybdenum is an essential element for all nitrogen-fixing microorganisms and, indirectly, for the vascular plants with nitrogen-fixing symbiont. This has ecological and agricultural significance, many soils being sufficiently deficient to limit fixation. Stewart *et al.* (1978) found about 25 % of Scottish soils showed increased nitrogenase activity on addition of molybdenum at concentrations as low as 0·1 $mg\,l^{-1}$. Iron, and also magnesium, which is necessary, in conjunction with ATP, for the fixation, are unlikely to act as primary limits, being needed in relatively small amounts compared with needs for other plant functions.

Nodulated roots with their high respiratory oxygen demand may exceed the oxygen transporting capacity of the internal air space system of roots in anaerobic soil (p. 318 ff) and it may be for this reason that few leguminous plants inhabit strongly waterlogged soil. One exception to this generalization is the genus *Alnus* of which some species live in very wet soil but nodules rarely, if ever, form on roots below the permanent water-level and may even be above soil level in leaf litter.

Nitrification and denitrification

Except for small additions of nitrate, ammonium and nitrogen oxides from precipitation and atmosphere, nitrogen enters the soil in organic form, largely as the amino-N of proteins and free amino acids. Plants, however, absorb most of their nitrogen as nitrate, ammonium or free ammonia (Haynes and Goh, 1978). A mineralization process thus stands between the plant and its nitrogen supply, a mineralization controlled by soil microflora, wetness, nutrient status and acidity.

A wide variety of bacteria and fungi are able to liberate NH_4^+ or gaseous ammonia from organic-N compounds and, in soils which are too wet or too acid to support nitrifying organisms, this is the end product of mineralization and the plants of these habitats are able to absorb nitrogen in this reduced inorganic form (Haynes and Goh, 1978).

In alkaline and near-neutral soils, nitrate is formed by the process of nitrification and, until recently, plant physiologists assumed that NO_3^- was the 'normal' source of nitrogen. Nitrate is formed by oxidation of NH_4^+ by chemo-autotrophic bacteria, firstly of the genus *Nitrosomonas* to form nitrite (NO_2^-) and secondly of the genus *Nitrobacter* which oxidizes NO_2^- to NO_3^-. At very high soil pH, the activity of *Nitrosomonas* outstrips that of *Nitrobacter* and NO_2^- may accumulate in the soil solution (Campbell and Lees, 1967). A few fungi are also able to oxidize NH_4^+ to NO_3^- but their quantitative importance is not really known.

Nitrogen may be lost from soil by the process of denitrification following the reaction sequence

$$NO_3^- \rightarrow NO_2^- \rightarrow NO \rightarrow N_2O \rightarrow N_2 \quad \text{(Focht, 1978).}$$

The most common denitrifiers belong to the genera *Achromobacter* and *Pseudomonas* which are normally aerobic but can develop under anaerobic conditions if a nitrate supply is available to be used as a source of oxygen in the oxidation of organic matter. At pH 7·0 the critical redox boundary for $NO_3^- \rightarrow NO_2^-$ is about 420 mV and for $N_2O \rightarrow N_2$, 250 mV. Figure 9.10 shows that values in this range may be encountered within a few hours of waterlogging and, equivalently, a small distance below the surface of a waterlogged soil or within the surface of a wet soil aggregate. Thus denitrification is common in waterlogged soils, in anaerobic microsites of wet soil aggregates and in agricultural soils where high nitrate and organic content enhance denitrification under wet conditions.

Soil physical chemistry predestines the system to behave as a denitrifier. Ammonium cations are held by the cation exchange complex and protected from leaching but once nitrified, the NO_2^- and NO_3^- anions are not bound and, with high rainfall, are carried down into the subsoil where oxygen concentrations are lower and denitrifiers are active (Figure 9.1). Gas chromatography has facilitated analysis of the soil atmosphere, permitting detection of low concentrations of

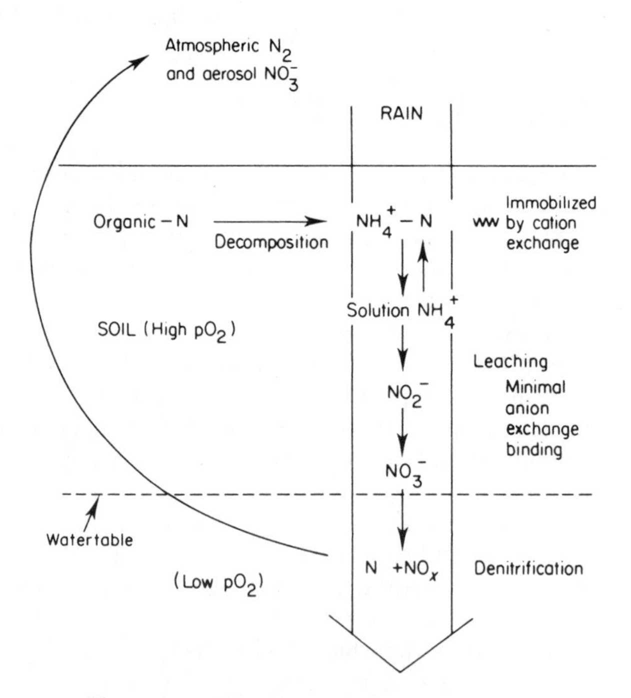

Figure 9.1 The soil as a denitrifying system.

nitrous oxide (N_2O), evidence of denitrifying activity. Values of 10–20 $\mu l\,l^{-1}$ are commonly found in fertile soils during the wetter months of the year.

Nitrogen fixation and soil nitrogen

N-fixation is the major source of N for plant growth though some small contribution is made by photochemical and electrical discharge fixation in the atmosphere which, added to nitrogen compounds from industrial smoke, appears in precipitation as nitrates. Some alkaline soils and waters contribute gaseous ammonia to the atmosphere while denitrifying soils may volatilize nitrogen oxides which are ultimately recycled in rain (see Chapter 12).

The symbiotic fixation of N may be very large, reaching 2–400 kg $ha^{-1}\,y^{-1}$ in some leguminous crops and, in optimum greenhouse conditions, rising to the equivalent of 1000 kg $ha^{-1}\,y^{-1}$ (Virtanen and Miettinen, 1963). Free-living bacteria produce less with a maximum of, perhaps, 70–80 kg $ha^{-1}\,y^{-1}$. In the case of the free-living organisms some excretion of organic-N compounds to the soil does occur but it is not large. This, in addition to limited organic substrate supply, probably accounts for the great difference in fixing efficiency between these and the symbiotic forms. In the latter case much of the fixed-N is removed, as it is formed, into the host's tissues or may even be excreted to the soil though there is some controversy in the literature concerning the latter point. Excretion to the soil is demonstrated experimentally in Figure 9.2: pea plants, innoculated with rhizobia, were cultured in nitrogen-free sand. During the experiment considerable amounts of nitrogen were fixed and, in the early stages of growth, about 30% of the fixed-N was excreted to the sand. As the plants continued

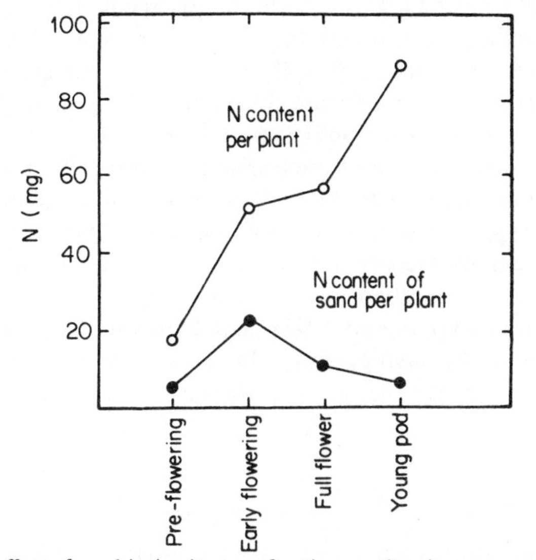

Figure 9.2 The effect of symbiotic nitrogen fixation on the nitrogen content of pea plants and on the nitrogen content of the originally nitrogen-free sand culture medium. Data of Virtanen and Miettinen (1963).

growth to maturity and pod formation, the excreted nitrogen was taken up again. In similar experiments Virtanen and co-workers showed that a non-leguminous plant grown in the same container as a nodulated leguminous plant also benefited from the excreted nitrogen. The results of these experiments are apparently indisputable but, in the field, it is much more difficult to demonstrate that plants are obtaining their N supply in this way, rather than through microbiological breakdown of dead nodule and root tissue. This may seem to be an academic point as the fixed-N ultimately reaches the soil pool of organic-N, but, in relation to competition for N, the availability of even a minimal supply could be very important.

Cultures of *Rhizobium* appear to have optima, for growth, between pH 6·5 and 7·5 while the same range is optimal for N-fixation in the field by species of *Trifolium* (clover) and *Medicago sativa* (medick). Most agricultural legumes fix nitrogen in the range pH 5·0 to 8·0 but many wild members are active below this value, for example *Ulex europaeus* (gorse) may extensively and successfully colonize acid, nutrient-deficient soils. Other wild legumes occur on very nutrient-deficient soils: *Ononis repens* (restharrow) and *Lotus corniculatus* (birds foot trefoil) are common as sand dune plants in northern Europe while various *Lupinus* (lupin) spp. have been utilized as pioneer N-fixers in the reclamation of poor quality and derelict soils. Non-leguminous N-fixers such as *Hippophaea rhamnoides* (sea buckthorn) also occur as colonists of sand dune soils while *Myrica gale* (bog myrtle) may fix at pH values as low as 3·8 with an optimum of pH 5·4. Stewart (1966) notes that *M. gale*, supplied with nitrogen in its rooting medium, has a growth optimum of pH 3·3 though its optimum for nodulation is 5·4. If, however, the pH is lowered after nodulation it continues to fix N, suggesting that the plant may have an efficient internal pH regulating mechanism.

To summarize: the greater part of the global reserves of soil nitrogen has been produced by symbiotic fixation though there is some evidence that free-living organisms may play a more important role than formerly suspected, particularly in natural ecosystems. Walker (1965) goes so far as to say that: 'In most cases the important colonizing plants are legumes, nodulated non-legumes known to fix nitrogen, or other plants with known or suspected mechanisms for fixing atmospheric nitrogen'. Excellent accounts of these topics may be found in a number of publications. General reviews are given by Burns and Hardy (1975), discussion of metabolic mechanisms by Shanmugam *et al.* (1978) while Stewart (1975), Nielsen and McDonald (1978), and Nutman (1976) have edited IBP synthesis volumes on free-living and symbiotic N-fixation. A useful agricultural account is given by Dobereiner *et al.* (1978).

Phosphorus

Natural soils usually contain between 0·02% and 0·5% of phosphorus, most of which is derived from the parent minerals (Bear, 1964). The surface layers of the soil are often enriched with P which has been absorbed by roots and passed to the surface in litter fall. As a result of this biocycling, much of the soil-P is usually organic; Bould (1963) suggests that it may range from 3 to 75% of the total.

Inorganic-P is often present as apatite (calcium phosphate) of which the commonest form is fluorapatite ($Ca_{10}F_2(PO_4)_6$). In alkaline soil, apatite may form 50% or more of the inorganic-P but, as weathering proceeds and pH falls, so the percentage of iron phosphate increases, to reach perhaps 40%. Other forms of inorganic-P are aluminium phosphates and occluded-P, the latter sealed into mineral grains with a coating of insoluble iron oxide. These two forms of soil-P usually represent some 10–20% each of the inorganic-P content. Occasionally, very high soil-P concentrations may be associated with unusual conditions, for example, vivianite (ferrous phosphate) may be precipitated under peat in bog soils while a few soils, derived from biogenic marine deposits or from guano, are immensely rich in calcium phosphate (Bear, 1964; Black, 1968).

The occurrence of other forms of inorganic-P is less well documented, but Jeffery (1964) has suggested that *Banksia ornata*, an Australian sclerophyll-heath plant, may synthesize and retain inorganic polyphosphate in its roots. Polyphosphate is a major storage pool in beech, *Fagus sylvatica*, mycorrhiza (Chilvers and Harley, 1980) but its stability in soil is not known. Under waterlogged, anaerobic conditions inorganic-P may be reduced to phosphine (PH_3) and lost by volatilization.

The common forms of inorganic-P in soil all have very low solubility products and, consequently, dissolved-P rarely exceeds 0·01–1·0 mg l^{-1} in soil solution. Fried and Broeshart (1967) cite data for a number of N. American soils, some of which had received P-fertilization, showing the P concentration range between 0·04 and 0·08 mg l^{-1} to be the most frequent. In the pH range of natural soils most of the dissolved-P is present as the ion species $H_2PO_4^-$ or, in very alkaline soils, HPO_4^{2-}. Because of these low solution concentrations plant uptake rapidly depletes the dissolved-P so that the rate-limiting process is solution from adjacent soil particles. Larsen (1967) also notes that P forms soluble complexes with many metal ions and it is, perhaps, an oversimplification to assume the bulk of the soluble-P to be $H_2PO_4^-$ and HPO_4^{2-}.

The organic-P entering a soil is largely derived from plant litter; though animal remains and faeces may be important under heavy grazing. The known groups of organic-P compounds in this material are inositol phosphates, phospholipids, nucleic acids, nucleotides and sugar phosphates, but only inositol phosphate persists in quantity in soil and even this may be of secondary, microbiological origin (Anderson, 1975b). Cosgrove (1967) notes that organisms occur, in the soil and rhizosphere, which are capable of dephosphorylating all known organic-P compounds of plant origin and that plant roots may also have similar phosphatase activity at their surfaces. Woolhouse (1969a) further investigated this possibility and found differential phosphatase activity between various ecotypes of *Agrostis tenuis* (bent grass). Despite this, a large proportion of the soil-P remains in the organic form, presumably because any inorganic-P which is produced by phosphatase activity is very rapidly mopped up by microorganisms. The fractionation of soil-P is fraught with difficulty and it may be that much of the soil-P, formerly thought to be organic-P, is held in living microbial cells and cell debris (Larsen, 1967).

In concluding discussion of the phosphorus status of soil, it may be noted that it shares with nitrogen the characteristic of being immobilized in organic form and is strongly dependent on soil microorganisms for its natural cycling. Unlike nitrogen, however, its inorganic compounds are rather insoluble and impose a further rate limitation on biological transfer processes and prevents significant leaching loss.

Potassium

Natural soils contain much more potassium than phosphorus or nitrogen. Black (1968) cites values between 0·3 and 2·5 %. Potassium in soils is generally derived from alumino-silicate minerals such as feldspars and micas and, consequently, finer textured soils with high silt and clay content tend to have a higher percentage of potassium.

Chemically, K is a mobile element, its simple compounds being very soluble in water and, were it not for its incorporation in the fairly stable lattice structure of the alumino-silicates, it would rapidly be lost by leaching. In soil only a very small proportion of the K is soluble or exchangeable, the remainder is a non-exchangeable component of the soil matrix. Black (1968) cites a mean figure of 99·6 % of the total –K in the latter form and, of the remaining 0·4 %, 0·17–0·87 me % were exchangeable and 0·003–0·02 me % were in solution. Fried and Broeshart (1967) record solution concentrations between 2 and 6 mg K l^{-1} (0·05–0·15 me l^{-1}) as being the most frequent in North American soils.

The cation exchange complex of most soils carries only a small proportion of K as Ca and Mg are relatively more abundant in the exchangeable form. Cation exchange capacity ranges between about 5 and 50 me % in all but the coarsest textured soils, but it is rare to find more than about 1 me % of K whereas Ca often ranges up to 30 me % and Mg to 20 me % in cation-saturated soils. In highly leached soils the exchangeable K is again low but the balance is made up in exchangeable hydrogen–aluminium. The K in solution is in equilibrium with that adsorbed on the exchange complex and its low concentration protects the soil against leaching loss. This protection is probably made more effective because rainfall contains a certain amount of K and more is leached from leaves in the canopy (see Chapter 12). The K content of the downward percolating water may thus be at a level which inhibits dissociation of K from the exchange complex and root uptake is likely to be from cyclic K in the soil solution, rather than at the direct expense of exchangeable K. In the context of K mobility, organic combination is not important as it is very easily leached from both dead organic matter and living plant parts such as leaves and roots.

Sulphur

Sulphur forms between 0·01 % and 0·15 % of temperate zone soils (Bould,

1963) and there is reason to believe that much of it is immobilized in organic combination (Freny, 1967), not more than 50–500 $\mu g\ g^{-1}$ being soluble sulphate. This contrasts with arid zone soils in which calcium or magnesium sulphates may accumulate after upward transport in capillary water, and evaporative concentration at the soil surface.

Sulphur, in soil, originates from the mineral matrix in which it may occur as various metals sulphides (e.g. FeS_2; ZnS) or as crystalline sulphates. Generally, sulphides occur in igneous rocks and sedimentary rocks which were laid down under reducing conditions and sulphates in sedimentary rocks produced in oxidizing conditions—often as evaporites. In industrial countries a significant amount of sulphur is now added to soil in precipitation, industrial production of SO_2 causing an annual input of up to 70 kg S ha^{-1}. There is a natural rainfall input of 5–6 kg $ha^{-1}\ y^{-1}$ from H_2S released by waterlogged soils, lake muds and, probably most important, continental shelf sediments (p. 304).

Like N and P, S is immobilized in the organic form so that rates of microbiological mineralization of the element are significant factors in pedogenesis and plant nutrition. It differs from P in that sulphate, its common inorganic form in soil, is moderately soluble and rapidly leached in high P/E conditions. Sulphate ions are not held by anion exchange in alkaline soils but some binding may occur in acid conditions. Availability from free sulphate and rainfall input is generally greater than from organic-S mineralization.

A wide range of organic-S compounds is formed by living organisms and may enter the soil in litter. These include S-amino acids, sulphonium compounds, sulphate esters, sulphur-containing vitamins and antibiotics, organic sulphides, sulphoxides and isothiocyanates (Freny, 1967). The majority of these are rapidly decomposed and become indetectable in soil. The amino acids cystine and methionine are produced on acid hydrolysis of soil, suggesting that they may occur in polypeptides or proteins incorporated in the humic complex or bonded to clay particles. About half of the organic-S is in this form while the remainder is possibly present as sulphate esters though there is considerable doubt about its true composition (Anderson, 1975a).

Plants absorb S as sulphate though there is some evidence that S-amino acids may also be assimilated. The mineralization process is, consequently, of considerable interest but no definite pathways for conversion of S-amino acids to sulphate (aerobic) or sulphide (anaerobic) have been identified (Freny, 1967; Anderson, 1975a). Some knowledge of environmental influences on mineralization of S has accrued, for example increasing pH and/or temperature (between 10 and 35 °C) increases SO_4^{2-} release, while drying of the soil slows the process. These observations suggest that mineralization is bacterially mediated.

Calcium

The calcium content of soils is widely variable: rendzinas and chernozems derived from Ca-rich parent material or enriched with evaporites may have

$CaCO_3$ contents of 40–50% (16–20% Ca), while acid sandy soils often contain less than 0·5% Ca and no free carbonate. In non-carbonate soils the Ca originates from the alumino-silicate and Ca-Mg silicate minerals with the consequence that Ca and clay content are often correlated.

In rich soils some 70–80% of the cation exchange complex may carry Ca ions and in non-carbonate soils this may be a relatively large proportion of the total soil-Ca. Black (1968) records a group of soils with a mean Ca content of 17·3 me % as having 23% of this Ca on the exchange complex. The equilibrium concentration in the soil solution is also high compared with other cations such as K. Fried and Broeshart (1967) suggest a mean value of about me l^{-1} for Ca compared with 0·1 me l^{-1} or less for K.

Calcium is, pedogenetically, a significant element in the sense that the presence or absence of calcium carbonate may be diagnostic of the P/E regime. This concept is applicable only to parent materials which are rich in free $CaCO_3$, but, as limestones are geographically ubiquitous and carbonate-rich loess deposits widespread in mid-continental areas, the concept of carbonate leaching is most useful in interpreting present-day soil conditions. Incoming rainwater is in solution equilibrium with the atmospheric CO_2 and, as it enters the surface layers of the soil, may dissolve more. The downward percolating water may be able to carry up to 250 mg l^{-1} of $CaCO_3$ by solution as $Ca(HCO_3)_2$ (Perrin, 1965).

If the P/E is much greater than unity, the surface layers of the soil may be completely leached of free $CaCO_3$ and an advancing front of decalcification passes down the profile at a rate governed by the P/E ratio, the original carbonate content of the soil, temperature regime and seasonal distribution of the rainfall input. If the P/E ratio falls much below unity during the summer months then upward movement of capillary water in response to surface evaporation causes $CaCO_3$ (and other salt) enrichment of the *A* horizon. Even if the winter precipitation is high it may not reverse this summer effect as the leaching process is less efficient at low temperatures or, more commonly in midcontinental areas, the soil is frozen.

Biocycling of calcium is important in restricting leaching loss by returning Ca to the soil surface but there is no question of serious organic immobilization of Ca as it is quickly released by the decomposition process. Its more important biological implication is in relation to the composition of the soil biota: both earthworms and bacteria are much reduced in numbers by low Ca status. Though Ca is not immobilized in organic matter, it does show chelation reactions with a number of organic acids, for example, citric and gluconic (Stevenson, 1967). Formation of Ca-chelates may have important effects on both pedogenesis and plant nutrient uptake.

Before leaving the discussion of Ca in soil it should be noted that $CaCO_3$ is the commonest soil constituent responsible for high levels of soil alkalinity, many $CaCO_3$ soils having pH values between 8·0 and 8·4. The relationship between such high pH values and plant nutrient availability is discussed later (pp. 263 ff).

Magnesium

Magnesium behaves similarly to Ca in soils being derived from alumino-silicate, silicate or sulphate minerals on non-carbonate parent materials or from dolomite ($MgCa(CO_3)_2$) and magnesite ($MgCO_3$). The total soil content of Mg, like Ca, is widely variable, ranging from 0·003 % to 0·6 % Mg in normal soils (Bould, 1963) to much higher values in dolomite soils where 1·2 % Mg may be exceeded (Bear, 1964).

Exchangeable Mg in soils is usually less than exchangeable Ca despite the fact that the total Mg content of non-carbonate soils often exceeds total Ca. Bear (1964) gives a mean value of 1·6 me % for exchangeable Mg and 40·7 me %g for total Mg in a number of North American soils. The concentration of Mg in the soil solution is similar to that of Ca at about 10 me l^{-1} (Fried and Broeshart, 1967).

Iron

Though iron is a micronutrient for plants it has a much wider pedogenetic and microbiological significance than its quantitative uptake would suggest. Iron makes up some 0·7–4·2 % of temperate zone soils and between 14–56 % of many tropical latosols. Small percentages of iron occur as lattice components of the clay alumino-silicates but the greater part is present either as hydrous ferric sesquioxide in particulate form or as coatings on other mineral particles. Under anaerobic conditions iron may become soluble in the divalent (Fe^{2-}) form, may form organoferrous complexes or precipitate as FeS, $FeCO_3$ or $Fe(OH)_2$ as shown in Figure 9.3. In normal neutral or alkaline soils the Fe^{3-} concentration in the soil solution is extremely low, but falling redox potential may allow Fe^{2-} to be produced, sometimes rising to toxic concentrations of several hundred mg l^{-1} (see p. 302). Similarly, exchangeable ferric iron is normally at rather low levels but reducing conditions permit Fe^{2-} to occupy some of the cation exchange sites. In these soils, iron availability is strongly dependent on redox status, but in acid soils ferric iron is more soluble, even under oxidizing conditions, so that iron deficiency is usually a characteristic of well-drained calcareous soils.

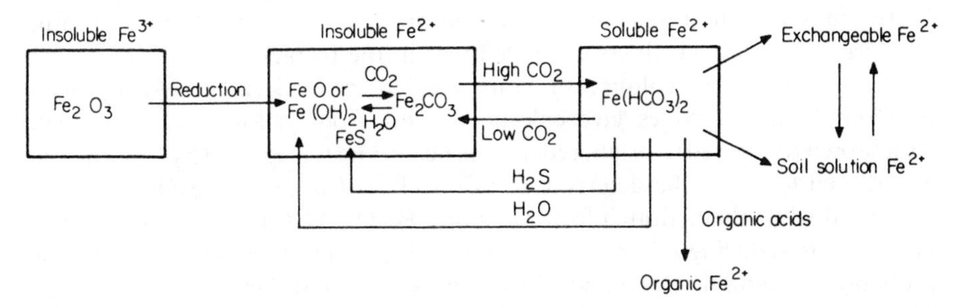

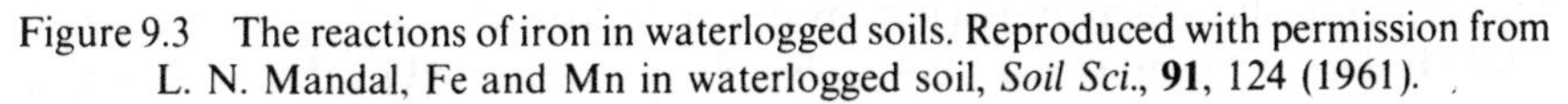
Figure 9.3 The reactions of iron in waterlogged soils. Reproduced with permission from L. N. Mandal, Fe and Mn in waterlogged soil, *Soil Sci.*, **91**, 124 (1961).

The distribution of ferric sesquioxide in soil is often a useful parameter of pedogenetic processes as its solubility, relative to that of other soil components, is widely variable. In alkaline to neutral conditions iron is not easily mobilized other than in waterlogged soils where low redox potentials permit its conversion to the more soluble ferrous form. In consequence, well-drained soils have silica/sesquioxide ratios close to those of the original parent material (Table 9.2).

Table 9.2 Silica: sesquioxide ratios ($SiO_2/(Fe_2O_3 + Al_2O_3)$) of various soils

Podsol	A_1 17·9	A_2 41·7	B_1 15·3	B_2 19·0	C 21·6	Mean $A_1A_2B_1B_2$ 23·5
Chernozem	A 4·6	B_1 4·9	B_2 3·8	C 4·1		Mean AB_1B_2 4·4
Oxisol	1 Parent rock 3·2	2 Inner weathered crust 0·16		3 Outer weathered crust 0·14	4 Laterite 0·23	Mean 2,3,4 0·18

Acid, temperate zone soils may show podzolization with iron loss from the *A* horizon and redeposition in the B horizon. Though the silica/sesquioxide ratio of the *A* horizon is increased and that of the *B* horizon reduced, the mean ratio for the whole profile is not much altered unless soluble iron is flushed away to the ground water. The only case of massive Fe_2O_3 loss from temperate zone soils occurs, with waterlogging, if ferrous iron is flushed downward to a *laterally moving* water table.

Under high temperature regimes and high rainfall in the tropics the situation differs as the silicate minerals are very unstable and a large loss of silica, probably as colloidal silicic acid, occurs. The ratio of silica/sesquioxide becomes very low, the Al and Fe sesquioxides of the parent minerals being resistant to decomposition and some synthesized *in situ* from ionic Al and Fe liberated during alumino-silicate breakdown. This process of *laterization* appears to be most common on base-rich parent materials and leaves a residue of *primary laterite* containing little else but iron and aluminium sesquioxides and a few resistant parent minerals (Mohr and van Bahren, 1959). Nearer to the water table, primary laterite may become resilicified, by deposition of silica from solution, to form *argillaceous laterite* which covers wide areas in the tropics.

The characteristic colours of soils are, in general, conferred by iron compounds. Sesquioxides give colours ranging from yellow-brown, through dark brown to rust red to bright red or pink according to degree of hydration. The dehydrated form, so characteristic of oxic soils, is *haematite* (Fe_2O_3) which is bright red while the hydrated form, *goethite* ($Fe_2O_3.H_2O$) is brown or reddish brown. Most reducing soils have a drab olive-grey, greenish or bluish tint which is probably attributable to organoferrous compounds and they are often mottled with ochreous spots of hydrated iron oxides or hydroxides. In the presence of free sulphide dark colorations from grey to black may be caused by the formation of FeS or FeS_2.

The migration of iron under podzolizing conditions has been the source of much controversy and various workers have considered it to be: (a) mobilization as inorganic ferrous iron in response to changing E_h–pH conditions, (b) movement of soluble organo-iron complexes formed from humic compounds or (c) movement of soluble complexes formed with organic leachates from fresh litter. It has also been suggested that iron may be eluviated in colloidal form, either alone or in association with organic materials.

Bloomfield (1965) maintained the view that humic compounds are not involved, except as an energy source for microorganisms, and postulated that iron is reduced to the ferrous condition by polyphenols leaching from fresh litter and then travelling down the profile as organo-iron complexes with polyphenols. Microbiological activity in the *B* horizon probably decomposes the organo-iron complex, initially liberating ferric hydroxide as coatings on mineral grains and intercalated between soil particles. The bacterium *Pedomicrobium* appears, commonly, to be associated with this process (Aristovskaya and Zavarzin, 1971).

The mobilization of ferrous iron in waterlogged soils is generally caused by a fall in redox potential and production of reducing organic compounds by a non-specific microbial population. The bacterium *Gallionella*, which oxidizes ferrous iron, is responsible, amongst other organisms, for the accumulation of ferric concretions and mottles in gley soils, for the formation of bog-iron deposits and for the precipitation of hydroxides in natural waters.

Manganese

In soil, manganese behaves similarly to iron and consequently shows similar patterns of distribution in profiles, related to pH–E_h variations and valency–solubility change. The total manganese content may vary from a few $\mu g\, g^{-1}$ to about 1 % though this is no index of its availability as it may occur in the di- tri- and tetravalent state of which only the first is particularly soluble. Manganese exists in soil as the higher valency oxides, as manganous carbonate and hydroxide or as exchangeable or soluble Mn^{2+} (Ehrlich, 1971). The manganous ion begins to appear in solution at a pH below 8·0 and a corresponding E_h of 600 mV.

Many microorganisms are involved in the oxidation and reduction of the soil Mn and are probably responsible for maintaining the availability of Mn^{2+} in well-oxidized, high pH soils. The reduction of Mn to the divalent form does not necessarily imply that it will appear in solution as $MnCO_3$ and $Mn(OH)_2$ may precipitate above neutrality. Microorganisms are also responsible for the oxidative production of manganese concretions in poorly aerated and gleyed soils. Manganese deficiency, like iron deficiency, is most commonly associated with high soil pH.

Copper

Copper, in soils, derives from various primary minerals, the commonest being chalacopyrite ($CuFeS_2$). Its concentration may range from 0·1–1000 $\mu g\, g^{-1}$ in mineral soils though less than 1 $\mu g\, g^{-1}$ is likely to be in solution. Its solubility is much reduced by high pH and by the presence of carbonates and sulphides (Ehrlich, 1971). In solution it may exist as the Cu^{2+} cation in equilibrium with the exchange complex and it may also appear in solution as an organic complex. Deficiency of Cu is usually associated with low concentration in the parent material.

Zinc

Zinc occurs in a wide range of primary minerals and is easily released by weathering either as the Zn^{2+} cation or as an organic complex. The normal soil content is between 10 and 300 $\mu g\, g^{-1}$. Above pH 5·0 its availability may be reduced by precipitation of $Zn(OH)_2$. In acid soils much of the Zn content is exchangeable but in calcium soils a large proportion may only be extracted by dilute acid treatment (Bould, 1963). Deficiency of Zn is associated with a low Zn content of parent minerals.

Boron

Boron in soils originates from borosilicates, calcium and magnesium borates and iron and aluminium boron complexes. The total content is usually between 2 and 100 $\mu g\, g^{-1}$ (Bould, 1963) and it may appear in solution as an equilibrium mixture of *ortho*-, *meta*- and tetraborate anions (Colwell and Cummings, 1944). Usually less than 3 $\mu g\, g^{-1}$ is soluble as borate but much higher values occur in arid zone saline soils where boron toxicity may occur. Boron deficiency is associated with light, sandy, easily leached soils.

Chlorine

Until comparatively recently the essential role of chlorine as a plant nutrient was not recognized (Hewitt, 1963). It occurs ubiquitously in soils as the Cl^- anion and its concentration is often similar to that of NO_3^- and SO_4^{2-} in the soil solution. In the saline soils of salt deserts and salt marshes the concentration of Cl, associated with Na, may be very high, reaching toxic concentrations and sometimes appearing as a crystalline efflorescence at the soil surface or within soil pores.

Molybdenum

Molybdenum occurs in natural soils as the sulphide or the molybdate, usually entering soil solution as the molybdate anion. It may, like phosphorus, become immobilized by the production of insoluble aluminium or iron molybdates. The usual soil content is 1–10 $\mu g\, g^{-1}$ but most of this is insoluble. Deficiencies may be associated with absence from the parent minerals or fixation, particularly in acid soils. A few calcareous, P-rich soils show excessive Mo availability which causes 'teart' disease in cattle grazing on such land. Molybdenum is also an essential element in bacterial N-fixation.

Other elements of ecological interest

Various non-essential elements play a part in soil formation and plant ecology. Quantitatively, the most significant of these is *Aluminium* which forms a part of the structural lattice of most clay minerals but also enters the exchange complex of some acid soils. Aluminium commonly forms 5 % to 15 % by weight of the soil but the Al^{3+} cation does not appear in solution above pH 5·0 (Moore, 1974). More acid soil solutions contain over 1 to 3 mg Al^{3+} l^{-1}, toxic concentrations for many plants. Differential tolerance is a part of the mechanism controlling the distribution of calcicole and calcifuge plants (p. 263 ff).

Silicon, as silica or silicates, makes up 20 % to 30 % of many soils but monosilicic acid in soil solution rarely exceeds a few mg l^{-1}. The Equisetales, the Gramineae and a few other plant groups accumulate silica to between 2 and 20 % of the dry weight. Despite this there are only a few indications that it is essential to higher plants (Hoffman and Hillson, 1979) though it is often a limiting element for the diatoms with their siliceous frustules. Accumulation in most terrestrial plants seems to be a passive accompaniment to the transpiration stream, thus in wheat (*Triticum vulgare*) the total Si content of lemmas and glumes is very strongly correlated with total water use (Hutton and Norrish, 1974).

Sodium is ubiquitous in soils, even ombrotrophic bog waters containing 10–20 mg l^{-1}. The soil solutions of mineral soils often contain many times this concentration and in salt-marshes the value may exceed 10,000 mg Na l^{-1}. In extreme conditions salt may crystallize and produce efflorescences in salt-desert conditions, demanding a solution equilibrium of more than 14 g Na l^{-1}. Despite its widespread occurrence and essentiality for animals, sodium is not demonstrably essential for other than a few halophytes and perhaps some CAM plants (p. 289). Its major ecological significance is at high concentrations when it may create the osmotic and toxicity problems associated with halophytism (p. 285).

A wide range of elements which are not essential to plants, but are taken up by them, has been studied in relation to toxic effects in particular where industrial use is spreading them through many ecosystems. Examples are lead, mercury and cadmium, all of which have a diversity of industrial uses. Lead, after combustion of lead tetra-ethyl and -methyl 'anti-knock' fuel additives in the presence of ethylene dibromide, is broadcast to the environment as microparticulate lead bromide or oxide which may accumulate as local concentrations on vegetation or in soil and may also enter atmospheric circulation. A similar problem of dissemination arises with mercury, the insoluble oxides and sulphides of which were once thought to be immobile and relatively non-toxic. In anaerobic systems such as lake and ocean sediments, particularly in the presence of hydrogen sulphide, volatile methyl mercury is generated by methanobacteria and may be widely transported. Cadmium, which appears as a particulate accompaniment to zinc-smelting dusts and in solution as a waste product of metal plating and other industries, is not so extensively mobile in the environment but it is very toxic to organisms and, behaving analagously to zinc, is very easily taken up by plants. Further consideration of these toxic metals is given on p. 280 and 369.

The elements nickel, chromium, arsenic and selenium have attracted ecological attention because they are associated with specific geological-soil environments. Ni and Cr are common in soils derived from the group of ultramafic 'serpentine' rocks and, worldwide, are associated with 'serpentine barrens' of which the depauperate vegetation has been explained in part by Ni, Cr and perhaps Co toxicity (Proctor and Woodell, 1975; and p. 281). Arsenic is frequently associated with igneous rocks and their metamorphic aureole, for example in Cornwall, South West England (Applied Geochemistry Research Group, 1978) where toxicity to plants and animals has been recorded. The fourth element, Se, is interesting for two reasons: firstly it is selectively accumulated by some species of *Astragalus* (locoweed) and poisons grazing stock. Secondly, it is associated with uranium deposits and its accumulation by plants has been used in biogeochemical prospecting (Cannon, 1971).

A non-metallic toxic element, *fluorine*, is currently attracting attention as an environmental hazard deriving from high-temperature firing of alumino-silicate clays, mainly in the brick industry and from the electrolytic manufacture of aluminium from bauxite with the addition of cryolite (Na_3AlF_6). It is essentially an atmospheric pollutant, emitted as gaseous fluorine or hydrogen fluoride: in this form it has free access to vegetation, unlike the calcium and aluminium fluorides which occur in natural soils but are very insoluble.

It is not only inessential elements which pose toxicity hazards. Many of the micronutrients, for example copper and zinc, at higher concentrations in naturally mineralized areas or in polluted sites, may be very toxic.

NUTRIENT AVAILABILITY AND ABSORPTION

The foregoing discussion has shown that only a small proportion of most nutrients is present in a rapidly assimilable form. The soil solution is the primary source of inorganic nutrients for plant roots and, in the case of the more important cations, the concentration of this solution represents an equilibrium with the cations adsorbed on the exchange complex. Thus, for many cations, the soluble + exchangeable component is a good index of availability for plant growth, though the situation is rather different for calcium (and perhaps magnesium) in calcareous soils where free carbonates are present.

Short-term availability of cationic nutrients will be governed by the solution concentration, and the rate of replenishment from the exchange complex, unless contact exchange plays any significant part. Overstreet and Jenny (1939) claimed that plants accumulated cations more rapidly from clay suspensions than from equilibrium dialysate solutions of the clays, because contact exchange of cations occurred due to overlapping of the ionic atmospheres surrounding the roots and the soil particles. This hypothesis is also supported by evidence from physical systems in which cation transfer between solid adsorbents is more rapid, if contact occurs, than if cation diffusion in liquid is the only possible pathway (Black, 1968). The evidence for contact exchange is, however, still tenuous and cannot be fully accepted without further experimentation.

The availability of some micronutrient cations is governed more by their solubility relationships than by exchange phenomena, thus iron may be very insoluble in well-aerated calcareous soil due to precipitation of ferric hydroxides and in a similar fashion manganese may disappear from solution at high pH.

Anionic nutrient availability is more difficult to specify as the anion exchange activity of soils is a rather variable parameter, and also because several anionic nutrients, such as sulphur, nitrogen and phosphorus, became involved in biological cycles which effectively remove them from the physicochemical equilibrium relationship.

Phosphorus shows a much stronger binding reaction with soil than does either nitrogen or sulphur and it may be removed almost quantitatively from solution if added as orthophosphate (Fried and Broeshart, 1967). This immobilization of P is due not so much to ion exchange as to formation of insoluble phosphates at the surfaces of Fe-, Al- and Ca-containing minerals. In theory, if organic matter is mineralizing fast enough to supply adequate N, it should also be releasing sufficient P for plant growth but, unfortunately, much of the phosphate-P so produced is precipitated before it can be absorbed and also, because the solution-P is at a very low concentration, P tends to be rather immobile. Unless the mineralizing organic matter is in the immediate vicinity of absorbing roots the released inorganic-P will not reach root surfaces and, for this reason, organic-P cannot be considered to form an immediate source of nutrient-P.

Sutton and Gunary (1969) note that the immediate uptake by roots depletes the soil solution of dissolved P and further phosphate ions will then be drawn into solution from the solid phase. The quantity which may be liberated in this way is called the *labile* fraction. To specify the total P-supplying potential of the soil it is also necessary to know the P-adsorbing capacity as some inorganic-P will be temporarily trapped by anion exchange, at positively charged sites, before entering solution. Sutton and Gunary describe the P-supplying behaviour of a soil by using a well analogy (Figure 9.4): the water in the well represents the immediately available solution-P; the water in the porous matrix, the labile-P and the volume of the porous material is a measure of the adsorbing capacity. If two soils have the same adsorbing capacity then the labile-P content of each is sufficient to specify their relative P-supplying power but, if the absorbing capacities differ, then both these and the labile-P contents must be specified. Because P is so immobile, the P-supplying power of a soil must be considered as its ability to supply P to an individual absorbing root; the total length of absorbing root thus becomes very important in the context of whole-plant supply. A soil which may contain adequate P to support an established plant with an extensive root system may, at the same time, be quite unable to support a seedling with a single, short root (Sutton and Gunary, 1969).

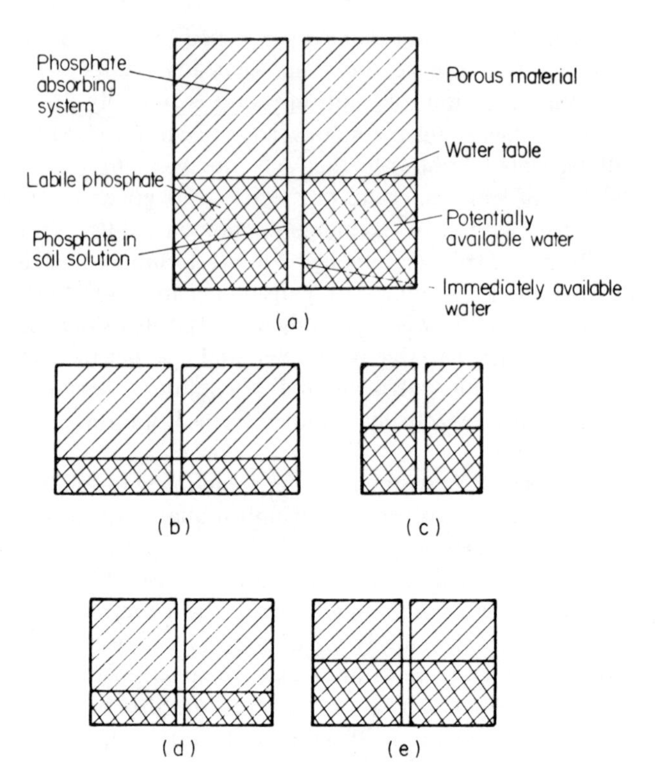

Figure 9.4 The well analogy of soil phosphorus behaviour. (a) depicts the analogy between the porous matrix surrounding a well and the phosphorus absorbing system. The amount of water in the porous medium represents the labile phosphorus supply. In comparing soils, if the absorbing system is of equal size as in (b) and (c), it is only necessary to specify the labile quantity but, if the adsorbing systems are unequal as in (d) and (e), both the size of the adsorbing system and the labile quantity must be specified. Reproduced with permission from C. D. Sutton and D. Gunary, *Ecological Aspects of the Mineral Nutrition of Plants*, Blackwell, Oxford, 1969, pp 127–34.

The soil–root interface

Nutrient absorption, like water absorption, is a function of the soil capacity-intensity factors, of the kinetics of release and movement in the soil, of surface area of root–soil contact and of internal transport characteristics of the plant. The root system of a plant is a much branched structure, the terminal parts of which intimately permeate the soil pore space and have their surface contact further increased by the development of root hairs which enter even very small pores and also become adpressed to the surfaces of soil particles. It is currently thought that the major pathway for water and solute mass flow to the root stele is through the microporous structure of the cortical cell walls and the free space (p. 148). This pathway is interrupted by the suberized Casparian strips of the endodermis but, beyond this barrier, the mass flow pathway is again available in cell wall micropores and in the xylem lumina. At the root apex where the Casparian strips are not fully suberized, and further up the root, where passage cells or endogenous root branches penetrate the endodermis, mass flow entry of the soil solution to the xylem can occur: perhaps up to 25% of the total intake (Weatherley, 1969).

Nutrient elements may move through the soil by mass-flow or by diffusion. In the case of elements which are at high concentration in the soil solution the transpiration stream may deliver sufficient to the root surface to satisfy demand but, if not, the solution adjacent to the root will become depleted of a particular element and establish a diffusion gradient from the surrounding soil. If the rate of diffusion is low then this will become the limiting factor of elemental uptake. In the cases of elements such as sodium which are actively excluded, the transpiration stream may carry more than the plant's requirement and the element may become concentrated at the root surface.

In natural soils it is likely that low solution concentration and slow diffusion will be the limiting factors. Prenzel (1979) investigated nutrient supply in a European *Fagus sylvatica* (beech) woodland on an acid brown-earth soil using suction plate lysimeters to extract soil solution. Uptake by the trees was measured by tissue analysis and potential mass-flow uptake computed as transpiration loss multiplied by solution concentration. The ratio of uptake to mass-flow transport (the mass-flow coefficient) was found not to differ significantly from unity for S, Fe and Mg, indicating that these three elements were delivered at an equivalent rate to plant demand. Chlorine and Na were strongly excluded (Figure 9.5) as might be predicted from physiological considerations while P, N, K, Ca and Mn were strongly accumulated. The exclusion of Al was just statistically significant and might, again, be expected in this acid soil where, with a solution concentration of more than 1 $mg\,l^{-1}$, it might be toxic.

Different results might be expected in other soil types, thus in an alkaline soil with more Ca there would be no need for exclusion of aluminium and, as found by Oliver and Barber (1966), Ca would arrive at a more than adequate rate in the transpiration stream. Similarly, in fertilized soils, solution concentrations of the major elements may be much higher than those in natural soils (Table 9.3). A typical hydroponic nutrient solution is included in Table 9.3 to illustrate its extreme departure from the composition of soil solution, a fact which is now always realized in ecological experimentation. Clement *et al.* (1978) have recently described a system for supplying nutrients from flowing solution at concentrations much closer to normal.

The interaction of mass flow and diffusive effects at the soil–root interface may be demonstrated for a range of elements using suitable isotopic labels followed by autoradiography of the soil and root. Suitable labels are ^{32}P, ^{33}P, ^{35}S, ^{45}Ca and ^{65}Zn. Results of such experiments show zones of depletion, for example of ^{33}P, adjacent to the root, indicating an inadequate supply of the element in the transpiration stream (Nye and Tinker, 1977). In the case of ^{35}S, soil solution SO_4^{2-} concentration was more than sufficient to meet plant demand with the

Table 9.3 Solute concentrations in soil solution, natural waters and nutrient culture solutions

	Ion (mg l^{-1})				
Element	Natural acid soil	Agricultural soil	Fen peat	Natural streamwater	Nutrient solution (modified Hoagland's)
N (NO_3^-, NH_4^+)	1.0	16	—	0·45	224 (NO_3^-)
P ($H_2PO_4^-$)	0·01	0·13	—	0·00075	62
K	1·0	8	0·8	0·23	235
Ca	2·4	52	100	1·65	160
Mg	0·5	12	10·9	0·38	24
S	3·0	11	—	2·1	32
Fe	0·13	—	—	—	1·1
Mn	0·9	—	—	—	0·11
Na	1·7	—	7·8	0·87	—
Cl	8·4	—	—	0·55	1·8
Al	1·3	—	—	0·24	—
pH	3 to 4 ($CaCl_2$)*	6·9	7·25	4·9	5·0
Extraction	Suction-plate lysimeter	Pressure extract	Filtration	—	—
Soil/vegetation	Acid brown earth. *Fagus sylvatica*. Germany	Fine sandy loam. Mixed arable–alfalfa, South California	Fen peat. *Carex paniculata*	Spodosol-runoff from 60–70 y forest. Mixed deciduous, New England	—
Reference	Prenzel (1979)	Burgess (1922)	Sparling (1967)	Likens *et al.* (1977)	Epstein (1972)

*pH measured in $CaCl_2$ slurry.

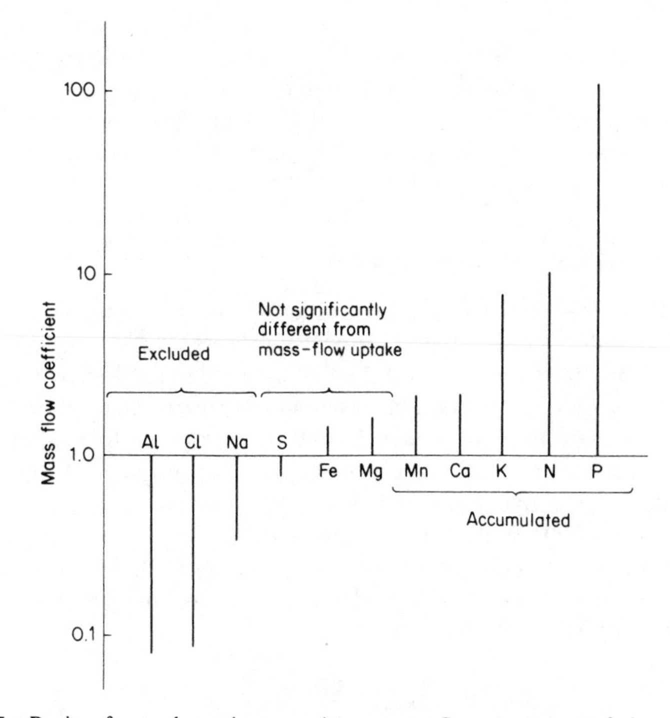

Figure 9.5 Ratio of actual nutrient uptake to mass flow transport of elements in the transpiration stream (mass flow coefficient) in *Fagus sylvatica* on an acid brown earth. Note the log scale, the very strong accumulation of N, P and K and the exclusion of Al and Cl. Data of Prenzel (1979).

consequence that a zone of ^{35}S concentration formed at the root surface (Barber *et al.* 1963).

Various workers have attempted to quantify the treatment of solute movement to root surfaces using derivatives of the equations governing diffusion and mass flow given in Chapter 4 (Equations 4.1 and 4.2). Nye and Marriot (1969) used the continuity equation:

$$\frac{\partial C_{\text{liq}}}{\partial T} = \frac{1}{a}\frac{\partial}{\partial a}\left(a\,D_{\text{if}}\frac{\partial C_{\text{soil}}}{\partial a}\right) + Ja_0C_{\text{liq}} \tag{9.1}$$

where C_{liq} = the concentration of solute in soil solution (mol m^{-3}); C_{soil} = the concentration of solute in soil (mol m^{-3}); T = the time (s); a_0 and a = the root radius and radius of surrounding concentric zone of soil (m); D_{if} = the diffusion coefficient for solute in soil (m^2 s^{-1}); and J = the water flux (m^3 m^{-2} s^{-1}).

The computed solutions are based on the boundary conditions:

(1) $T = 0$; $a > a_0$; C_{liq} = initial concentration in bulk soil solution.
(2) $T > 0$; $a = a_0$; C_{liq} given a new value related to the Michaelis relationship for solute uptake from solution.

The treatment makes no allowance for 'root interception' caused by root growth during the time period: Nye and Marriot considered that root growth displaced soil and solution and solute movement would still have to occur. The solution of this equation quite accurately predicts the type of nutrient depletion effects which have been shown autoradiographically under relatively low transpiration conditions (Nye and Tinker, 1977).

Root hairs, mycorrhizas and the rhizosphere

The root may be treated as a solid cylinder of radius a_0 in the above equation (9.1) but this neglects the fact that the most actively absorbing zone is surrounded by a halo of root hairs which are sufficiently crowded to diffusively interfere with each other. Calculation may best be based on the premise that absorption takes place through the wall of a cylinder defined by the tips of the root hairs, i.e. of radius a_0 + root hair length. The validity of this approach is supported by experiment, for example Bhat and Nye (1973) measured the density of an autoradiographic ^{33}P depletion zone around *Brassica napus* (rape) roots and calculated profiles of depletion using both a_a and a_0 + r.h. length in the above equation. Figure 9.6 shows that the best fit is given by assuming that the root hairs increase the radius of the absorption cylinder.

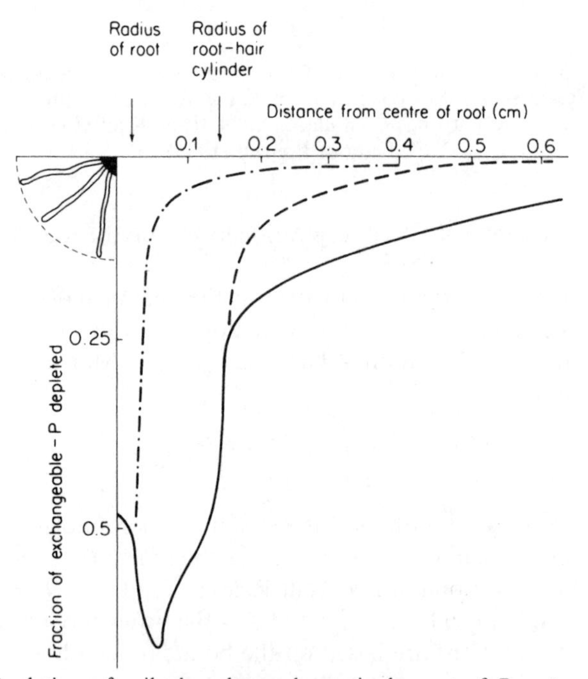

Figure 9.6 Depletion of soil phosphorus by a single root of *Brassica napus.* ——— measured ^{33}P-depletion; –·–·–·–·– calculated depletion profile assuming root hairs inactive; – – – – – calculated depletion profile assuming root hairs active in a soil cylinder of the indicated radius. Modified from Bhat and Nye (1973).

Roots are associated with a spectrum of microorganisms ranging from loose associations with the rhizosphere zone, through ectomycorrhizas, endomycorrizas to root-nodule organisms and finally on to the more extreme interactions with invading pathogens. All of these associations have consequences which influence vegetation ecology (Newman, 1978).

There is now strong evidence that mycorrhizae increase the efficiency of nutrient absorption (Marks and Kozlowski, 1973; Sanders *et al.* 1975; Nye and Tinker, 1977). Ectomycorrhizas, mostly of the class Basidiomycetae, form a sheath, external to the root with hyphae radiating into the surrounding soil. They are associated with a limited number of plants, mostly trees and at least in some cases the host plants fail to thrive in the absence of the fungal symbiont. Endomycorrhizas may belong to any fungal class, are more common but difficult to observe for which reason they were formerly a neglected group. The most widespread are the vesicular-arbuscular mycorrhizas which, apparently, infect most of the world's vegetation. There is no external sheath; the hyphae form vesicles and arbuscles within the root cortical cells and ramify several cm into the surrounding soil, often forming connections between adjacent roots.

Most of the early work on nutrient uptake was with *Fagus sylvatica* (beech) ectomycorrhizas (Harley, 1969a and b). The infected roots are considerably more effective in phosphorus uptake, partly because their life is prolonged, possibly because P may be absorbed and transported to the root through the external hyphae and possibly by direct mobilization of P from organic sources (Nye and Tinker, 1977). It has also been suggested that ectomycorrhizal infection may assist water uptake in difficult environments.

During the past two decades considerable research has been devoted to the ecology and physiology of endomycorrhizas. It is now suspected that most of the world's vegetation carries v-a mycorrhizal infection, for example almost all species of British grassland, scrub and woodland were found to be infected (Read *et al.* 1976). Infection improves the nutrient absorbing ability of roots: in the case of phosphorus a two- to fourfold increase (Sanders and Tinker, 1973; Nye and Tinker 1977; Barrow *et al.* 1977).

Both endo- and ectomycorrhizas store inorganic polyphosphate which may be seen as metachromatic granules or identified histochemically (Chilvers and Harley, 1980) and by electron probe microanalysis (Cox *et al.* 1980). The latter workers measured rapid transport of the granules by hyphal cytoplasmic streaming, a fact which may part-explain the increase in root absorbing efficiency when surrounded by a halo of hyphae attached to the internal symbiont.

Clarkson (1974) described these hyphae as 'super' root hairs which permit more thorough exploration of soil volume. Went and Stark (1968) made the same suggestion and also implied that direct recycling of nutrient elements might occur if the external hyphae were involved in mineralization of organic material. This would be of particular importance in very nutrient deficient habitats, for example tropical rain forest on extreme oxisol or spodosol soil types. Even in temperate environments the consensus is that mycorrhizal infection plays its major role in low fertility sites (Sparling and Tinker, 1978).

Clarkson (1974) argues against direct mobilization, citing evidence from ^{32}P-labelling that mycorrhizal and normal roots both draw on the same resource of soluble-P. There is no doubt however that infected roots are very efficient scavengers of soluble nutrients. Stark and Jordan (1978) sprayed the surface of a tropical forest soil with inorganic nutrient labelled with ^{45}Ca and ^{32}P in amounts simulating mineralization input. Less than 0·1 % was leached through the mycorrhizal root mat and all loss ceased within 2 months. The experiments were, unfortunately, not undertaken with unforested controls but the oxisol and spodosol soils of these forests would be unlikely to prevent leaching loss and the authors' conclusion that the mycorrhizal root mat acts as a nutrient trap is valid.

In addition to the rather specific microorganism associations of the mycorrhizal symbiosis, plant roots have a range of interactions with rhizosphere organisms which benefit from leakage of metabolites and, to a lesser extent, from sloghed-off tissues. Rovira (1979) cites 5–10 % of total plant photosynthesate being lost to the soil in sterile systems, increasing to 12–18 % in the presence of rhizosphere organisms. Bacteria are likely to receive such exudates only when they are close to the root surface: Newman (1978) has offered a mathematical model prediction that the equilibrium population will have a biomass density at the root surface which is more than 700 times that at 1·8 mm. The soil solution concentration of soluble metabolite is unlikely to exceed $10\,mg\,l^{-1}$ and, if it moves by diffusion, will be utilized within a very short distance of the root surface.

Fungi in the rhizosphere may extend much further from the root surface and, with rapid hyphal transport by cytoplasmic streaming, will be able to short-circuit the high resistance diffusion pathway of nutrients and metabolites in the soil. At least for mycorrhizal external hyphae there is evidence that their presence permits nutrient transfer between adjacent plants.

Many of the rhizosphere organisms are symbiotic, the plants benefitting from locally enhanced N-fixation (Hassouna and Wareing, 1964) and increased nutrient supply from the mineralization of organic materials. Bacterial dissolution of insoluble phosphates has been suggested as a further benefit but firm evidence does not exist (Tinker, 1975; Nye and Tinker, 1977). The rhizosphere also contains non-symbiotic organisms some of which may be potential pathogens and there must be a gradation of relationships between symbiosis and pathogenicity and also between root exudate dependence and independence.

The effects of microorganisms on nutrient uptake were largely ignored until the early 1970s (Barber, 1969) since when a substantial amount of work has been reported (Clarkson, 1974; Nye and Tinker, 1977; Harley and Russell, 1979).

Comparison of ^{32}P-labelled phosphate absorption by sterile and non-sterile roots shows that slightly less P is absorbed in sterile conditions but a much larger proportion of it is transported to the shoot (Barber, 1969). Autoradiographs reveal that the bacterial colonies on the surface of non-sterile roots immobilize the phosphorus (see Clarkson, 1974) but it is not known whether this would affect the long-term nutrition of a wild plant. Newman's (1978) model predicts that the rhizosphere biomass will reach equilibrium with metabolite exudation and this would presumably limit the amount of P-immobilization which could occur unless there was an accumulation of undecayed, dead bacterial tissue which is unlikely.

Phosphorus has been the focus of most attention in the root–microorganism interaction but some other elements are also involved, for example sulphate, rubidium and potassium may be immobilized by root surface organisms (Clarkson, 1974) while mycorrhizas may be involved in sulphur and zinc absorption (Clarkson and Hanson, 1980). Further study is almost certain to reveal involvement of microorganisms in the uptake of most other nutrients but it is too early to be certain of the full ecological significance of many of these interactions.

Behaviour of ions at the absorbing surface

It is difficult for the ecologist to define the root absorbing surface except to say that it lies somewhere between the outermost layer of the cortex and the exterior of an indefinite concentric soil-cylinder related to the ramification of root hairs and fungal hyphae. By contrast most experimental studies of ion absorption kinetics have been made in solution culture where the absorptive surface is easily defined. Most of the discussion of mechanisms of ion uptake is still speculative (Epstein, 1972; Clarkson, 1974; Hewitt and Smith, 1975) and, for the ecologist it will usually suffice to accept the experimentally demonstrated kinetic relationships without attention to details of postulated carrier mechanisms.

In dilute nutrient solutions the absorption of ions may be represented by the scheme shown in Figure 9.7. Epstein and his co-workers in the early 1950s first treated this situation by quantitative analogy to enzyme kinetics. An ion species ('substrate') combines with a carrier ('enzyme') and is transported to the cell interior where it remains immobile as a 'product'. The Michaelis–Menten representation of the ratio between the opposing transport processes is formalized as:

$$K_m = \frac{K_2 + K_3}{K_1} \tag{9.2}$$

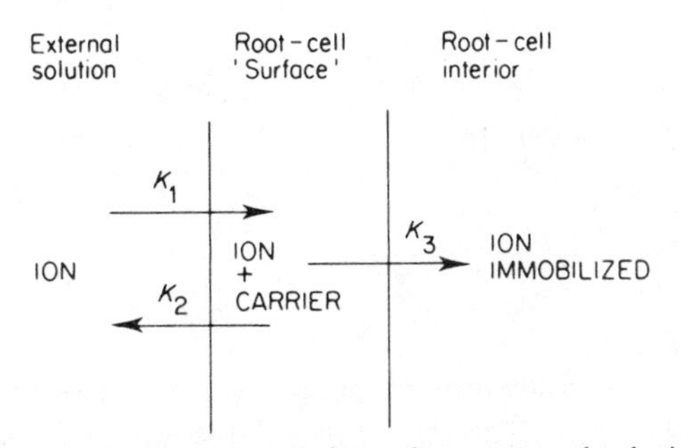

Figure 9.7 A scheme for the absorption of ions from external solution by roots. K_1, K_2 and K_3 are the rate constants used in the kinetic treatment of Equation 9.2.

where K_m is the Michaelis constant and $K_{1,2,3}$ are the rate constants shown in Figure 9.7.

The rate of ion uptake into the root is related to K_m:

$$V = V_{max} \frac{C_{liq}}{K_m + C_{liq}} \tag{9.3}$$

where V = the rate of ion uptake ($mol\,kg^{-1}\,s^{-1}$); V_{max} = the maximum rate of ion uptake which is approached asymptotically as the external solution concentration increases and the carrier mechanism saturates ($mol\,kg^{-1}\,s^{-1}$); and C_{liq} = the external solution concentration ($mol\,m^3$). If the special case of $V = V_{max}/2$ is substituted in Equation (9.3) it will be found that $K_m = C_{liq}$ when absorption is at half maximum rate. This relationship is of great value in practically determining K_m and V_{max}: roots are allowed to absorb a radio-isotopically labelled ion species from a suitable range of concentrations. After a standard time period the roots are washed to remove ions from the free space, their radioactivity is measured and used to estimate ion uptake into the root cells. The raw data may be plotted as in Figure 9.8a, but replotting the double reciprocal relationship of Figure 9.8b permits graphical estimation of both V_{max} and K_m.

Published values suggest that, for most ions, V_{max} is reached with a solution concentration between 0·1 and 0·5 mM. If there is a very high affinity between ion and carrier, K_m will have a small value with a minimum of about 0·006 mM (*Zea*

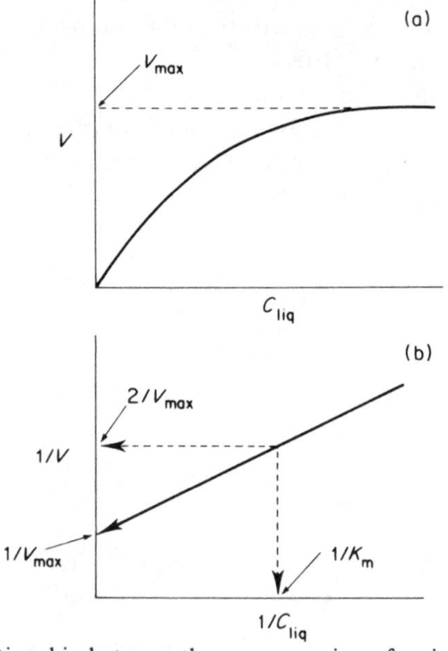

Figure 9.8 (a) The relationship between the concentration of an ion in external solution (C_{liq}) and the velocity of its uptake (V). Maximum uptake velocity = V_{max}. (b) Double reciprocal (Lineweaver–Burke) plot of the same data showing the calculation of the Michaelis constant (K_m).

mays (sweetcorn) root for phosphate). If the affinity is low there is a relative exclusion of the ion, K_m having a high value with a maximum of about 0·32 mM (*Hordeum vulgare* (barley) root for sodium). Epstein (1972) gives a range of values, though mainly for agricultural species.

If the concentration of the external solution increases substantially beyond the value which gives V_{max} it is found that ion absorption does continue to increase with a very shallow slope, often rather irregularly and ultimately reaching a new, ill-defined V_{max} in the region of 50 mM or more. The dilute solution mechanism is usually named 'system I' and it is operative over the range of solution concentration which is found in natural soils (p. 254). The concentrations which initiate 'system II' are usually encountered only in saline soils and occasionally with very high fertilizer dressings. There is, in any case, considerable controversy concerning system II, which some workers consider to be a definite mechanism (Epstein, 1972), but others dismiss as a passive diffusion of ions from the highly concentrated external solution (Nye and Tinker, 1977; Clarkson, 1974).

It is rather difficult to relate this theoretical treatment of ion absorption kinetics to the real world situation in which the absorptive behaviour of roots will be a mixed function of inherent plant characteristics and the imposed effects of microorganisms, particularly mycorrhizal fungi. It is likely that K_m values of non-sterile roots in soil will not be the same as those determined in nutrient solutions.

The presence of a soil mineral matrix as a barrier to mass-flow and diffusion is likely to reduce rates of ion uptake by formation of depletion zones. Even in unstirred nutrient *solutions* the formation of a diffusion depletion zone around absorbing roots may reduce the rate of ion uptake by almost 50% at an external C_{liq} which would give V_{max} in stirred solution. In these conditions roots with a very low K_m would be at a competitive advantage and it is of course in depleted nutrient conditions that the additional absorbing capacity given by mycorrhizal hyphae, confers its greatest benefit.

Further problems arise in the extrapolation of laboratory studies of ion uptake kinetics to field conditions because experimental solution concentrations are often well in excess of those in soils. Hoagland's solution, which is widely used for solution culture (Table 9.3), contains N, K and P at 16, 6 and 2 mM, the first two concentrations between 10 and 100 times those of many soil solutions, and that of P up to several thousand times more!

Soil solutions are continuously replenished from exchangeable or other labile stores and it seems likely that ecologically realistic experiments should attempt to duplicate such conditions. Plants are much more efficient in scavenging nutrient elements from low concentration sources than they are in uptake from high concentration solutions. Wild *et al.* (1979) compared K-uptake from a continuously flow-replenished solution of 1·2 μm K with uptake from a 100 μM K solution which was allowed to deplete to less than 1·2 μM K as the plants grew. The root-entry fluxes were 3 to 6 times greater in the low concentration conditions and uptake was not affected by concentration above 1·2 μM K in replenished solutions. This concentration is only 0·0002 of that afforded by Hoagland's solution!

In the same experiments K_m values were measured by the usual technique. Plants which were grown in the 100 μM K depletion mode had K_m values between 15 and 20 μM K compared with those from the low concentration solutions in which K_m was less than 1·0 μM K. It would appear that any attempt to measure K_m and relate it to ecological circumstances should be made using plants which have been suitably pretreated in solution concentrations matching those of the soil. In these experiments K-uptake by *Lolium perenne* (ryegrass) and *Raphanus sativus* (radish) was compared. The two species showed markedly different K-absorbing characteristics in the depletion experiments but very little difference in the replenished solutions.

The measurement of K_m by short-term uptake at different external solution concentrations is, in any case, not satisfactory for characterizing field behaviour in which ion-uptake satisfies the demands of newly growing tissue rather than that of existing but nutrient-depleted structures. Natural demand will be governed by at least three factors: prevailing relative growth rate, ratio of root surface area to plant weight and internal nutrient status. This contrasts sharply with most physiological studies of uptake kinetics in which nutrient status alone dominates the characteristics of the absorption isotherms.

Selective absorption of ions

Ions may enter cells passively by diffusion or actively by ion-pumping mechanisms, such as the adenosine triphosphatase membrane-bound systems already demonstrated in animal and bacterial cells. Both mechanisms, either through selective permeability, selective internal ion-metabolism or selective functioning of pumps, result in some ions being absorbed and others excluded.

There is still much controversy concerning mechanisms of uptake (Lutge and Pitman, 1976) and the only concrete information concerns the concentration of ions and the electrical potentials in different parts of the plant and the influx–efflux behaviour of different ion species thus enabling us to recognise the location and direction of ion-pumps (Baker and Hall, 1975). There are enormous voids in physiological knowledge of energy sources, their coupling to ion pumps and the biochemistry of pumps. Fortunately the ecologist has less need of this missing information, being more concerned with the direction and quantity of ion movement from soil to plant system.

Terrestrial mesophytes usually absorb ions to concentrations far in excess of those in soil solution but they must also be selective, since the ion-balance of the external solution rarely, if ever, matches that of the internal ion 'atmosphere' of the cell which is controlled by fairly critical biological 'chemo-stats'. In temperate soils for example, the molarity of Ca usually exceeds that of Mg and, similarly, Na of K. In the mesophyte cell vacuole these relationships are reversed. (Table 9.4a) indicating a selective uptake or exclusion mechanism. Non-succulent halophytes growing in soils with excess Na^+, Cl^-, SO_4^{2-} or other ions, face very severe problems of concentrating essential elements while excluding others of which the concentration may be well in excess of system-I control (Table 9.4b).

Table 9.4 Concentrations of common elements in vacuolar fluid and soil solution

	Element (mmol l^{-1} ion)					
	K	Na	Ca	Mg	Cl	SO_4
Vacuole: nine mesophytes*	174	44	50	81	28	31
Typical soil-solution†	0·8	3·0	42	3·6	4·6	11·8
Concentration ratio	218	15	1·2	23	6·0	2·6
Vacuole: nine semi-arid Chenopodiaceae*	75	370	38	34	50	18
Typical saline soil-solution‡	3·5	224	90	66	296	51
Concentration ratio	21	1·7	0·4	0·5	0·17	0·4

* Cram (1976).
† Fairbridge and Finkle (1979).
‡ Fried and Broeshart (1967).

MINERAL NUTRITION AND PLANT ECOLOGY

The diversity of soil, freshwater and marine habitats occupied by plants has encouraged ecologists to compare the physiological behaviour of 'normal' plants with that of extreme habitat species. The results have been particularly fruitful in the field of mineral nutrition, providing physiological-ecological explanations of some species distribution patterns. The comparisons have included calcicoles and calcifuges of, respectively, alkaline and acid soils; halophytes of saline soils; species of nutrient-poor (oligotrophic) and nutrient-rich (eutrophic) soils and plants which are particularly tolerant of high concentrations of specific phytotoxic elements.

Calcicoles and calcifuges

A few moments with a flora will reveal such statements as 'absent from calcareous soils' or 'usually on basic soils' while experimental attempts to grow such plants reveal considerable differences in physiological response or competitive behaviour when they are grown on soils differing in reaction from those which they normally inhabit.

An early example of such experiments was Tansley's (1917) comparison of the bedstraws *Galium saxatile* and *Galium sterneri* syn. (*G. sylvestre*) or (*G. pumilum*). He found that germination and growth of the calcifuge *G. saxatile* was inhibited on calcareous soil and the plants were not able to compete with those of *G. sterneri*. On acidic soil this relationship was reversed, the calcicolous *G. sterneri* being suppressed and giving *G. saxatile* a competitive advantage. No attempt was made to establish the cause, but Tansley commented that it would make 'an interesting investigation in physiological ecology'.

The most obvious chemical characteristics of calcareous soils are associated with the high content of calcium carbonate and the near saturation of the cation exchange complex with calcium ions. As might be expected, various workers have found ecologically significant differences in response to calcium supply. Jefferies and Willis (1964) grew the calcifuge *Juncus squarrosus* (heath rush) and the calcicole *Origanum vulgare* (marjoram) in nutrient solutions with differing levels of calcium supply. *J. squarrosus* proved to have a very low calcium requiement and was intolerant of high concentrations. By contrast the growth of *Origanum vulgare* was strongly limited by low calcium supply. Clarkson (1965) found a similar relationship by comparing calcicole, neutrophile and calcifuge species of *Agrostis* (bent grasses) grown in solution culture with differing additions of calcium. The growth of the calcicolous and neutrophilous species, *A. stolonifera*, *A. canina* and *A. tenuis* was much improved by increased calcium supply but *A. setacea*, a calcifuge, was little affected (Figure 9.9).

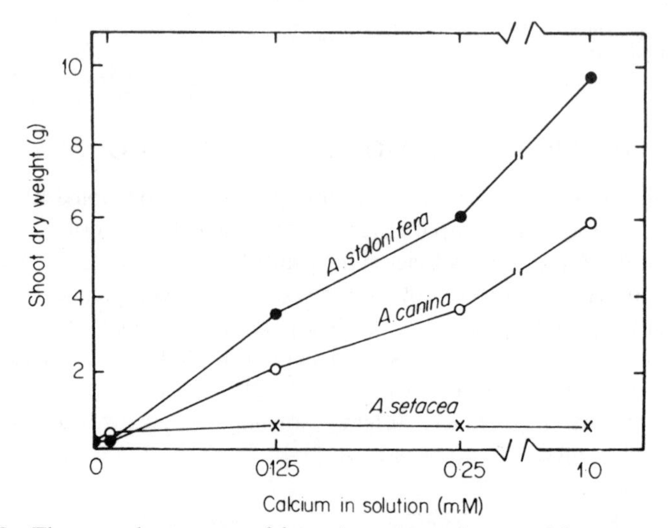

Figure 9.9 The growth response of four *Agrostis* species to calcium concentration in solution culture. Data of Clarkson (1965).

Similar results have been achieved with ecotypic populations of single species derived from contrasting habitats; for example, Snaydon and Bradshaw (1961) found that *Festuca ovina* (sheep's fescue grass) populations from acidic soils have a much lower calcium requirement than those from calcareous soil. Ramakrishnan (1968; 1970) showed similar differentials in calcium requirement of ecotypes of *Melilotus alba* (melilot) and, in the second paper, showed that the differentials of growth response were enhanced when the ecotypes were grown in competition with each other on acid and on calcareous soils.

Some European workers have gone so far as to suggest that calcicoles have a much higher content of soluble calcium and malate than calcifuges in which the calcium is precipitated as oxalate. The molar ratio of soluble K/Ca affords a means of comparison, ranging from below one in calcicoles up to about 100 in calcifuges (Lauchli, 1976).

Such work might suggest that the relationship between pH and available calcium is, alone, sufficient to explain the differentiation of these broad types of plant. Unfortunately this is not so: despite its ecological usefulness as an easily measured single parameter of soil chemistry, pH is a function of a host of interacting soil characteristics (chapter 4, p. 90 ff). Many of these are implicated in control of physiological function and growth, thus showing an inevitable correlation with pH.

Figure 9.10 presents a synopsis of these pH-related soil conditions from which it may be seen that calcium availability is only one of a large number of factors which exert selective pressure according to the relative acidity of the soil.

The concentration of hydrogen ions *per se* appears not to influence plants in the range pH 4 to pH 8 providing that sufficient Ca is available and iron or aluminium does not reach toxic concentration. The evidence for this comes mainly from the experiments of Arnon and Johnson (1942) and Olsen (1958) using agricultural species in solution culture. Below pH 4·0, direct H^+ injury may interfere with the selective permeability of root cells, for example causing potassium loss, particularly in the absence of Ca (Moore, 1974) but it would be useful to know if wild plants of acid soils are sensitive in this way. Rorison (1973) has suggested that, because there is difficulty in controlling solution pH in the face of active root ion uptake, flowing nutrient experiments will be necessary to answer these questions.

In agricultural soils receiving irrigation water rich in bicarbonate, toxicities have been reported (Bear, 1960) but these generally arise from other pH-linked effects such as iron deficiency or increases in the exchangeable sodium level of the soil. Woolhouse (1966) found that bicarbonate impeded iron uptake and translocation in the calcifuge *Deschampsia flexuosa* (wavy hairgrass) but had no effect on plants of less acid or alkaline soils such as *Holcus mollis* (soft grass), *Arrhenatherum elatius* (false oatgrass) and *Koeleria cristata* (crested hairgrass).

The solubility relationships of iron and aluminium have now been strongly implicated in the behaviour of calcifuge and calcicole species. Hartwell and Pember (1918) first recognized the differential effects of aluminium in acid soils, the toxicity strongly depressing the growth of barley but having little influence on rye, reflecting its ability to grow on more acid soils than barley can. They also noted the interaction of high soil aluminium with phosphorus status; phosphate fertilization suppresses aluminium toxicity, presumably by precipitation of insoluble phosphates. Until the early 1960s aluminium toxicity was not extensively studied as an ecological factor, possibly because its agricultural effects are easily overcome by liming. Jones (1961), however, raised ecological questions concerning the relationship of aluminium to wild plants and suggested that calcifuges tend to be tolerant of high aluminium concentrations compared with the great sensitivity of calcicoles.

Clymo (1962) investigated the growth and distribution of two sedges species, *C. lepidocarpa* and *C. demissa*, which are morphologically similar and both of which inhabit wet soils, thus eliminating the problem of consistent physical differences between acidic and calcareous habitats. By growing plants in nutrient solutions he was able to show that the calcicolous *C. lepidocarpa* was excluded from the habitat of the calcifuge *C. demissa* by Al^{3+} concentrations of above *c.* 1 p.p.m. and

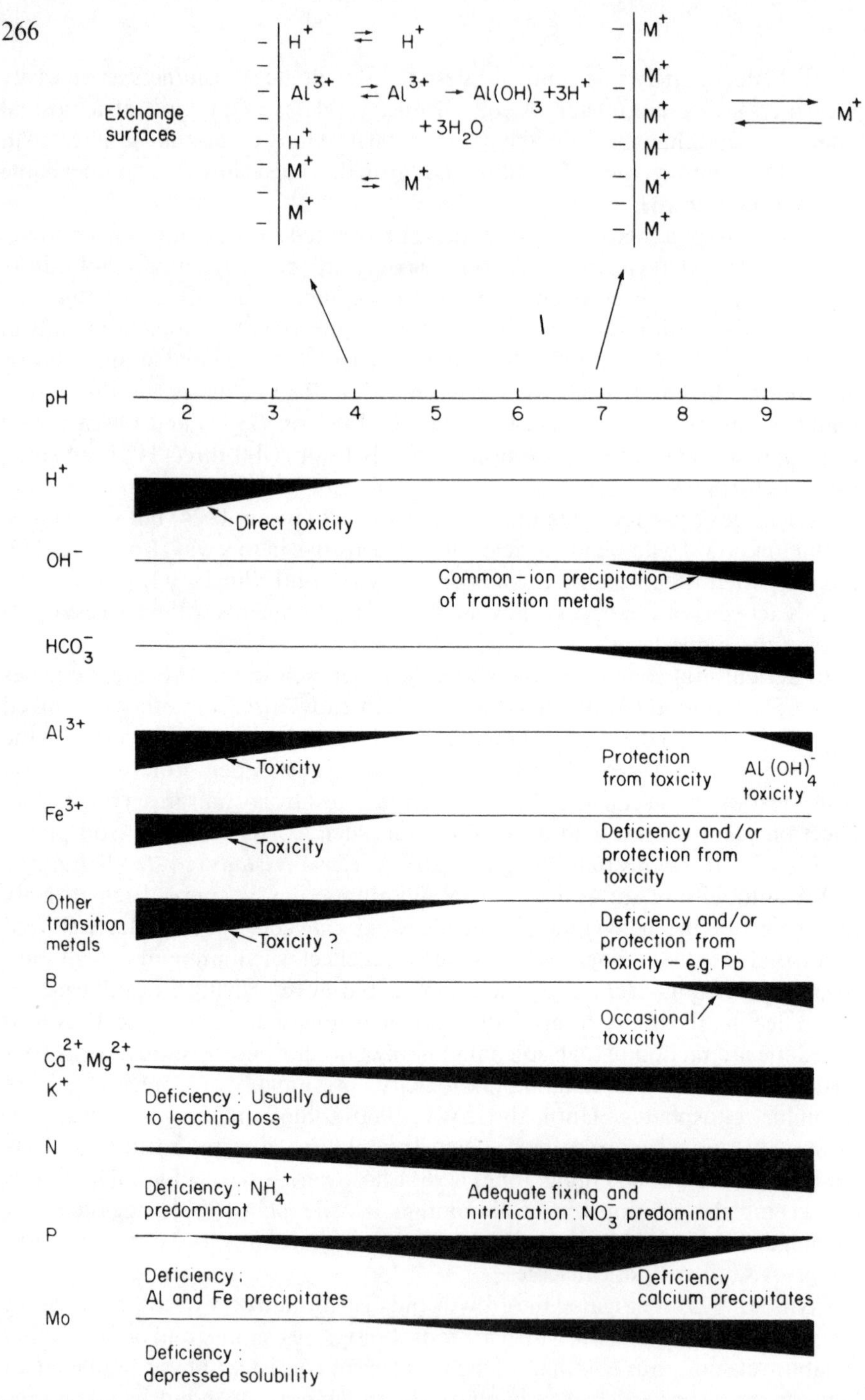

Figure 9.10 The relationship of soil chemical characteristics to soil pH. Increasing height of the shaded band indicates increasing soil solution concentration and/or availability to plants.

Ca^{2+} concentrations below *c.* 30 p.p.m. This conclusion supports Jones' suggestion, as also do Rorison's findings (1960a and b) for *Scabiosa columbaria* (dovesfoot scabious), a calcicole which showed growth inhibition in acid soil extracts, caused by high concentrations of Al^{3+}. Hackett (1965) investigated the mineral nutrition of the strongly calcifuge grass, *Deschampsia flexuosa*, and found it to be insensitive to a wide range of nutritional conditions but tolerant of high levels of Al^{3+}. It has an experimental pH optimum of 5·5–6·0 which is above the range of its field occurrence, suggesting that the natural distribution does not reflect its habitat 'preference'. It is more likely that tolerance of high aluminium concentrations permits it to use acid habitats as a 'refuge' from the competition of faster growing species which are more aggressive in their demands on mineral nutrient supply.

Clarkson (1969), in reviewing the aluminium toxicity problem, defined three attributes of the calcifuge plant which may be summarized as: (i) ability to cope with a low calcium supply and its impaired absorption and translocation; (ii) ability to withstand a low phosphorus supply, and fixation of phosphorus by aluminium at the root surface and (iii) tolerance of high aluminium concentrations conferred either by cytoplasmic sites at which the element may be harmlessly accumulated, specific methods for chelating aluminium or a mechanism for precipitating it at the cell wall, thus preventing entry into the cytoplasm.

Recent work has divulged the possibility that the mechanism for coping with aluminium toxicity may also explain the need for high soil iron concentrations shown by most calcifuges. It has been known for many years that calcifuge plants when grown in a calcareous soil develop a 'lime-induced chlorosis' and cease to grow actively. The chlorosis may be cured by foliar application of ferrous iron salts or chelated iron compounds suggesting that it is due to an induced iron deficiency (Hewitt, 1963). By contrast, calcicolous plants do not show such chlorosis and it was assumed that they either have a lower iron demand or they are more efficient at extracting soil iron than calcifuges. However, Grime and Hodgson (1969) have suggested that the mechanism responsible for immobilizing aluminium and rendering it harmless in calcifuges may be non-specific to the extent that it could also immobilize iron, possibly by a chelation process. As a result, in calcareous soil, when aluminium ions are absent, the chelating system absorbs the small amount of iron in the incoming soil solution and induces iron deficiency. In acid soil with a high aluminium concentration the chelating system shows a higher affinity for this than for iron and no deficiency arises.

A major mechanism of aluminium toxicity appears to be inhibition of root meristematic cell division (Clarkson, 1969) and tracer experiments using ^{46}Sc as an Al-analogue show its very free movement amongst and into meristematic cells compared with its confinement to cortical intercellular space and binding to the cell walls elsewhere (Clarkson and Sanderson, 1969). The first sympton of Al-toxicity is the formation of stunted, peg-like lateral roots and Yeo and Flowers (1977) have suggested an interesting explanation of the stimulation by very low Al-concentrations which has occasionally been reported (e.g. Hackett, 1965): the meristem is damaged, even by very low concentrations of Al, but this is followed by the initiation of new lateral roots which are produced faster than they also are damaged. The consequence is a slight enhancement of plant growth.

Widespread occurrence of lime induced chlorosis in North American agricultural soils has prompted an extensive research programme. Jones (1976) was able to list at least 16 factors which cause or contribute to chlorosis and it is now known for certain that absorption and transport mutants of several species exist. Iron inefficient varieties of *Glycine max* (soyabean) which develop chlorosis on calcareous soil were first reported in the 1940s. Similar iron inefficient varieties exist for *Zea mays* (maize), *Sorghum vulgare* (sorghum) and other crops, while Grime and Hutchinson (1967) found widespread lime chlorosis in wild English vegetation.

Brown (1961) and his co-workers have shown with an elegant technique using ferricyanide-ferrichloride as an iron source, that iron efficient varieties have a greater iron reductive capacity at the root surface than iron inefficient varieties, suggesting that plants probably absorb iron in the more soluble divalent form. Further work (Brown and Ambler, 1970) also reveals the ability of iron efficient roots to lower the pH of the external medium and to maintain lower phosphorous and higher citrate concentrations in the root sap. Lower pH favours divalent iron in the $Fe^{3+} \rightleftharpoons Fe^{2+}$ equilibrium, citrate complexing maintains internal mobility (Tiffin, 1966) and lower phosphorus concentration prevents formation of insoluble phosphates.

The solubility products of iron and aluminium phosphates are so low that it has been usual to suggest P-immobilization as another hazard of acid soils. Experiments in which phosphorus and iron supply are varied independently certainly suggest that high phosphorus availability may depress iron uptake (Patel *et al.* 1976). The situation may be more complicated in soil than in nutrient solution, for example Maschner and Azarabadi (1979) found that P inhibited Fe uptake from solution but in soil they suggested that P was sufficiently depleted by uptake at the soil–root interface for iron to become adequately available.

Ecological aspects of calcicole-calcifuge behaviour

The suggestion that vegetation may be categorized by its response to soil acidity is very old. In 1900 Roux was able to review work which included not only plant geographical studies but transplanting and liming experiments with wild and cultivated species. Rayner (1913) suggested that the failure of *Calluna vulgaris* (ling) on calcareous soil was due to lack of mycorrhizal infection. By 1921 Salisbury was aware of almost all those characteristics of calcicoles and their soils which have subsequently been investigated. In addition to the chemical characteristics of the soil, Salisbury noted that many obligate calcicoles of moist climates are much less edaphically restricted in arid conditions. In Britain and northern Europe the so-called calcicoles are often plants which are at the northernmost point of a much more general distribution in central and southern Europe.

Within limited geographical areas the relationship between calcicoles and the substratum is very striking (Figure 9.11) but interpretation is complicated by the much wider distribution of such plants in the drier parts of mainland Europe and also by the occasional occurrence of limestone-heath, an apparently anomalous vegetation type containing mixtures of calcicole and calcifuge species. In the earlier literature (Tansley and Rankin, 1911) limestone heath was interpreted as a consequence of surface leaching and stratification of root systems but this has proved not to be the case (Grubb *et al.* 1969). Etherington (1981) has suggested that limestone heath on shallow soil may be a recent relict vegetation in which calcifuges have been 'stranded' on an alkaline surface by soil erosion of acid loess cover, managing to persist only be vegetative means.

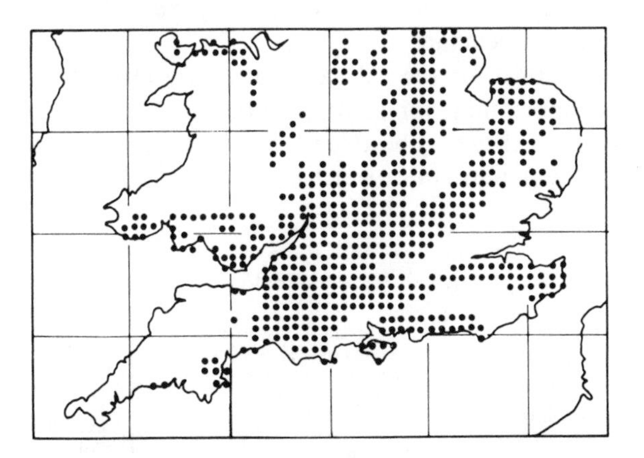

(a) Distribution of limestone outcrops

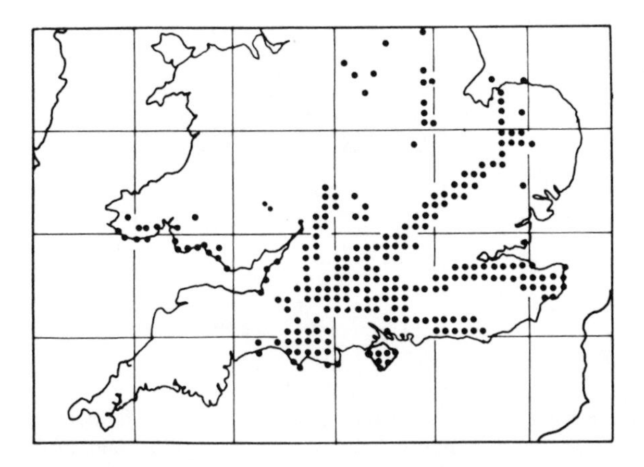

(b) Distribution of *Asperula cynanchica*

Figure 9.11 The distribution of *Asperula cynanchica* in relation to limestone outcrops in southern Britain. Each dot represents occurrence in a 10 km grid square. Redrawn with permission from F. H. Perring and S. M. Walters, *Atlas of the British Flora*, Botanical Soc. of the British Isles, 1962, Map no. 483/2.

Grubb and his colleagues suggested that pH 5 is a critical lower limit for seedling growth in most calcicoles (Figure 9.12) which would permit their establishment in a great range of soil types but, between pH 5 and pH 6, many large and aggressively competitive grasses exclude the slower growing calcicoles *if* the nutrient and water supply is adequate and *if* grazing pressure is low. Consequently the calcicoles are competitively excluded from many soils, but with heavy grazing, prevalence of nutrient deficiency or drought they form a characteristic limestone association. Superimposed upon this general pattern are other divisions caused by extreme calcicoly, waterlogging relationships and the potential substitution of Mg for Ca in some calicole species. Extreme calcicoly may simply reflect a greater than normal sensitivity to competition, the very alkaline soil serving as a refugium, or it may be a function of excessive physiological calcium demand.

Just as pH is critical for the calcicole so it may also be for the calcifuge, representing an upper limit for seedling survival. After establishment in the surface acidified layer they may be able to tolerate a higher pH, between 5 and 6, provided that grazing, nutrient or water deficiency eliminates competition. The

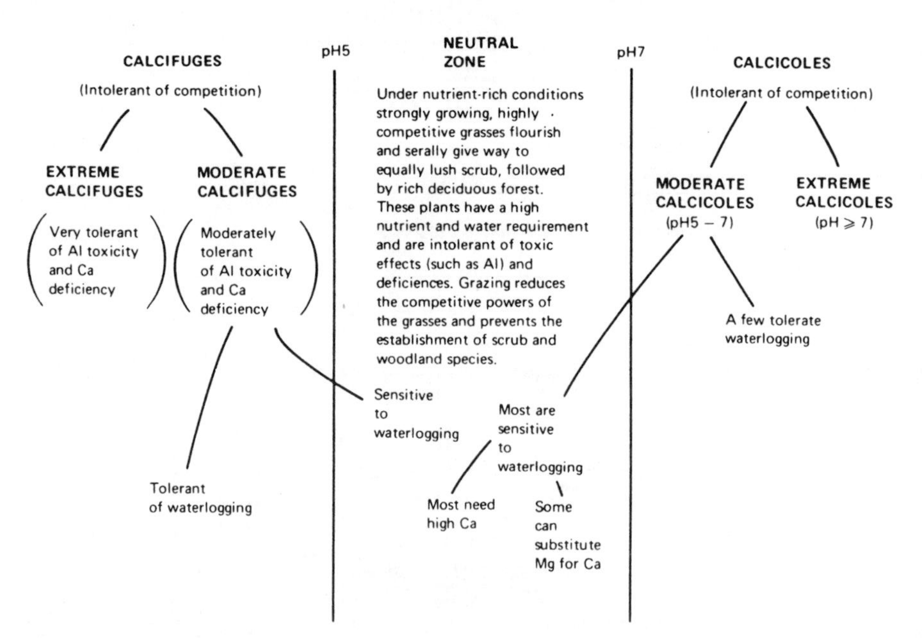

Figure 9.12 A classification of calcicoles and calcifuges. The neutral zone between pH-5–7 may be occupied by the highly demanding but strongly competitive species which exclude the moderate calcicoles and calcifuges. Under heavy grazing, nutrient deficiency or water shortage, the moderate calcicoles and calcifuges may cohabit in this pH range. pH 5 seems to be the upper boundary of Al toxicity effects while pH 7 is probably near the lower boundary of lime-induced iron deficiency. With high rainfall, transient waterlogging and anaerobiosis of soil aggregate centres may extend the pH range by calcifuges by making iron temporarily available in the reduced, ferrous form.

problem for these calcifuges is iron supply and it is under these circumstances that lime-induced chlorosis is often encountered in the wild. Localized or short periods of waterlogging may make iron available in the ferrous form, which could explain why some of the more extreme calcifuges are commoner in limestone heath in the higher rainfall areas of Britain.

Grime (1963a) noted that a number of calcifuges could occur on limestone rendzinas even without superficial acidification, but these observations were all made in northern England or western Ireland under high P/E regimes, suggesting that the temporary formation of ferrous iron might be significant in these soils. This could be a partial explanation of the observed distribution but, also, many of the species cited are not particularly strict calcifuges, and those which are, were not noted above pH 5–5·5. Grime (1963b) also described the invasion of a rendzina soil by the calcifuge grass *Deschampsia flexuosa* (wavy hairgrass) at a surface pH of *c.* 5. At this pH it was able to compete with the existing calcicole flora, which it replaced, but burning or drought caused reversion to normal limestone rendzina grassland and Grime suggested that the calcifuge cover and accumulation of acid mor humus might be considered a transient phenomenon caused by a sequence of abnormally humid seasons.

From the available data it seems that calcicoles and calcifuges can coexist in a narrow pH range near pH 5 providing that the interaction of nutritional, moisture and grazing factors gives each type an equal competitive status and excludes the vigorously growing neutrophilic species which normally flourish in this pH range. The problem of aluminium toxicity does not appear to arise as concentrations in the soil solution are fairly low at this pH (Grime and Hodgson, 1969). In the northern limestone areas the calcicole niche is generated by the complex of factors already discussed but, in southern Europe and other dry areas, the single factor of water shortage protects them from taller, vigorously growing species. As might be expected, the NPK fertilization of calcareous grassland leads to the massive growth of a number of nutrient-demanding grasses which eliminate the calcicoles (Willis, 1963; Bradshaw, 1969). Similarly, the removal of grazing pressure causes an increase in the rankly growing palatable grasses, followed often by invasion of woody plants. Thomas (1960) describes such changes, caused in British chalk grassland in the late 1950s by the myxomatosis killing of the rabbit population.

Nutrient availability

Several nutrient elements are required by plants in quite substantial amounts by proportion to dry weight, at least in herbaceous tissues (Figure 9.13). The total demand for some of these elements may exceed supply, and plant growth will thus be limited. Nitrogen and phosphorus are most frequently limiting as they are not abundant in parent materials. Potassium and calcium are less often limiting because they occur as common components of aluminosilicate minerals and also, though calcium is often taken up in large quantity, the actual plant requirement is rarely greater than that for some trace elements.

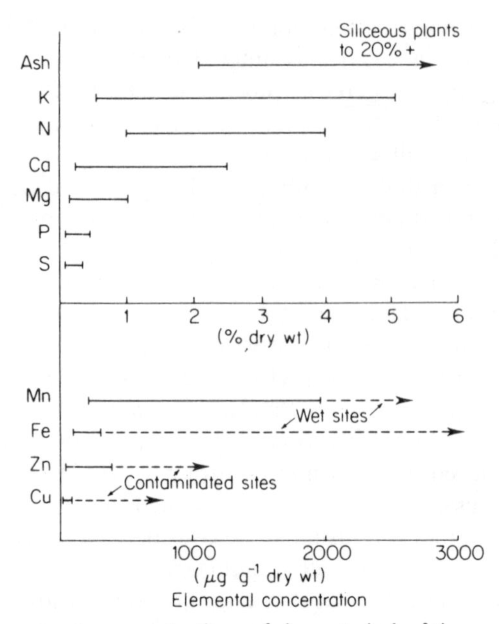

Figure 9.13 Approximate concentrations of elements in leaf tissues. Woody tissue often contains an order of magnitude less of most elements. Data of Rodin and Bazilevich (1967) and Likens and Bormann (1970).

The fact that N and P are the commonest limiting elements is revealed by the surge of growth and change of species composition consequent on NP addition to almost any natural or semi-natural vegetation. The classic Park Grass experiment, started in 1856 at Rothamsted Experimental Station, England, provides good examples (Brenchley, 1958; Thurston, 1969): NPK fertilization gives very large hay yields of robust grasses such as *Dactylis glomerata* (cocksfoot) which reduce diversity by competitively excluding small grassland species. Addition of KP without N does not affect diversity but it does promote the growth of N-fixing Leguminoseae, a consequence which is reminiscent of the aquatic situation where P-eutrophication produces blooms of N-fixing blue-green algae but NP together allow green algae to bloom and the blue-greens are out-competed.

The pattern of change following NP addition varies according to the nature of the original vegetation. In 1930, working in Central Wales, Milton started an agricultural experiment in which species-poor hill grasslands of either *Molinia caerulea* (purple moor grass) or *Festuca-Agrostis* (fescue-bent) were fertilized, limed and exposed to controlled or uncontrolled grazing. During the following 16 years NPK–lime addition caused dramatic increases in species diversity, exclusion of the original dominants and development of a sward resembling lowland pasture (L. I. Jones, 1967). If the original vegetation is species-rich, as in the Park Grass experiment, nutrient addition tends to reduce diversity by enhancing the growth of a few robust species. Willis (1963) confirmed this suggestion by NPK fertilization of calcareous sand dune grassland which permitted *Poa pratensis* (meadow grass), *Festuca rubra* (red fescue) and *Agrostis stolonifera* (creeping bent) to make such lush growth that many creeping and rosette plants were excluded within 2 to 3 years.

All of these experiments reveal the great sensitivity of total production and species composition to nutrient status. Natural habitats vary very widely in their nutrient status and it might be suggested that wild plants may show an equally wide variation in their adaptation to these conditions. The very fact that man has been able to select and breed crop plants which are enormously responsive to fertilization is a consequence of this variation.

Specific adaptation to oligotrophy and eutrophy

The most sophisticated response to nitrogen deficiency is the acquisition of the N-fixing symbiosis but, at a less obvious level, it is now apparent that plants of different habitats vary in their ability to assimilate inorganic nitrogen (Gigon and Rorison, 1972). Lee and Stewart (1978) present analyses for numerous species which show that the nitrate reductase enzyme content of leaf tissue is very low in species of acid soils compared with those of nitrogen rich, disturbed habitats (ruderals). The mean values were, respectively for 20 acid species and 13 ruderals, 0·3 and 5·1 μmol NO_2 h^{-1} g^{-1} fresh weight.

Nitrogen assimilation by plants involves the reduction of nitrate, via nitrite, to ammonia which is then incorporated in organic materials catalysed by glutamine synthetase. Plants of fertile, well drained soils encounter available nitrogen mostly as nitrate but in acid or seriously waterlogged soil no nitrate is formed and nitrogen is assimilated directly as ammonium, thus the low nitrate reductase activity of the acid soil species correlates well with the soil conditions which they endure. Some plants of acid soils which assimilate NH_4–N are harmed by NO_3–N, an unusual situation in which a normally essential nutrient is toxic at physiological concentrations. Various Ericaceae, including the cultivated North American blueberry (*Vaccinium angustifolium*), show this effect (Foy *et al.* 1978). These plants are totally lacking in root nitrate reductase activity.

Lee and Stewart also show results of induction experiments in which plants were exposed to a nitrate source for some hours before the nitrate reductase assay. Plants of very wet, acid soils showed virtually no additional reductase activity but those of dryer acid sites showed increases of up to eightfold. All ruderal species showed the induction effect but never more than about threefold, a predictable result as their soils will already have been nitrate-rich. Herbs of fertile *Quercus robur* (oak) woodland in the field assay had relatively low nitrate reductase activity but manifested large inducible changes, in one species of $\times$ 15.

In this case the interface between the soil inorganic environment and the plant's biochemical pathways seems remarkably well adapted to environmental conditions: energy would be wasted in synthesizing an enzyme which is never needed. Some habitats show marked variation in seasonal nitrate availability, for example as temperatures increase in spring: species which encounter periodic flushes of nitrate are provided with a quickly inducible enzyme system.

Wet acid soils not only rarely form nitrate but also have limited rates of nitrogen mineralization due to their acidity and calcium deficiency. For this reason they are nitrogen deficient habitats for plants and often also deficient in other elements. The remarkable plant adaptation of insectivory is almost confined to such soils and provides plants such as *Drosera* spp (sundews) with sufficient nitrogen to ensure optimum growth in a totally N-deficient soil (Chandler and Anderson, 1976a). It also provides additional phosphorus which Chandler and Anderson found to be advantageous in permitting better vegetative propagation during subsequent growth. Treating the leaves with a fungicide/antibiotic inhibited the insect enhancement of growth (Chandler and Anderson, 1976b) suggesting that carnivory is yet another plant–microorganism symbiosis in which the plants reap the benefit of the bacterial extracellular hydrolytic enzymes. Tracer experiments using insects, feeding-labelled with ^{35}S, have given evidence of the direct incorporation of whole amino acids or polypeptides from the prey into the plant tissue (Chandler and Anderson, 1976c).

The orthophosphate concentration in soil solutions and natural waters is very low (often 10^{-3} to 10^{-2} mg P l^{-1}) and it is to be expected that plants, as for nitrogen, will have acquired adaptive characteristics to different P-supplying environments and to permit pre-emption of limited supplies in competition. Microorganisms, particularly mycorrhizal fungi, play an important role in P-acquisition which has already been discussed on p. 275 ff.

Various workers have suggested that root-surface or other extracellular production of phosphatase enzyme would give the plant access to external organophosphorus compounds. Phytoplankton organisms utilize internal stores of polyphosphate when external orthophosphate is limited and as the reserve is depleted, alkaline phosphatase activity is induced (or de-repressed) permitting them to absorb exogenous organophosphorus (Taft and Taylor, 1976).

Woolhouse (1969a) suggested that acid phosphates activity at root surfaces influences the rate of P uptake and that production of the surface enzymes is a genetically controlled trait varying from one species or population to another. Tinker (1975) questions this as inositol hexaphosphate and other P-esters are strongly adsorbed by soil, consequently little organic-P could reach root surface enzymes. This conclusion might be modified if the ramification of root hairs or mycorrhizal hyphae brought cell surfaces into very extensive and intimate contact with soil particles. The fact that normal and mycorrhizal roots both draw on the same resource of soluble P (Clarkson, 1974) suggests that this is unlikely. Rhizosphere organisms themselves have a demand for P (and other nutrients): it is possible that root-surface phosphatases could play a significant part in competition for organic P between the root and its uninvited guests but there appears to be no information concerning this possibility.

A rather limited number of species are adapted to highly eutrophicated habitats in which nutrient elements, particularly nitrogen and phosphorus, are very freely available. In today's disturbed environment these mainly appear as a group of rather specialized ruderals but their former habitats must have been soils enriched with animal urine and faeces; some of them comprise the 'lair flora' of the Victorian naturalists and others probably colonized the mats of decomposing vegetation cast-up on river, lake and sea shores.

The high levels of nitrate reductase found in such plants has been mentioned but some of them also have a remarkably high phosphorus requirement. An excellent example is *Urtica dioica* (stinging nettle) which is now found in disturbed habitats throughout the world. It has long been described as a 'nitrate plant' both because of its habitat and because of the high nitrate concentration in its tissues (Olsen, 1921). Pigott and Taylor (1964) investigated its growth on natural soils and concluded that its response to phosphate, rather than its nitrophily, control its distribution. In most natural soils its seedlings will not grow much beyond the cotyledon stage unless the soil is phosphorus enriched.

Rorison (1968) compared the response of *U. dioica* and three other species, to P concentrations between 10^{-7} M and 10^{-3} M (Figure 9.14). At the highest concentration *U. dioica* was still responding strongly but plants of oligotrophic acid or alkaline habitats were showing signs of saturating between 10^{-5} M and 10^{-4} M. It is hardly surprising that *U. dioica* is so characteristic of bird roosts, rabbit warrens, dunghill and sewage seepages, not to mention its more recent spread to roadside car-parking areas!

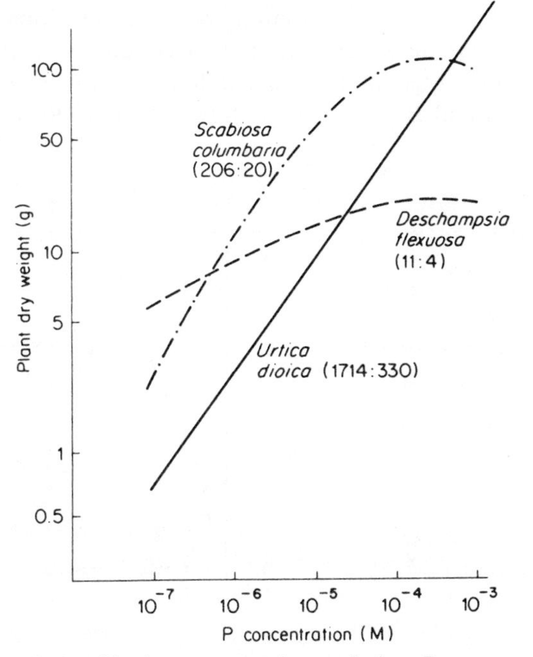

Figure 9.14 The relationship between nutrient solution P-content and dry weight increase of three plant species. The appended figures are the ratios of P-content and dry weight of the 10^{-3}-M P treatment to the 10^{-7}-M P treatment. Modified from Rorison (1968).

Many wild plants would grow very successfully at the high nutrient concentrations associated with these ruderals and it is obvious that disturbance must play a crucial part in their lives, protecting them from competition which would otherwise oust them (p. 384).

General adaptation to variation in nutrient supply

No habitat exactly resembles another in the availability of all essential elements, and either species or local populations thus face selective pressures deriving from these differences, the consequences of which have been demonstrated by innumerable experiments.

The investigations of various grasses by Bradshaw, Lodge *et al.* (1958; 1960), and Bradshaw, Chadwick *et al.* (1960, 1964) showed very clearly the strong, specific differences in response to varying calcium, phosphorus and nitrogen supply which, in most cases, correlated strongly with ecological behaviour. Plants of low nutrient potential habitats, such as *Nardus stricta* (mat grass), were insensitive to increasing nutrient supply, at low levels, and at high levels growth began to be inhibited. By contrast, plants of fertile soils such as *Lolium perenne* (perennial ryegrass) and *Agrostis stolonifera* (creeping bent) showed much improved growth with increasing nutrient supply, particularly at the lower levels (Figure 9.15).

Genetic flexibility permits the selection, over a period of time, of local ecotypic populations which are suited to specific habitat conditions. Such selection may be imposed by a wide range of environmental variables including nutrients. The work of Snaydon and Bradshaw (1961), previously cited, with *Festuca ovina*

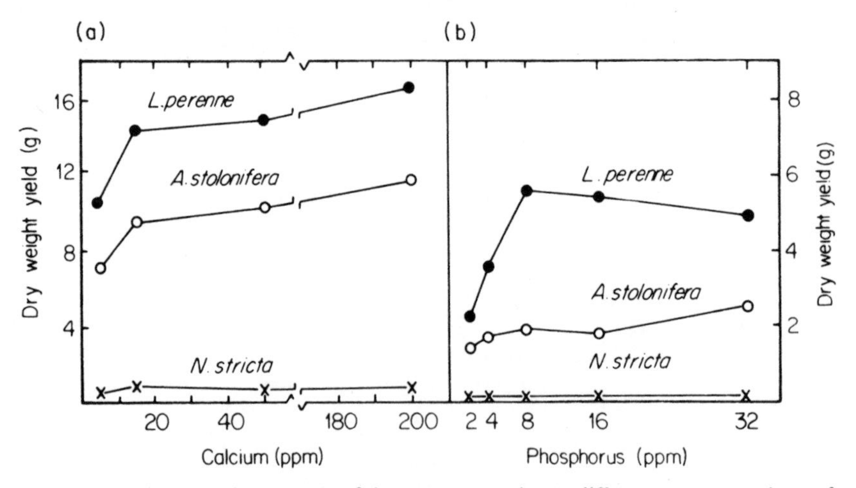

Figure 9.15 The growth response of three grass species to different concentrations of (a) calcium and (b) phosphorus in nutrient solution. Data of A. D. Bradshaw *et al.*, *J. Ecol.* **46**, 749–57, part of Figure 1 (1958) and A. D. Bradshaw *et al.*, *J. Ecol.*, 631–37, part of Figure 1 (1960).

(sheep's fescue) ecotypes illustrates the point. Upland populations had a much lower calcium requirement than lowland populations, suggesting that sites of low nutrient potential are likely to support individuals with low nutrient demand. Goodman (1969) studied a collection of *Lolium perenne* populations and concluded that virtually every combination of growth response to N, P and K could be found. The selection of populations with generally low nutrient demand is probably also favoured by heavy grazing pressure which promotes the survival of smaller, slow growing individuals.

The rapidity of selection for optimum response to particular nutrient levels and the localization of such selection is dramatically illustrated by the work of Davies and Snaydon (1973a, b; 1974) and Davies (1975) with *Anthoxanthum odoratum* (sweet vernal grass) populations collected from the various treatments of the Rothamsted Park Grass experiment. Growth response to calcium in nutrient culture was strongly positively correlated with pH and calcium status of the source-plot soil. The difference was of the same magnitude as some between-species differences of Figure 9.15 and must have evolved in less than 50 years since the start of liming treatments in sites less than 30 m apart. Responses to P and K were similarly positively correlated with the P and K status of the source-plot soil.

A complex of characteristics related to growth and development may be interpreted as adaptive to soils of low nutrient status thus the extent of root hairs or mycorrhizal hyphae, degree of branching of the root system, root/shoot ratio, rate of growth compared with the rate of absorption and mechanisms for recycling or storage may all make important contributions to survival and competition in oligotrophic conditions.

The present consensus is that the major role of root hairs and mycorrhizas is to increase the exploration of soil volume (e.g. Clarkson, 1974). In some species there is a correlation between the frequency of root hairs and extraction of phosphorus from deficient soils while mycorhizal hyphae certainly permit P-transport over several cm in the soil (Clarkson and Hanson, 1980). Experimentally induced nutrient deficiency often results in more profuse branching of roots (Clarkson and Hanson, 1980) and sometimes a more radical redistribution of growth has been shown, thus Wallace *et al.* (1976) showed more than 60% reduction in leaf/root ratio of *Phaseolus vulgaris* (bush bean) in 10^{-5} M P compared with 10^{-3} M P. This ability to modify shoot/root ratio is under genetic control, different varieties of *P. vulgaris* varying widely in response (Gerloff, 1976).

Ecological information for wild plants is sadly lacking but the enormous root systems of sandhill and sand-dune species (Weaver, 1926; Salisbury, 1952) which have usually been interpreted as adaptive to water stress would equally well serve to 'mine' nutrients in these very deficient habitats.

The effect of modified local nutrient supply may also be significant. Dimbleby (1953) found that rotting tree stumps and roots were explored by the mycorrhizal roots of *Pinus sylvestris* (Scot's pine) and *Betula* spp. (birch) Experimental work is much needed but, for *Hordeum vulgare* (barley), Drew and Saker (1978) found that local enhancement of P-supply to deficiency-cultured plants causes compensatory root development sufficient to bring plant growth back to non-deficient control levels. The same workers previously showed a similar response to nitrogen.

The annual loss of leaves to the soil surface and the laying-down of valuable nutrient elements in effectively 'dead' woody tissues are both obvious profligacies. Storage and recycling would be the obvious adaptive responses to this problem in the perennial plant. Goodman and Perkins (1959) showed very considerable accumulation of phosphorus and potassium in the tussocks of the moorland species *Eriophorum vaginatum* (sheathed cotton grass), translocation from dying leaves preventing loss in leaf litter. The leaf-bases of this and many other tussock-forming species of oligotrophic habitats thus represent biological accumulations of nutrients over many years of scavenging from the surrounding environment.

Trees and shrubs which annually accumulate a layer of xylem which remains as physiologically dead tissue for the remainder of the life of the plant usually show substantial export of material, presumably via vascular rays and xylem parenchyma. Table 9.5 shows analyses which indicate that heartwood may contain no more than 5% to 10% as much of the potentially limiting elements as the newly formed sapwood.

Table 9.5 Phosphorus and potassium concentration in sapwood and heartwood of three tree species

Species, age and organ	P($\mu g\ g^{-1}$)		K($\mu g\ g^{-1}$)	
	Sapwood	Heartwood	Sapwood	Heartwood
*Hippophaë rhamnoides**				
Trunk 18 y	365	19	—	—
Trunk 27 y	542	26	—	—
Branch 18 y	182	21	—	—
Branch 27 y	333	25	—	—
*Ulmus glabra** trunk	488	146	—	—
Pinus rigidus† trunk	70	20	490	240

* Vovides and Etherington, unpublished data.
† Woodwell *et al.* (1975)

The retention of leaves by sclerophylly or evergreenness has been suggested as a mechanism for P-conservation (Lovless, 1962; Thomas and Grigal, 1976), limiting the rate at which inorganic-P can re-enter the soil, potentially to be lost. Further studies are required, particularly of long-lived herbaceous perennials such as *Pteridium aquilinum* (bracken) in which individual clones may persist for more than 1000 years (Oinonen, 1967) and must continually remobilize nutrients which will be carried forward with new growth, possibly accounting for the cyclic patterns of invasion, maturity and senescence described by Watt (1947) which can sometimes result in the formation of very large open-centred circular clones of *P. aquilinum* and other creeping perennials.

Longevity and storage may assist plants in oligotrophic habitats but a further adaptation may be an inherently slow growth rate which, effectively gives the plant more time to obtain essential elements and prevents limitation by slow mineralization or diffusion. Hackett (1967) suggested that the slow growth rate of *Deschampsia flexuosa* (wavy hair grass), coupled with its tolerance of aluminium in acid soils, permitted it to find refuge from competition with more vigorous species which, even without the problem of toxicity, would be unable to absorb nutrients sufficiently fast to keep pace with their rapid tissue growth.

Many species show this slow growth adaptation, for example the *Nardus stricta* (matgrass) of Figure 9.15 shows almost no response to phosphorus or calcium supply and maintains a very low growth rate compared with *Lolium perenne* (ryegrass) of fertile soils, which is very responsive to nutrient supply but maintains a greater growth rate than *N. stricta* at any P or Ca level. Similarly, the response of the three *Agrostis* spp. (bentgrasses) to Ca shown in Figure 9.9 indicates that *Agrostis setacea*, a species of acid poor soils, shows no growth response to increasing Ca and has a constant, low growth rate at all external Ca. concentrations. In a later paper Clarkson (1967) showed a similar pattern of response to phosphorus supply and suggested that the slow growth of *A. setacea* was adaptive to low nutrient conditions. Similarly to *Deschampsia flexuosa*, *A. setacea* is also much more aluminium tolerant than the remaining *Agrostis* spp of less acid and deficient soils (Clarkson, 1966).

INTERACTIONS

In the field a number of deficiencies may occur simultaneously and with strong interactions. This is particularly apparent in the relationship of the three major elements, nitrogen, phosphorus and potassium, as illustrated by Figure 9.16 which shows the effects of factorial additions of N, P and K on leaf number in *Ipomoea caerulea* (morning glory). The plants were grown in a basal medium which supplied sufficient NPK to prevent deficiency symptoms but caused a considerable limitation of growth. Addition of N to the basal medium caused a growth increment but P, K and PK, in the absence of N, had little influence. N shows a slight interaction with P, K and PK, its response line becoming steeper with each further interaction. Much stronger interactions are apparent when K is added in the presence of N and NP or P in the presence of N and NK. This particular set of interactions is fairly simple and may be interpreted as the result of N limiting the response to P and K. Less markedly K or P act as limits in the NP and NK treatments, thus reducing the response to N in these two cases. Much more complex interactions may occur and the involvement of synergism, antagonism or chain-linked physiological effects makes interpretation and investigation very difficult. In natural ecosystems a single nutrient element addition may often elicit a response but usually it also reveals the existence of other limiting deficiencies. Furthermore, in mixed species associations, differential responses often lead to changes in the vigour and representation of individual species. The very great differences in floristic composition and productivity which exist between eutrophic and obligotrophic habitats, and which often overrule climatic considerations, may be interpreted as the outcome of differential competitive effects in relation to nutrient supply.

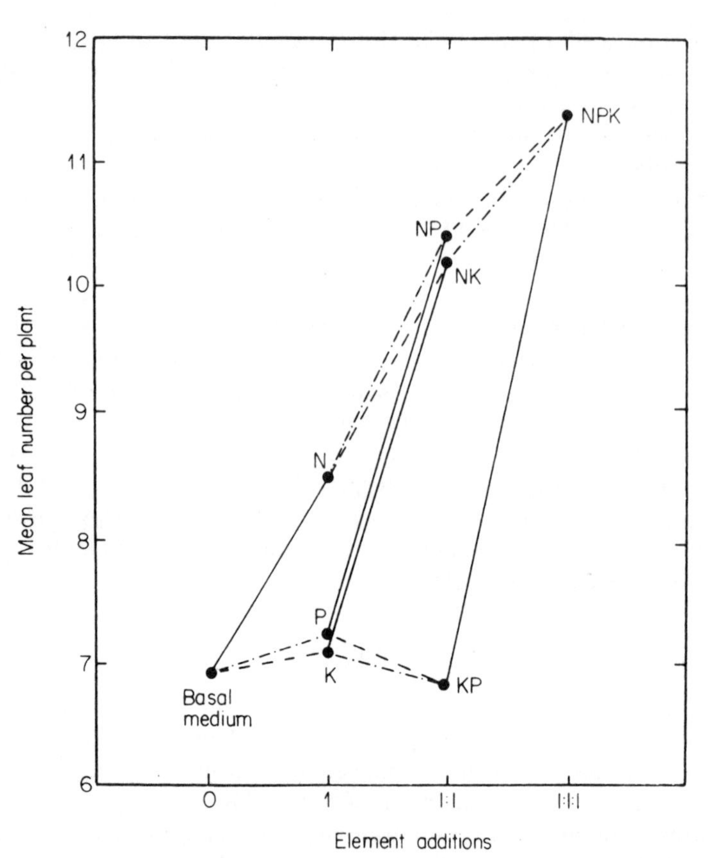

Figure 9.16 The interaction of leaf number with factorial NPK addition in *Ipomoea caerulea* (Data of Njoku, 1957). The method of plotting is that of Richards (1941) which permits interactions to be identified visually. If there are no interactions, the set of response lines for a single element addition will be parallel. In this case N shows a slight interaction with P, K and PK, but the K interaction with P, N and NP is very much stronger. ——— N response; –·–·–·–·– P response; – – – – – K response.

TOXIC SOILS

Some of the most inhospitable environments on earth are those which contain high concentrations of toxic materials such as heavy metals, aluminium, compounds of arsenic, boron, selenium and sulphur, fluorine and various organic toxins. The rather more widespread effects of saline soils will be considered in the next section.

The heavy metals include some 38 elements of density greater than $5\,\mathrm{g\,cm^{-3}}$. Some of these, mainly those which are commoner in the environment, were utilized during biological evolution, thus Fe, Mn, Cu, Zn and Mo are required by plants. Others which are normally present at low concentrations include Ag, Cd, Hg and Pb and may be toxic in very small quantities. The reasons for both essentiality and toxicity are similar: the heavy metals are transition elements which are able to form stable coordination compounds with various inorganic and organic ligands, the metal cation acting as a Lewis acid and being donated an electron from the electron-rich ligand.

Examples of such coordination binding are with the unshared electrons on a sulphur atom, for example in an amino acid such as cysteine, and the coordination of iron or magnesium to four nitrogen atoms in the porphyrin structure of cytochrome or chlorophyll. Thus the coordination structures may be essential parts of the organism's functioning or they may be disruptive, for example interfering with an enzymatic protein configuration.

Geological fortuity, biological concentration during soil formation or mining, smelting and industrial emissions may expose plants to a high concentration of a single toxic material whilst all other environmental factors are normal. The toxin may be either an inessential heavy metal or a required metal at higher than normal concentration. In either case the plants are exposed to a very strong, highly specific selection pressure.

The earliest interest in this subject centred around species which occur in association with ore-bodies and other rocks with a high heavy metal content and rarely, if at all, in other habitats. Antonovics *et al.* (1971) have reviewed the topic and note that higher plants characteristic of metal-contaminated soils have been recognized since the sixteenth century at least. Some have even been used as indicators in prospecting for heavy metals and owe their specific epithet to the association, for example *Viola calaminaria* (pansy) has been used to locate zinc deposits in various parts of Europe.

The earliest descriptions of metal-tolerant plants associated with toxic soils were of a subjective nature and, subsequently, various phytosociological techniques have been used to describe the assemblages of plants found on toxic soils. Such correlative work does not critically prove the existence of tolerant species of ecotypes and it is only since the 1950s that good experimental evidence has been produced. Comparative experiments in which 'normal', and supposedly tolerant, populations are grown in nutrient media containing different levels of the toxic metal have shown that the association of certain plant species with metal toxic soils is due to their ability to evolve metal-tolerant races. Thus Kruckberg (1954) found the serpentine endemic, *Streptanthus glandulosus* (jewel flower), to be a serpentine-tolerant race. Populations from the few non-serpentine localities were much less tolerant. Plants with a more widespread occurrence, such as *Agrostis tenuis* (bent grass) are also frequent colonizers of toxic mining spoil, and selection for tolerant ecotypes has again proved to be the cause.

Antonovics *et al.* (1971) suggest that the appearance of metal-tolerant ecotypes is one of the best documented examples of evolution in action. *A tenuis* populations on toxic spoil have a greater metal tolerance and a lesser variability of tolerance than that of the seed which they produce. This suggests that natural selection is acting to preserve the high tolerance of the adult populations and further investigation shows that this is indeed the case: if non-tolerant seed is sown on contaminated soil, most germinates, but the seedlings fail to root and die. Studies of mine spoil and pasture populations of *A. tenuis* have revealed a higher proportion of tolerant individuals in the mine seed though the level of tolerance is less than that of the adult population because of dilution by non-tolerant pollen from outside the toxic area.

The evolution of tolerant races is extremely rapid, as evidenced by their occurrence on spoil heaps between 50 and 100 years old, while Snaydon (cited in Bradshaw *et al.* 1965) has shown zinc tolerance in *Festuca ovina* (sheep's fescue) and *A. tenuis* under galvanized fences less than 30 years old. Tolerant individuals may also be selected in one generation from the seed of normal populations (Antonovics *et al.* 1971). Thus the initial colonization of a toxic soil relies upon selection of tolerant seedlings from the surrounding normal populations followed by a continuous selection for the tolerance characteristic in the face of the diluting effect of gene flow from surrounding populations and a high rate of population turnover. Species which are described as endemic to, or frequently associated with, toxic soil are those in which the normal gene pool permits the occasional appearance of tolerant recombinants. Tolerant individuals are only present at low frequency in normal habitats as they have less competitive vigour than their normal counterparts (Antonovics *et al.* 1971). There is some difficulty in interpreting this point because mine spoil populations may also have been selected for their ability to survive a physically harsher and chemically more oligotrophic environment than pasture populations. Where tolerant ecotypes are significantly less competitive than normal plants, the toxic habitat may be interpreted as a refugium in the same way that extreme calcifuges and calcicoles escape the competitive pressures of the more normal mesotrophic environment.

The revegetation of toxic waste poses a serious problem which is only totally solved by burying the material under a deep layer of non-toxic soil, though short-term improvements have been made by very heavy liming to precipitate the metals, or organic amendments such as sewage sludge which complex the metals so reducing availability. Recently the potential for selection of tolerant plants has been exploited and Walley *et al.* (1974) give details of a seedling screening process. A number of varieties are now commercially available and Smith and Bradshaw (1979) describe successful trials with *Agrostis tenuis* cv. Goginan (acid Pb and Zn) and cv. Parys (Cu) and *Festuca rubra* cv. Merlin (calcareous Pb and Zn). These permit revegetation at a fraction of the conventional cost and with NPK fertilization give a long-term stable sward (Gemmel and Goodman, 1980). Some care must be taken in grazing management as the grasses take up significant quantities of the metals.

The tolerance mechanism is usually metal specific: selection for one tolerance is not automatically accompanied by another. Tolerance is not 'all or nothing' but continuously variable, being polygenically controlled. In *Anthoxanthum odoratum* and *Agrostis tenuis* the genetic control of Zn-tolerance is directionally dominant with a high additive genetic variance, the mine populations showing constant renewal by gene-flow from surrounding populations while retaining Zn-tolerance (Gartside and McNeilly, 1974).

Phytotoxicity of heavy metals is complex and relatively little is known of the essential physiological processes which are affected (Foy *et al.* 1978). The commonest symptoms are stunting and chlorosis, the latter probably due to induced iron deficiency. Coordination binding to the sulphur atoms in protein S-amino acids is an obvious source of biochemical disruption and it has been suggested that most tolerance mechanisms in resistant plants are related to protection of active enzymes by various exclusion mechanisms.

Explanation of tolerance must account for its heritability, range within a species and metal-specificity. Tolerance is usually positively correlated with the soil metal content and adaptation to one metal rarely confers tolerance to another though Gregory and Bradshaw (1965) did find a population of *Agrostis tenuis* in which zinc tolerance was accompanied by a linearly related nickel tolerance.

Tolerant plants may exclude metals by cell wall binding, presumably a metal specific chelation process, and there is evidence for this (Wainwright and Woolhouse, 1975) but in other cases tolerant plants accumulate as much, if not more metal than intolerant ones. In such cases binding to walls may be in cells other than of the root cortex or some other compartmentation mechanism must be active. Mathys (1977) has suggested that malate may act as a complexing agent for metals, permitting their non-toxic transport through the cytoplasm, and that oxalate may act as a terminal acceptor when the malate complex is transported to the vacuole for storage. It is difficult to see how this mechanism could be metal specific but if it operates it may throw further light on the accumulation of malate in waterlogged root systems (p. 324 ff) as they may also be exposed to toxic concentrations of Fe^{2+}.

The alternative to compartmentation is tolerance and again there is some evidence for this interpretation. Enzyme adaptation to high metal concentrations has been shown *in vitro* (Wainwright and Woolhouse, 1975), and the same authors review information suggesting that mitochondrial membranes and plasma membranes may have similar adaptive tolerance.

Plants may be used as indicators of metal deposits either by studying the distribution of tolerant species and assemblages of species, or by analysis of plant organs in which metals may have become concentrated. Rune (1953) presents a summary of plants which are known to be associated with specific minerals: some examples follow. *Viola calaminare* (pansy) and *Thlaspi calaminare* (penny cress) occur on the calamine (zinc carbonate and silicate) soils of Belgium, Poland, Germany and Austria: Rune considers them to be, respectively, chemomorphoses (ecotypes) of *Viola lutea* and *Thlaspi alpestre*. A whole assemblage of copper-indicating plants has been described for the Congo, amongst them *Buchnera cupricola* and *Guttenbergia cupricola*. Howard-Williams (1970) has investigated the ecology of *Becium homblei* in Central Africa where it is believed to be a copper indicator and though he showed that it does occur on other soils, it is tolerant of up to 15,000 p.p.m. of soil copper and is generally competitively restricted to contaminated soils. The topic of geobotanical prospecting has been well reviewed by Cannon (1960) and Malyuga (1964).

Some further comment is required on the rather widespread, naturally toxic habitat of serpentine. This is a magnesium-iron silicate rock formed by metamorphosis of peridotite which, in many cases, contains chromite ($FeCr_2O_4$) and garnierite (Ni, Mg)$SiO_3 . nH_2O$) (Walker, 1954). Most of the world's serpentine areas are 'barrens' with a very sparse vegetation of unusual species composition, the cause of which may be general infertility due to shortage of major nutrients, nickel and chromium toxicity or imbalance of calcium:magnesium ratio. Kruckberg (1954) established that species associated with serpentine were able to produce races which were tolerant of the soil conditions and Proctor (1971) showed, with *Avena sativa* (oat) as an indicator plant, that toxicity may be due to the high ratio of magnesium to calcium in the soil. He also cites previous evidence, derived from indicator plants, that nickel and chromium toxicity occur and that general infertility may be attributed to a major nutrient deficiency. Proctor and Woodell (1975) give an extensive review.

Aluminium is rarely toxic to animals but plants frequently suffer in acid soils of less than pH 5·5 if the concentration exceeds 0·1 to 1·0 $mg\,l^{-1}$ in soil solution. This has been discussed in more detail in the previous section (p. 267) and Foy (1974) gives an extensive review. Very acid soils may also produce toxicities of iron and manganese though these are more frequent in anaerobic soils where reduction to the more soluble divalent condition may occur (Foy *et al.* 1978). Species differentials of aluminium tolerance have been found and may evolve rapidly, for example in response to acidification of soil by repeated ammonium sulphate fertilization as in some plots of the Rothamsted Park Grass experiment (Davies and Snaydon, 1973b).

Boron may reach phytotoxic levels in a few natural soils, for example where it becomes concentrated by evaporation in salt-desert soils. Reclaimed sites on pulverized fuel ash (PFA) may also contain toxic amounts of boron and sometimes, also, arsenic (Bradshaw and Chadwick, 1980). Some desert plants are remarkably tolerant of boron: *Salsola nitraria* (saltwort) is able to tolerate over 30 % in desert alkali soils (Cannon, 1971). Genetic control of B-tolerance and uptake is known in agricultural plants and presumably exists in wild populations exposed to either deficiency or excess levels.

Sulphide toxicity occurs in waterlogged soil containing free H_2S or S^{2-} (p. 304). The sulphate anion is not toxic except at the very high concentrations associated with sulphide oxidation in mining spoils or drained alluvial soils ('cat clays') but the extreme acidity ($<$pH 2·5) causes extreme aluminium, iron and manganese toxicity which probably mask any sulphate effect. Sulphur dioxide has attracted wide attention as a pollutant and contributor to the sulphur cycle (DOE, 1976 and p. 367), but as a natural phytotoxin it only occurs locally as a volcanic emission.

Fluorine occurs as a toxic component of soils derived from fluorspar-containing materials, for example the gangue-mineral of many limestone lead-zinc ores is calcite with fluorspar which may create land reclamation problems if grazing animals are able to absorb sufficient fluorine to cause fluorosis.

The occurrence of allelopathy, if the evidence is accepted (p. 404) implies a whole range of specific responses to soil-borne toxins. Ethylene is generated in waterlogged or temporarily anaerobic soil (A. M. Smith, 1976) and there is no doubt of the specifically variable responses to this potent growth controlling substance. Many other organic compounds occur as plant and microorganism products in soil and may be responsible for phytotoxicity responses. A few organic compounds of geological origin, for example bituminous materials, may also be toxic (Cannon, 1971).

Table 9.6 gives a synopsis of chemical analyses of plants and soils from both normal and toxic sites together with information on accumulator species where this is relevant.

Halophytism

Another long-recognized ecological group of plants is that of salt affected soils such as the solonchaks of the arid-zones, and the sea-coast salt marshes and mangrove swamps vegetated by a specialized group of halophytes of which the universal attribute is the ability to grow in soils so saline that they are either physiologically toxic to normal plants or impose so serious an osmotic water stress that water absorption is impeded (Waisel, 1972; Reimold and Queen, 1974).

Sodium chloride is not essential for the healthy growth of most halophytes though, in non-saline soils, they usually conompete with normal plants. A few, however, show a strong growth response to additions of sodium chloride well in excess of the trace required by most plants. This response is particularly characteristic of the Chenopodiaceae (Pigott, 1969; Binet, 1978) many of which contribute to the world's halophilous flora.

The concentration of solutes in sea water is about 33–38 g/l giving a solute potential of *c.* $-2{\cdot}4$ MPa (Pigott, 1969). Freshly flooded salt-marsh soil will have this osmotic potential but the upper marsh may have lower values after dilution by rain, or higher values following water loss in dry weather. Salt concentrations in arid zone soils may be very much higher, often reaching saturation, with production of crystalline efflorescences of salt at the soil surface and including sulphates, carbonates and bicarbonates as well as chlorides of Na, K and Ca.

Plant cells which are bathed in a medium of high osmotic potential must maintain a higher vacuole osmotic potential or else lose water. This has caused the evolution of a transport mechanism which permits sodium and chloride accumulation, in halophytes, which contrasts with the exclusion of these ions by normal plants. The internal osmotic environment is thus balanced against the external one and, contrary to past opinion, the plant does not suffer 'physiological drought' by osmotic loss of water to the soil medium.

The halophyte has to cope with other consequences of this accumulation mechanism. Firstly, it must have an internal tolerance of sodium and chloride levels which would be directly toxic to normal plants and secondly, it must concentrate essential ions such as potassium from a low environmental concentration in the presence of high concentrations of potentially competitive ions. The transport mechanism must possess sufficient specificity for the required ion to prevent its competitive exclusion by a high concentration of a chemically related ion.

Table 9.6 Approximate concentrations (mg kg^{-1}) in soils and plants of some trace elements which may occur at toxic levels

	Heavy metals							Semi-metals		Non-metals
	Cd	Cr	Cu	Hg	Ni	Pb	Zn	As	Se	B
Normal soil (*1*)	0·5 +	5 +	5–30	0·01–0·05	2–20	2–10	1–90	0·1–1	0·01 +	2·5
Toxic soil (total)	10–200 + (*7*)	10^2–10^3 (*2*)	10^2–10^3 (*9*)	0·5–16 (*9*)	10^2–10^3 (*2*)	to 4 × 10^4 (*3*)	to 5 × 10^4 (*3*)	10^2–10^4 (*5*)	0·1–80 (*6*)	to 3·7 × 10^4 (*4*)
Normal plant (*1*)	0·1–0·5	0·1–0·5	4–5	0·02–0·2	1–5	1–15	10–50	to *c.* 0·002	to *c.* 0·1	· 10–50
Tolerant plants	10–10^3 (*7*)	10–10^2 (*2*)	10^2–10^3 (*4*)	1–3 (*10*)	10^2–10^5 (*2*)	10–10^3 (*8*)	10^2–10^4 (*8*)	to *c.* 30 (*5*)	10^2–10^4 (*6*)	to 200 (*1*)

1. Bowen (1979).
2. Proctor and Woodell (1975).
3. Etherington, unpublished.
4. Cannon (1971).
5. Wild (1974).
6. Lewis (1976).
7. Denayer de Smet (1974).
8. Johnston and Proctor (1977).
9. Thornton and Webb (1980).
10. Jorgensen (1979).

Halophytes in general cope with the problems of osmotic water stress, cation nutrition and salt toxicity by a series of mechanisms detailed in Figure 9.17 which may be summarized as:

(1) Limitation of uptake or transport coupled with synthesis of organic osmotica.
(2) Unlimited uptake combined with compartmentation or tolerance of high internal salt concentration.
(3) Control of internal concentration and ion-balance by excretion.
(4) Control of Na and K selectivity at root or organelle surfaces.

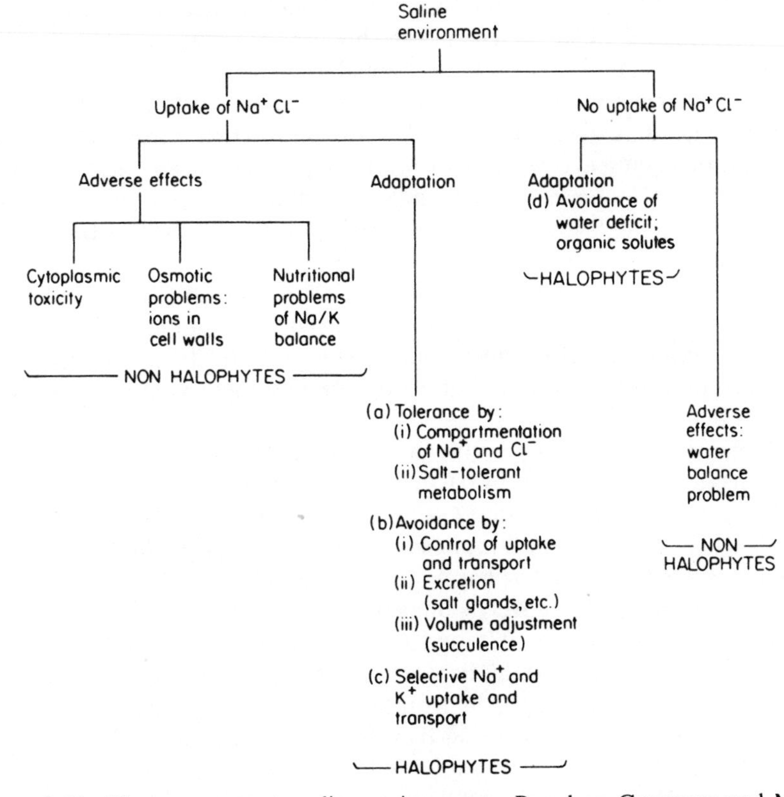

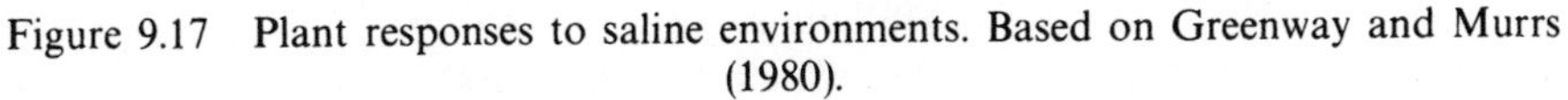

Figure 9.17 Plant responses to saline environments. Based on Greenway and Murrs (1980).

Organic solute synthesis and compartmentation may be discussed together as they are often interrelated in situations where inorganic solutes are sequestered in the vacuole: the cytoplasm is then freed from toxicity problems but requires a balancing osmoticum which is furnished by nitrogenous compounds or carbohydrates. Stewart and Lee (1974) showed enormous accumulation of proline in many salt-marsh halophytes which often reaches 30% of the total amino acid pool and, if confined to the cytoplasm would act as a significant balancing osmoticum. Betain and other quaternary ammonium compounds have been recorded in the same role (Binet, 1978) while marine brown algae accumulate the carbohydrate mannitol. The halophilic green alga *Dunaliella viridis* will grow in 4.25 M NaCl and produces internal glycerol levels approaching 4.4 M (Rains, 1979) but the solute potentials of most terrestrial halophytes are lower than *D. viridis* (equivalent to almost −10 MPa) with a common range from −2·0 to −5·0 MPa.

Regulation of osmolality and ion balance through limitation of uptake or transport shows a complete range from glycophytes (non-halophytes) which specifically exclude Na^+ and Cl^-, to halophytes which both absorb and translocate Na^+ so freely that they have a much higher internal concentration of which over 90% is often present in the shoots (Greenway and Munn, 1980). Maritime plants in non-saline soils occasionally encounter salt if blown sea spray impacts on their leaves, causing either toxicity or osmotic problems. There is now some evidence that differences of leaf wettability deriving from amount and configuration of cuticular waxes may be an adaptive characteristic. Ahmad and Wainwright (1976) for example found that inland populations of *Agrostis stolonifera* (creeping bentgrass) were markedly less water repellent than coastal populations.

Upward movement of Na from the roots is limited in glycophytes which consequently have a much more uniform internal distribution of Na. The ionic balance is variable in halophytes, in some cases Na and Cl accompanying each other in equimolar proportions, but in other cases synthesis of inorganic anions permits exclusion of Cl.

Halophytes which do not control salt uptake may accumulate NaCl surplus to osmotic-control requirements by passive entry with the transpiration stream. Salt may be disposed of *via* specialized salt glands (Lutge, 1975), through modified epidermal hairs (Osmond, 1979) or by shedding of salt-laden organs (Waisel, 1972) but each of these mechanisms has a metabolic cost.

Salt glands resemble actively pumped hydathodes to which they may be evolutionary related (Lutge, 1975) and are universal in plants of saline soils. They occur in temperate salt marsh species such as *Limonium vulgare* (sea lavender) and *Spartina anglica* (cord grass), in tropical mangroves, for example *Aegitalis annulata* and in salt-desert shrubs such as *Tamarix aphylla*. Lutge (1975) suggests that the biochemical control of their ion-pumps shows a fine control relationship with their ecology. *Aegitalis annulata* pumps salt so fast that crystals may accumulate on the leaves, each of which may lose 100 μmol day^{-1}! Not all mangroves rely on salt-regulating glands, for example *Rhizophora mucronata* is an uptake-controlling species.

A number of halophytes secrete salt into bladder-like epidermal hairs, prominent amongst them being many of the Chenopodiaceae. Osmond (1979) gives examples of the concentrations which may be achieved in several species of *Atriplex* (oraches and saltbushes). In these cases the salt is out of contact with living cell contents and may subsequently be removed by rainwater leaching. It may also act as an efficient reflector of long-wave infrared, protecting such leaves from thermal overloading.

Many halophytes are succulent, and succulence may also be promoted in some mesophytes by salinating the growth medium (Jennings, 1968). This is another facet of compartmentation, the large vacuoles of the succulent tissue's cells acting as both salt storage and water reservoirs in which dramatic concentration changes will not accompany small changes of water content. In this function they resemble the malate-storing vacuoles of CAM plants but, surprisingly, only a relatively small number of CAM plants are strongly halophytic though in some cases CAM may be initiated by exposure to salt, for example in *Mesembryanthemum crystallinum* (ice plant) (Kluge and Ting, 1978). Synthesis of proline as an osmoticum is less important in succulent halophytes as might be expected if the vacuole serves as an osmotic 'buffer' (Cavalieri and Huang, 1979).

The halophyte, usually rooted in a sodium rich medium, faces a problem of univalent cation uptake while working in the System II concentration range (Epstein, 1972, and p. 261) which is not particularly ion-selective. A species which does not discriminate against Na will, despite this, increase the K/Na ratio within the plant compared to the soil solution while those plants which exclude Na and synthesize organic solutes as osmotic balancers must withdraw K^+ from soil solution in competition with Na^+ which is present at enormously higher concentration. The solution of this physiological problem is a characteristic of a well adapted halophyte but one which so far eludes the research worker!

Tolerance of high salt concentration in the cytoplasm is rare in vascular plants but halophilic bacteria have organelles which are modified in this way though showing a preference for K rather than Na as the osmoticum, consistent with their tendency for preferential K-uptake (Lanyi, 1979).

Soil salinity is a major limitation to much agricultural crop production and the present research drive is toward selection of salt-tolerant crops. Recent results suggest that genetic variability of salt-tolerance is rather large in crop plants and varieties of *Hordeum vulgare* (barley) have been selected which will germinate and establish in 75% seawater (0·4 M NaCl), while similarly successful pilot results have been achieved with *Triticum vulgare* (wheat) and *Lycopersicon esculentum* (tomato) (Epstein *et al.* 1979).

CONCLUSION

The differentiation of ecological groupings of plants, in relation to nutritional and other soil chemical conditions, is one of the strongest factors promoting localized variability in vegetation, and it reflects both the physiological status of the individual plant species and their competitive relationship with each other. The acquisition of adaptive characteristics suiting species or ecotypes to specialized soil environments provides a very suitable experimental opportunity for the investigation of mineral nutrition at both the physiological and the genetic control level. This opportunity is now beginning to be exploited and may lead to rapid advances in this particular field of eco-physiology.

Chapter 10

Waterlogged soils

Contributed by Dr. W. Armstrong

INTRODUCTION

High rainfall, topogenic water accumulation or poor surface drainage may permit the development of waterlogged conditions and the pedogenesis of the hydromorphic soil types discussed in Chapter 3.

Waterlogged soils are of worldwide distribution. They range from subaquatic sediments, estuarine marshes and swamp through to ombrogenous organic peats. Flooding may be frequent, seasonal or permanent, but even among non-waterlogged soil types there may be occasional temporary flood periods. Rice, which is the staple diet for so large a proportion of the earth's population, is cultivated chiefly as a wetland crop.

Waterlogging and the ability of plants to cope with such conditions is thus of wide ecological and agricultural importance, and as much of our knowledge in this area is still confined mainly to specialist literature, a full chapter dealing with a single soil condition seems to be justified.

PHYSICOCHEMICAL CHARACTERISTICS OF WETLAND SOILS

This section gives an outline of the more fundamental characteristics of wet soils. More detailed accounts have been given by Ponnamperuma (1972) and Gambrell and Patrick (1978).

Gas exchange

In the absence of rapid temperature and pressure fluctuations gas movement in soils occurs chiefly by diffusion (Grable, 1966). Diffusion is essentially a passive process which 'seeks' to redress inequalities in concentration. It is the name given to the *net* transfer of matter from one part of a system to another as a result of random molecular movement. Matter is thus effectively transported from places of higher to lower concentration so long as differences in concentration exist.

The mathematical treatment of the diffusion process originates with Fick (1855: See Nobel 1974) who established that the quantity of substance Q (moles) diffusing through unit area of surface A (cm^2) per unit time t (s), i.e. the flux, is dependent upon the concentration gradient $\partial C/\partial x$ (mol cm^{-3}/cm), normal to the surface, and upon the velocity of the random walk process. The latter, which depends upon the molecular size of the diffusing species and upon the characteristics of the medium in which diffusion occurs, finds expression in a conductivity constant which quantifies diffusivity: the diffusion coefficient D ($cm^2 s^{-1}$). States of quasi-equilibrium in diffusion are common in plant and soil where for respiratory purposes the atmosphere provides both an oxygen source of constant and high concentration (*c.* 8500 nmol cm^{-3}) and a carbon dioxide sink, again constant, but at low concentration (*c.* 12 nmol cm^{-3}). Fick's first law defines diffusion rate at equilibrium as follows:

$$\frac{Q}{t} = -DA\frac{dC}{dx} \tag{10.1}$$

where Q/t is the diffusion rate (mol s^{-1}), and dC/dx the concentration gradient, in which dC represents the concentration or 'pressure' difference of diffusate, and dx, diffusion path length. The negative sign indicates that diffusion is in the direction away from increasing concentration. Diffusion is thus analogous with electricity flow as expressed in Ohm's law, viz. $I = \Delta V/R$, where I, the current (coulombs s^{-1} or amperes) $\equiv$ diffusion rate; ΔV, the potential difference or 'driving force' (volts) $\equiv$ the concentration difference (mol cm^{-3}); and R, the resistance (ohms) $\equiv dx/Da$ (s cm^{-3}).

Resistance to diffusion thus increases as the value of D becomes smaller, and since the diffusivities of oxygen and carbon dioxide in air (0·205 $cm^2 s^{-1}$ and 0·162 $cm^2 s^{-1}$ at 23 °C) are approximately 10,000 × greater than the respective values in water ($2{\cdot}267 \times 10^{-5}$ $cm^2 s^{-1}$ and $1{\cdot}86 \times 10^{-5}$ $cm^2 s^{-1}$), it follows that the resistance of aqueous diffusion paths will normally be of the greatest importance. As a consequence of this we find that diffusion through the extensive gas-filled pore space of the well-drained soil is usually more than adequate to sustain the respiratory activity of the soil population: the rate of respiratory gas exchange may be dependent more upon the thickness of water films which surround the soil organisms than upon distance from the soil surface.

The values of diffusivity given above are unmodified by any structural peculiarities of the diffusion path and are usually denoted as D_0. When due account is taken of the proportion of soil area A which is porous and when allowance is made for any lengthening of the diffusion path due to tortuosity, D_0 is modified and we obtain the effective diffusion coefficient, D_e where $D_e = D_0\tau\varepsilon$, ε being the fractional porosity and τ a tortuosity term, <1. The physical resistance to diffusion in equation (10.1) is thus more accurately defined by the expression dx/D_eA, and if the equation is rearranged to define flux, the resistance becomes dx/D_e (s cm^{-1}).

Whilst x/D_eA (or x/D_e) may satisfactorily express the physical impedance to diffusion in the soil–plant system there is another factor equal in importance which can significantly limit the transport of respiratory gases: the respiratory process itself. Where respiratory sites are distributed along the length of the diffusion path, as through a soil profile, the activities of sites proximal to the oxygen source (the atmosphere) will tend to deprive of oxygen those more remote from it. At the same time the escape of CO_2 from the more remote parts will be hindered by CO_2 efflux in the proximal sites since these lie closest to the CO_2 sink; CO_2 will therefore tend to accumulate at the remote sites. The greater the physical resistance to diffusion the more is the effect accentuated and it is this synergism between respiration and physical resistance which leads to the establishment of anaerobiosis in wet soils. The interaction of these two factors causes what may be termed the effective diffusive resistance (Armstrong, 1979), and anaerobiosis may be thought of as the product of an infinite effective diffusive resistance. The electrical analogue of diffusion in the soil is a 'population' of constant current (i.e. voltage independent) devices (the respiratory sites) tapping electricity at intervals along a linear resistor (the diffusion path).

In situations where effective diffusivity is constant, and respiration uniformly distributed along the diffusion path (x), the synergism between respiration and diffusive resistance finds expression in the following equation:

$$C_x = C_0 \pm \frac{Mx(2l - x)}{2D_e} \tag{10.2}$$

(Greenwood, 1967) where l is the total length of the diffusion path and M, which it is assumed is indifferent to oxygen concentration, is the respiratory rate ($\text{mol cm}^{-3}\,\text{s}^{-1}$); for CO_2 output the right-hand expression will be positive, for oxygen uptake it will be negative (Currie, 1961). From this equation the concentration (C_x) of respiratory gases at all distances ($x < l$ to $x = l$) along the diffusion path may be found, C_0 being the atmospheric gas concentration.

The expression $Mx(2l - x)/2D_e$ represents the oxygen deficit between the surface and point x. However, written in this way rather obscures its origins. In electrical terms it can readily be seen to be the sum of two components. Consider a uniform electrical resistor (length l) from which current leaks away uniformly along its length. It can be shown that the voltage drop across the resistor could equally be achieved if the current was removed in total at a point half-way along the resistor (Armstrong, 1979). Hence from Ohm's law the voltage drop from $x = 0$ to $x = l$ (analogous with soil oxygen deficit) would be given by

$$V = I\tfrac{1}{2}R$$

where I is the total current being lost along the resistor. However at a point x along the resistor the deficit ΔC_x will arise in two ways: the oxygen loss between $x = 0$ and $x = x$ will give a deficit $\Delta_1 = I_x\tfrac{1}{2}R_x$, and a further deficit is brought about by the current flowing through R_x to the more remote parts, $l - x$. This component, Δ_2, is given by $I_{l-x}R_x$. Consequently for a cylinder of soil (radius r) having uniform respiratory activity and diffusivity, the deficit at point x can be expressed as follows:

$$\Delta C_x = \Delta_1 + \Delta_2 \tag{10.3}$$

$$= I_x\tfrac{1}{2}R_x + I_{l-x}R_x \tag{10.4}$$

where I represents the oxygen consumption per unit time. I is given by MV where M is the oxygen uptake per unit volume per unit time and V the volume. Since cylindrical volume is given by $\pi r^2 h$ and R_{soil} by $x/D_e A$ (or $x/D_e \pi r^2$) equation (10.4) may be expanded to give:

$$\Delta C_x = M\pi r^2 x \left(\frac{1}{2} \cdot \frac{x}{D_e \pi r^2}\right) + M\pi r^2 . l - x \left(\frac{x}{D_e \pi r^2}\right) \qquad (10.5)$$

which simplifies to

$$\Delta C_x = \frac{Mx^2}{2D_e} + \frac{M}{D_e}(l - x)(x) \qquad (10.6)$$

and finally to give equation (10.2).

For soils in which diffusivity and respiration vary along the diffusion path the electrical approach can be pursued further and should prove valuable for the non-mathematician.

In the well drained soil x/D_e may never be sufficiently large for the respiratory activity to bring about a state of infinite effective resistance (Currie, 1962), but in wet soils the situation is very different: with the effective diffusivities found in wet soils ($D_e \leqslant 1 \times 10^{-5}$ cm^2 s^{-1}) and with quite unexceptional soil respiratory rates one may predict aerobic conditions within a few millimetres of the soil surface (see Table 10.1). Experience confirms that in the fully flooded soil anoxia approaches

Table 10.1 Surface oxygenation expressed as depth of aeration (cm) in saturated soils: predictions. If in equation (10.2) we put $C = 0$ on $x = l$ we obtain $C_o = (M/2D_e)(l)^2$ from which l, the aerated path length, may be calculated. The table shows how the aerated path length in a saturated soil would vary with different combinations of soil oxygen consumption M_S, and soil oxygen diffusivity $D_{e(s)}$. The solution oxygen concentration, C_o, at the air/soil interface has been taken as $8{\cdot}57 \times 10^{-6}$ g cm^{-3} (= 20·41 % oxygen in the gas phase).

Soil oxygen consumption	Oxygen diffusivity in the wet soil (10^{-6} cm^2 s^{-1})		
(g cm^{-3} s^{-1})	0·56[a]	3·54[a]	10[b]
$5{\cdot}27 \times 10^{-8}$[c]	0·013	0·034	0·057
$5{\cdot}27 \times 10^{-9}$	0·042	0·107	0·180
$5{\cdot}27 \times 10^{-10}$	0·135	0·339	0·570
$5{\cdot}27 \times 10^{-11}$	0·426	1·073	1·803
$5{\cdot}27 \times 10^{-12}$	1·349	3·393	5·702

[a] Computed from the data of Currie (1965).
[b] Greenwood and Goodman (1967).
[c] From data of Teal and Kanwisher (1966).

the soil surface, and that in freshly flooded soils the respiration of aerobic organisms will reduce the oxygen concentration to zero within a few hours (Scott and Evans, 1955). With deeper water tables there will be a fringe of anaerobiosis above the free water level, the depth of which will be related to capillary rise (Boggie, 1972). The soil will be totally anaerobic below the water table unless it is being laterally flushed with oxygen-containing water (Armstrong and Boatman, 1967).

Once the soil is depleted of oxygen and the diffusion rate is no longer sufficient to maintain the supply for aerobic organisms, a new population of anaerobic microorganisms builds up, the already lowered redox potential continues to fall and chemically reducing conditions are established (Ponnamperuma, 1972; Gambrell and Patrick, 1978).

Anaerobic conditions are not always confined to fully saturated soils or soil horizons: anaerobiosis may occur within the crumbs of wet but unsaturated crumb-structured soils. The extent to which this will occur can be theoretically predicted for different levels of soil oxygen demand (Currie, 1962; Smith, 1980) or measured experimentally (Greenwood and Goodman, 1967). The porosity and tortuosity of crumbs can be such that when saturated, oxygen diffusivity can be as low as $5{\cdot}6 \times 10^{-7}$ $cm^2 s^{-1}$. Currie's calculations show that a soil crumb just small enough to remain wholly aerobic (i.e. having the critical dimensions) would, if its radius were doubled, have 30% of its volume made anaerobic; at four times the critical radius the predicted anaerobic volume approaches 85%. An interesting consequence of crumb anaerobiosis is that the oxygen concentration in the intercrumb gas-filled pore space rises since the overall oxygen consumption of the soil falls. Greenwood and Goodman (1967) both predicted and experimentally measured a critical diameter of about 0·35 cm for water-saturated spheres of a clay-loam soil. Figure 10.1 shows the profile of oxygen concentration in such a sphere.

The commonest technique for the assessment of soil oxygen status is based upon polarography and uses a platinum microelectrode in place of the usual dropping mercury electrode. The method was originally described by Lemon and Erickson (1952, 1955) and has recently been excellently reviewed by McIntyre

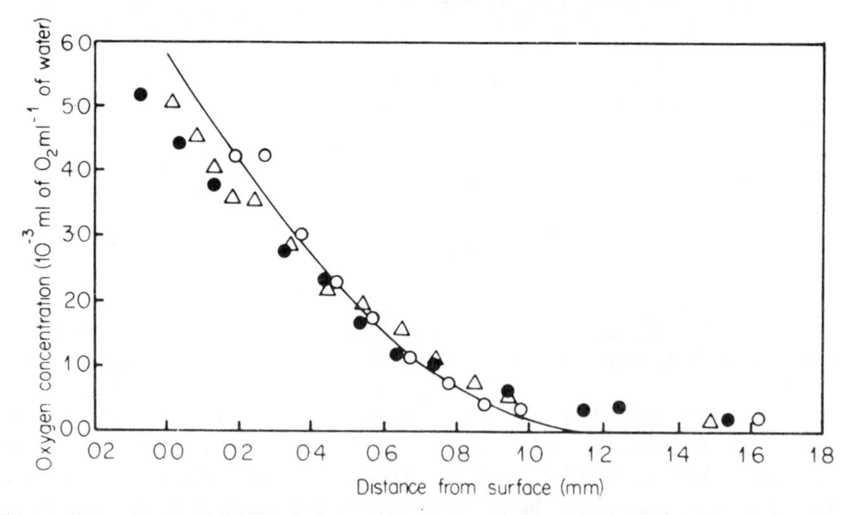

Figure 10.1 Oxygen profiles in a water saturated soil sphere. First, second and third replicated electrode insertions, ▲, ○, ● respectively. Predicted relationship ———. Reproduced with permission from D. J. Greenwood and D. Goodman, *J. Soil Sci.*, **18**, 190, Figure 2 (1967).

(1970). Electro-oxidizable or electro-reducible substances give unique current–voltage curves which permit their identification and assay. At pH values above 3·5 Oden (1962) considers that the electrolytic reduction of oxygen at the platinum microelectrode will follow the equation

$$O_2 + 2H_2O + 4e^- = 4OH^-$$

and for each molecule of oxygen reduced there is a current transfer of $4e^-$. Figure 10.2 shows the typical electro-reduction curves for oxygen at a platinum electrode in both air-saturated and oxygen-deficient waterlogged loam; the measuring circuit is illustrated in Figure 10.3. Oxygen is scarcely reduced when the applied potential is low, and in the absence of oxygen this 'residual current' extends out to point E on the current–voltage curve. In the presence of oxygen, at about point B, the curve steepens, rising sharply with voltage between B and C as the rate of oxygen reduction increases. Between C and D the curve forms a plateau, the rate of reaction being independent of applied voltage and governed entirely by the rate at which oxygen can diffuse to the electrode. The value of this plateau current at equilibrium (minus the residual current if known) is used to calculate the oxygen diffusion rate (ODR) to the electrode using the equation:

$$i_t = nFAf_{x=0,t} \tag{10.7}$$

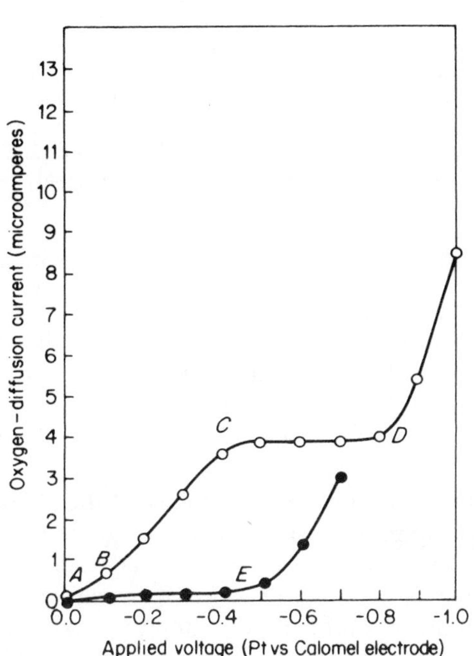

Figure 10.2 Current–voltage curves obtained with platinum versus calomel electrodes used in the circuit shown in Figure 10.3. Letters are referred to in text. Curve for air-saturated waterlogged loam soil (○); curve for the same soil approaching oxygen depletion (●). Adapted with permission from W. Armstrong *J. Soil Sci.*, **18**, 29–31, Figures 1 and 3 (1967b).

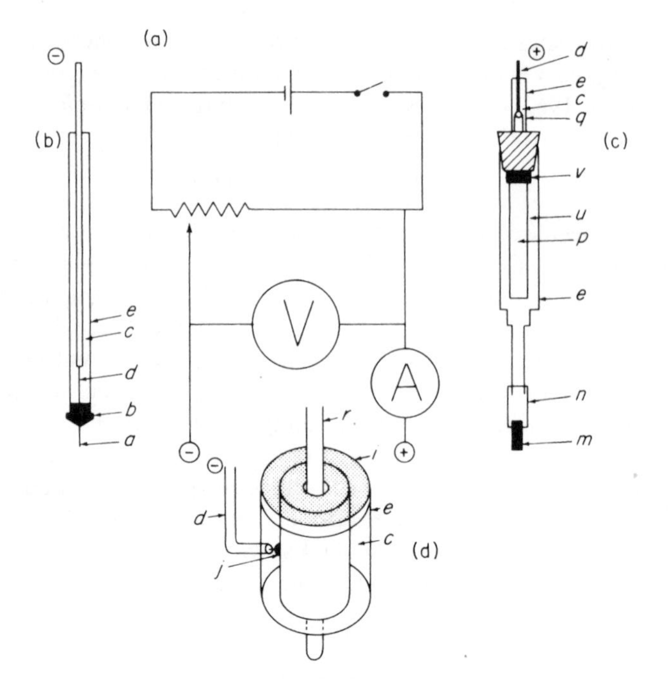

Figure 10.3 Polarographic circuit for assessment of oxygen diffusion rate (a), with platinum wire electrode (b) for measurement of oxygen diffusion rate in soil, and Ag/AgCl reference anode (c). Radial oxygen loss (ROL) from plant roots may be measured by substituting the cylindrical platinum electrode (d) for the soil electrode (b). Key: *a*, 18 swg Pt wire; *b*, Pt–Cu wire junction cemented with conducting 'Araldite' and set in household 'Araldite'; *c*, epon-epoxy resin; *d*, sleeved Cu wire; *e*, perspex tube; *f*, Pt cylinder; *i*, celluloid guide (one at each end); *j*, soldered joint; *r*, root; *m*, glass tube with ceramic or fibre plug; *n*, polythene extension; *p*, Ag/AgCl core; *q*, Ag/AgCl to Cu wire junction; *u*, saturated KCl solution; *v*, household 'Araldite'.

where i_t = diffusion current (amperes) at the time of equilibration t; n = number of electrons required for the reduction of one molecule of oxygen, assumed to be 4; F = the Faraday (96,500 coulombs); A = surface area of the platinum electrode (cm^2); $f_{x=0,t}$ = oxygen flux at zero distance (x) from the platinum surface at time (t).

Because the dimensions and oxygen affinity of the platinum-wire electrode resemble those of a respiring root it has been suggested that the oxygen flux (weight/unit area/time) to the electrode simulates the potential respiratory oxygen supply to the root. This idea has been questioned for well-drained (three-phase) soils, but the analogy probably does apply to a two-phase system such as a saturated soil. The technique has been widely used to specify soil oxygenation status and the results are likely to be trustworthy in wet soils. It must be recognized, however, that certain soil factors may cause complications, for example, plateau shifts accompany the lowering of oxygen status (Figure 10.2) or increase in soil acidity (Armstrong, 1967a; Black and West, 1969). For this reason, great care must be taken to ensure that oxygen diffusion determinations are made at a realistic applied potential which may vary from soil to soil or even within a profile. This caution has not always been observed and the literature abounds with data in which the diffusion values may be erroneously inflated.

As oxygen depletion occurs in waterlogged or wet soils so carbon dioxide concentration increases. Recent work by Greenwood (1970) and Greenwood and Nye (unpublished) seems to confirm the general opinion that CO_2 concentrations in aerobic soils never become sufficiently high to inhibit plant growth. Despite the fact that the diffusion coefficient of CO_2 in water is less than that of oxygen, its solubility is *c.* 27 times greater so that dispersion from sites of production is more rapid than the opposite movement of oxygen. In waterlogged soils, high CO_2 concentrations may be encountered, particularly after addition of fresh organic matter as happens under rice cultivation. However, soil CO_2 levels rarely, if ever, reach lethal values (15–20 %), even in rice cultivation conditions. Dilution by methane, produced from decomposing organic matter, is probably responsible for keeping CO_2 concentration down under these circumstances.

Oxidation–reduction potential

When the oxygen supply is limited, a proportion of soil microorganisms make use of electron acceptors other than oxygen for their respiratory oxidations. This results in the conversion of numerous compounds into a state of chemical reduction and is reflected in a lowering of the oxidation–reduction (redox) potential, a physicochemical property of the soil (Pearsall, 1938).

The redox potential (E_h) of a system is a measure of its tendency to receive or supply electrons and is governed by the nature and proportions of the oxidizing and reducing substances which it contains. For example, a common redox couple in soil is the reversible ferrous:ferric system $Fe^{2+} \leftrightharpoons Fe^{3+} + e^-$. In pure solutions, when the ferrous and ferric ions are at equal concentration, this system has an E_h of +771 mV relative to the standard hydrogen electrode of $E_h = 0{\cdot}00$ mV ($\frac{1}{2}H_2 \leftrightharpoons H^+ + e^-$).

The tendency of systems to gain or supply electrons can be observed quantitatively by measuring the potential at an unattackable electrode, e.g. platinum, which when immersed in a system takes on the electrical potential of that system. If the half-cell formed by the immersed platinum is electrically coupled to a standard half-cell, e.g. the hydrogen electrode, also in contact with the redox system, a cell is thus formed. The potential of the redox system is effectively the electromotive force of this cell and can readily be measured as the imposed potential required to prevent the passage of current through the circuit. (This potential will be equal and opposite to the electromotive potential of the system.) In practice, however, the redox potential is generally measured using a Pt/standard half-cell electrode pair in conjunction with a high-resistance amplifying millivoltmeter. Most modern pH meters have a millivoltage scale which can be used for this purpose. Because of its very high input impedance an electrometer of this type will directly measure the 'cell' e.m.f. with no significant passage of electricity which could change the potential of the system under investigation. In normal usage the standard hydrogen electrode is replaced by other standard half-cells with their own potentials relative to the hydrogen electrode (e.g. the saturated calomel electrode = +250 mV; the saturated Ag/AgCl electrode = +199 mV) and consequently $E_h = E + 250$ (or 199) mV where E is the meter reading.

Redox measurement is slightly affected by temperature (Hill, 1956), but more important is its dependence on pH which complicates comparisons between soils. To overcome this problem, redox potentials have been converted to standard pH values, for example E_6 or E_7, on the assumption that E_h increases by 59 mV for each unit of pH decrease. This may introduce error as the correction value is variable according to the nature of the redox couples involved (see Ponnamperuma *et al.* 1966). It is more desirable to specify both E_h and pH separately.

Redox measurements are chiefly of value in characterizing negative soil aeration, since, in spite of the chemical heterogeneity of soils, it has been possible, particularly for wet soils, to define the potentials at which a number of important chemical changes in equilibria occur. Nevertheless, it is important to realize that redox potential is a measure of intensity level and not capacity. In this it resembles temperature and pH, and just as temperature and pH give no information as to heat capacity and buffering power respectively, so redox potential is independent of the poising effect, the capacity term in oxidation–reduction potentials (Hewitt, 1948).

Figure 10.4 illustrates, in relation to redox potential, the sequence of events which follows the flooding of a mineral soil within a closed system. The concentration of oxygen declined first and was then accompanied by nitrate reduction. (Oxygen and nitrates are usually undetectable in soils at E_7 values below +250 mV.) Reduction of insoluble but easily reducible manganese compounds caused an increase in exchangeable manganese which persisted for the whole of the 7 days period. By contrast, the reduction of iron did not begin until the 4–5th day when redox potential (E_7) had fallen to about 150 mV. This delay was probably due to the reserves of reducible manganese present in the soil. The iron concentration in solution then increased very rapidly to 300 μg/ml on the seventh day when redox potential had fallen to −150 mV.

Reactions which appear as a time sequence in the above examples usually occur sequentially with increasing depth in permanently flooded soils and this is reflected in both oxygen and redox potential measurements.

Comparative studies of soil aeration are more easily conducted by redox assessment than by the measurement of oxygen flux. Permanently installed electrodes, which function adequately for an indefinite period in some soils, can also record the aeration dynamics of field sites. The contrasting aeration patterns shown in Figure 10.5 were obtained in this way. Each is illustrative of soil aeration under a distinctive plant community in the lower reaches of a Humber salt-marsh. *Spartina anglica* (cordgrass) which dominates the lowest elevations receives very

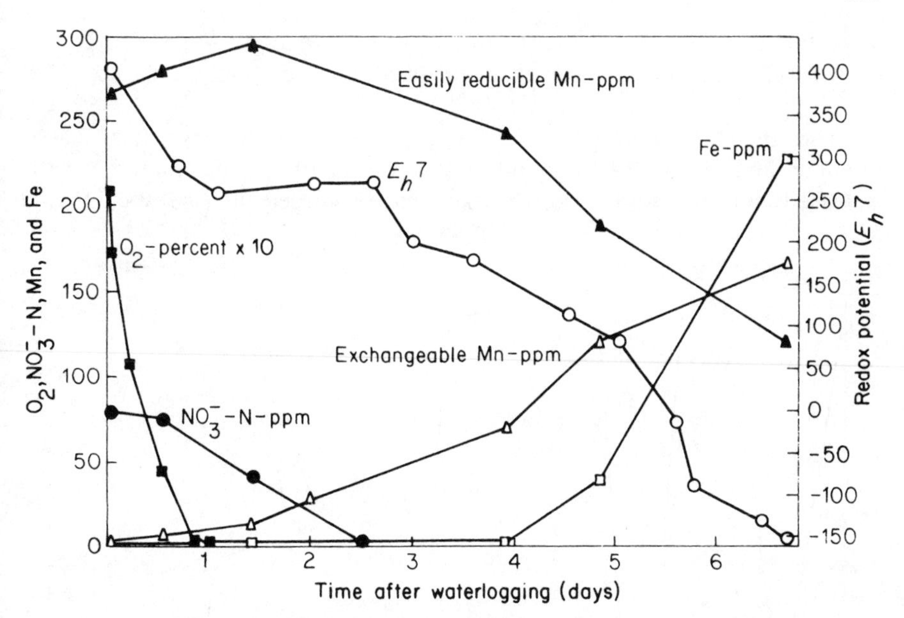

Figure 10.4 Changes in oxygen, nitrate, manganese, iron and redox potential of a silty clay during 7 days following waterlogging. Reproduced with permission from W. H. Patrick and F. D. Turner, *9th. Int. Congr. Soil Sci. Trans.*, Vol IV, paper 6, Figure 4 (1968).

frequent tidal cover and the redox potential shows little tidal response: the soil profile remains totally anaerobic and reducing save during a few neap periods. At these times during what was a particularly dry summer the mud surface became fissured and conditions oxidizing to $\leqslant 5$ cm. At slightly higher elevations flooding is restricted to days of the high spring tides (monthly) and is sufficient to afford periods of very adequate aeration to the marsh grass *Puccinellia maritima*. Neap and low spring cycles give particularly high potentials to at least 10 cm. Rainfall effects brought about by exceptional precipitation during a neap-low spring cycle can be seen in this example. By virtue of its position lining the banks of drainage creeks, the best aeration is afforded to the woody Chenopodiaceous shrub *Halimione portulacoides* (sea purslane). Better drainage is associated with less fluctuation in potential and conditions are rarely reducing even at 30 cm. The data in each example show how responsive is redox potential to soil flooding. The removal and reappearance of oxygen itself probably accounts in large measure for the broad and rapid fluctuations in potential.

Soil nitrogen

Certain facultative anaerobic microorganisms can use nitrate as an oxygen source in respiration causing denitrification by the liberation of gaseous nitrogen or nitrous oxide (N_2O). Ammonia may also be produced after submergence although, contrary to popular belief, little of this is formed from nitrate but originates from anaerobic breakdown of organic matter. At high pH, high temperature and low cation exchange capacity ammonia may volatilize from waterlogged soils.

Losses by denitrification may be very rapid if an energy source is available. Bremner and Shaw (1958) found that addition of organic matter greatly increased NO_3^- loss and various workers have recorded rates between 15 and

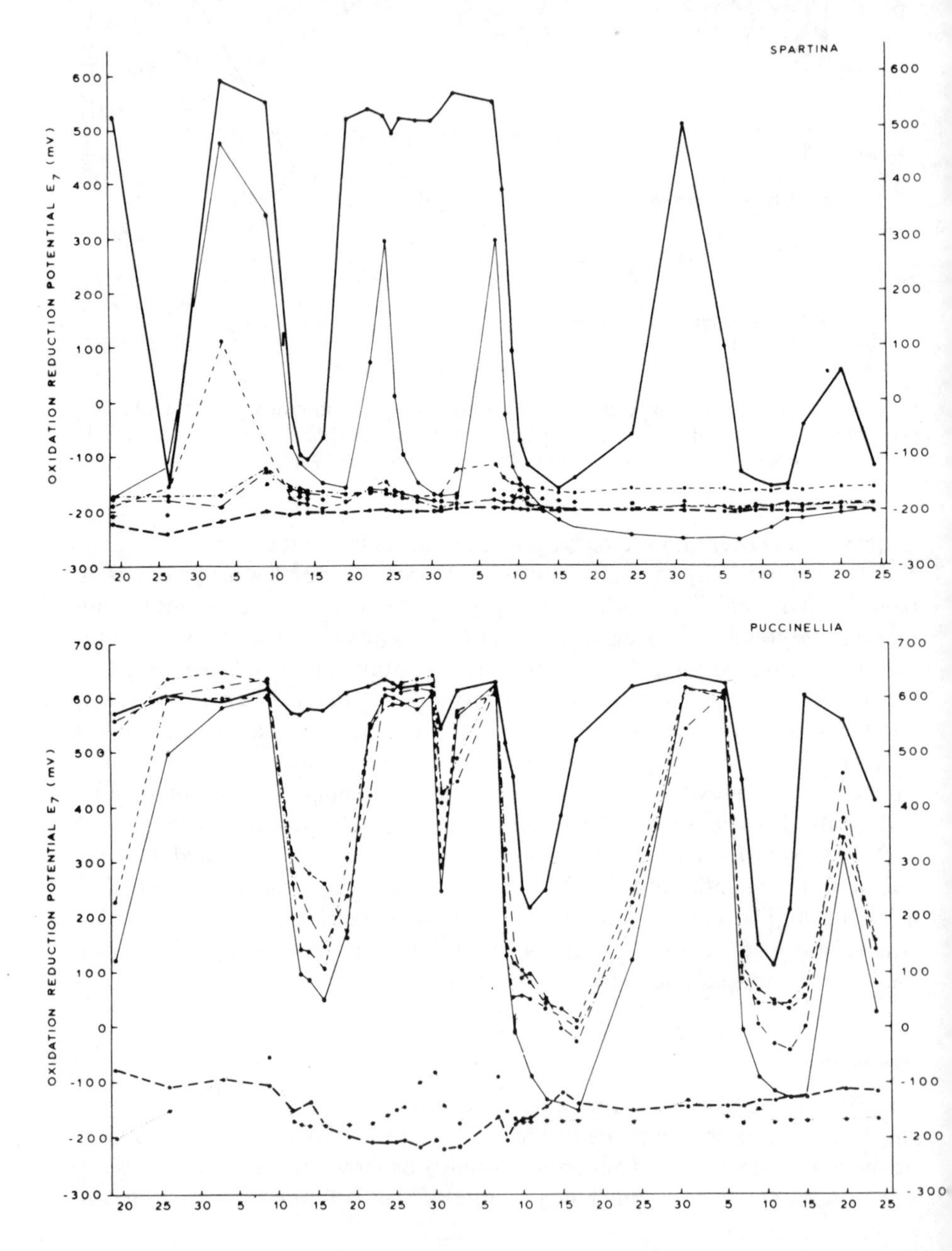

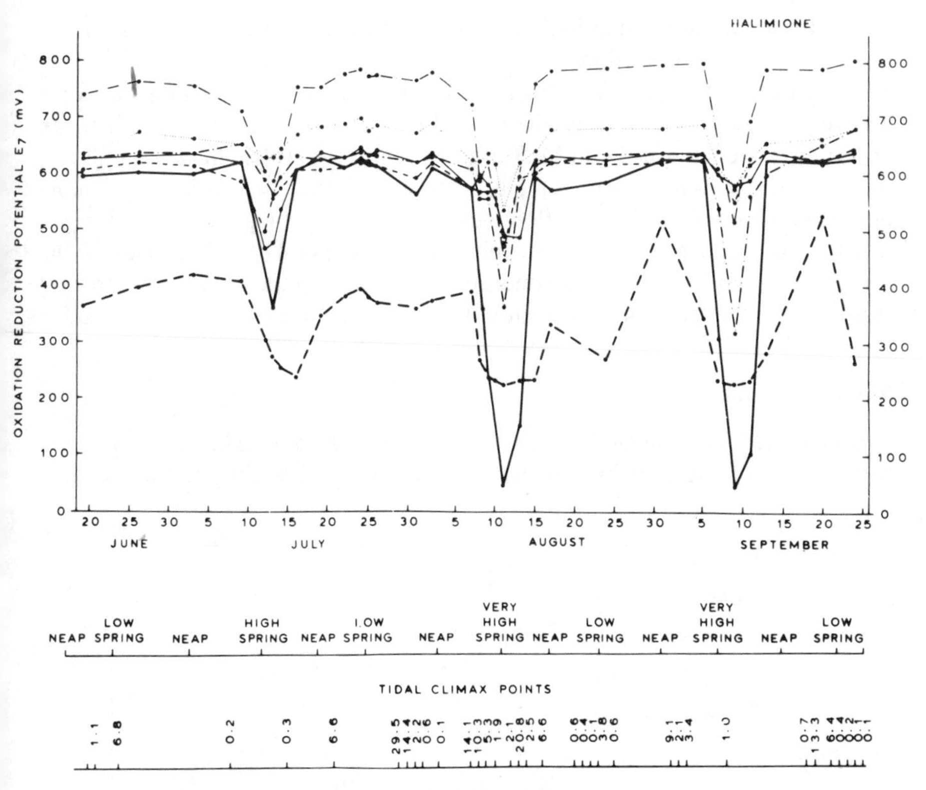

Figure 10.5 Measurements of oxidation-reduction potential in a saltmarsh (River Humber, E. England) during three summer months with permanently installed electrodes at the following depths: ——— 1.5 cm; ——— 2.5 cm; - - - - - 5.0 cm; – · – · – · – · – 7.0 cm; · — · — · 10 cm; .... 20 cm; and - - - - - 30 cm. The aeration pattern is chiefly governed by the frequency and duration of tidal flooding with occasional effects of rainfall. This is particularly noticeable at the *Puccinellia maritima* site where heavy rain at the end of July caused a sudden fall in potential in the 0–10 cm zone and the high spring tides after August 8 imposed an extended period of low potential. W. Armstrong, T. J. Gaynard and S. Lythe (unpublished).

55 μg/g/day from soils to which no additional organic matter was added. In rice cultivation, serious losses from ammonium fertilizers may be caused by rapid nitrification followed by leaching of the nitrate into the underlying anaerobic layers where it is denitrified. Nitrogen usually reaches plants in waterlogged soils as the ammonium ion derived by slow release from decaying organic matter. Under these circumstances it is prone to leaching loss due to displacement by ferrous or manganous ions.

Denitrification is generally associated with oxygen deficient soils. When it does occur in apparently aerobic soils it is likely to be associated with oxygen depletion in the centres of structural aggregates. Greenwood (1962) has shown that nitrification may proceed at half of its aerobic rate in oxygen concentrations as low as 1/80 of air saturation.

Nitrogen fixation by free living microorganisms is important in some waterlogged fertile soils such as rice paddy where blue-green algae form an important nitrogen source but McRae and Castro (1967) also showed significant fixation in dark-incubated samples, suggesting that bacteria are equally important in these conditions. Ponnamperuma (1972) suggested that nitrogen transport through the internal atmosphere of the rice plant (p. 307) and loss to the rhizosphere may provide ideal conditions for fixation in a situation where soil resistance to diffusive movement otherwise limits fixation.

Manganese and iron

Because soils contain more iron than manganese, the dominant redox system is usually that of the iron hydroxides (Figure 10.6) rather than the manganese

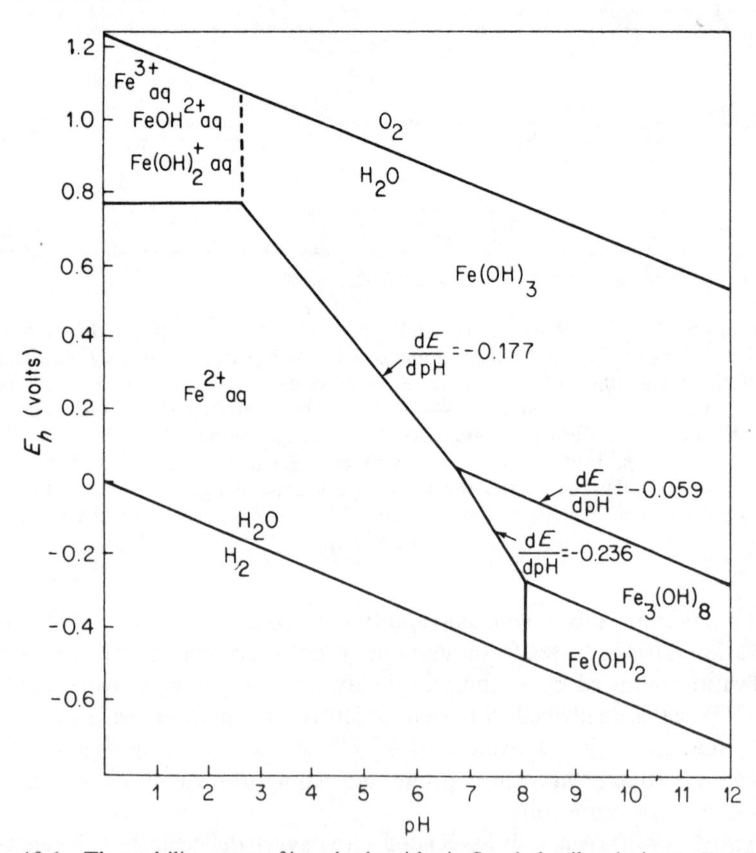

Figure 10.6 The stability areas of iron hydroxides in flooded soils relative to E_h, pH and an aqueous Fe^{2+} activity of one millimolar at 25 °C. Reproduced with permission from F. N. Ponnamperuma, T. A. Loy and E. M. Tianco, *Soil Sci.*, **103**, 380, Figure 1 (1967).

system of oxides and carbonate (Figure 10.7). However, just as manganese reduction does not occur until all free nitrate has disappeared (Takai and Kamura, 1966), so the presence of manganese dioxide or manganic compounds may delay or prevent iron reduction to the ferrous state. Additions of manganese dioxide have been used to buffer rice soils against the development of the extremely reducing conditions injurious to crops. These two Figures indicate the wide range of soil pH and E_h in which aqueous solutions of divalent iron or manganese may occur. More detail of the behaviour of iron in waterlogged soil is given in Figure 9.3. High concentrations of iron and manganese are toxic to plants and some protective mechanism is required by plants which inhabit waterlogged soils (p. 313 and following). With organic matter addition, paddy soils may become extremely reducing and even rice, which normally has a high iron requirement, may develop symptoms of iron toxicity (Ponnamperuma *et al.* 1955).

High concentrations of divalent iron and manganese are not confined only to agriculturally manured paddy soils; for example R. Jones (1967) measured the

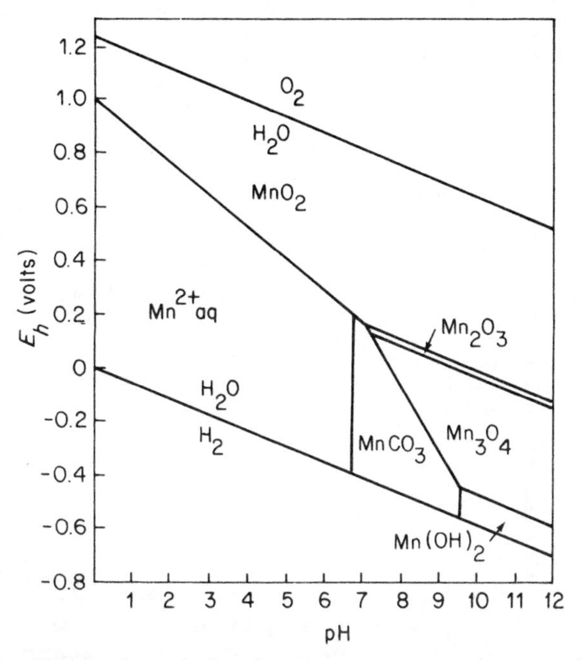

Figure 10.7 The pH/redox boundaries of manganese compounds assumed to be present in manganiferous soils. The E_h of the solutions of flooded soils high in manganese is usually 0·1 to 0·2 volt at pH 6·5 to 7·0, and P_{CO_2} is 0·05 to 0·1 atm. In this region, the stable solid phases are MnO_2 (at E_h higher than 0·17 volt), Mn_2O_3, and $MnCO_3$. If the activity of Mn^{2+}_{aq} and/or P_{CO_2} increases, the $MnCO_3$–Mn^{2+}_{aq} boundary will move left, enlarging the $MnCO_3$ area and excluding Mn_2O_3 and Mn_3O_4. The situation will be reversed if Mn^{2+}_{aq} or P_{CO_2} decreases. For soils low in manganese, in which the less reactive forms of the manganese oxides are involved, the MnO_2 – Mn^{2+}_{aq} boundary will be depressed 0·1 to 0·2 volt, narrowing the Mn^{2+}_{aq} area. Reproduced with permission from F. N. Ponnamperuma, T. A. Loy and W. M. Tianco, *Soil Sci.*, **108**, 55, Figure 1 (1969).

seasonal fluctuation of exchangeable iron and manganese in wet sand dune slacks, finding very high values during the late spring and early summer months. These high concentrations coincided with the period of maximum growth and were presumably associated with increasing temperature and enhanced microbiological activity (Figure 10.8).

Sulphate–sulphide

As redox potential falls, the next inorganic reduction following the formation

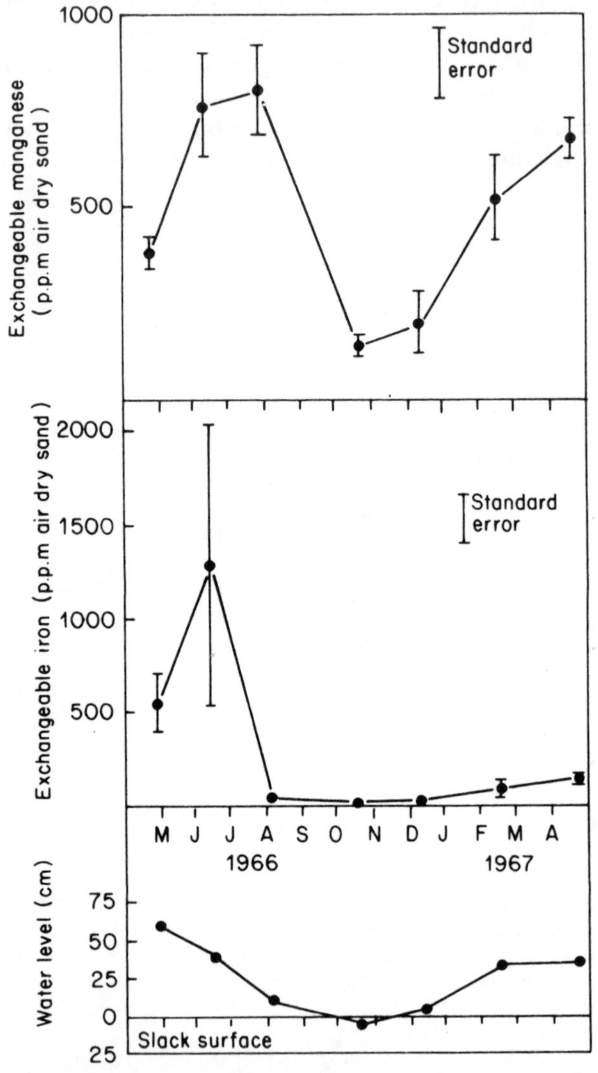

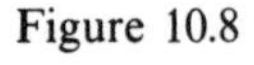
Figure 10.8 Seasonal changes in exchangeable iron and manganese in a dune slack dominated by *Agrostis stolonifera*. From Jones (1967).

of ferrous iron is the reduction of sulphate. This is generally caused by a group of obligate, anaerobic bacteria of the genus *Desulphovibrio* which use SO_4^{2-} as an electron acceptor in respiration. These bacteria are largely responsible for the accumulation of sulphide as iron sulphide in waterlogged soils, bogs and marine sediments.

Sulphate reducers are more exacting in their requirement of anaerobic conditions than most anaerobes and, consequently, require very low ambient redox potentials. They function best near neutrality and have an ecological range from pH 5·5–9·0 (Starkey, 1966). They have a high sulphide tolerance, some species producing up to 2000 μg/ml of hydrogen sulphide in culture solutions. Many bacteria and other microorganisms can produce sulphides from other reduced sulphur compounds but these are normally insignificant compared with the sulphate reducers. Alexander (1961), Postgate (1959), Starkey (1966) and Zobell (1958) have reviewed the physiology and ecology of sulphate reduction.

Various workers have found that sulphate reduction starts at redox potentials between zero and -190 mV (Postgate, 1959; Takai and Kamura, 1966) while Connell and Patrick (1968) demonstrated very strict pH limits for the accumulation of sulphide in soil, reflecting the normal pH tolerance limits of the bacteria concerned. They also found that sulphate became unstable below $E_h = -150$ mV suggesting that sulphide may accumulate abiotically in very reducing soils.

The soluble sulphides (S^{2-}, HS^- and H_2S) are considered highly toxic to plants and other soil organisms. Allam (1971) has reviewed the influence of soluble sulphide on such enzyme systems as catalase, peroxidase, ascorbic acid oxidase, polyphenol oxidase and cytochrome terminal oxidase which affect root oxidative capacity in rice. However, Connell and Patrick (1968) point out that ferric iron is reduced to ferrous at higher potentials than that of sulphate reduction. In iron-containing soils, any sulphide formed by bacteria is likely to be precipitated by the ferrous iron already present. Ponnamperuma (1965) suggests that in normal soils the presence of Fe^{2+} will keep the concentration of hydrogen sulphide below 1×10^{-8} M (0·0034 μg/ml) in solution. Allam (1971) also suggests that the clay fraction may reduce soluble sulphide concentration by sorption.

The concentration of dissolved hydrogen sulphide which can exist in equilibrium with a particular concentration of ferrous iron increases as pH decreases. However, at the same time, the solubility of ferrous iron increases with the decrease of pH. Thus, if the soil has a high iron content the sulphide in solution will remain low, but if the soil is iron-deficient it is possible for dissolved hydrogen sulphide and low concentrations of ferrous iron to co-exist.

Phosphorus

The status of soil phosphorus is discussed in Chapter 9, p. 240, where it is noted that it may occur either as iron phosphate or as occluded phosphate surrounded by a sheath of iron oxide. In both forms the solubility product of the phosphate is very low but iron reduction on waterlogging may result in its release to the soil solution. The transformations of phosphorus in waterlogged soil have been reviewed by Patrick and Mahapatra (1968).

Silica

Increased levels of dissolved silica are characteristic of waterlogged conditions, apparently due to the reduction of iron in ferrisilica complexes. Evidence from rice cultivation suggests that increased availability of silicon in waterlogged soils is beneficial, apparently decreasing the uptake of iron and manganese from solution. Okudu and Takahashi (1964) have suggested that it promotes the oxidizing power of the root (p. 314), probably by improving internal diffusive oxygen transport rather than by any biochemical mechanism.

Organic products of anaerobic metabolism

Among the organic products of anaerobic microbial metabolism are methane, ethane, ethylene, propylene, fatty acids, hydroxy and dicarboxylic acids, unsaturated acids, aldehydes, ketones, alcohols, monoamines, diamines, mercaptans and heterocyclic compounds. Organic materials may, in fact, dominate the redox conditions of many waterlogged soils. For example, the soil solution of many rice soils may contain up to 20 me/l of permanganate oxidizable material of which more than half is organic in nature.

Although it is possible that some organic products of waterlogging may have a beneficial effect as growth promoters (Wang, Chen and Tung, 1967), many are demonstrably phytotoxic (Takai and Asami, 1962; Wang, Yong and Chuang, 1967; Takijima, 1963; Chandrasekaran and Yoshida, 1973; Sanderson, 1977). Acids of the aliphatic monocarboxylic acid homologous series are particularly harmful and increase in potency in the order:

formic < acetic < propionic < butyric < valeric,

which suggests that higher lipid solubility renders the acid more toxic.

Soil pH

Redman and Patrick (1965) studied the pH behaviour of a large number of soils on submergence and noted that those originally above pH 7·4 decreased in value while those below pH 7·4 increased. The reasons for this are discussed in Chapter 4, p. 94.

Specific conductance

With the onset of reducing conditions the specific conductance of the soil solution usually rises. This is partially due to the increase in ferrous iron concentration and also due to the displacement of other cations such as magnesium and calcium from the exchange complex into the solution. Soils initially high in nitrates or sulphates may show a reduction in conductance. In calcareous soils, calcium and magnesium will be mobilized by the increase in carbon dioxide in solution. These cation displacements and CO_2 solvent effects usually cause increases in cation concentration following waterlogging even though they are not directly involved in the soil reduction process. Cations such as potassium may become prone to leaching or lateral drainage loss under these conditions, particularly on agricultural, coarse-textured soils with drains.

HIGHER PLANTS AND THE WETLAND ENVIRONMENT

Many of the characteristics of wet soils which have just been described create problems for plant growth and survival.

Regarding soil oxygen deficiency there have been many attempts to determine the optimum and limiting soil and solution culture oxygen regimes for plant growth. It is not now difficult to understand why the results obtained have been so variable. Oxygen requirements at root surfaces depend upon demand, oxygen permeability of the root wall, radial diffusivity through the tissues and root radius. Each of these factors can vary from root to root and even along the same root; also demand, permeability, and diffusivity are considerably influenced by temperature. Indeed temperature must exert such considerable effects on soil and plant activity in relation to soil wetness that it is a factor which should never be overlooked (Cannel, 1977). It has not been generally appreciated that oxygen can diffuse into the roots from the shoot system and so reduce the need for a radial intake from soil or solution culture (see below). Neither has it been appreciated that stationary moisture films can surround roots and add resistance to the radial diffusion path even in bubbled solution cultures. Further to this, errors of technique have beset most attempts to relate soil ODR to plant performance. Perhaps the least controversial field study yet published has been that of Williamson (1964) who found that for a number of non-wetland crop species optimum growth occurred at a soil ODR of *c.* 150 $ng\ cm^{-2}\ min^{-1}$ and that yield reductions ranged from 25–75% according to species at ODRs ranging from 0–30 $ng\ cm^{-2}\ min^{-1}$.

Internal aeration notwithstanding, waterlogging can quickly effect profound changes in the non-wetland plant, some injurious, some concerned with adaptation to the wetland condition. However, as Drew (1979) has been careful to point out, it would be misleading to ascribe the early effects of transient waterlogging principally to the effects of soil toxins. The accumulation of potentially toxic substances to concentrations injurious to well-aerated plants may take days or weeks depending on temperature, the availability of substrates and the relative rates of production and utilization by microorganisms. Also the presence of oxides of nitrogen and metal oxides can delay the onset of strongly reducing conditions. More and more evidence now points to oxygen deficiency itself as the primary trigger for early plant response. Oxygen deficiency is the first major change effected in the flooded soil and whilst roots can apparently function normally at very low oxygen pressures (Armstrong and Gaynard, 1976; Holder and Brown, 1980; Hopkins *et al.* 1950; Huck, 1970), total anoxia can, within hours, lead to root-tip death in wetland and non-wetland plants alike (Huck, 1970; Webb, Gaynard and Armstrong, unpublished). It should be noted that the immediate cause of death is now thought to be the result of C-starvation (Vartapetian *et al.* 1978) rather than a poisoning by products of anaerobic metabolism such as alcohol. Since non-wetland plants are furnished with less effective systems for internal aeration, their root systems are more readily made anoxic by waterlogging (see later).

Disturbed root function attributable to oxygen deficiency *per se* has now been implicated in many of the symptoms of shoot damage which characterise the non-wetland plant following soil waterlogging. These include reduced transpiration, early leaf senescence, epinasty, wilting, and a slowing of shoot and leaf extention and dry matter accumulation (Kramer, 1951). At the chemical level increased ethylene and ABA synthesis and auxin accumulation are noteworthy.

Free abscisic acid can increase in concentration manyfold within a few hours of oxygen exclusion from roots (Hiron and Wright, 1973). Although the triggering mechanism for this important response is unknown the suggestion that it might be a transient water stress in the leaves now seems less plausible in view of the low water potentials that must develop in excised wheat leaves before there is an acceleration of ABA synthesis (Wright, 1977). The release of ABA provides an explanation for the early closure of stomata and reduction in transpiration which have been observed (Holder and Brown, 1980; Pereira and Kozlowski, 1977); the inhibition of shoot growth and early leaf senescence are possibly conditioned by increased levels of ABA in stressed shoots. However, Drew *et al.* (1979) have shown that the state of plant nitrogen can respond much more rapidly to soil waterlogging than had been thought and that nitrogen deficiency induced by malfunctioning roots might also play a role in the onset of senescence and contribute to the decline in shoot growth. Net rates of uptake of N, P and K to the shoot of barley were reduced shortly after waterlogging and, because dry matter accumulation was little affected initially, the average concentration of these nutrients declined markedly during the first 48 h of waterlogging. Shoot nitrate dropped by $\frac{2}{3}$ and evidence was found of a redistribution of nitrogen from older to younger leaves and tillers preceding the premature senescence of the older leaves. Foliar sprays of the synthetic cytokinin benzyladenine (BA) improved the retention of chlorophyll, suggesting that early senescence might be due in part to an interference with the natural cytokinin supply of the root system. The addition of calcium nitrate to the soil as a surface dressing prevented any signs of waterlogging damage provided that there was present both a skin of oxidized soil and some aerated roots. Since a continuous supply of nitrate is held to be necessary for the synthesis of cytokinins in roots and their transport to the shoots (Wagner and Michael, 1971; Yoshida and Oritani, 1974), it is possible that the nitrate/cytokinin effects were linked in this way. Availability of nitrate for the nitrogen metabolism of the growing shoot tissues may explain their sustained growth. The possibility that added nitrate could serve the root systems as an electron acceptor in place of oxygen has been examined and discounted by Lee (1978) and Drew *et al.* (1979).

Giberellin synthesis in the roots and its translocation to the shoot is also restricted by poor root aeration. Jackson and Campbell (1979) found that the inhibition of stem extension by waterlogging can be lessened by applying benzyladinine (BA) and giberellic acid (GA), or by the presence of aerobic roots. This supports the view of Went (1943) and Reid and Crozier (1971) that hormones from aerobic roots are necessary to sustain normal rates of stem extension.

Waterlogging leads also to increased ethylene levels in shoots (Kawase, 1974; Jackson and Campbell 1975a, b) and accumulation of auxin in the basal parts of the shoot (Phillips, 1964a, b). Both of these plant hormones have been mutually implicated in causing stem hypertrophy and leaf epinasty. Bradford and Dilley (1978) and Jackson and Campbell (1979) have concluded that the higher levels of ethylene in the shoot arise primarily from *in situ* synthesis but triggered by root disorder or ethylene precursors from the damaged root system. Ethylene and auxin are also thought to be involved in the adventitious root replacement which follows waterlogging. Wample and Reid (1979) found that soil flooding which submerged only the apical 2·5 cm of the root system was sufficient to increase shoot ethylene, cause IAA to accumulate, and new roots to be initiated.

Although unable to survive in many situations where wetland roots remain healthy, most non-wetland species probably have some capacity for internal oxygenation. This transport of oxygen from the atmosphere through the intercellular spaces to the roots may be expected to provide some degree of buffering against soil oxygen deficiency. Troughton (1972) showed that the growth of *Lolium perenne* in culture solution was related to the amount of intercellular air-space in the roots even where the solutions were artificially aerated.

The degree to which internal aeration renders non-wetland roots independent of soil oxygen largely remains to be discovered. Recently, however, Healy and Armstrong (1972) have evaluated the effectiveness of internal transport in Pea. In aseptic, anaerobic 3 % agar jelly roots grew to 20 cm or more, but in anaerobic liquid culture periodically gassed with pure nitrogen, growth ceased at 8–9 cm. Mesophyte tissues have low internal air porosities and it was found that oxygen loss by diffusion from the root surface into the liquid medium was sufficiently rapid to deplete the limited internal supply and cause oxygen starvation at a root length of 8–9 cm. By contrast, diffusion into 3 % agar is less rapid and in the absence of degassing permits the build-up of an oxygenated zone around the roots which further reduces the rate of oxygen loss from the root surface. A larger proportion of internally diffusing oxygen is thus available for root respiration and accounts for the prolonged root growth. It follows that because of the high oxygen demand in waterlogged soils, oxygen 'drainage' from the root's limited internal supplies should be a more immediate factor in root failure in mesophytes than the lack of soil oxygen. This may be sufficient to prevent many mesophyte species from occupying a wetland niche, although other soil factors can also be shown to play a part.

The reduced products of more permanently wet soils may play an important part in determining plant distribution. Tolerance of, or susceptibility to, high iron and manganese concentrations affects the natural distribution of many wetland species and may be responsible for the exclusion of some mesophytic species from waterlogged soils. Martin (1968) found that *Mercurialis perennis* (dogs mercury), a woodland plant of well-drained soils, was very sensitive to ferrous iron toxicity in culture solution, compared with *Deschampsia caespitosa* (tufted hairgrass) which inhabits much wetter, heavier woodland soils. He also found other woodland species to have a graded tolerance of ferrous iron which corresponds with the wetness of their natural habitat soils (Table 10.2).

Table 10·2 Tolerance of species to ferrous iron in water cultures. (Reproduced by permission from M. H. Martin, *J. Ecol.*, **56**, 786 (1968))

Species	p.p.m. Fe^{2+} (observations on root system)	
	Survival	Death
Mercurialis perennis	2	4
Endymion non-scriptus	5	10
Brachypodium sylvaticum	–	15[a]
Geum urbanum	10	10–20
Circaea lutetiana	10	15
Primula vulgaris	10–20	20 +
P. elatior	20	30
Carex sylvatica	30	30–40
Deschampsia caespitosa	50	80–100

[a] Not tested below 15 p.p.m. Fe^{2+}

Bannister (1964c) found that the distribution of the two heathers *Erica cinera* and *E. tetralix* was governed by water availability (p. 187) and waterlogging. Fairly short periods of waterlogging were lethal to *E. cinerea* but left *E. tetralix* unharmed. Jones and Etherington (1970) and Jones (1971a, b) studied the problem further and suggested that iron toxicity was a key factor governing this differential response. On waterlogging pot-cultured plants, *E. cinerea* died quickly following the development of a characteristic waterlogging syndrome which included leaf discoloration and massive leaf water loss. On analysis, the *E. cinerea* plants were found to have taken up more iron than the *E. tetralix* plants which were unharmed by the waterlogging.

A second experiment showed that cut shoots of the two species differed in their sensitivity to iron, supplied in solution. *E. cinerea* was quickly killed by concentrations which did not harm *E. tetralix*. The appearance and water status of the *E. cinerea* shoots before death almost exactly simulated the waterlogging syndrome. In a final experiment *E. cinerea* was grown in two peat soils, one of high and one of low iron content. On waterlogging, the plants in the high iron peat developed the waterlogging syndrome but those in the low iron peat survived, confirming the iron toxicity hypothesis.

The influence of chemical processes in waterlogging may be studied, as above, by contrasting various plant species or by comparing waterlogged and well-drained habitats. The sand dune and dune slack catenary system (Jones and Etherington, 1971) is an ideal habitat for this purpose. Willis *et al.* (1959a and b) had already demonstrated the marked relationship between the duration of winter flooding and the distribution of many dune and slack species. Jones and Etherington extended this type of study by comparing the waterlogging response of a number of dune and slack species. The slack species, *Carex nigra* (black sedge), recorded by Willis *et al.* as requiring a long period of flooding, was slightly stimulated by, or insensitive to, waterlogging compared with the dry dune species, *Festuca rubra* (red fescue). The slack species, *Agrostis stolonifera* (creeping bentgrass), was only slightly impeded by waterlogging and *Carex flacca* (carnation sedge), which inhabits both slacks and dune slopes, responded best to a partial waterlogging regime.

Jones (1972a) compared the iron and manganese contents of the plants in this experiment and found that uptake of both metals was greatly enhanced by waterlogging in most cases. Under waterlogged conditions the iron content of roots was much greater than that of shoots. Jones suggested that this might be caused by internal precipitation of iron hydroxides (see p. 313). *C. nigra*, the species requiring the wettest slack habitat, showed a smaller increase in root iron, with waterlogging, than the remaining species. In a further series of experiments Jones (1972b) showed that the dune grass *F. rubra* was harmed by high concentrations of manganese (200 μg/ml) in culture solutions whereas the slack plants *C. nigra* and *A. stolonifera* showed increased or unchanged growth compared with controls containing 2 μg/ml of manganese. Figure 10.8 gives some indication of the natural levels of iron and manganese to which these plants may be exposed.

Iron and manganese toxicity effects have also been documented for agricultural conditions and particularly for rice culture where soils are not only flooded for long periods but organic matter additions result in very low redox potentials. Ferrous iron is, for example, involved in the 'bronzing' disease of rice described by Baba *et al.* (1964) and Ponnamperuma (1965) has discussed the role of divalent manganese in relation to rice nutrition and toxicity effects. In the case of rice there are fewer instances of proven toxicity for manganese than for iron.

In extremely reducing soil, hydrogen sulphide is also responsible for a number of toxicity effects which have been most widely observed in rice culture. Sulphide or ferrous iron toxicity are suspected of causing 'suffocation disease' in Taiwan (Takahashi, 1960a, b), 'straighthead' in the U.S.A. (Atkins, 1958), 'akagare' and 'akiochi' in Japan (Baba *et al.*, 1963; Park and Tanaka, 1968) and 'brusone' disease in Hungary (Vamos, 1959). Hollis (1967) has reviewed toxicity diseases in rice and also drawn attention to the occurrence of symptomless disease caused by soil reduction but manifested only as yield reduction. Tanaka *et al.* (1968) have suggested that the simultaneous occurrence of dissolved hydrogen sulphide and ferrous iron under low pH conditions may cause 'bronzing' disease as high sulphide concentrations interfere with the oxidizing activity of normal roots and allow them to take up excessive amounts of ferrous iron.

Root rot, often accompanied by blackening, is of fairly common occurrence in wet soils and is suggestive of sulphide damage. The 'die-back' disease of *Spartina anglica* (cord grass) involves rotting of the rhizome apex and Goodman and Williams (1961) reproduced the symptoms by addition of sulphide to water cultures. 'Die-back' soils have a very low redox potential and high sulphide content. Armstrong and Boatman (1967) suggested that *Molinia caerulea* (purple moorgrass) was excluded from some valley-bog sites in Northern England by the presence of up to 8 μg/ml of sulphide in soil solution. At the same site, *Menyanthes trifoliata* (bogbean) which has a greater root oxidizing activity (p. 315) was unaffected by a 2 μg/ml dissolved sulphide but a proportion of its roots which had entered deeper horizons with 7–8 μg/ml of sulphide were stunted or dead.

Among the organic waterlogging products which have been considered phytotoxic, butyric acid is the most commonly described, particularly with reference to rice (Mitsui *et al.* 1954; Baba *et al.* 1964). Wang *et al.* (1967) investigated waterlogging injury to sugar cane and found the monocarboxylic acids to be toxic but hydroxy- and dicarboxylic acids were growth-promoting at low concentrations.

Ethylene may sometimes accumulate in soils after waterlogging and whilst it may not be a lethal agent it can depress growth. Smith and Scott-Russell (1969) found that barley root elongation was strongly inhibited by 10 μl/l and reduced to 50 % by 1 μl/l. Soil solutions from waterlogged fields gave equilibria of 0·1 to 8 μl/l but, because of the variable effects of ethylene on plant tissues, according to carbon and oxygen concentrations (Burg and Burg, 1965) Smith and Scott-Russell were cautious in interpreting waterlogging damage as a consequence of ethylene toxicity (see also Smith and Restall, 1971). It must be stressed also that ethylene effects in culture solution are readily reversible upon removal of the ethylene (Drew, 1979).

ADAPTATIONS TO THE WETLAND ENVIRONMENT

Although the wetland environment is unusually hostile to plant life there is nevertheless a vast assemblage of species either endemic to wet sites or tolerant of some degree of soil anaerobiosis. Many plants can produce extensive and healthy root systems in all but the most reducing of soils.

The degree to which tolerance is achieved may depend upon one or more of a number of recognizable features characteristic of endemic wetland species. These will be discussed in the remaining sections and include: (i) the ability to exclude or tolerate soil-borne toxins, (ii) the provision of air-space tissue, (iii) the ability to metabolize anaerobically and tolerate an accumulation of anaerobic metabolites and (iv) the ability to respond successfully to periodic soil flooding.

Exclusion of soil toxins

Extraction of roots from waterlogged soil or observation of such roots through a glass screen almost always reveals red-brown deposits of ferric compounds associated with part of their length (Figure 10.9). Sectioning of these roots may even show oxidized iron deposits on the walls of the cortical intercellular space (Armstrong and Boatman, 1967; Green and Etherington, 1977). The iron II oxides Goethite (α-FeOOH) and Lepidocrite (τ-FeOOH) have been identified as major components of deposits on rice roots (Bacha and Hossner, 1977; Chen *et al.* 1980). Chen *et al.* found that cultivar, growth stage, and soil type were significantly related to the quantity of FeOOH deposited on and in rice roots. Accumulation varied from $<2\%$ of the dry weight of roots at 7 days flooding to a maximum of 10% at maturity. The role of iron as a toxin in wet soils has already been mentioned and these deposits indicate that the plant may have the ability to exclude significant quantities of iron from its roots by re-oxidation processes (see

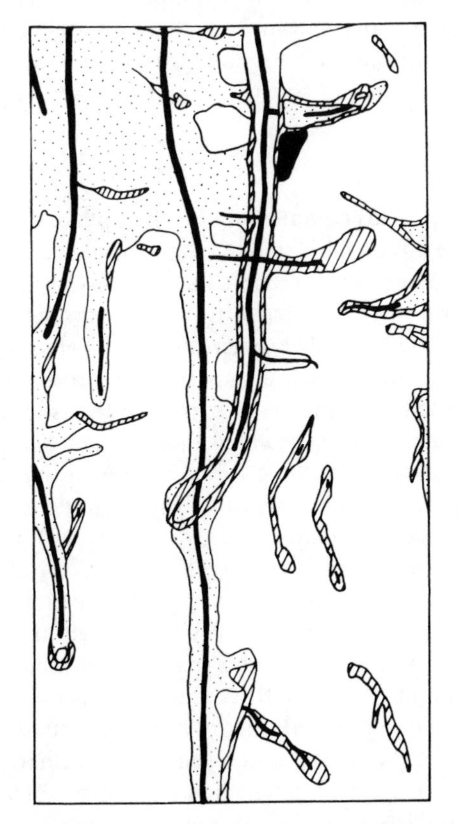

Figure 10.9 Deposits of iron oxides around the roots of *Eriophorum angustifolium* in a waterlogged soil of pH 6·2–6·4. (Reproduced to scale from a colour transparency). The diameters of the main roots range from 0·06 to 0·1 cm. Reproduced with permission from W. Armstrong, *Physiol. Pl.* **20**, 924, Figure 1 (1967). Key: Roots ■ ; Hydrated ferric oxide light deposit ⚄, heavy deposit ▨ ; soil remaining in a reduced state □.

also Engler and Fitzpatrick, 1975). As early as 1888, Molisch showed root oxidizing activity toward substances such as pyrogallol, and Raciborski (1905) stated that while no phanerogams lack such powers, there are specific differences in effectiveness.

Such differences have since been correlated with tolerance of waterlogged soil and Fukui (1953) related the ability of roots to oxidize α-naphthylamine with their ability to penetrate reduced paddy soil. Bartlett (1961) investigated the oxidation of iron compounds adjacent to plant roots in soils and culture solution and showed that high oxidizing capacity conferred the ability to limit iron uptake from a reducing environment. Rice, which is of such economic importance and is associated with very reducing soils, was claimed by Doi (1952) to have the highest root-oxidizing capacity of any known plant. Goto and Tai (1957) found that resistance to rice root rot was correlated with the ability to oxidize the redox dye, aesculin.

The exclusion of soil-borne toxins by oxidation can be related to radial oxygen loss (ROL) from the roots causing direct oxidation of the rhizosphere, to enzymatic oxidation within the root and at its surface, and to microorganism-dependent oxidations adjacent to the root surface.

Radial oxygen loss

Cannon and Free (1925) commenting on the uptake of oxygen from soil by roots suggested: 'It is conceivable that ... in appropriate conditions movement of oxygen may be in the opposite direction, that is to say, from the plant into ... the atmosphere of the soil'. Van Raalte (1941, 1944) showed, by redox measurement, that rice roots were able to oxidize anaerobic media. He suggested that oxygen, which he detected in the cortical air space of the roots at concentrations of *c.* 8%, was transported from the aerial parts and that soil oxidation ultimately depended on this supply. He speculated that oxygen was probably, itself, the oxidizing medium diffusing from the roots, but noted that some other organic redox-couple in its oxidized state might diffuse from the root surface. He called this oxidized substance a 'bio-indicator' by analogy with redox indicators.

Van Raalte's work established not only the concept of radial oxygen loss from roots but also the need for internal diffusion from the aerial parts to replenish the supply. Isotopic techniques subsequently confirmed that the movement of oxygen in the stem and root intercellular space can be modelled as diffusion in a hollow tube. Evans and Ebert (1960) established this point using ^{15}O in *Vicia faba* (broad bean) seedlings and Barber *et al.* (1962) found that the build-up of ^{15}O activity in the roots of rice far exceeded that in barley, a finding which correlated well with the larger intercellular air space of rice. These workers suggested that the large volume of internal atmosphere was one factor determining the ability of rice to withstand waterlogging.

Radial oxygen loss from roots was first demonstrated using polarographic techniques (Heide *et al.* 1963; Armstrong, 1964; Vartapetian, 1964). Armstrong (1964 and 1967b) measured root oxygen losses using a hollow cylindrical platinum electrode insulated on its outer surface (Figure 10.3). The root is threaded through the platinum cathode and the oxygen diffusion current measured using the circuit and technique previously described (Figure 10.3). If the root is in a de-oxygenated medium the oxygen diffusing from it to the platinum cathode is the only source of an oxygen diffusion current and the radial oxygen loss from the root may be calculated from equation (10.7) where A = the surface area of root within the electrode. The oxygen flux is usually given as $ng\ cm^{-2}$ root surface, min^{-1} (see Armstrong, 1979).

Measurements of radial oxygen loss have shown that the lowest values are associated with non-wetland species such as *Mercurialis perennis* (dog's mercury) (Martin, 1968) and Pea (Healy and Armstrong, 1972). Oxygen flux from the roots of wetland species is usually higher and, corresponding with general oxidizing activity, is highest near the apex. Table 10.3 shows some of the interspecific or

Table 10.3 Radial oxygen losses from the roots of some wetland species ($ng\ cm^{-2}\ min$). All values refer to the apical centimetre of root. Root lengths approximately 6–10 cm. (Reproduced with permission from W. Armstrong, *Physiol. Pl.*, **20**, 540–53 (1967b); **25**, 192–7 (1971))

Species	Value
Menyanthes trifoliata	163
Eriophorum angustifolium	128
Schoenus nigricans	128[a]
Juncus effusus	71
Spartina townsendii (s.l.)	67
Molinia caerulea	14
Rice cv Norin 36	
—waterlogged	150
—non-waterlogged	75
cv Norin 37	
—waterlogged	183
—non-waterlogged	71

[a] Over the ½ cm segment behind the meristem the value rose to 190.

intervarietal differences which have been found. In the cases examined, increasing oxygen diffusion rate correlates with improved ability to root in wetland soil (Martin, 1968; Armstrong and Boatman, 1967; Boatman and Armstrong, 1968; Keeley, 1979). Teal and Kanwisher (1966) concluded that the dominance of *Spartina alterniflora* (cordgrass) in intensely reducing salt-marsh muds is related to internal aeration which is sufficient to supply both root respiration and also the oxygen demand of the reduced mud.

Protection by radial oxygen loss implies the formation of an oxygenated zone in the rhizosphere which forms a buffer between the cells of the root and the hostile soil environment and the pattern of iron deposition adjacent to roots (Figure 10.9) appears to support this view. Microorganism activity may be aerobic in this zone (Read and Armstrong, 1972). Armstrong (1970) has predicted the size of the oxygenation zone for particular species and known soil oxygen consumption using a mathematical model. The model is based on Fick's law of diffusion and relates the diffusion of oxygen from a cylindrical source (root) to consumption by the soil oxygen sink. It defines the thickness of the oxygenated root sheath in terms of two soil variables, the oxygen consumption rate and the oxygen diffusion coefficient, and two plant variables, the root radius and the oxygen concentration at the root surface. This concentration is assumed to be the solution equilibrium which corresponds with the oxygen concentration in the cortical gas space, when the root is in a steady state of oxygen loss to a reducing soil.

In Figure 10.10 the thickness of the oxygenated sheath, predicted from the model, is plotted against experimentally obtained value of radial oxygen loss for three British bog plants and two rice varieties. At higher values of ROL a higher root surface oxygen concentration can be maintained and the diameter predicted for the sheath becomes greater. This is important as many reduced inorganic toxins including forms of ferrous iron, manganous ions and sulphide are only slowly oxidized. Increased thickness of the oxygenated zone increases their

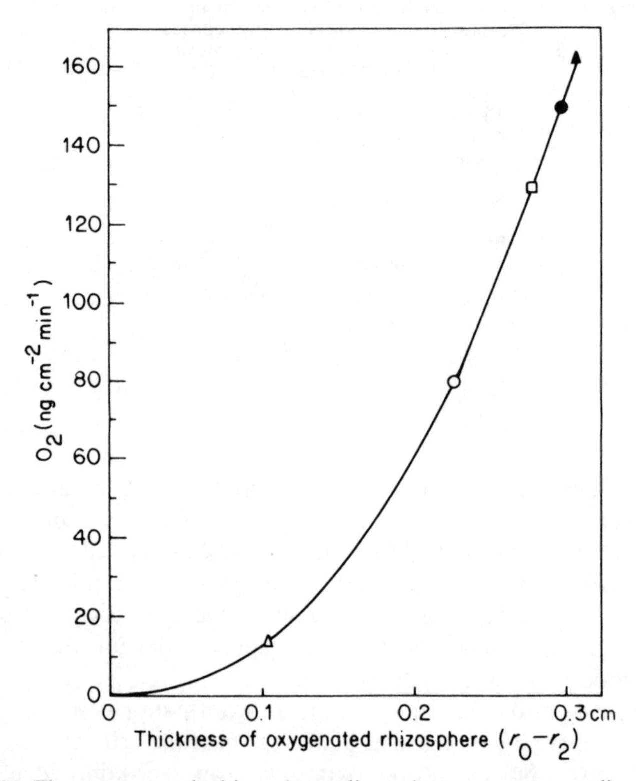

Figure 10.10 The oxygenated rhizosphere dimensions, $r_0 - r_2$, predicted for rice cultivars Yubae (●) and Norin 36 (○), *Menyanthes trifoliata* (▲), *Eriophorum angustifolium* (□) and *Molinia caerulea* (△) plotted against the values of ROL previously measured for these species. Reproduced with permission from W. Armstrong, *Physiol. Pl.* **23**, 623–30 (1970).

residence time as they move in with the transpiration stream, and makes it more likely that they will be made harmless before reaching the root surface.

Figure 10.11 shows the predicted thickness of the oxygenated sheath in relation to root radius and for specified values of soil oxygen consumption. With soils of average oxygen consumption rate, narrow roots, e.g. laterals, may maintain an oxygenated rhizosphere which is many times their own diameter. The model still predicts an oxygenated zone for narrower roots in soils of high oxygen consumption. However, narrow lateral roots may lack the gas space provision necessary to provide oxygen concentrations within the root at a sufficient level to bring about the predicted rhizosphere oxygenation.

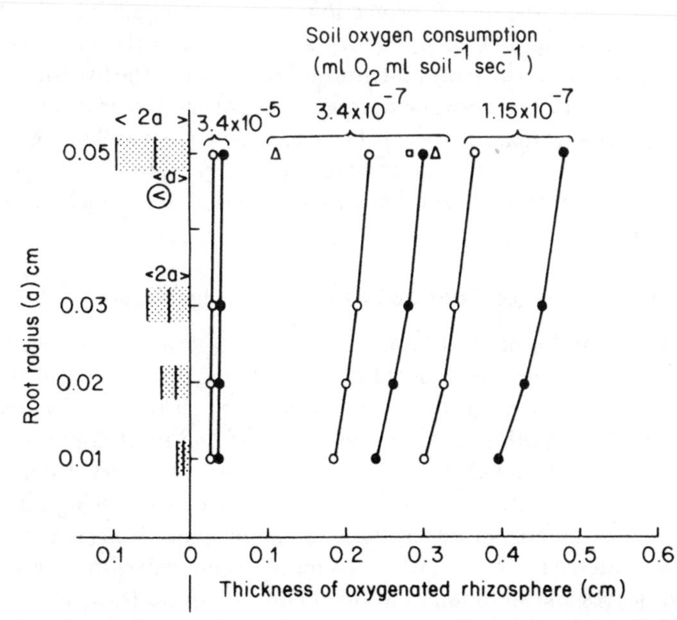

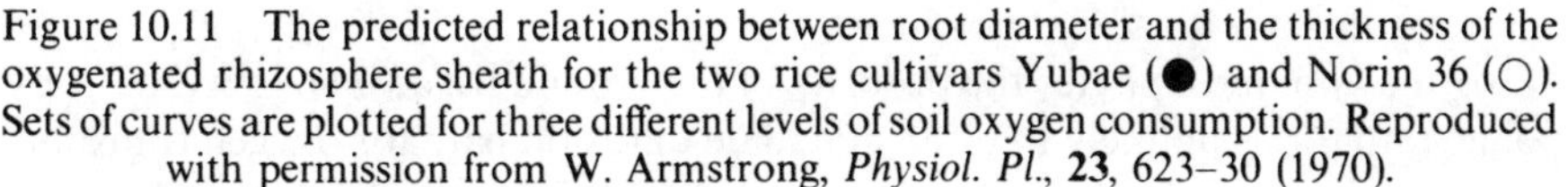

Figure 10.11 The predicted relationship between root diameter and the thickness of the oxygenated rhizosphere sheath for the two rice cultivars Yubae (●) and Norin 36 (○). Sets of curves are plotted for three different levels of soil oxygen consumption. Reproduced with permission from W. Armstrong, *Physiol. Pl.*, **23**, 623–30 (1970).

Enzymatic oxidations

It seems certain that direct oxidation by diffusing molecular oxygen only account in part for observed root oxidizing activity. Yamada and Ota (1958) found that extracts prepared from rice roots actively oxidized iron in the presence of molecular oxygen. Intact root systems performed similarly. They attributed the activity to a new kind of peroxidase or new iron enzyme closely related to peroxidase, and the ratio of Fe^{2+} oxidized to oxygen consumed indicated that in the reaction $Fe^{2+} = Fe^{3+} + e^-$, $\frac{1}{2}O$ was used as an electron acceptor. Concentrations of $FeSO_4$ which proved to be very toxic to the rice plant retarded oxidation by the extract to the level of a boiled control.

Armstrong (1967c) attempted to compare the contribution made by oxygen *per se* and other oxidants to the oxidizing activity of *Molinia caerulea* (purple moorgrass) and *Menyanthes trifoliata* (bogbean) roots. By comparing reduced dye-oxidation by living and artificial roots with assessments of radial oxygen loss he concluded that enzyme-mediated oxidation might account for up to 90% of the total.

Microorganism-induced oxidations

Pitts (1969) and Hollis (1967) have shown that the bacterium *Beggiatoa* may have a complex ecological relationship with rice and, perhaps, other marsh plants, which gives the plant protection from sulphide toxicity. *Beggiatoa* oxidizes hydrogen sulphide to sulphur, intracellularly. It is also known to need an external supply of catalase to prevent auto-intoxication by the hydrogen peroxide which it produces. Pitts demonstrated the release of catalase from rice roots and also found that *Beggiatoa* isolates, from a paddy soil, grew well in company with rice roots. There is thus a symbiosis in which *Beggiatoa* protects the rice roots from high sulphide concentrations and rice provides the catalase which maintains the bacterial system.

The provision of air-space tissue and its functional significance

Under normal soil water conditions, the oxygen entering the root from the soil will probably satisfy the oxygen demand of the cortex. Whether it will also supply the stele requirement is less clear (Fiscus and Kramer, 1970). As soil water content increases, the liquid path to the root will increase and reach a point where the external supply will be insufficient to meet the root respiratory demand. Under fully waterlogged conditions there will be no external oxygen supply at all. In nearly all plants there is, however, the alternative route of gas exchange in the ground tissue intercellular space which connects, ultimately with the atmosphere, through the leaves and stomata. In woody plants the lenticels usually provide the nearest contact with the atmosphere.

Normally in mesophytes the porosity of the ground tissue is so low (2–7% by volume) that it is unlikely to be an adequate diffusion path for oxygen transport to the underground parts (see also pp. 314 and 322): the synergism between respiration and low porosity can severely limit the length of aerated root. A rough guide to the relationship between length of aerated root (l), effective porosity ($\tau\varepsilon$) and root respiratory activity (M) may be obtained by appropriate substitution in equation (10.8), a modification of an earlier equation (see Table 10.1).

$$l = \sqrt{\frac{2D_0\tau\varepsilon C_0}{M}}. \tag{10.8}$$

Results so obtained are independent of root radius, and examples are illustrated in Figure 10.12.

It must be noted that contrary to popular belief, roots will continue to function normally at internal oxygen concentrations $\geqslant$2–3 per cent (Armstrong and

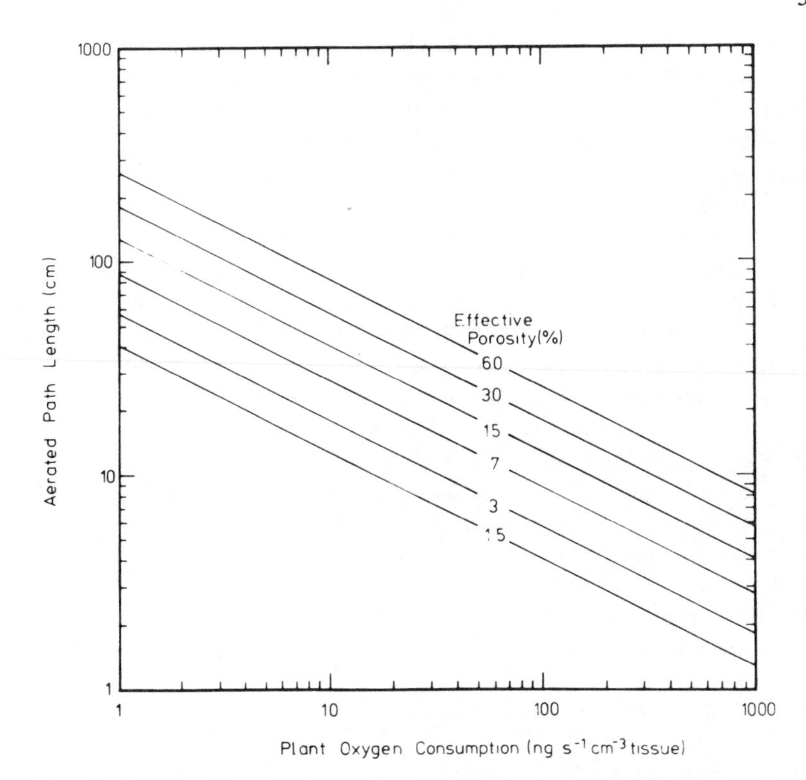

Figure 10.12 The predicted distance to which oxygen will diffuse through plant organs of different porosity at specified rates of respiratory oxygen consumption. Computation using Equation 10.8 taking C_0 as the concentration of oxygen in moist air at 23 °C—$268{\cdot}7 \times 10^{-6}$ g cm^{-3} (20·41 %).

Gaynard, 1976) but for convenience it has been assumed that M is unaffected by all concentrations >0. At an effective porosity of 2 *per cent* and the not unusual respiratory rate of 100 ng O_2^- s^{-1} cm^{-3} root tissue, the length of aerated root will be no greater than 3 cm. A reduction in the respiratory rate or an increase in porosity will naturally increase this figure but with a reduction in respiratory rate to 50 ng s^{-1} cm^{-3} and an increase in $\tau\varepsilon$ to 7 *per cent* the predicted length barely exceeds 10 cm. In wetland species up to 60 *per cent* of the plant body is pore space and consequently diffusive resistance will be much lower. At the same time the respiratory rate per unit volume of highly porous organs is much below that of their low porosity counterparts and internal ventilation will be much improved (see again Figure 10.12). The increased porosity is a result of exceptionally loose packing of cortical parenchyma or the occurrence of much larger air spaces or lacunae. These air spaces are formed either by cell separation (schizogeny) during maturation of the organs, or by cell breakdown (lysigeny) (Arber, 1920; Sculthorpe, 1967). Cell collapse frequently accompanies schizogeny (e.g. in grasses and sedges).

Air-space formation is not confined to primary tissue and may be conspicuous in secondary tissues arising from a phellogen or from a normal cambial layer. Schrenk (1889) in fact originally proposed the use of the term aerenchyma to describe air-space tissue arising from a phellogen. It has subsequently become customary to describe all air-space tissues as aerenchyma.

Although more porous than normal parenchyma, lacunate tissues often contain sites of much lower porosity. For example, in a honeycomb type of structure the gaseous exchange between individual chambers is confined to the normal intercellular spaces of the bounding walls. Similarly, the gas-filled cavities in many hydrophyte stems are interrupted by watertight cellular diaphragms of comparatively low porosity. Tissue of restricted porosity also occurs in the compacted zone at the root and stem cortical junction. These zones must act to some extent as rate-limiting barriers in gaseous transfer.

Plants vary in their ability to form aerenchyma depending on species and degree of soil anaerobiosis (Arikado and Aduchi, 1955; Arikado, 1959; Martin, 1968; Armstrong, 1971a, b). Figure 10.13 illustrates this point. Plants which cannot form aerenchyma appear to be intolerant of wet habitats, give low ROL values and have poor ability to oxidize inorganic toxins in the rhizosphere. It is interesting to note that certain mesophytes have at least a limited capacity to increase root porosity in response to soil wetness (Yu *et al.* 1969), and the value of

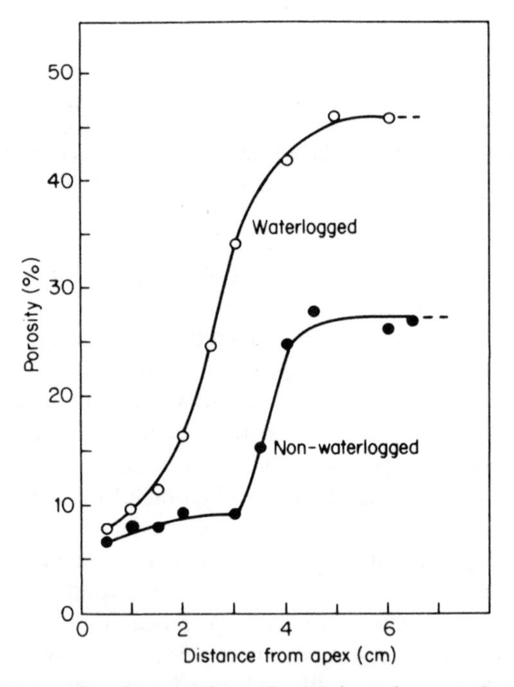

Figure 10.13 Root porosity changes along rice roots under waterlogged (○) and non-waterlogged conditions (●). Reproduced with permission from W. Armstrong, *Physiol. Pl.*, **25**, 192–7 (1971b).

this response would seem to be greatest when a feature of the new adventitious roots which frequently grow to replace those killed by the initial flooding. The ability to rapidly produce adventitious roots of increased porosity may be the most important adaptive response to flooding in plants (see also Jackson, 1955).

Williams and Barber (1961) reassessed the role of aerenchyma and concluded that its function was not primarily one of oxygen transport, or one of storage, but that it fulfilled the necessary requirements for a mechanical-cum-metabolic compromise in the wetland plant body; a compromise involving a considerable reduction of respiratory oxygen demand while at the same time retaining adequate mechanical strength. The formation of lacunae they regarded as necessary primarily to reduce the respiratory demand of the plant body. The size of the lacunae was unnecessarily large for an oxygen pathway. The honeycomb structure of aerenchyma provides maximum strength with the deployment of the minimum of tissue, consequently reducing the ratio of respiratory oxygen uptake to volume; obviously of survival value in an oxygen-deficient environment.

Williams and Barber's interpretation was not derived from experimental evidence and no account was taken of the oxygen transport requirements to maintain an oxygenated rhizosphere. Evidence now available indicates that in terms of rhizosphere oxygenation alone the provision of aerenchyma to a maximum porosity of 60 *per cent* may not be excessive (Armstrong, 1972, 1979). Whilst the cellular partitions do increase the diffusive resistance of aerenchyma they are probably essential in the event of injury to prevent the spread of internal flooding; their thinness and low frequency are thought to minimize their detrimental effects on aeration (Teal and Kanwisher, 1966, Armstrong, 1972; Gaynard—see Armstrong, 1979).

In the living plant the assessment of the gas-space requirement for respiration and radial oxygen loss in combination is complicated by the interaction of many factors. These include: effective porosity of the internal path, respiratory activity (and its distribution), root wall permeability, root radius, and the oxygen diffusivity and demand in the rhizosphere. Model building provides the best approach to this multivariable system and Luxmoore *et al.* (1970, 1972) have produced a computer model of the soil–plant oxygen diffusion system. More simply, an electrical analogue may be built in which resistors simulate root porosity, root wall permeability and diffusion path length. Voltmeters and milliammeters indicate, respectively, oxygen concentration and oxygen flux while a variable resistance leakage to 'earth' simulates respiratory oxygen consumption in root and rhizosphere (Figure 10.14; Armstrong and Wright, 1976; Armstrong, 1979).

A bank of 10 analogue units is used in this model to simulate the oxygen diffusion activity of roots 10 cm long. The units may be programmed to predict the oxygen relations of either aerenchymatous or non-aerenchymatous species in conditions of soil oxygen deficiency and mixtures of characteristics may be introduced. Figure 10.15 clearly shows that root porosity is the overriding factor governing internal oxygen concentration and radial oxygen loss. Waterlogged rice, for example, has an apical oxygen concentration of 11 % in the model (see

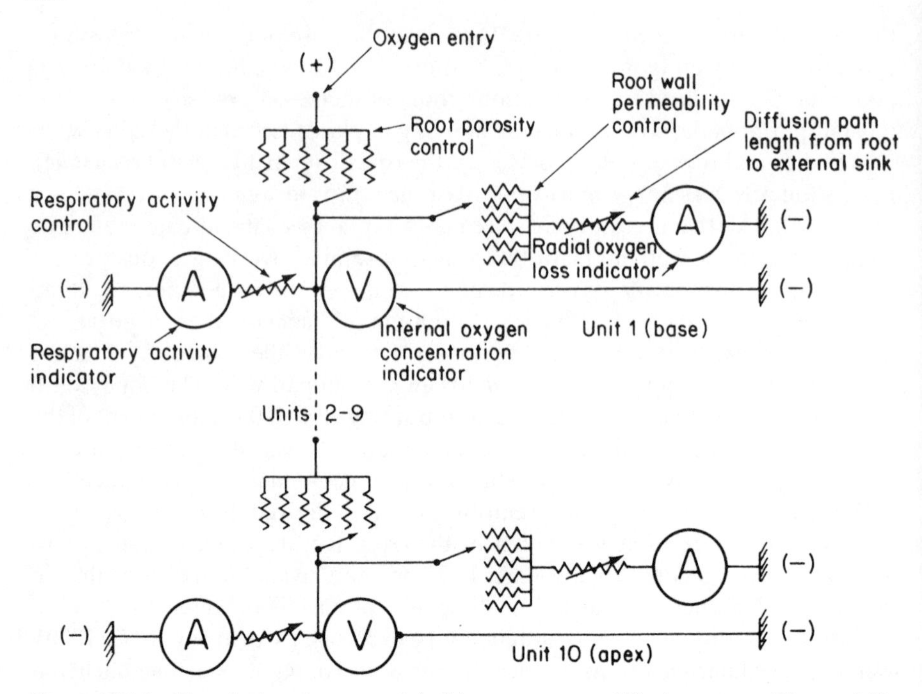

Figure 10.14 Electrical analogue model of the root oxygen diffusion system. The model is based on ten analogue units each of which represents a 1 cm segment of the root. Each unit may be programmed to representative values of simulated porosity, wall permeability, respiratory activity and diffusion path length to external oxygen sink. The equilibrium oxygen concentrations, respiratory oxygen consumption and radial oxygen diffusion loss may then be read from the three meters of each unit. It should be noted that the internal oxygen concentration indicator (meter V) has a very high impedance compared with the meter A. The current drawn by meters V has negligible effect on the system. From Armstrong and Wright (1976) and see Armstrong (1979).

curve 1). If the aerenchymatous structure is lost the concentration falls to 5 % in a root only 7 cm long (curve 5). If in addition, the respiratory and leakage characteristics are those which typify non-wetland plants (curve 8) this concentration declines below 2 % which is lower than the critical oxygen pressure (COP) for rice root respiration (Armstrong and Gaynard, 1976). These modelled data are in keeping with available experimental results (Armstrong, 1971b; Healy, 1975).

The foregoing discussion suggests that the oxygen transport requirement is probably the overriding factor governing the ecological need for aerenchyma development but that the structure also accords with the requirements of a mechanical-metabolic compromise. What little evidence there is does not seem to favour a reservoir hypothesis and it has been found recently (Gaynard and Armstrong, unpublished work) that the internal oxygen supply in the aerenchymatous bog species *E. angustifolium* (cottongrass) is almost exhausted after 1 hour's submergence in the dark.

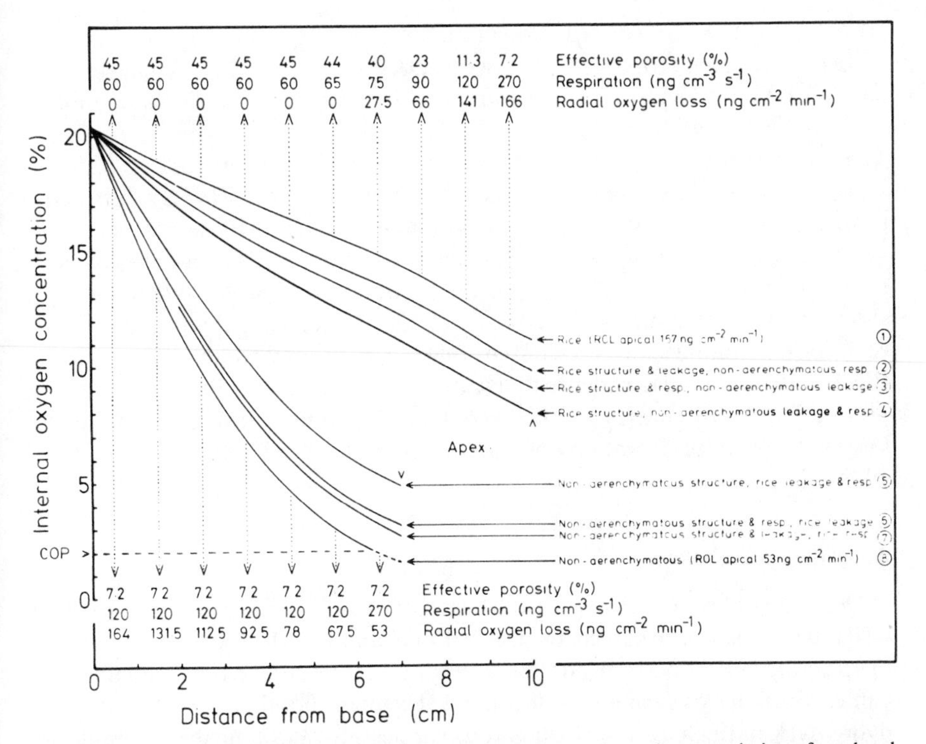

Figure 10.15 An analogue analysis shewing how the various characteristics of wetland (aerenchymatous rice root) and non-wetland (non-aerenchymatous) roots contribute to the oxygen status of the root in the wetland condition. The data were compiled on the assumption that the wetland soil, where aerated, consumes oxygen at the uniform rate of $5{\cdot}27 \times 10^{-8}\,\mathrm{g\,cm^{-3}\,s^{-1}}$, and that oxygen diffusivity in the soil was a uniform $1 \times 10^{-5}\,\mathrm{cm^{2}\,s^{-1}}$. It was assumed also that wall permeability to oxygen of the rice root declined from a maximum of 100% at the apex, to zero at 5 cm from the apex; in the non-aerenchymatous root the minimum value (60%) was attained at 6 cm.

Although the functional significance of aerenchyma is now better understood, important questions remain concerning its physiological origin. Until quite recently the literature revealed a paradoxical situation: the general belief that aerenchyma is triggered by oxygen stress in the tissues (McPherson, 1939; Kawase, 1979) running counter to the foregoing evidence of the highly aerobic character of, and yet sustained aerenchyma production in, the aerenchymatous root.

The prospects for an early solution to this problem now seem hopeful. The earliest signs of aerenchyma formation in roots have been identified as planes of middle lamella discontinuity which arise at an early stage of growth. Lacuna formation, involving cell tearing and collapse, have been attributed to a lack of cell adherence along these planes during subsequent growth (Boeke, 1940). An ethylene induced separation of stem cell walls leading to aerenchyma production, and attributed to a stimulation of cellulase activity has been identified by Kawase

(1979). Kawase considered enhanced ethylene concentrations to be a consequence of anaerobiosis in the tissues. Drew *et al.* (1979) have similarly noted an association between ethylene and aerenchyma formation, this time in corn roots, but point out that ethylene biosynthesis in plants usually requires oxygen. They suggest that ethylene concentrations rise simply because escape from the tissues is hindered by submergence (see also Kawase, 1976). This is an attractive proposition since it accords with the continuing production of aerenchyma in the aerobic root. It is also consistent with other observations that root porosity seems more related to immersion than aeration (Luxmoore and Stolzy, 1969). Finally it should be noted that nitrate deficiency may also induce aerenchyma formation in corn roots (Konings and Verschuren, 1980), as also may conditions of drought (Beckel, 1956). There is evidence that water stress can increase the rate of ethylene production (McMichael *et al.* 1972; Wright, 1977), whilst nitrate deficiency is also known to produce symptoms of waterlogging damage in shoots (Drew *et al.* 1979).

Anaerobic metabolism

Occasions may frequently arise when tissues in the underground organs of wetland species will be subject to anoxia. For example, if the root systems develop in an aerated interflood period they may contain little aerenchyma and therefore will easily suffer oxygen stress during a subsequent flood period. The seasonal dying back of the leaves of rhizomatous species provides a further example, for in high water table conditions this can lead to prolonged periods of oxygen stress.

Anaerobic respiration will naturally predominate under these circumstances and in some species ethanol accumulates to quite a high concentration, higher perhaps than the levels reached in non-wetland plants (Hook *et al.* 1971). In other cases lactic acid may be a principal end-product of anaerobic respiration (Chirkova, 1968; Hook *et al.* 1971; Brown *et al.* 1969). However, wetland species also exhibit forms of anaerobic metabolism in which the accumulated end-products are much less toxic than ethanol, and this may better enable them to withstand prolonged periods of oxygen stress without injury. An example is the formation of shikimic acid in the rhizomatous species *Iris pseudacorus* (yellow iris) and *Nuphar lutea* (yellow waterlily) (Boulter *et al.* 1963; Tyler and Crawford, 1970). This acid is relatively non-toxic and reaches high concentrations during the winter when loss of the aerial parts, coupled with raised water tables, severs gaseous connection with the atmosphere.

Many species, both wetland and non-wetland, are capable of the anaerobic dark fixation of carbon dioxide by carboxylation of phosphoenolpyruvate to oxaloacetate. In the presence of the malic dehydrogenase enzyme the oxaloacetate may subsequently be converted into malic acid. Mazelis and Vennesland (1957) were of the opinion that malic acid should be considered as a principal end-product of anaerobic respiration in many, if not all, plant tissues; of greater importance perhaps than ethanol, lactate and carbon dioxide. However, Crawford and co-workers (1966, 1968, 1969) have recently suggested that malic acid accumulation may be a characteristic feature of wetland species only. They consider that in non-wetland species accumulation will be prevented by the presence of 'malic enzyme' which will recycle excess malic acid to pyruvate and thence to ethanol and carbon dioxide. McMannon and Crawford (1971) have outlined these ideas as follows. When intolerant plants are flooded their respiration is blocked and glycolysis proceeds to the formation of acetaldehyde and then ethanol. Acetaldehyde induces alcohol dehydrogenase activity which, together with a reduction in the apparent Michaelis constant of the enzyme–substrate reaction, accelerates glycolysis. Carboxylation of phosphoenolpyruvate can form malate via oxaloacetate but the malate is rapidly decarboxylated by 'malic' enzyme to form pyruvate and further contribute to acetaldehyde and ethanol accumulation which ultimately reaches toxic levels. The flooding of tolerant plants partially blocks respiration to cause acetaldehyde and ethanol accumulation but the former fails to induce alcohol dehydrogenase, the Michaelis constant of the reaction is not changed and glycolysis does not accelerate. The malate which is formed via carboxylation of phosphoenolpyruvate cannot be decarboxylated again as these plants lack the 'malic' enzyme and consequently malate accumulates. It is non-toxic and remains in the plant until aerobic conditions are restored.

Despite the attractiveness of these ideas it has to be acknowledged that there is little experimental evidence to support them: ethanol clearly accumulates to high concentrations in a number of wetland species (Boulter *et al.* 1963; Hook *et al.* 1971; John and Greenway, 1976), malate accumulation is more associated with aerobic respiration in some species (Effer and Ranson, 1967), and highly active malic enzyme has now been found in some flood-tolerant species (Davies *et al.* 1974) including those studied by McMannon and Crawford (1971); Avadhani *et al.* (1978) have found that ethanol formation in rice seedlings during a single day of anoxia can be 20–30-fold greater than the highest amounts of malate recorded in the seedlings. Keely (1978), Keely and Franz (1979) and Smith and ap Rees (1979) have strongly rejected the malate hypothesis. Keely has suggested that malic acid may accumulate as a balance for ionic disequilibrium within the cells. Survival of the flood-tolerant *Nyssa sylvatica* var *biflora* (blackgum) could be correlated with the capacity to produce new roots after flooding. Existing secondary roots initially increased their potential for ethanol production but eventually senesced as did the roots of the intolerant species, *N. sylvatica* var *sylvatica*. New roots in var *biflora*, present 1 month after flooding, had the capacity for substantial ethanol production but also accumulated malic acid. Malic acid levels remained high in the younger parts of the root system after a whole year's flooding but the capacity for ethanol production had fallen below that of drained controls and could be correlated with improved internal aeration.

Smith and ap Rees (1979) made a careful study of anaerobic metabolism in the roots of the flood-tolerant *Ranunculus scleratus* (celery-leaved water crowfoot), *Senecio aquaticus* (marsh ragwort) and *Glyceria maxima* (reed sweetgrass). Estimates of enzyme activity indicated a capacity to convert the product of anaerobic glycolysis to ethanol, lactate and malate. Of all the enzymes studied

ADH had the highest activity in unaerated roots and it showed the greatest rise in activity when the supply of oxygen to the roots was decreased. Lactate was also a product of fermentation but not a major one; malate did not accumulate under anaerobiosis. The pattern of fermentation was precisely that followed in pea roots, which are intolerant of flooding.

Forms of anaerobic metabolism may be of benefit to submerged overwintering (non-absorbing) roots and rhizomes. Whether they ever enable morphologically ill-adapted roots (non-submerged, non-aerenchymatous) to survive the rigours of a reduced soil in which phytotoxins abound remains to be shown. However, active growth seems always to depend upon a readily available source of oxygen (Armstrong, 1979).

Vartapetian (1978) has suggested that plants should be classified according to the character of their resistance to anoxia into the following categories: (1) truly resistant to anoxia; (2) resistance apparent rather than true; and (3) non-resistant to anoxia. At present there is little known of category (1), but the coleoptile stage of rice appears to be representative (Vartapetian, 1978; Kordan, 1976a, b) and even the first adventitious root primordia may be established anaerobically. However the further development of rice which includes chlorophyll formation and adventitious root growth will not proceed under anoxia. The adult rice plant (most probably in common with the other wetland species) must be placed therefore in category (2), the apparent resistance to anoxia being a property of the internal oxygen transport. Most mesophytes should probably be placed in category (3) but ought not to be considered any less resistant to anoxia than the post-coleoptile stage in rice. Perhaps even the converse might be true since it has been shewn that following the onset of anoxia, mitochondrial disintegration in rice precedes that in pumpkin, bean and tomato (Vartapetian, 1970; Vartapetian *et al.* 1977). An extracellular carbon source can delay such disintegration in the excised roots of both wetland and non-wetland plants (Vartapetian *et al.* 1976, 1977, 1978; Morriset, 1978) and in whole plants (Webb, Gaynard and Armstrong, unpublished data), and has led Vartapetian to suggest that carbon-starvation is the immediate cause of death from anoxia rather than the build-up of toxic materials (see Vartapetian *et al.* 1978).

Other adaptations to waterlogging

A feature common to many wetland plants is the xeromorphic nature of their leaves and stems, despite the fact that they may be rooted in flooded soil. Good examples may be found in the genus *Juncus* (rush) and in wetland species of the Ericaceae such as *E. tetralix* (cross-leaved heath). It seems likely that the primary function of xeromorphism in these plants is not the usual one of water conservation *per se* but is rather to reduce the velocity of water movement to the root surface. In this way, the time available for the oxidation of phytotoxins in the rhizosphere zone of oxygenation will be increased. Jones (1971a) showed that the use of antitranspirants on *Erica cinerea* (bell heather) could improve survival of waterlogging and also limit iron uptake. She suggested that the more xeromorphic nature of *E. tetralix* probably improves its ability to tolerate high levels of ferrous iron in the soil solution.

A change in root orientation in poorly aerated or anaerobic soil horizons may be frequently observed and the dwarfed plants of *Molinia caerulea* (purple moor grass) found in the surface water channels of valley bogs provide excellent examples of this obviously beneficial phenomenon (Armstrong and Boatman, 1967). In these places the adventitious roots become orientated into a horizontal growth pattern along a tolerable redox plane. In this way they avoid the deeper, more intensely reducing and hostile horizons. Under these circumstances the lateral roots of this species invariably grow upwards towards more oxidizing conditions.

The development of the 'shaving brush' effect in root systems exposed to a fluctuating water table appears to be an outward sign of yet another survival mechanism. The root apices are killed in the rising water table but the root bases remain healthy and new laterals rapidly grow from these and adjust to the new level. If the water table oscillates frequently, the cycles of death and regrowth produce a mass of brush-like roots. The roots are obviously harassed by these conditions but the mechanism which confines death to the apices, and allows rapid basal regrowth, must assist the plant to tolerate these marginal conditions. The 'shaving-brush' effect is common in conifers and other species not fully tolerant to the extremes of soil anaerobiosis, e.g. *Molinia caerulea*.

WETLAND TOLERANCE IN WOODY SPECIES

Few tree species or other woody plants adapt well for growth in permanently flooded soil. Dormant trees may survive weeks of winter flooding, but a single day of flooding during the growing season may be very harmful (Kramer, 1969). The response of most trees in wet soils is probably similar to that noted for Lodgepole pine (*P. contorta*) in some recent field experiments (Boggie, 1972). Young trees remained alive for little more than one season when the water table was permanently maintained at the soil surface. Survival increased enormously with a slight lowering of the water table while growth continued to improve with progressively lower water levels.

Because of their scarcity those species which are successful in the wetland habitat are relatively well known. They include the mangroves such as *Rhizophora*, *Avicennia* and *Sonneratia*, the swamp cypress (*Taxodium distichum*), as well as species of *Nyssa*, *Salix*, *Alnus* and *Myrica* (tupelo willow, alder and bog myrtle).

A study of the swamp cypress and the tupelo gum (*Nyssa aquatica*) (Dickson and Broyer, 1972) shows that they grow best in waterlogged conditions, as do many wetland herbaceous species. Again, as with herbaceous species the provision of an adequate ventilating system appears to be essential for the successful wetland growth of woody species. Aeration in trees has recently been reviewed by Hook *et al.* (1972), Hook and Scholtens (1978) and Coutts and Armstrong (1976).

Special biochemical pathways may be available to cope with periods of anoxia brought about by any sudden raising of water tables, an example being the production of glycerol in *Alnus* (Crawford, 1971).

Perhaps the best known study of the ventilation of woody species is that of Scholander *et al.* (1955) who examined the gas exchange of *Rhizophora mangle* and *Avicennia nitida* (mangroves) in the tidal swamps of Florida. *Rhizophora* perches on arched stilt roots, which are richly provided with lenticels above ground and in the mud terminate in bunches of long, spongy, air-filled roots about finger-thick. The stilt roots are above the tide level so that submerged roots remain in gaseous contact with the atmosphere through the intercellular space system and the lenticels. The oxygen concentration in roots submerged in the reducing mud remained continuously high, 15–18 %, but if the lenticels were blocked this concentration fell to 2 %, or less, in two days. *A. nitida* (the black mangrove) is a bush or tree up to 20 m tall which produces prolific numbers of air roots (pneumatophores) protruding from the mud. A single tree may produce several thousand of these which are usually 20–30 cm high, one centimetre thick, soft and spongy, and studded with numerous white lenticels. In the mud they arise from radially-running main roots, which are also soft and spongy, and contain large amounts of air. Root aeration is influenced by tidal rhythm as the air-roots and their lenticels are covered by each tide. Following submergence the root oxygen concentration falls, to rise again as the tide retreats. The authors noted that there was a progressive pressure drop in the internal gas space while the air-roots were covered and that air would be drawn in when the lenticels were again exposed. As the internal oxygen diffusion mechanism is so efficient in highly porous organs, it may be that this pressure change is not a necessary mechanism.

A number of other woody species produce structures which behave like air-roots, for example Bond (1952) observed that the nitrogen-fixing nodulated roots of *Myrica gale* grew upwards and he suggested these might improve the oxygen supply to the nodule tissue. The well known 'knees' of the swamp cypress, produced by upward proliferation of root xylem, have long been thought to improve gas exchange, though Kramer *et al.* (1952) have cast some doubt on this.

Oxidation of reduced dye solutions has been used to show root-oxidizing activity in the woody species *Betula pubescens* (hairy birch), (Huikari, 1954), *Salix atrocinerea* (sallow) (Leyton and Rousseau, 1957), *Nyssa sylvatica* and *N. aquatica* (black tupelo, water tupelo) and *Fraxinus pennsylvanica* (ash) (Hook *et al.* 1972). Hook was unable to detect root-oxidizing activity in the non-swamp species *Liquidambar styraciflua*, (sweetgum), *Liriodendron tulipifera* (tulip tree) and *Platanus occidentalis* (American sycamore), but concluded that tolerance of flooding in swamp tupelo is achieved by the combined adaptations of rhizosphere oxidation (ultimately dependent upon oxygen entering through the lenticels), anaerobic respiration at low oxygen concentrations and the tolerance to high carbon dioxide concentrations of the new roots produced after flooding (Hook *et al.* 1971; see also Keely, 1979).

Armstrong (1968) has demonstrated internal oxygen transport and radial oxygen loss with rooted cuttings of *Salix* (willow) species and the heath shrub

Myrica gale (bog myrtle). In these cases oxygen entered the stems through the lenticels, and if these were progressively blocked, from the base of the stem upwards, radial oxygen loss gradually declined. Radial oxygen loss ceased in *S. atrocinerea* and *S. fragilis* when only 3 cm of stem had been treated in this way.

If the lower stem lenticels form the main point of oxygen entry, sudden floods may severely limit the root oxygen supply, thus causing death, but in those species prone to flooding the rise in the water table is usually accompanied by a rapid proliferation of adventitious roots. These new roots arise from hypertrophied lenticels just below the new water surface.

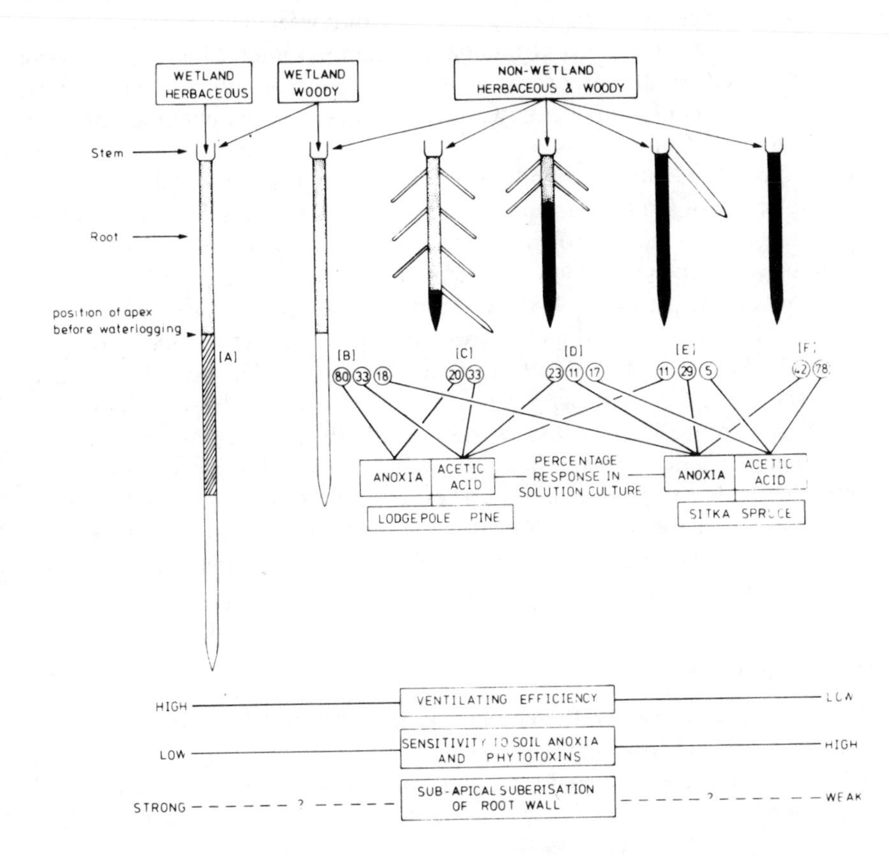

Figure 10.16 Suggested relationship between internal root ventilation and the responses of wetland and non-wetland plants to a limited period of soil waterlogging. Root growth before and during, and regrowth after the end of, the waterlogged period is indicated as follows: before (and surviving) *dotted*; before but killed by the waterlogging *black*; during *shaded*; regrowth *white*. The figure also records how Sitka Spruce and Lodgepole Pine respond when exposed for a limited period to anoxia or to acetic acid (100 ppm) in solution culture. N.B. Mature woody roots may at times substitute for the stem in the above scheme, and it must be noted also that in some non-wetland species new adventitious roots of greater porosity may develop during a flood period.

The path of oxygen movement through woody species is not yet established with certainty. In young shoots and seedlings it is almost certainly the primary cortex while in a few plants, such as the legume *Aeschymomene aspera*, xylem aerenchyma is produced. The difficulty is with the remainder which in the adult condition have no cortical parenchyma. The most obvious route for gas transfer is through the empty conducting elements of the stele but until recently the cambium was held to act as a barrier to gas exchange with the xylem.

There is now evidence that the cambium may be adequately pervious to oxygen in wetland trees. Hook and Brown (1972) have reported the presence of intercellular spaces and free gas exchange through the cambia of both swamp tupelo and green ash. In sycamore, tulip poplar and sweet gum, species which can withstand only short periods of inundation, cambial intercellular spaces were found to be either absent or so small that their continuity could not be established. Gas could only be drawn through these cambia under a tension of between 30 and 80 mm Hg.

More recently Coutts and Philipson (1978) have attributed the rhizosphere oxidizing power of *Pinus contorta* (lodge-pole pine) roots to oxygen transported via the stele and the limited tolerance of soil wetness found in *P. contorta* would seem to be very dependent upon the xylem path. This was the conclusion also of Sanderson and Armstrong (1978) who studied differential sensitivity to oxygen deficiency, Fe^{2+}, and the organic toxins acetic and butyric acid, in the two species, *P. contorta* and *Picea sitchensis* (sitka spruce). This investigation also indicated that the oxygen deficiency which rapidly ensues in the soil after sudden flooding is a more likely cause of root death than are the later produced phytotoxins, and this observation may well hold true for most non-wetland plants whether herbaceous or woody. These and related observations have been used to construct the scheme shown in Figure 10.16 which outlines for both wetland and non-wetland plants the possible relationship between internal root ventilation and the response to a limited period of soil flooding.

Acknowledgements

I would like to thank Mr. R. Wheeler-Osman, Miss S. Lythe, Mr. P. Meaker, and Miss A. Lyon for drawing the illustrations, Miss E. M. Sharpe for typing the manuscript and my wife, Jean, for reading and criticizing this chapter during its preparation.

Plants in Ecosystems

Chapter 11

Production ecology

INTRODUCTION

Life on earth has evolved over the past $3{\cdot}5 \times 10^9$ years and, during this time the metabolism of the biosphere has grown from an originally minute carbon budget to an annual cycle which exceeds man's mineral extractive capacity by at least two orders of magnitude and is exceeded only by the cycle of water, which is mainly physical in character. Primary photosynthetic production is thus the most massive of the earth's surface chemical processes.

The trapping of solar radiant energy by the biochemical activity of the chlorophyll molecule now provides almost all of the energy which fuels the global ecosystem. Chemo-autotrophic energy conversions provide only a tiny proportion of the primary energy harvesting. *Net primary production* is the surplus organic production remaining after plant respiration has utilized some of the gross primary production:

$$\begin{array}{ccccc} \text{NET} & & \text{GROSS} & & \text{RESPIRATION} \\ \text{PRIMARY} & = & \text{PRIMARY} & - & \text{(DARK + PHOTO)} \\ \text{PRODUCTION} & & \text{PRODUCTION} & & \\ P_{net} & = & P_{gross} & - & R_{gross} \end{array} \qquad (11.1)$$

(kg dry matter m^{-2} $time^{-1}$ or, for energy budgets, MJ m^{-2} $time^{-1}$)

Net primary production is ecologically a valuable characteristic as it represents the maximum available energy source for consumer organisms and so governs the whole community food-web. Gross primary production is a theoretically useful value but it is difficult to determine accurately, in particular for C-3 plants (p. 128), as there is no simple means of measuring photorespiratory carbon use.

Net primary production is expressed in its raw form as *kg plant dry matter* m^{-2} $time^{-1}$ but, because different tissues differ in specific energy content and because whole-ecosystem energy budgets are often needed, it is better to convert the dry matter data to energy equivalents (J kg^{-1}) and to express the production figures as J m^{-2} $time^{-1}$. These are then compatible with measurements of solar radiant flux density (J m^{-2} s^{-1} = W m^{-2}). Table 11.1 presents some useful conversions and equivalents for use in production studies.

Table 11.1 Units and equivalences in production ecology

A. Terms	Fundamental unit	Most convenient unit, conversion, etc.
Specific energy content (dry matter*)	$J\ kg^{-1}$	$MJ\ kg^{-1}$
Biomass density (*B*) (dry matter*)	$kg\ m^{-2}$	$1\ kg\ m^{-2} = 10\ t\ ha^{-1}$
Biomass density (*B*) (energy)	$J\ m^{-2}$	$MJ\ m^{-2}$
Net primary production (P_{net})	$kg\ m^{-2}\ s^{-1}$	$kg\ m^{-2}\ time^{-1}$ or $t\ ha^{-1}\ time^{-1}$ where time is in days or years
Net primary production (P_{net}) (energy)	$J\ m^{-2}\ s^{-1}$ ($= Wm^{-2}$)	$MJ\ m^{-2}\ time^{-1}$ where time is in days or years
Energy content (dry matter*)	$J\ kg^{-1}$	$MJ\ kg^{-1}$

B. Energy content and other characteristics of biological materials (mainly from Lieth, 1975).

Material	Energy content ($MJ\ kg^{-1}$)
Plant dry matter: Leaves	*c.* 16 } 14–21
Plant dry matter: Woody tissue	*c.* 18 } 14–21
Animal dry matter (Callow, 1977) (mean)	*c.* 24
Cellulose	17·6
Lignin	26·4
Crude protein	23·0
Fat	38·9

1 g dry plant material ≡ 0·5 g C ≡ 1·8 g CO_2

C. Conversions.

1 t = 1000 kg
1 ha = $10{,}000 m^{-2}$
W = $J\ s^{-1}$
1 kcal = $4{\cdot}1868 \times 10^3$ J
100 Wm^{-2} = 4·32 $MJ\ m^{-2}$ 12-hour-day^{-1}

* Plant and animal material is dried to a constant weight at 80 °C in a ventilated oven. This prevents charring but inhibits initial respiratory loss.

The primary producers are the first of a series of *trophic levels*, the remainder of which comprises the *secondary producers*, all heterotrophic organisms. The herbivores occupy the second trophic level, the carnivores the third and subsequent levels. In many cases this is a classification only of convenience as some organisms may feed at more than one trophic level. Detritivores and decomposer microorganisms take the dead remains of all other trophic levels and close the cycle of production–decomposition by returning materials to the environment in the inorganic, plant-available form.

The relative quantities of organisms in the different trophic strata may be expressed simply by *biomass density* (*B*):

$$B_{total} = B_{1ary} + B_{2ary} + B_{decomposer} + B_{dead\ organic} \qquad (11.2)$$

The subscripts 1ary and 2ary denote, respectively, primary and secondary producer organisms. All units $kg\,m^{-2}$ or $J\,m^{-2}$.

It has been customary to represent the biomass density at different trophic levels by pyramid diagrams such as those of Figure 11.1a and if these pyramids are shown alongside the equivalent pyramids for energy flow (Figure 11.1b), they draw attention to a number of fundamental characteristics of ecosystems. The energy transfer pyramids *always* indicate a primary production which is very much greater than the subsequent energy flows through the consumer trophic levels, and each transfer to subsequent levels is accompanied by a massive reduction in the handing-on of energy. The biomass density pyramids show the same pattern, each trophic level having a considerably larger biomass than the succeeding one but, in this case, it is not an invariable rule and occasionally the pyramid may even be inverted (Figure 11.1c). The ratio of primary producer to herbivore biomass is much more variable than the equivalent ratio of energy flow because of the wide variation in morphology and physiology of the primary producer organisms. If the plant is large, for example a tree with a great deal of mechanical support tissue, it will inevitably have a larger biomass per unit of photosynthetic tissue than a herbaceous plant with its leaves near ground-level.

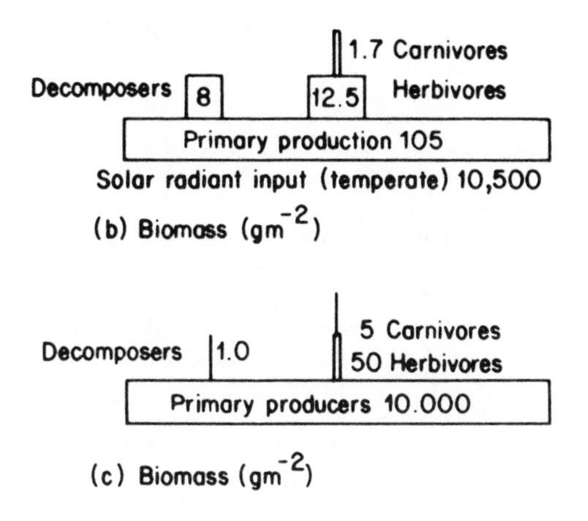

Figure 11.1 Energy flow and biomass pyramids. (a) Energy flow in immature summer deciduous woodlands. Radiant input in the photosynthetic range (0·4–0·7 μm). (b) Biomass distribution in the same woodland. (c) Inverted biomass pyramid for phytoplanktonic primary producers and zooplanktonic and benthic faunal consumers in the English Channel. Data for (a) and (b) are hypothetical values based on Rodin and Bazilevich (1967), Edwards, Reichle and Crossley (1970) and Odum (1971) and (c) Harvey (1950).

Even less supporting tissue is needed by aquatic plants which are buoyed by water and the smaller investment in perennial support systems gives the advantage of a lower respiratory rate, the ratio of net to gross primary production is consequently greater and the ecosystem can support a herbivore biomass which is relatively large compared with its plant biomass.

Figure 11.2 dramatically shows the impact of biomass density on the P_{net}/B ratio and clearly separates vegetation types into groups such as terrestrial woody, terrestrial herbaceous, shallow water aquatic and deep water aquatic ecosystems.

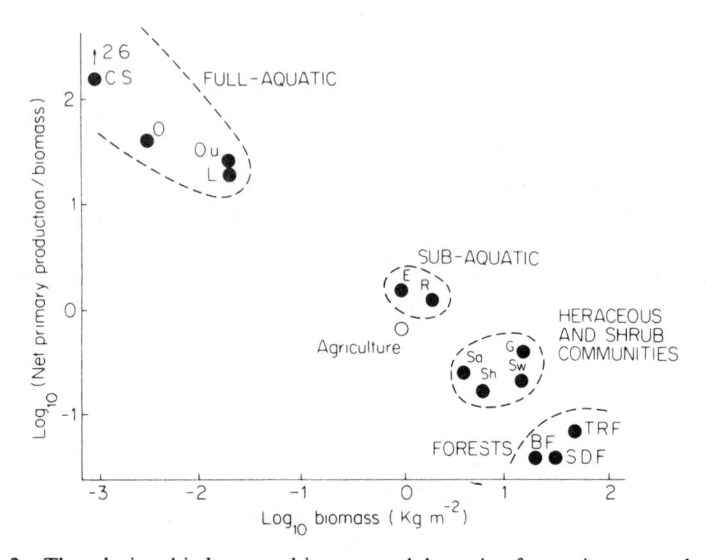

Figure 11.2 The relationship between biomass and the ratio of net primary production to biomass (P_{net}/B). The plotted values are derived from world means given by Whittaker and Likens (1975). Note the logarithmic scales. Key: C.s., continental shelf; O., ocean; O.u., ocean upwelling; L., lake; E., estuary; R., river; Sa., savanna; Sh., shrub; G., grassland; Sw., swamp; B.F., boreal forest; S.D.F., summer deciduous forest; T.R.F., tropical rain forest.

The most efficient of the non-woody plant ecosystems, in terms of P_{net}/B ratio are those such as grasslands, shallow water reefs or estuaries and deeper water continental shelf systems. In all of these cases constant rejuvenation of the vegetation by browsers or filter feeders maintains a high ratio of photosynthetic cells or tissues to energy-costly storage or support systems. Cropland shows an anomalously high P_{net}/B ratio, again because grazing has a constantly rejuvenating effect or, in the case of arable land, the majority of cultivated plants originated from pioneer species which devote little of their production to support or storage tissue.

Biomass density estimates also draw attention to the dynamics of ecosystem function as they represent the balance of production versus consumption and decomposition. The values are never constant but change with time, either cyclically with the seasons, less predictably from season to season, and in the long term (decades or centuries) may increase, decrease or remain constant. It has been suggested that such increase or decrease indicated either successional build-up or degeneration of the ecosystem while long-term constancy, at least of total, or plant biomass, is found in climax conditions (Odum, 1969; Colinveaux, 1973; and p. 393).

HISTORY

The Aristotlean idea that plants derived their organic substance from soil persisted for virtually 2000 years and was not seriously challenged until about 1600 when van Helmont's willow tree grew 150 pounds of wood while removing only 2 ounces of material, other than water, from the soil. Another 200 years passed before de Saussure gave the correct equation for the photosynthesis: carbon dioxide + water = plant matter + oxygen; but the humus theory of nutrition was not killed for almost another half-century (Liebig, 1840).

Liebig fought the case for photosynthetic carbon assimilation and also realized that production was limited by factors other than those which directly affect photosynthesis, for example mineral nutrition. He was also the first to speculate quantitatively about the effect of vegetation on the atmosphere (Liebig, 1862): 'If we think of the surface of the earth as being entirely covered with a green meadow yielding annually 5000 kg/ha, the total CO_2 content of the atmosphere would be used up within 21–22 years if the CO_2 were not replaced'. He calculated this transfer to be 230–240 $\times$ 10^9 t CO_2 y^{-1}, a figure which is within 10 % of the most recent estimate of world primary production (Whittaker and Likens, 1975)!

Lieth's (1975) review of production-ecological history shows that the years which separated Liebig's first speculation from the systematic work of the International Biological Programme (1964–74) saw a limited number of advances in which most estimates of terrestrial primary production were drawn from measurements by foresters and agronomists while extrapolations to the global scale were gradually refined by the improved mapping of the distribution of vegetation types. Measurements of aquatic production, difficult even now, were not seriously made until after the Second World War. This is slightly surprising as the Winkler method of the light-dark oxygen bottle or CO_2 titration had long been available, though ^{14}C fixation, which is free from the problem of respiratory interference, could not be widely used until the post-war years (p. 340).

Liebig's and other early speculations on world production led, by the early 1940s, to the posing of more critical, quantitative questions about ecosystems. The efficiency of the vegetation in collecting and storing solar energy, the manner in which the energy is re-used by consumer organisms, the relationship of such use to successional maturity and the impact of the hydrarch succession from oligotrophic to eutrophic all formed subject matter for seminal publications at this time.

Transeau (1926) used data from northern central Illinois for sweet corn (*Zea mays*) yield to construct a carbon budget and thence to estimate photosynthetic CO_2 and solar energy trapping efficiency. Taking a total yield (net primary production) of 14 t plant dry matter ha^{-1} and assuming a respiratory consumption of 23 % of the total carbon fixation, Transeau calculated that gross primary production accounted for 1·6 % of the total solar energy income during the growing season. This was one of the first attempts to assess the *efficiency* of primary production. Using Transeau's data and assuming 50 % of the solar radiation to be photosynthetically available, the net primary productive efficiency would have been 2·4 % for the 100 day growth period, very close to the average of values cited by Cooper (1975) for *Z. mays.* It is relevant to the discussion on p. 15 ff that Transeau also calculated a water budget which showed that 45 % of the incoming radiation was dissipated by transpiration of 3700 t ha^{-1} (37 cm) of water from the crop.

Transeau's work concerned a single species during its growing season. The next step, which was to prove crucial in ecological energetics, was the development of detailed annual budgets for multi-species systems. Juday (1940), at the University of Wisconsin, produced such a budget for an inland Lake and influenced the research of Lindeman (1942) which established the trophic-dynamic model of ecosystem structure and function, the foundation upon which all subsequent production ecology has rested. Juday showed that plant production harvested no more than 1 % in physical processes such as evaporative and other heat losses.

Lindeman (1942) produced a set of definitions of trophic levels and studied the efficiency of energy transfer from one level to the next in lake ecosystems. His suggestion that transfer efficiency gradually improved as energy passes along a food chain has been generally accepted. Lindeman also discussed the initially rapid increase in productivity during succession, levelling to a more or less constant value and then showing a senescence decline. He also related such successional interpretations of energy flow to the natural eutrophication of water bodies as hydrarch succession proceeds.

The classic papers of the Odum brothers (Odum and Odum, 1955; Odum, 1957) on trophic structure and productivity of a coral reef community and an extensive freshwater spring system were of direct lineage with Lindeman's work and have served to mould the ecosystem approach of a generation of research workers. Both papers drew attention to the very high productivity of shallow, flowing-water systems in which light, rather than nutrients, is the limiting factor. They also noted the relationship between high gross primary productive efficiency (5–6 %) and organisms of low biomass which have a minimal maintenance respiration and rapid turnover time. The use of an energy flow diagram in the 1957 paper illustrated the use which could be made of the first two laws of thermodynamics to partition energy flows between various compartments of the ecosystem. Such diagrams have formed the conceptual model which underlies all subsequent attempts at ecosystem analysis, not only of energy flow but also many other aspects of materials movement and change with time (Figure 11.3).

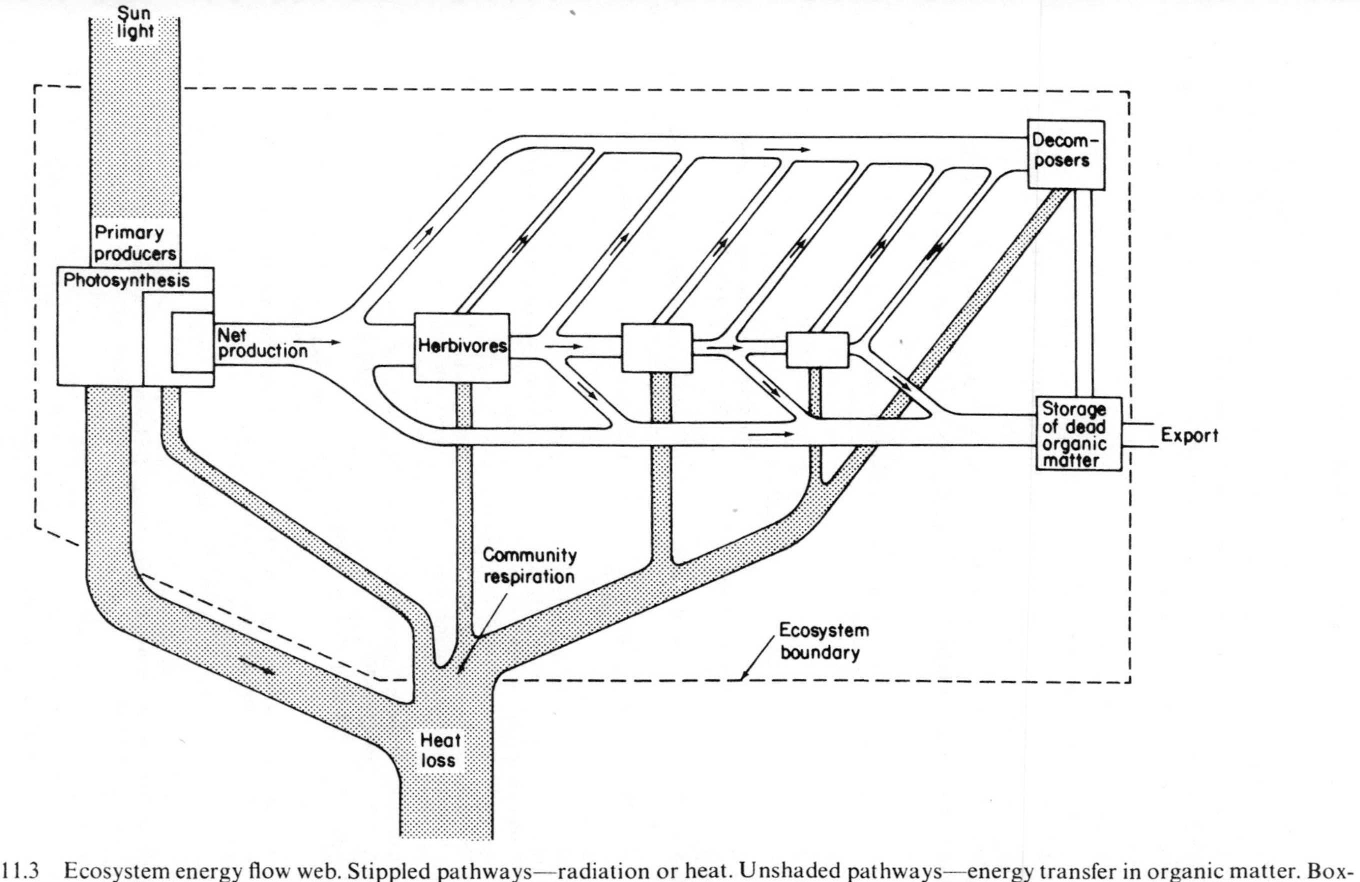

Figure 11.3 Ecosystem energy flow web. Stippled pathways—radiation or heat. Unshaded pathways—energy transfer in organic matter. Box-sizes indicate relative biomass. The diagram cannot be drawn to scale as the transfers at top-carnivore level may be several orders of magnitude less than net primary production.

By the early 1960s a fairly large fund of information had become available though Newbould's (1963) review shows that much of this was still derived from either forestry and agriculture or from physiological experiments concerned with the photosynthetic process. Despite the contribution made by aquatic ecologists to the development of the ecosystem approach there was also a dearth of freshwater and marine information. Throughout the IBP decade, methods were tested and refined, culminating in the reviews of terrestrial and aquatic techniques given, respectively by Whittaker and Marks (1975) and Hall and Moll (1975). Table 11.2 gives a synopsis of the recommendations, some of which are discussed in more detail below.

Most of the early data, relevant to uncultivated ecosystems, were derived from forest yield studies in which it is possible to sample either even-aged stands or to

Table 11.2 Methods of estimating primary productivity

Method	Problems and limitations
A. Terrestrial.	
1. Biomass measurement at sequential harvests.	Difficult in large scale vegetation. Corrections for grazing, litterfall and decomposition.
2. Dimensional relationships	
a) Mean tree biomass × tree density.	Calibration—Inaccurate unless even-aged and few species.
b) Regression of easily measured parameter (e.g. dbh) on sample tree biomass.	Calibration—Assumption of constant ratios.
c) Production ratios. Increment borings	Calibration—Assumption of constant
c) Production ratios. Increment borings on stratified clips regressed against measured annual production.	Calibration—Assumption of constant ratios.
3. CO_2 exchange in a cuvette or by micrometeorological calculation (IRGA measurement).	Modification of environment by cuvette and inaccurate extrapolation. Micrometeorological requirement for uniform stand.
4. Leaf area or chlorophyll per unit area and light extinction measurement.	Requires knowledge of photosynthetic rate versus light intensity. Inaccurate extrapolation.
5. Underground production: direct sampling as (1).	Problems of separation from soil: difficult for large-scale vegetation.
6. Underground production. Regression of an easily measured above ground characteristic against root weight. E.g. root/shoot ratio.	Calibration—Assumption of constant ratios.
B. Aquatic.	
7. Diurnal monitoring dissolved O_2, CO_2 or CO_2^- related pH change in: a) Free system. b) An enclosure (bottle or larger). Light and dark bottle may be used.	Enclosure modification of environment. Assimilation and leakage of organic solutes interferes with calculation. Diffusive O_2 and CO_2 exchange in open system. Correction for respiration (see Hall and Moll, 1975).
8. Incorporation of ^{14}C from $H^{14}CO_3^-$ label.	Short-term: involves time multiplication errors. Problems of respiratory loss.
9. Biomass harvesting as in (1).	Only feasible for attached or easily filterable organisms.

measure age by annual-ring counting. The techniques of even-age stand sampling are described in detail in Ovington (1957, 1959, 1961). The field sampling for fresh weight and subsampling for fresh/dry weight ratios, energy and nutrient content is very laborious in the forest environment (Figure 11.4). It is further complicated in mixed-age stands where trees of different age-size classes must be adequately sampled and their frequency of occurrence related to total tree density in the forest stand.

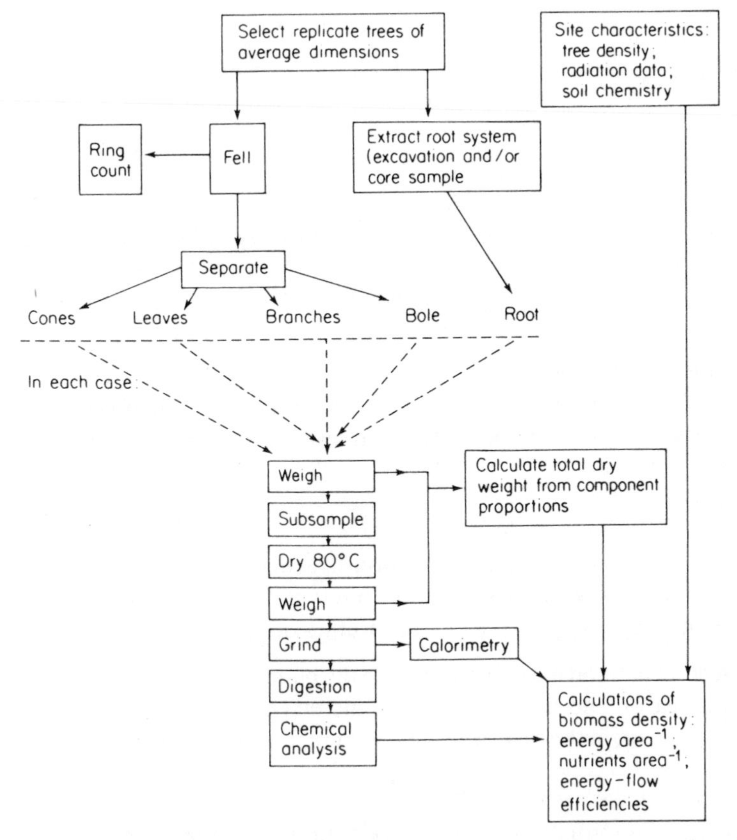

Figure 11.4 Sampling procedure for energy flow and nutrient cycling studies of forest and other large-scale vegetations (see, for example, Ovington 1957, 1961).

Because of these difficulties, dimensional analyses are attractive: an easily measured standard forestry characteristic such as diameter-at-breast-height (DBH) is regressed against tree biomass for a suitable range of samples. The regression line may then be used to convert DBH to biomass estimates in more extensive sampling. If the tree ages are determined from increment cores, annual primary production may be estimated in the same way. A useful example is given in Figure 11.5 showing the species-independence of the regression line for North American deciduous forest trees.

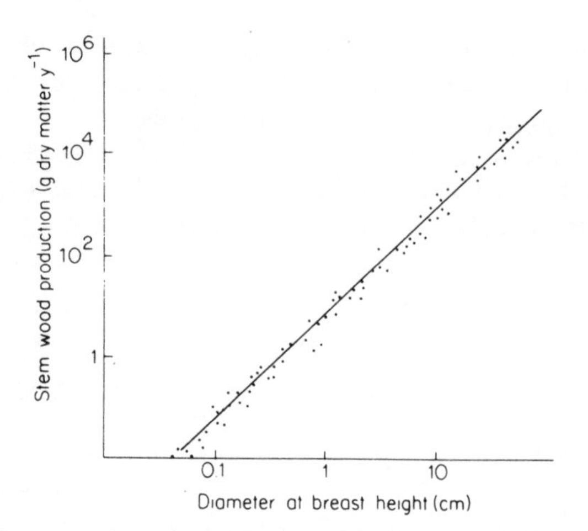

Figure 11.5 The regression of stem wood productivity on diameter at breast height (DBH) for several species of shrub and tree. Note logarithmic scales. Adapted from Whittaker and Marks (1975).

If plant age cannot be measured, two or more samples must be taken, separated by suitable growth periods. Newbould (1967) explains the techniques very clearly. The first involves two harvests separated by a measured time interval:

$$P_{net} = (B_2 - B_1) + L_F + G \tag{11.3}$$

where P_{net} = the net primary production over time T; B_1 and B_2 are biomass densities at harvests separated by a time interval T; L_F = the litterfall per time T; and G = grazing-loss per time T; all in units of $\mathrm{kg\,m^{-2}}\,T^{-1}$.

The second method may be used where the plants can be harvested at the end of a growing season and the current year's increment separated from the tissues of previous years or, in the case of annuals, where there is no previous year's growth:

$$P_{net} = B_c + L_c + G_c \tag{11.4}$$

where P_{net} = the annual net primary production; B_c = the current year's biomass density increment; L_c = the current year's litterfall; and G_c = the current year's grazing loss; all in units of $\mathrm{kg\,m^{-2}\,y^{-1}}$.

Both of these techniques require a suitable method of measuring biomass density, either clipping from quadrats of known area or sampling of individual plants and conversion to biomass density from measurements of plant spacing.

The sampling of root biomass is extremely difficult and inaccurate. The usual techniques involve extraction of cores, known-volume monoliths or digging out of intact root systems. Even when a mechanical coring device is used (e.g. Wellbank and Williams, 1968), the sampling represents by far the greatest part of harvesting time consumption. The roots may be removed from the soil by hand, or automatic-washing, through sieves of suitable size. Problems arise in both sampling and washing if small roots are lost: this source of error is more serious in herbaceous ecosystems than in forest or scrub where much of the root biomass is woody. In tundra, desert and some grassland ecosystems, root/shoot ratios are very large and may change rapidly as reserves are mobilized at the beginning of the growing-season. Productivity estimates in these biomes rely heavily on root biomass estimates and, if net annual production is small, may be seriously in error. A further problem arises from soil contamination which confounds dry weight determination and contaminates tissue analyses, particularly with elements such as iron. Newbould (1968) and Lieth (1968b) have reviewed methods for sampling root biomass and both authors suggested that ecologists are in need of a suitable labelling technique for measuring root production but no satisfactory developments have emerged despite a few attempts to use ^{14}C distribution after photosynthetic assimilation (e.g. Caldwell and Fernandez, 1975).

Dimensional analysis has been attempted, particularly in forest ecosystems, to reduce the sampling difficulty. Regressions of above ground on below ground biomass are especially useful (Figure 11.6) but give little assistance in the estimation of short-term changes which are, however, less marked in forest, compared with herbaceous communities.

Harvest sampling of biomass density change is so difficult in the forest environment that attempts have been made to measure net production by monitoring carbon dioxide exchange of the canopy by micrometeorological techniques (p. 30) or by using suitable cuvettes enclosing either whole trees or component branches. The main disadvantage of both techniques is the need to extrapolate short-term metabolic gas exchanges to the much longer periods over

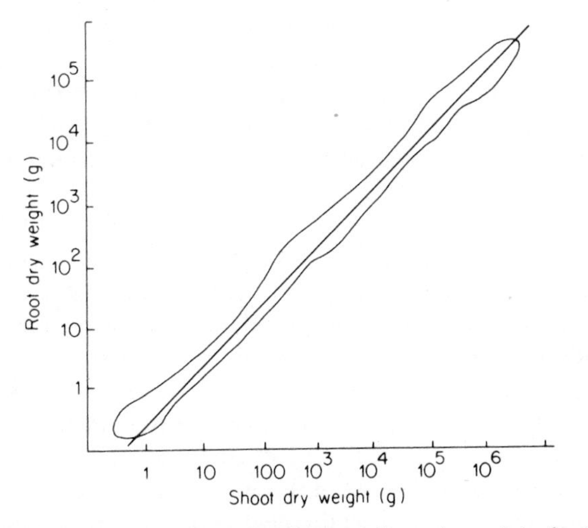

Figure 11.6 The regression of root dry weight on shoot dry weight for five tree species sampled in New Hampshire, USA, and ranging from seedlings to adult trees. The outline encloses the scatter of sample points from over 100 samples. Note logarithmic scales. Adapted from Whittaker and Marks (1975).

which productivity is estimated. Micrometeorological techniques are, furthermore, fairly inaccurate especially in non-uniform canopies, while cuvettes modify plant microclimate. The volume edited by Monteith (1975) gives extensive accounts of micrometeorological approaches to different types of forest and to other vegetation types. The best known enclosure technique was used in a study of tropical rain forest in Puerto Rico (Odum and Jordan, 1970). An open-topped giant cylinder of nylon-vinyl plastic 20 m in height could be hoisted on an aluminium frame to enclose a group of trees 18 m in diameter. Air-flow was maintained by a 2 m diameter extractor fan at the base of the cylinder and air-stream monitored by infra-red carbon dioxide analysis. Similar measurements have been made in herbaceous ecosystems, for example Trlica *et al.* (1973) described an air condition cuvette which encloses a substantial portion of grassland sward.

Aquatic systems pose special problems, except for rooted macrophytes and bottom-attached algae which may be treated similarly to terrestrial plants. Planktonic communities are swept along by water currents or, on death may settle below the photic zone, sometimes into deep anaerobic water, to be permanently lost from the ecosystem. Assessment of planktonic production, grazing and detritivore consumption by swimming organisms is consequently difficult.

Three approaches to planktonic production have been developed depending on: (i) direct counts or pigment (chlorophyll measurements; (ii) metabolic gas exchange (O_2, CO_2 or consequent pH change); and (iii) ^{14}C fixation by feeding $^{14}CO_2$, usually from labelled bicarbonate. Details of the methods are given by Vollenweider (1969) and a critical review by Hall and Moll (1975). Most applications of the gas exchange and ^{14}C methods have involved 'light' and 'dark' bottles filled with water and organisms from a given depth and re-exposed at that depth, in the case of ^{14}C, with the addition of an aliquot of labelled bicarbonate. After a suitable time, the change in O_2 content (Winkler titration or electrometric), CO_2 content (electrometric, titration or pH change) or ^{14}C incorporation into organic matter, is measured. Scintillation counting is most often used for ^{14}C assay.

All of these methods give short-term estimates so that monthly, seasonal or annual production data are subject to multiplicative error. The gas exchange techniques are confounded by changes in respiratory rate caused by illumination, though it is usually assumed that the light-dark bottle technique estimates gross production if dark respiration is added to light photosynthesis. In the case of ^{14}C fixation it is not certain whether respiratory CO_2 loss is from existing unlabelled reserves or from recent, labelled assimilates. It is also difficult to compensate for dark fixation of CO_2 (p. 128) which does not involve energy trapping. Most workers have assumed that fixation in the light measures net photosynthesis (see Hall and Moll, 1975, for details) but this neglects the problems detailed above, the possible photo-assimilation of organic compounds (Fogg, in Vollenweider, 1969) and the leakage of large amounts of recently formed assimilate to the water as glycollic acid (p. 129) (Coughlan and Al-Hasan, 1977).

Litter production and grazing estimates are needed to compute primary production using equations (11.3) and (11.4) for terrestrial vegetation or for aquatic communities if counting or chlorophyll techniques are used to measure biomass of plankton. A variety of techniques have been developed for application in different ecosystems. Forest litterfall may be measured by exposing suitable traps such as trays, suspended bags or nets below the canopy but precautions are needed to prevent wind-drift and, if chemical analysis is intended, contamination from soil-splash and bird faeces must be considered. Newbould (1967) and Phillipson (1971) describe suitable techniques. Sinking planktonic detritus may be trapped in an analagous fashion using suitable containers if care is taken to avoid trapped material being washed out by water currents.

Grassland and other herbaceous systems in which leaf-death is continuous are more difficult to sample, especially when dead material remains attached for long periods of time before reaching the soil surface. Williamson (1976) noted that production in such systems is poorly correlated with living biomass, and devised a leaf-marking technique to measure 50 % mortality rates so that biomass change estimates of productivity could be corrected for leaf turnover. More detailed methods of grassland productivity measurement may be found in Milner and Hughes (1968).

Litter production and grazing are very difficult to assess in aquatic systems as planktonic organisms are ingested by filter feeders and, after death, detritivores may consume the slowly sinking remains before they reach detritus traps exposed below the photic zone. In running water the detritus may be swept out of the system or carried in from external, perhaps terrestrial sources. Wetzel (1975) notes that some streams and rivers receive so much terrestrial leaf litter that it exceeds aquatic primary production and the overall metabolism of the ecosystem is heterotrophic. Odum and Heald (1975) record the export of more than 50 % of mangrove (*Rhizophora* spp.) litter to adjacent coastal waters: an intense coupling of terrestrial primary to aquatic secondary production with far-reaching consequences for the management of tropical coastal waters.

The measurement of grazing is difficult and often inaccurate, in particular when it is partitioned between a large number of herbivorous species. Most estimates are made indirectly, the herbivore population being assessed by direct counts, capture-recapture, marking or other methods (Southwood, 1978) coupled with controlled environment measurements of feeding rates. The techniques of such studies are described by Waters (1977) and Edmondson and Winberg (1971), aquatic secondary production; Golley *et al.* (1975), small mammals; Duvigneaud (1971), forest secondary producers; and Grodzinski *et al.* (1975), general bioenergetics techniques.

MEASUREMENT OF ENERGY CONTENT

The conversion of biomass data to caloric values is a necessary step in constructing energy budgets for different organisms or ecosystems. The most common measuring technique is oxygen bomb calorimetry. The sample is ground to a powder and either compressed to form a tablet or loaded into a gelatine capsule after weighing. The sample is then ignited in oxygen using a suitable bomb calorimeter and the temperature rise may then be used to calculate caloric content which may also need correction for the generation of nitric and sulphuric acids during the ignition. The necessary techniques are described by Lieth (1975), and information is also given which permits calculation of caloric values from chemical analyses of biological materials. Paine (1971) summarizes various other techniques for caloric determination. Many measurements of the caloric values of organic compounds, tissues and organs have now been published and, in some studies, it is possible to use these values to convert productivity data to energy flows. A few generalized values are cited in Table 11.1 and further information may be found in Lieth (1975) and Golley (1961).

POTENTIAL AND ACTUAL PRODUCTION

If solar energy is the only limiting factor, a vegetation canopy with optimum leaf area index and leaf display will manifest its maximum daily rate of dry matter production. The sum of such daily totals, over the whole growing season (or year), represents the *potential net primary production* of the ecosystem. Optimal leaf area index will, of course, be greater during the summer months and this must be incorporated in the calculation. The energy converting efficiency of such a canopy may approach the theoretical limit of 5–6 % conversion of photosynthetically available radiation (PAR) (Loomis and Gerakis, 1975).

In the temperate zone, annual solar PAR input is about 1500 MJ m^{-2} (50 % of the estimate of total short-wave income given by Monteith, 1973). Assuming a conversion efficiency of 5 % and dry matter energy content of 18 MJ kg^{-1} this would give an annual production of $(1500 \times 0{\cdot}05)/18 = 4{\cdot}2\,kg\,m^{-2}$. This is approximately twice the observed maximum yields for uncultivated terrestrial vegetation (Whittaker and Likens, 1975). If the same calculation is repeated for equatorial regions the predicted potential production of *c.* 8 kg m^{-2} is again about twice the observed natural maximum.

Potential production derived in this way is a direct function of the latitudinal distribution of solar radiant energy input (Figure 11.7). Actual production shows considerable departure from this pattern for a host of reasons related to both plants and environment. Some of these such as drought and cold are overriding, but despite their effects there remains a substantial geographical component in the distribution of actual production. Figure 11.8 shows data extracted from Lieth's Innsbruck Productivity Map (1975) on a midline transect from southern South America, northward through the U.S.A. to Canada: over most of this transect water deficit is not a major limiting factor: the distribution of productivity closely mirrors the radiant energy distribution and consequent temperature gradients. By contrast, Figure 11.9, a south to north transect through North Africa and Europe on longitude 15 °E, shows the massive departure from expected latitudinal value caused by water deficits in the Sahara. The pattern of marine productivity differs from that of land: equatorial and mid-

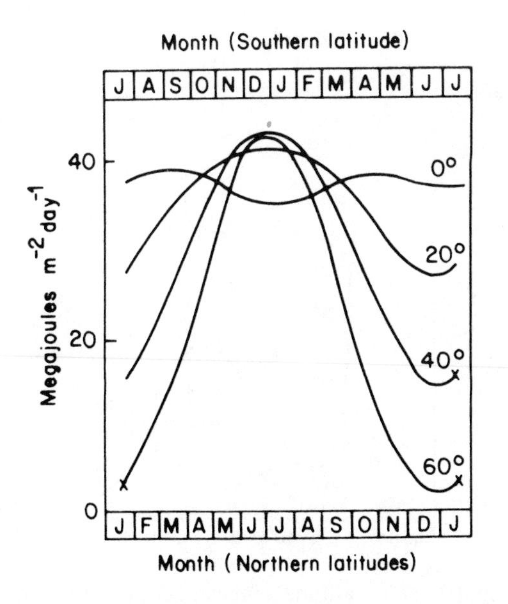

Figure 11.7 Variation in extraterrestrial solar irradiance with latitude and time of year. Modified with permission from R. O. Slatyer, *Plant-water Relationships*, © Academic Press, New York, 1967.

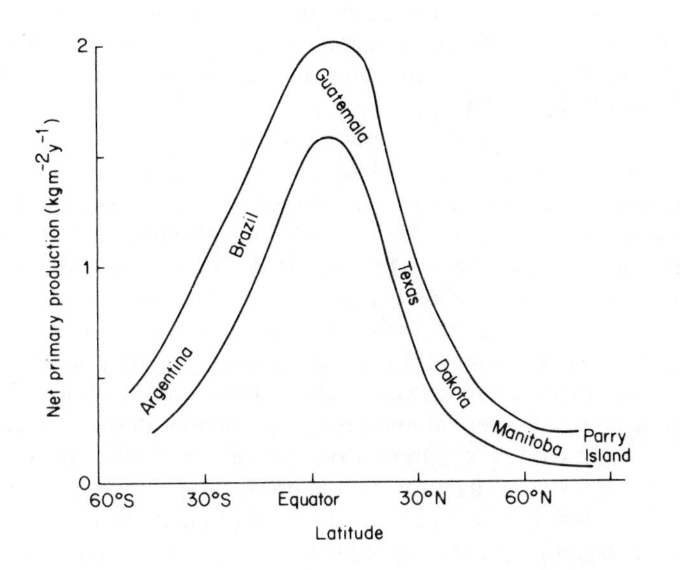

Figure 11.8 Distribution, with latitude, of mean net primary production on a transect forming the mid-line of S. America, USA and Canada. The spacing of the two curves represents the range. Data of Lieth (1975).

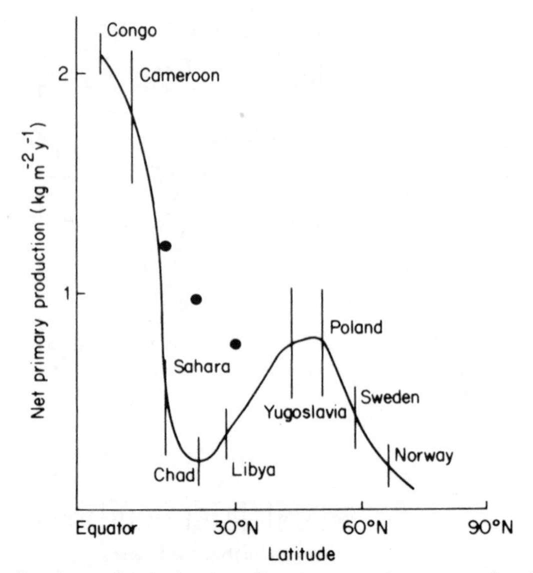

Figure 11.9 Distribution, with latitude, of mean net primary production on a transect comprising the 15 °E line of longitude running northward from mid-Africa to N. Europe. Vertical bars represent range and the three plotted points are the expected P_{net} values in the absence of drought limitation, their values being taken from Figure 11.9. Data of Lieth (1975).

latitude waters being poor, large productivites occurring only at high latitudes where winter thermal overturn brings nutrients such as phosphorus, nitrogen and silica to the surface. Equatorial and adjacent waters are permanently thermally stratified and consequently cut off from deep-water sources of nutrient replenishment (Figure 11.10).

Actual production fails to reach potential levels for a variety of reasons: (i) as described above, environmental factors may be limiting; or (ii) inherent characteristics of the plant canopy may prevent it from achieving maximum yield. An early productivity study by Black (1964) strikingly illustrates the contribution of environmental and inherent biological limitation. *Trifolium subterraneum* (subterranean clover) is widely grown in Australasia as pasture legume, and factors limiting its agricultural performance are of great interest. Black calculated potential productivity from maximum daily rates and, by a series of harvests, related actual production to these values. Figure 11.11 shows the marked discrepancy between potential and actual production; during the five summer months, drought entirely suppressed production and then at the beginning of autumn (April) water availability allowed production to increase, but initially it was limited by low leaf area index and failure to intercept all of the incoming radiation. Production transiently equalled the potential value at mid-winter (June) but quickly fell during July–November as LAI became supra-optimal and canopy self-shading reduced photosynthesis of lower leaves to a rate less than the compensation point.

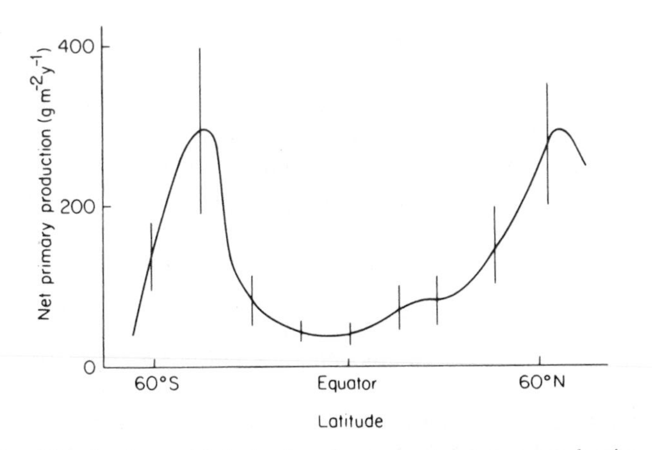

Figure 11.10 Distribution, with latitude, of mean net primary production on a mid-Atlantic transect comprising the 30 °W line of longitude. Vertical bars represent range. Note the expansion of the scale (× 4) compared with Figures 11.8 and 11.9. Data of Bunt, in Lieth and Whittaker (1975).

Black's figures show that 60 % of the annual production is lost to the 5-month drought and, of the 27 t ha^{-1} which could be produced during the wet season, 63 % is lost to inherent limitations by leaf area index leaving an actual production of only 1·25 t ha^{-1} y^{-1}. The study provides a theoretical basis for agricultural practice, as irrigation and manipulation of LAI by defoliation and increased sowing density would recover a substantial proportion of the loss.

Blacks's study, undertaken at Adelaide, illustrates drought limitation in a Mediterranean climate where natural vegetation is a xeromorphic sclerophyll scrub with a high drought tolerance and much lower inherent productivity than radiant input would predict. The *T. subterraneum* results may be considered as an

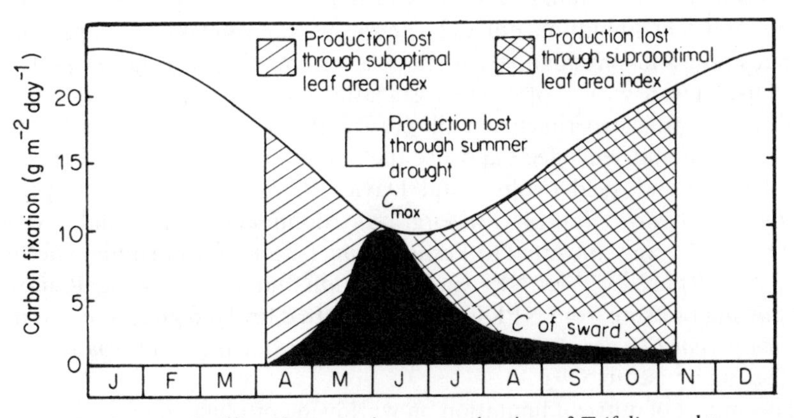

Figure 11.11 Factors limiting the net primary production of *Trifolium subterraneum* in arid-zone conditions in S. Australia. C_{max}, annual potential production; C, actual production. Reproduced with permission from J. N. Black, An analysis of potential production..., *J. appl. Ecol.*, **1**, 3–18 Figure 9 (1964).

ecological 'bioassay' of this situation. In other parts of the world cold is the overriding limitation: the tundra environment, with an annual mean temperature below 0 °C or presence of permafrost (p. 54), is an extreme case. During the few months when light intensity is above compensation point, air *temperatures* are often below freezing. 0 °C may be considered a physiological lower limit for the metabolic processes of most higher plants to proceed at any significant rate, though it does appear that some arctic-alpine lichens have photosynthetic optima near zero and continue carbon assimilation at temperatures as low as −20 °C. This compares with normal lichen optima of 10–15 °C (Kallio and Karenlampi, 1975) and higher plant optima of 10–25 °C (C-3) and 30–40 °C (C-4) (Troughton, 1975). Arctic plants generally fall in the lower part of the C-3 temperature optimum distribution (Tiezen and Wieland, 1975).

In addition to direct limitation by low temperature, tundra plants are often exposed to high intensity sunlight while the soil is still frozen. This causes tissue warming and the initiation of a transpiration stream which cannot be met from the soil. Such winter drought stress may affect wintergreen species unless they are protected by snow cover. If the soil is not frozen, radiant heating may permit photosynthesis in air temperatures well below zero.

Next to water deficit and cold, nutrient deficiency must be the commonest limitation to terrestrial primary productivity; the universal increase of agricultural yield by fertilization and in particular the increases which may be obtained from semi-natural grassland, well illustrate this point. Bradshaw (1969) cites the case of chalk grasslands in South-East England which were once thought of as summer drought limited, but now give heavy agricultural yields after fertilization, especially with phosphorus. The Park Grass Plots at Rothamsted Research Station, England, afford another example. They were laid down in 1856 as fertilizer trials which have been continued to the present day. The original grassland, which persists in untreated control plots, yields about $2\,t\,ha^{-1}\,y^{-1}$ which is increased by liming and NPK fertilization to $8{\cdot}7\,t\,ha^{-1}\,y^{-1}$ (Thurston, 1969). Nutrient limitation not only depresses yield but also controls species composition and physiognomy of the ecosystem: the Park Grass plots have developed a wide variety of species associations and soil types (Thurston, 1969), while comparable experiments in hill-grassland, started by Milton in 1930, have shown similar nutrient effects and also the dramatic impact of animal grazing on composition and yield (Milton and Davies, 1947). Over a 16-year period a *Molinia caerulea* (purple moorgrass) dominated grassland, yielding $2{\cdot}6\,t\,ha^{-1}\,y^{-1}$, was converted to *Festuca ovina* (sheep's fescue) dominance and a yield of $10{\cdot}9\,t\,ha^{-1}\,y^{-1}$ by NPKCa fertilization and heavy grazing. Removal of the grazing pressure from NPKCa plots permitted lowland grass-species such as *Lolium perenne* (ryegrass) to assume dominance but the yield was reduced to $5{\cdot}4\,t\,ha^{-1}\,y^{-1}$ (see p. 379).

The impact of nutrient limitation on worldwide productivity is highlighted by the growth of the fertilizer industry during the past century. At the end of the Second World War, total NPK use was less than $1{\cdot}0 \times 10^7\,t\,y^{-1}$ but this had grown to more than $5 \times 10^7\,t\,y^{-1}$ by the mid-1960s (Brown, 1970). The

Haber–Bosch process of nitrogen fixation was not developed until 1914 but the total industrial output had reached $3 \times 10^7\,t\,y^{-1}$ by 1970 (Delwiche, 1970) and now represents more than twice the global total of biological fixation. In England and Wales alone during 1968–9 the consumption of fertilizer elements was N: 6·6, P: 1·6 and K: $3{\cdot}1 \times 10^5$ t (MAFF, 1970). The energy-cost of industrially produced ammonium sulphate is about $42\,GJ\,t^{-1}$, at high fertilization rates a very substantial proportion of the energy content of the crop! Little wonder that H. T. Odum (1971) wrote: 'potatoes partly made of oil'.

Manufacturers claim that nitrogen fertilization up to $450\,kg\,ha^{-1}\,y^{-1}$ produces a yield increase in temperate grasslands and that $300\,kg\,ha^{-1}\,y^{-1}$ is profitable in short-term leys which are sown to varieties which have been selected for nitrogen responsiveness. The primary productivity may be increased from less than $1{\cdot}0\,t\,ha^{-1}\,y^{-1}$ to $10\,t\,ha^{-1}\,y^{-1}$ or more. Nitrogen fertilization has a similar impact on arable crop yield which has increased massively in consequence: Postan (1972) cites medieval records of wheat production in England equivalent to $400–450\,kg\,ha^{-1}$ (6–8 bushels per acre) compared with modern yields exceeding $3000\,kg\,ha^{-1}\,y^{-1}$ (Milthorpe and Moorby, 1974). Some of this increase is a consequence of varietal improvement but most derives from nitrogen and other fertilizer input. At these high yield levels the ratio of grain to straw is about 0·6 indicating a total above ground net primary production of about $8\,t\,ha^{-1}\,y^{-1}$ and total P_{net} of $10–12\,t\,ha^{-1}\,y^{-1}$. Most of the substantial yield improvement given by 'green revolution' maize, rice and wheat varieties stems from very heavy use of nitrogenous fertilizer: the breeder's role has mainly been to provide dwarfed plants which will withstand such nitrogen fertilization without the wind 'lodging' to which tall varieties are susceptible. General response of terrestrial communities and crops to nitrogen input are shown in Figure 11.12.

Most aquatic and terrestrial herbaceous ecosystems are nutrient-limited unless seriously stressed by acidity, cold or some other single adverse factor. Nitrogen is commonly the primary limitation and if this is corrected by fertilization other deficiencies may be induced, phosphorus frequently being the next limitation. In aquatic systems nitrogen is less often limiting as it is readily fixed by blue-green algae, phosphorus frequently being the overriding production limiter. It is for this reason that inland waters, estuaries and continental-shelf seas are so prone to eutrophication by sewage-born phosphorus, for example Schelske (1975) showed an enhancement of algal growth in Lake Superior by increasing the total phosphorus concentration from 0·0025 to less than $0{\cdot}01\,mg\,l^{-1}$. This effect was found without addition of extra nitrogen and Schelske suggested that phosphorus eutrophication ultimately causes a shift to blue-green algal nitrogen-fixing species when diatoms and green algae become nitrate limited. Lund (1965) similarly showed that the spring peak of diatom growth which occurs in both ocean and lakewater is ultimately limited by silica deficiency, and green or blue-green algae then become the major summer producers according to the nitrogen and phosphorus status of the water. Diatoms are unique in needing very large quantities of silica for frustule synthesis; 26 to 63% of their dry weight (Lund, 1965), and such silica limitation cannot occur in terrestrial ecosystems.

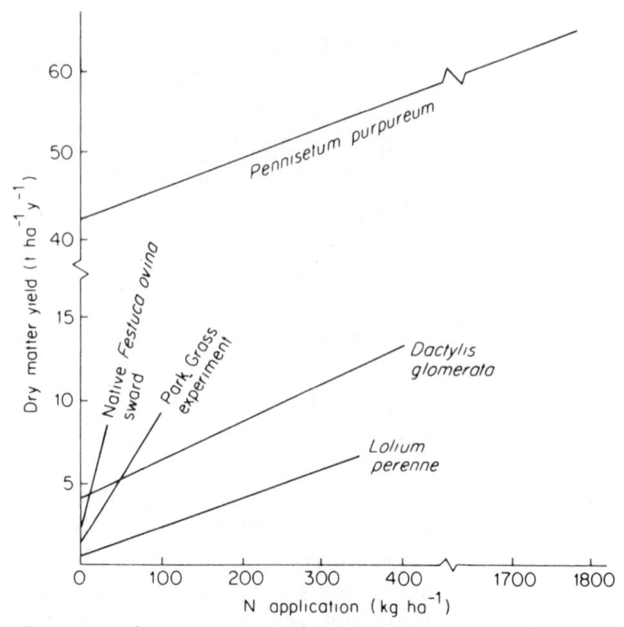

Figure 11.12 Response of net primary production to nitrogen fertilization. *P. purpureum*: Puerto Rico (Vincente-Chandler *et al.* 1964). *F. ovina*: Wales—experimentally fertilized, upland pasture (Milton and Davies 1947). Park Grass: Rothamstead, England—experimental, long-term fertilization of mixed-species permanent pasture (Thurston 1969). *D. glomerata*: N. America (Rhyked and Woller 1973). *Lolium perenne*: England (Spedding 1975). Lines joining points are for clarity and are not intended to model intermediate situations.

Aquatic and terrestrial ecosystems differ strongly in their CO_2 relationships, partly because some aquatic plants utilize dissolved bicarbonate (HCO_3^-) in addition to dissolved CO_2, producing a complicated, pH-related relationship, but mainly because the epilimnion of a productive lake is often CO_2 undersaturated at mid-day and primary production is limited by the rate of invasion of CO_2 from the atmosphere above (Schindler and Fee, 1975). The lake epilimnion might thus be compared to a single giant mesophyll cell in this context! Just as solution of CO_2 at the wet mesophyll surface may limit plant photosynthesis, so lake production may be limited by the diffusion resistance of the air–water boundary.

Because of low inorganic solubilities, slow diffusive exchanges at phase boundaries and lack of rapid turbulent mixing, deep lakes and oceans are very sensitive to thermally induced turnover such as the autumn and winter mixing of epilimnic and hypolimnic water in temperate lakes or to current induced upwellings such as that on the Pacific west coast of South America. The appearance of nutrient-rich deep water at the surface causes an increase in the primary production. Permanent thermal stratification in tropical waters maintains such areas as phytoplanktonic deserts, while the winter overturn is responsible for the productivity of high latitude waters (Figure 11.10) and for the spring and early summer 'blooming' of lakes.

Productivity of eutrophic aquatic systems may be limited by self-shading similarly to the development of supra-optimal leaf area index in terrestrial canopies (Black, 1964, p. 349). Planktonic algae in the lower part of the water column may be below light compensation point but, unlike leaves, turbulent mixing or active migration causes them at some time to experience higher irradiances near the surface, though the net photosynthetic effect will be similar. Talling (1975) shows variable patterns of algal biomass distribution with depth which are analagous to the non-uniform stratification of leaf area index and he described the depth range from the surface 100% PAR to 1% PAR as the euphotic zone in which photosynthesis is likely to be positive.

Algal suspensions offer the advantage, in culture, that they can be maintained at optimal areal density (analagous to optimum leaf area index) by filtration or centrifugal separation of large mature cells. Filter-feeding organisms similarly maintain relatively low algal densities. An algal concentration representing *c*. 300 mg chlorophyll-*a* m^{-2} in the euphotic zone gives maximal absorption of PAR (Talling, 1975) and represents a biomass density of *c*. 0·015 kg m^{-2}. This may be contrasted with the enormously larger biomass densities of forest vegetation, ranging from 20 to 45 kg m^{-2}. This variation in P_{net}/B ratio has already been discussed (Figure 11.2 and p. 336).

Whittaker and Likens (1975) tabulate the areal density of chlorophyll in the world's major ecosystems. The range is from 0·03 g m^{-2} in infertile ocean water to a maximum of 3·5 in some forest systems, a span only just exceeding two orders of magnitude. The corresponding range of biomass density is 0·001 to 45 kg m^{-2}, well over four orders of magnitude, again suggesting that primary production is unrelated to biomass density which mainly represents the storage of organic materials as structural components or energy reserves. In the aquatic planktonic systems, primary production is almost simultaneously consumed, decomposed and mineralized-recycled without accumulation of stored organic products.

Distribution of productivity in the biosphere

Some indication of the world-wide range of net primary production may be gained from Figures 11.2 and 11.8–10, and more detail appears in Table 11.3. Maximal P_{net} values are associated with tropical swamps and marshes, closely followed by cultivated land and then by tropical rain forest. There are a few records of greater P_{net} values for cultivated tropical grassland, for example Vincente-Chandler *et al.* (1964) obtained an above-ground maximum yield of 71 t ha^{-1} y^{-1} from *Pennisetum purpureum* (Napier grass) in Puerto Rico: with a realistic ratio of shoot to root yield this must be close to a total P_{net} of 100 t ha^{-1} y^{-1}, a value which Milthorpe and Moorby (1974) cite for the same species in Kenya. It is no surprise to note that *P. purpureum* is a C-4 plant or that other C-4 crops give similarly high yields under optimum conditions with long growing seasons. *Saccharum officinale* (Sugar cane) in Hawaii gave 78 t ha^{-1} y^{-1} (Burr *et al.* 1957) and *Zea mays* 30 t ha^{-1} y^{-1} in the cooler climate of Illinois (Milthorpe and Moorby, 1974). With their high temperature optima, high maximal photosynthetic rates and adaptation to water stressed habitats, the C-4

Table 11.3 Biomass and net primary productive characteristics of the biosphere and of specific vegetation types. Derived from Whittaker and Likens (1975) and other cited authors

Biosphere totals		
	Biomass (10^9 t)	P_{net} (10^9 t y^{-1})
Marine	3·9	55·0
Continental	1837	117·5
Total biosphere	1841	172·5
Efficiency		
Energy equivalence of 172·5 × 10^9 t dry matter (see Table 11.1)	2·9 × 10^{15} M J	
Annual solar PAR input	1278 × 10^{15} M J	
Efficiency	0·23 %	
Ecosystem types: normal ranges		
	Biomass (kg m^{-2})	P_{net} (kg m^{-2} y^{-1})
Tropical forest	6–80	1–3·5
Temperate deciduous forest	6–60	0·6–2·5
Boreal conifer forest	6–40	0·4–2·0
Temperate grassland	0·2–5	0·2–1·5
Tundra	0·1–3	0·1–0·4
Desert	0·1–4	0·1–0·25
Swamp	3–50	0·8–6·0
Freshwater	up to 0·1	0·1–1·5
Open ocean	up to 0·005	0·2–0·4
Ocean upwelling	0·005–0·1	0·1–0·4
Maximal P_{net}—some examples	(kg m^{-2} y^{-1})	
Algal mass culture, e.g. *Scenedesmus* sp. (Wassink, 1975)	5	
Marine littoral brown algae, e.g. *Laminaria* spp. and *Macrocystis* spp. (Mann and Chapman, 1975)	0·5–5	
Temperate emergent swamp, e.g. *Phragmites communis* and *Typha* spp. (Westlake, 1975)	4–6	
Tropical grassland—cultivated, irrigated fertilized and planted with *Pennisetum purpureum*, a C-4 grass (Milthorpe and Moorby, 1974)	10	
Temperate grassland—cultivated and fertilized (Loomis and Gerakis, 1975)	2–2·5	
Tropical forest (Wassink, 1975)	4	

species, both agricultural and wild, tend to excel in the low latitude, dry sub-tropics.

It is of interest to compare the maximum photosynthetic rates of different vegetation types (Table 5.4) with the maximum recorded P_{net} values (Figure 11.13). Predictably, there is a fairly strong correlation, confirming the intuition that photosynthetic rate must be determinant of production in the absence of limiting factors other than light and CO_2. The lower limits of productivity for different vegetation types show no such correlation with optimal photosynthesis as the low values may be caused by a variety of factors including cold, seasonal shortening of day-length, water and nutrient limitation. Further information relating the distribution of biomass, P_{net}, vegetation and soil types to climatic limitation is given in Figure 11.14.

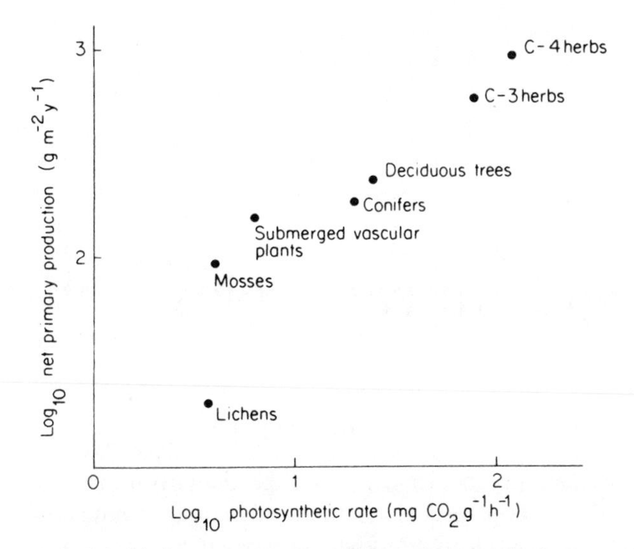

Figure 11.13 Relationship between mean maximum photosynthetic rate and net primary productivity of various types of plants. Note logarithmic scales. Data from Whittaker and Likens (1975) and, for lichens, Kallio and Karenlampi (1975).

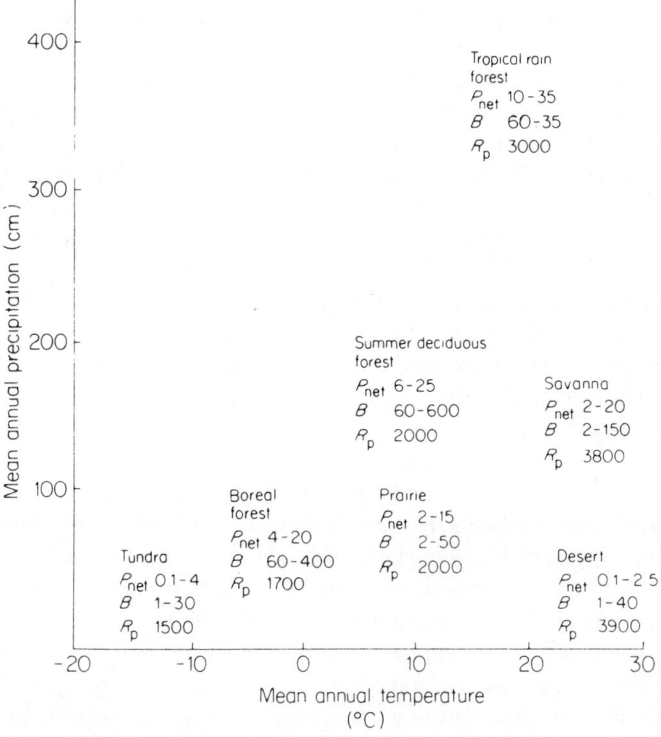

Figure 11.14 The climatic-geographical distribution of primary productivity, biomass and annual irradiance presented in the same framework as the distribution of soils and vegetation types shown in Figure 3.2. P_{net}—net primary production (t ha^{-1} y^{-1}); B—biomass density (t ha^{-1}); R_p—photosynthetically available radiation (MJ m^{-2} y^{-1}, 0·4–0·7 μm). Data from Lieth and Whittaker (1975) and Budyko (1974).

Chapter 12

Biogeochemical cycles

INTRODUCTION

Certain reactions in geochemical cycles, particularly those involving redox changes, would not be expected on thermodynamic grounds. This is not to suggest that they contravene thermodynamic laws, it is simply that the harnessing of solar energy by photosynthetic organisms must be considered as a part of the geochemical process, hence the unwieldy but useful noun, biogeochemistry, which was coined in the knowledge that unaided physics and chemistry could not fully explain the present state and functioning of earth's surface environment.

When life originated, the atmosphere was substantially anoxygenic: there is some dispute concerning oxygen production by photochemical dissociation of water (Towe, 1978) but geochemical evidence and physico-chemical calculation suggest that it probably did not exceed 10^{-4} of the present level when the first organisms appeared ($>3{\cdot}5 \times 10^9$ y BP). These first microorganisms are believed to have been heterotrophs subsisting on abiotically synthesized organic compounds and would initially have had a quantitatively negligible effect on the environment.

The event which triggered the production of the present-day disequilibrium of N_2 and O_2 in the atmosphere and NO_3^- and H^+ in seawater was the evolution of the photosystem II pathway of photosynthesis which releases free oxygen. B. N. Smith (1976) has argued that the inhibition of C-3 photosynthesis by the modern atmospheric oxygen concentration (21 % v/v), the Warburg effect (p. 128), is circumstantial evidence for the process having evolved in a lower oxygen environment. He also suggests that the more recent evolution of the C-4 pathway, which is not oxygen-inhibited, is the consequence of natural selection in the contemporary oxygen-rich atmosphere. Similarly the present day sub-optimal level of CO_2 for photosynthesis suggests that the process may have its origins in the CO_2 richer atmosphere suggested for the early earth.

The anoxygenic early atmosphere, which is almost certainly a prerequisite of the origin of living organisms, seems most likely to have been a mixture of N_2 and CO_2 with much smaller quantities of reduced carbon compounds and NH_3 originating from degassing of the earth and photochemical effects. Hart (1978) has produced a convincing simulation based on these assumptions which accounts for the present state of the atmosphere and sedimentary rock chemistry, provides for adequate duration of an anoxygenic atmosphere for life to appear and then models the correct time-scale of oxygen evolution. It also avoids producing a runaway glaciation or 'greenhouse effect', fairly difficult constraints as the distance from the sun giving high enough and low enough radiant inputs to avoid inevitable glaciation or atmospheric heating is a band only 0·05 astronomical units wide.

The photosynthetic production of oxygen not only affects atmospheric composition and redox-related equilibria but also has an indirect impact on a host of microorganism-mediated redox reactions which are involved in the geochemical cycles of elements such as C, H, O, N, Fe and S amongst others. Carbon compounds are formed during photosynthesis and, for equilibrium, equal quantities must be oxidized to CO_2 by respiration. However, during geological time, if oxygen has accumulated in the atmosphere this implies a disequilibrium and carbon must have been stored out of reach of microorganism respiration, presumably in marine sediments.

This carbon storage has indeed occurred, in the obvious form of the fossil fuels and in the less obvious form but much greater quantity of interstitial carbon, the 'kerogen' of sedimentary rocks. Holland (1978) estimates an annual storage of $1{\cdot}2 \times 10^{11}$ kg which would correspond to an oxygen doubling time of 4×10^6 y, geologically a very short period. It is oversimplistic to think of oxygen accumulation in terms only of carbon sequestration, as both iron and sulphur in igneous rocks are in the reduced condition so that weathering also forms a sink for atmospheric oxygen. During the early years of photosynthesis it is likely that all of the oxygen produced was 'sinked' to divalent (ferrous) iron and, for organisms which had evolved in anoxic conditions this may well have been protection from oxygen toxicity and the cause of banded ironstone deposition which was common at this time.

Carbon deposition had reached substantial rates by the early Archaean: Reimer *et al.* (1979) estimated an annual carbon preservation of $0{\cdot}32\,g\,m^{-2}$ in shales of $3{\cdot}4 \times 10^9$ y old. They were unable to differentiate between anoxygenically reduced carbon (as in modern photosynthetic bacteria and some Cyanophyta) and carbon produced by oxygenic photosynthesis, but suggested a most probable mixture of both, the banded ironstone formations in these South African 'Fig Tree' deposits being interpreted as forming by oxidation of ferrous iron during the 'sinking' of photosynthetic oxygen.

If the carbon dioxide compensation point of Devonian land plants was similar to that of modern C-3 plants, uniformitarianism would suggest that the CO_2 concentration of the atmosphere has probably never been below the present level for the past 400 million years. Both fish and very large insects existed during the Devonian and the same principle would suggest, if respiratory O_2-demand of these organisms was at modern levels, that the atmospheric O_2 was likely to be near the present day 21%.

Considering the large impact of photosynthesis and other biological processes these facts imply that there may be a considerable homeostasis of atmospheric composition by living organisms. Lovelock and Margulis (1974) have suggested that this includes stabilization of many biogeochemical cycles as well as atmospheric temperature control and that the process is also subject to natural selection so that organisms, by a feedback process, may have maintained their own environment within tolerable limits.

TYPES OF CYCLE

The photosynthetic plants of the earth's surface inhabit an atmosphere which they, themselves, have produced and a soil system which is a reflection of the dynamic balance between the input of plant materials and the catabolism of heterotrophs. The consequence of these relationships is the establishment of a number of cycles of essential elements, some of which are essentially local and some global in scale (Figure 12.1). The local cycles involve the less mobile elements in which there is no mechanism for long-distance transfer but the global cycles have a gaseous component which links all of the world's living organisms to form the giant ecosystem which we know as the biosphere. In relation to geological time scales the functioning of all the essential element cycles is extended by the various processes of continental denudation and sedimentation or other redeposition of materials. Within this reference framework, biological activity is responsible for the process of biogeochemical cycling.

These cycles may be divided into three global types.

(i) The gaseous cycles of carbon, oxygen and water, elements which are not normally regarded as nutrients, but are either utilized or produced in photosynthetic and respiratory processes.
(ii) The gaseous cycle of the nutrient element, nitrogen.

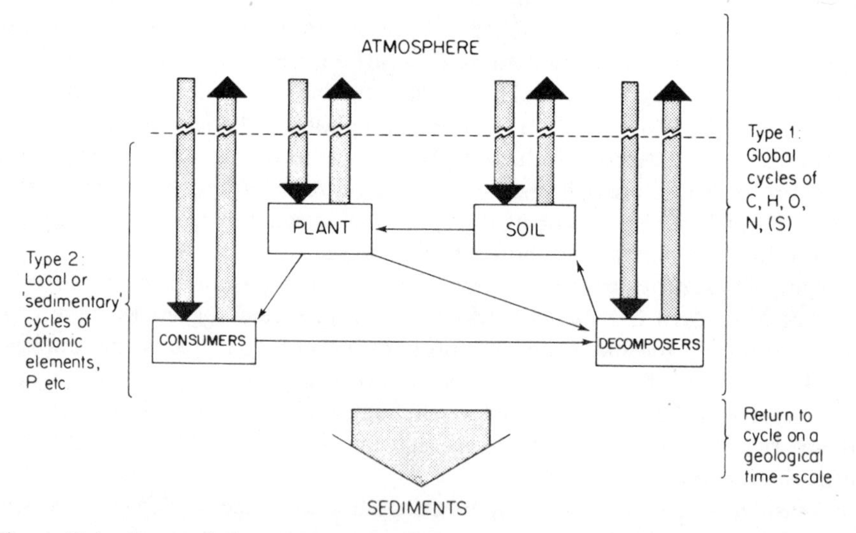

Figure 12.1 Types of elementary cycles. Breaks in arrows indicate that Type 1 cycles involve the atmosphere but Type 2 do not.

(iii) The non-gaseous, sedimentary cycles of the remaining nutrient elements. Sulphur is to some extent intermediate, since H_2S or SO_2, formed under some circumstances, add a gaseous component to its normally sedimentary cycle.

The sedimentary cycles are, in the short term, localized within ecosystems and the transport of their components by sedimentary processes permits them to become fully closed cycles only with the passage of geological time. The localized portions of the cycles, which are directly involved with biological activity, are strongly influenced by ecosystem energy flow so that an increased flow usually causes more rapid cycling. The strongest nutrient fluxes become associated with the main energy flow pathways in the producer-consumer food web; Figure 12.2

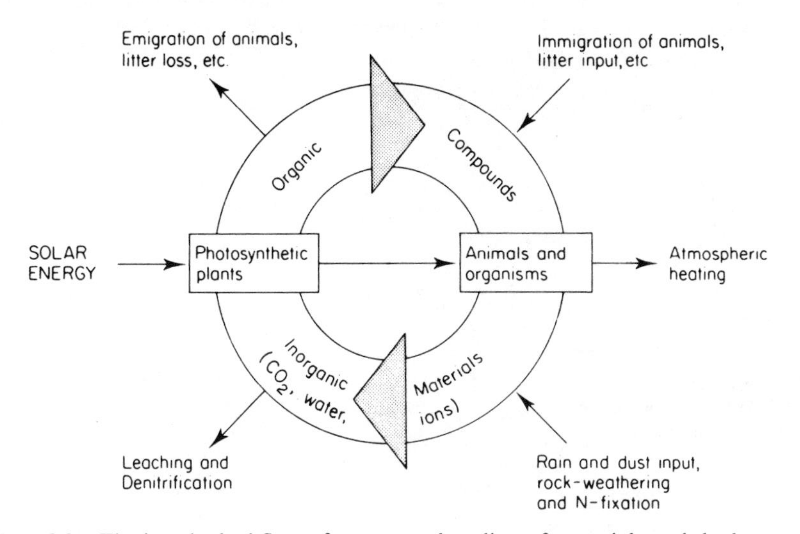

Figure 12.2 The interlocked flow of energy and cycling of materials and the losses and gains of materials which may brake or accelerate the system. Based on Etherington (1978).

shows this relationship and also indicates the biological concentration of nutrients which may accompany the growth of an energy flow pathway in an evolving ecosystem. It should be appreciated that the concentration of nutrients at soil surfaces, caused by litter fall, and the accumulation of nutrients in living organisms, from a low ambient concentration, are energy-demanding processes.

The relationship between energy flow and mineral cycling has been one of the more fruitful fields for the application of the systems analysis method, the classic Liebig limiting factor approach being too inflexible to cope with the complex interactions which are now known to occur in plant nutrition. Furthermore, as the flux of materials through an ecosystem gives some measure of its continuity and stability, and also follows the pathway of energy flow in food webs, Pomeroy (1970) suggests that the analysis of mineral cycling is a useful strategy with which to approach ecosystem analysis.

THE CARBON CYCLE

The primary consequence of the photosynthetic process is the storage of energy in reduced carbon compounds. These may then be exploited by the consumers and decomposers of the ecosystem so that the throughput of energy is accompanied by a cyclic exchange of CO_2 with the atmosphere. This exchange takes place over the whole land and ocean surface of the earth wherever temperature is high enough, or water supply adequate, to support life. It may be considered to be the most significant of the biogeochemical cycles as it is through the synthesis of carbon compounds that ecosystem energy flow is linked to the cycling of all other materials. Figure 12.3 illustrates the global carbon cycle and indicates the quantities involved in the various storages and annual fluxes. Though these estimates cannot be particularly accurate, they do indicate relative magnitudes; on a geological time scale it may be seen that photosynthetic carbon consumption would rapidly exhaust easily available resources if it were not for the respiratory return of CO_2 to the atmosphere.

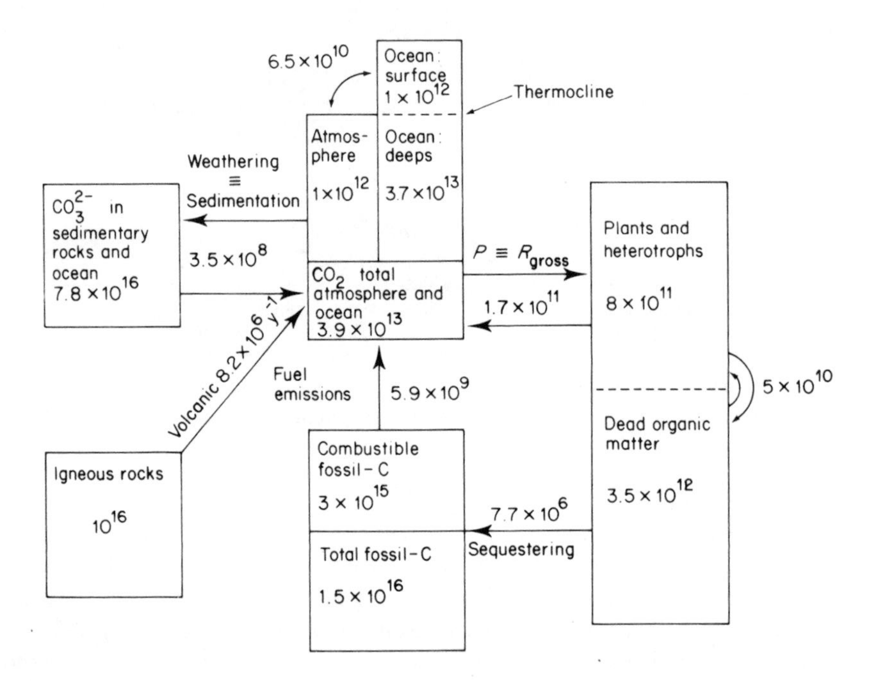

Figure 12.3 The global carbon cycle with quantities based on data collations of Lieth and Whittaker (1975) and Bowen (1979). Masses (in boxes): tonnes. Fluxes (arrows): $t\,y^{-1}$. These figures are the best estimates currently available from environmental measurements: some are relatively accurate, for example atmospheric and oceanic content and emissions from fuel combustion. Photosynthetic and respiratory fluxes are less accurate, while geological storage is little more than informed guesswork. Some workers have suggested a much larger ratio of fossil-C to fossil fuel-C (Brooks and Shaw 1973: 200 × more).

The majority of ocean-dissolved CO_2 (HCO_3^-) is below the thermocline and inaccessible for rapid exchange with the atmosphere. The immediate photosynthetic source is thus confined to atmospheric and epilimnic CO_2, a relatively small quantity which, without respiratory return, would be exhausted within one to two decades. The rate of exchange between ocean and atmosphere is such that the mean residence time of a carbon atom is only 3 years in either reservoir.

Several of the transfers shown in Figure 12.3, such as formation and dissolution of carbonates, photosynthesis and respiration, the production of organic materials and their oxidation by decomposers, must be assumed close to equilibrium for long-term stability of the cycle. The 'guesstimate' of leakage to fossilization in sediments and peat-beds is thus only $4{\cdot}5 \times 10^{-5}$ times the annual production.

The amount of CO_2 in the easily available reservoir is small: some 300 times the annual production by fossil fuel burning, which may have an impact on atmospheric concentration if the homeostasis by increased photosynthesis and by equilibrium with the ocean surface is inadequate to cope with this rate of release. There is evidence that the atmospheric concentration is increasing at nearly 10 v.p.m. per decade (Figure 12.4) and that there has been an increase of about 35 v.p.m. from a baseline of 290 v.p.m. in the pre-industrial era.

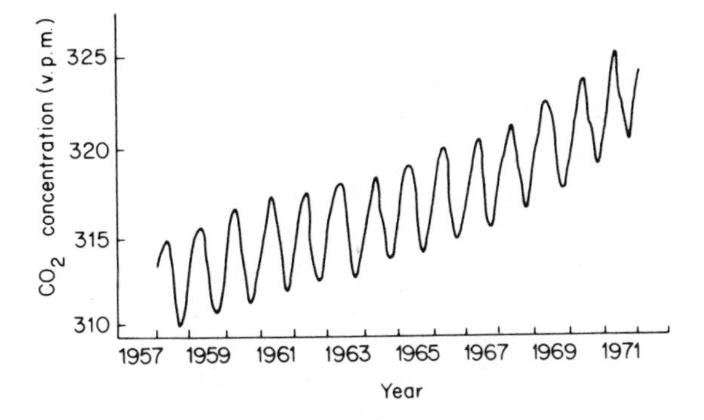

Figure 12.4 The time course of atmospheric CO_2 concentration in the northern hemisphere from measurements at Mauna Loa, Hawii, 1957–1971. Data of Ekdahl and Keeling cited by Holland (1978). This statistically fitted line is very close to the mean monthly analyses plotted in the original graph.

H. T. Odum (1971) has speculated that the return of biogenic fossil carbon to the cycle was inevitable, evolution ultimately providing an organism which can exploit an available energy source. When fossil fuel approaches exhaustion early in the next century it may be presumed that this pulse-increment of CO_2 will re-equilibrate back to the previous level, assuming the continued functioning of the biosphere, but some concern has been expressed that enhanced atmospheric CO_2 may lead to long-wave re-radiation being absorbed by the atmosphere (greenhouse effect) to cause atmospheric heating, climatic change and ocean-level rise. It may also be that a small amount of atmospheric heating will increase cloud-cover so homeostasing air-temperature by increasing short-wave albedo: at the present time it is impossible to assess these conflicting suggestions.

Global photosynthesis certainly has impact on the atmospheric CO_2 content as may be seen from the seasonal oscillation recorded on Hawaii (19·5 °N) in Figure 12.4. The rate at which the industrial pulse could re-equilibrate is dependent partly on sequestration into marine sediments, a process probably currently limited by surface water phosphorus concentration (p. 348). A further imponderable is the rate of evolution of soil and vegetation-stored carbon during current and future disafforestation (Woodwell, 1978).

The carbon cycle cannot be left without some mention of the element's isotopic composition, which has proved so useful in assessing not only the recent history of ecosystems but also some of the physiological and chemical processes which may have been involved in isotopic fractionation. Carbon is predominantly of atomic weight 12 but atmospheric CO_2 is continuously enriched with the radioactive ^{14}C produced in tiny amounts in the upper atmosphere by cosmic-particle interactions with nitrogen (^{14}N) nuclei. Once carbon has been photosynthetically assimilated, the radioactive decay of the incorporated ^{14}C gradually depletes dead tissue of the isotope with the consequence that its quantity may be used as a measure of the age (time of death) of the organ.

Carbon-14 forms only a tiny fraction of natural C but the stable ^{13}C is represented at just over 1 %. This has proved fortunate as a number of biological processes show differential discrimination with the result that the biogenic origin of fossil carbon can be established with some confidence and it is also possible to ascertain whether photosynthetic carbon assimilation has taken place directly through ribulose bisphosphate trapping (C-3 plants) or *via* phosphoenol pyruvate trapping as in C-4 or stressed CAM plants (Kluge and Ting, 1978). By comparison with an arbitrary standard (fossil carbonate), C-3 plants contain carbon which is depleted of about 27 parts per thousand ^{13}C ($\delta^{13}C = -27$ per thousand) but C-4 plants show less than half of this discrimination with a mode at $\delta^{13}C = -11$ per thousand. Brooks and Shaw (1973) show coal and modern land plants (C-3) to have the same ^{13}C composition. Similar analyses have been possible for a wide range of geological carbons.

OXYGEN, HYDROGEN AND THE WATER CYCLE

For each molecule of CO_2 assimilated in photosynthesis one molecule of oxygen will be released. Using the value from Figure 12.3 for C-assimilation, about $4{\cdot}5 \times 10^{11}$ t O_2 will be generated each year and, assuming an atmospheric O_2 mass of $1{\cdot}2 \times 10^{15}$ t (Bowen, 1979), this gives a total cycling time of two to three thousand years. The seasonal effects of photosynthesis and the effects of fossil fuel burning will thus be negligible, but the slow leak of carbon to fossilization is important in the geochemistry of oxygen as it probably just balances the oxidation of Fe and S produced in contemporary rock weathering.

The present day oxygen concentration of 21% v/v must be a compound product of the balance between photosynthesis and total respiration, the limitation of marine photosynthesis by phosphorus, the effect of oxygen concentration on the oxidation of sinking marine organic carbon and the liberation of reduced iron and sulphur by rock weathering.

The hydrogen cycle is dominated by that of water, in which some $4{\cdot}4 \times 10^{14}$ tonnes of water are precipitated and evaporated, annually, over the earth's surface. This is by far the largest global process in which plants are involved, most of the water evaporated from fully vegetated land surfaces being lost via the transpiration stream. Comparatively, only a very small proportion of the water taken up by plants is used in the photosynthetic process. The larger proportion is lost by transpiration from leaves and the latent heat exchange during its evaporation is a very significant contribution to the ecosystem energy budget.

The water cycle also plays a transporting role in geochemical cycles. More than 75% of global evaporation is from ocean surfaces, but less than 75% returns directly to the oceans as precipitation. There is a consequent run-off from land surfaces which removes soluble and particulate materials and deposits them in ocean basins, thus forming the transport mechanism for the sedimentary cycles of the less mobile elements.

Water is a lynch-pin of terrestrial life, a fact which is owed to its anomalous behaviour compared with that of the dihydrides of the related group-6 elements and the hydrides of adjacent period-2 elements. The anomaly is in its high melting point and boiling point which cause it to be liquid over the common range of terrestrial temperatures. The anomaly is attributable to hydrogen bonding which gives even liquid water a quasi-crystalline structure (Hutchinson, 1957). This confers other properties of which natural selection has also taken advantage: both its high specific heat and latent heat of vaporization are important to the thermal balance of organisms and ecosystems, while its characteristic of reaching maximum density at 4 °C above its melting point is responsible for the thermal stratification of the oceans and their activity as a global solar energy transducer. It is also because ice floats that the world's oceans do not become totally frozen in high latitudes.

The majority of the earth's water is contained in very deep ocean basins but, at the present time, the surface level slightly overtops the gently sloping continental shelves with the consequence that shore-lines will be very sensitive to ocean level changes. The polar ice caps contain only about 1·2% of the total ocean water volume but their melting would represent a sea-level rise in excess of 20 m. The possibility of global temperature change through greenhouse effect and cloud cover alteration consequent on human perturbation of the carbon cycle has been seriously considered as a factor which could trigger such melting, with the inundation of the majority of the world's coastal cities.

THE NITROGEN CYCLE

Gaseous nitrogen is the most abundant element in the atmosphere, comprising 78% by volume. Its global circulation provides almost inexhaustible replenishment of raw material for the nitrogen fixing microorganisms which supply almost all of the nitrogen used by plants. This very large reservoir takes 4×10^8 years to biologically cycle through sedimentary deposits (Holland, 1978) and because of the diversity of distribution of N-fixing organisms the global cycle is geochemically stable and not so sensitive to perturbation as the C-cycle. Local enhancement, for example by addition of fertilizer—or sewage-nitrogen—does have dramatic short-term consequences but these are rapidly damped by denitrification.

The nitrogen cycle has a complex soil-based component of small magnitude and a very large but functionally simple atmospheric component (Figure 12.5). In many ecosystems microbial N-fixation is rate-limiting to the cycle and for this reason coupled with slow N-mineralization and its offsetting by denitrification, plants are usually in strong competition for soil-N.

It is again very difficult to assess quantities in global cycles and the only value in Figure 12.5 which is known with any certainty is the atmospheric mass of nitrogen. Annual industrial nitrogen fixation is also fairly accurately known, thus both Holland (1978) and Bowen (1979) cite $4 \times 10^7\,t\,y^{-1}$ compared with Delwiche's (1970) earlier value of 3×10^7. However, to this must be added the formation of NO_x during fossil fuel combustion which may well equal industrial fixation. In the developed countries about equal contributions are made by motor vehicles and industrial sources of NO_x pollution.

It must be assumed that fixation by microorganism is generally in equilibrium with denitrification. The denitrifying organisms are limited by availability of nitrate and, consequently, surplus denitrifying capacity exists and is responsible for the rapid return to equilibrium after addition of nitrogenous fertilizers. It is this homeostasis which prevents progressive eutrophication of near-shore waters as man-made fixed nitrogen is now approaching half of the previous microbial fixation rate. Rapid loss of fertilizer nitrogen has also tempted agricultural researchers to suggest the use of either nitrification or denitrification inhibitors to reduce the loss-rate. The problem of nitrate leaching extends beyond the problem of eutrophication as its consumption may be harmful to humans causing, for example, methaemoglobinaemia in young babies: the WHO maximum permitted level of $22{\cdot}6\,mg\,l^{-1}$ is often approached and frequently exceeded in waters receiving agricultural drainage (Holdgate, 1979).

Comparably with carbon, no more than a tiny fraction of annual N-fixation is lost to fossilization in sediments because the anaerobic sedimentary environment is favourable to denitrifying bacteria (p. 238). Environments which are unfavourable for nitrogen fixation, such as acid rain-fed peats, are almost reliant on airborne fixed-N and, at least in pre-industrial times must have depended on photochemixally and electrochemically and volcanically produced NO_x arriving in rain as nitrate. Since industrialization the appearance of NO_3^- in precipitation has been part of the 'acid rain' phenomenon to which fossil-fuel derived SO_4^{2-} largely contributes. With modern attempts to control SO_2 emissions the proportion of NO_3^- has increased. In the north-east U.S.A. this increase was 300% in the decade ending 1973–74 (Likens *et al.* 1977).

THE SULPHUR CYCLE

Several sulphur compounds are gaseous: hydrogen sulphide and dimethyl sulphide are commonly formed by anaerobic bacteria and sulphur dioxide is

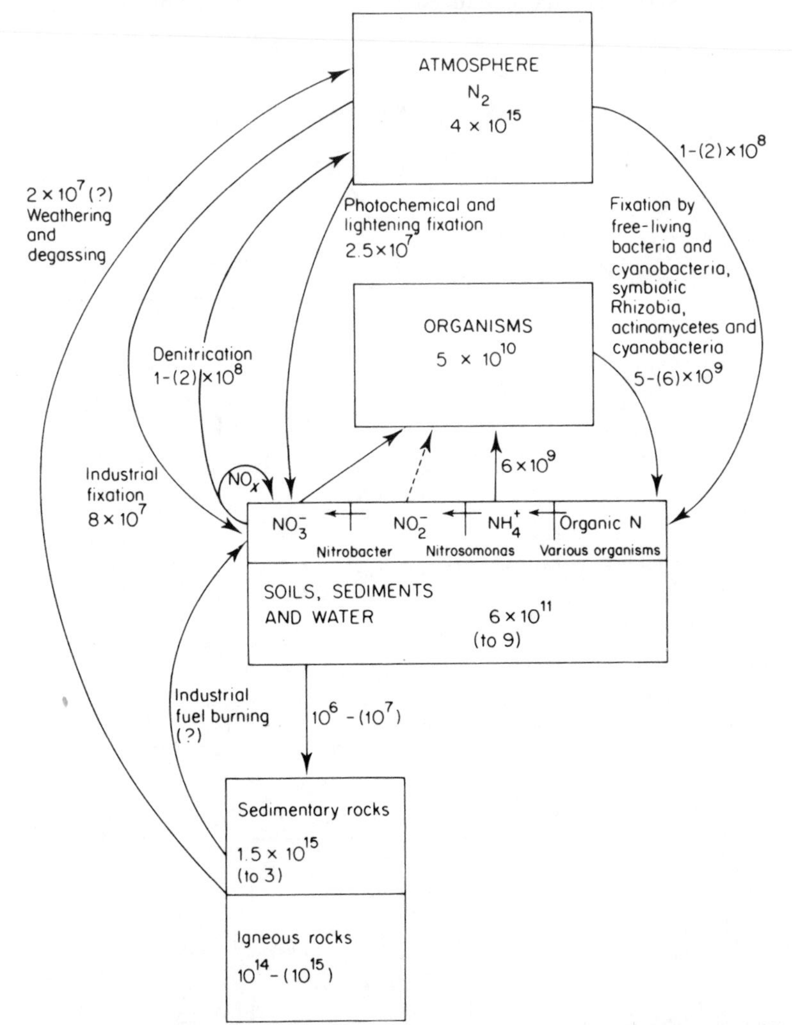

Figure 12.5 The global nitrogen cycle. From data collations of Holland (1978) and Bowen (1979). Masses (in boxes): tonnes. Fluxes (arrows): $t\,y^{-1}$. Atmospheric mass and industrial fixation are relatively accurate but the remainder is little more than order of magnitude estimation. Bracketed values indicate wide discrepancies in published values.

produced volcanically, by burning of vegetation and, now, in copious quantities by oxidation of sulphides and organo-S in fossil fuels. The cycle thus has a gaseous component but this is of limited size compared with that of the major gaseous cycles. Sulphur resembles nitrogen in having a rather complex set of redox-related soil microbiological transformations including the conversion of organic S compounds to SO_4^{2-} in oxidizing conditions and to elemental S in sediments and anaerobic soils (Figure 12.6).

Most sulphur in organisms exists as –SH and consequently must be reduced during assimilation of SO_4^{2-} by plants. During decomposition in aerobic conditions most of this –SH is oxidized back to SO_4^{2-} though most heterotrophic

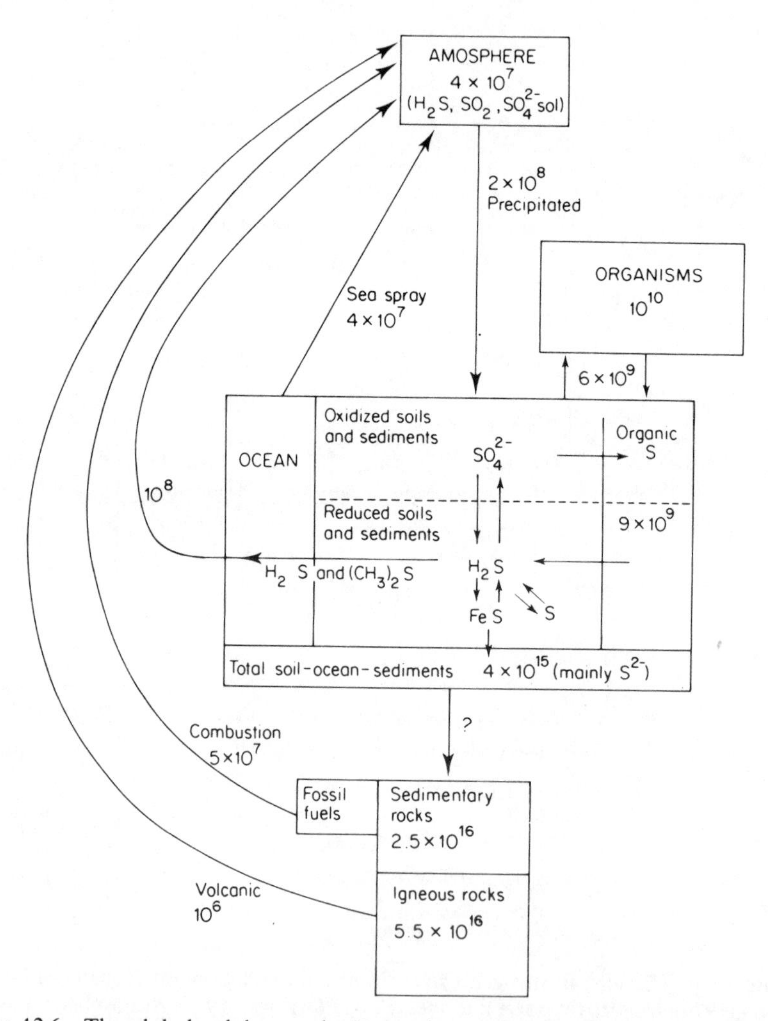

Figure 12.6 The global sulphur cycle. From data collations of Holland (1978) and Bowen (1979). Masses (in boxes): tonnes. Fluxes (arrows): $t\,y^{-1}$.

bacteria can also liberate H_2S from organic materials. In anaerobic soils and sediments H_2S is formed by this mechanism and also very freely by sulphate-reducing bacteria of which *Desulphovibrio desulfuricans* is the best known. In iron-rich materials, much of this H_2S is scavenged by ferrous iron to produce the very insoluble, black FeS which gives the characteristic colour to gyttja muds and other very reducing soils and sediments. Over a long period of time FeS is converted to pyrite (FeS_2) of which crystals may sometimes be found growing in high pH, mineral-rich peats and sediments.

The photosynthetic red and green sulphur bacteria may play an important role in shallow water and tidal mud-flat systems as they photosynthesize anoxygenically using only photosystem I, with H_2S or S as the hydrogen or electron donor and producing either S (red and green) or SO_4^{2-} (red only) as an end-product (Hutchinson, 1957). Various colourless sulphur bacteria such as *Beggiotoa* spp. may also oxidize H_2S to S or S to SO_4^{2-} when the H_2S supply is exhausted.

When sulphide-rich anaerobic sediments are exposed to air, either by drainage, for example of coastal marshes, or by seasonal fluctuations of water-tables, S-oxidizing bacteria such as *Thiobacillus thiooxidans* may produce very large quantities of SO_4^{2-} and strongly acidify the soil. Bacterial sulphate formation not only harms plants but also causes corrosion of buried pipes and submerged concrete and metal structures. River impoundments have also caused fish-kills when sulphide-rich water is released from dams and oxidizes to cause acidification downstream.

Fossil fuel burning has, as with carbon, perturbed the S-cycle which has a rather small atmospheric reservoir. SO_2 emissions may cause direct toxicity to organisms close to the source but, at large distances (and across national boundaries) its oxidation causes the acidification of rain with harm to natural waters and poorly buffered soils (Hutchinson and Havas, 1980). In the north-eastern U.S.A. during the mid-1960s the sulphate contribution to precipitation acidity was above 80% but with the introduction of industrial pollution control measures this had fallen to about 65% in the mid-1970s (Likens *et al.* 1977).

The amount of industrially contributed sulphur forms a substantial proportion of the pre-fossil fuel burning atmospheric budget but, as with C, it is a transient pulse caused by restoration of formerly available sulphur to the biogeochemical cycle and presumably will be sequestered again to the sedimentary sulphides from which it came.

Biological conversions of sulphur are, as with carbon, responsible for fractionating the stable isotopes ^{32}S (95% natural abundance) and ^{34}S (4·2%) so that biogenic sulphide deposits are measurably depleted of ^{34}S and ocean sulphates enriched, a fact which has been useful in the study of ancient sedimentary sulphates and their origin.

Balancing the budget of atmospheric S-transfer is difficult because H_2S is easily oxidized in air and aerobic water, little reaching the atmosphere where it is present at less than 10^{-4} v.p.m. (Lovelock and Margulis, 1974). These workers suggested that dimethyl sulphide might be a less easily oxidized vehicle for S-transport and recent measurements now suggest that it may be responsible for more than 30% of the budget (Nguyen *et al.* 1978).

LOCAL CYCLES

All of the metallic elements which are absorbed by plants as cations, and phosphorus or other anionic elements with no naturally common gaseous compounds, have sedimentary, essentially local cycles (Figure 12.7). Depending on the solubility in water of their compounds, their ionic mobilities and the avidity with which their organic compounds are formed, some of these elements are essentially immobile in the environment over long time-periods while others show very rapid transport through soils and ultimately to ocean basins.

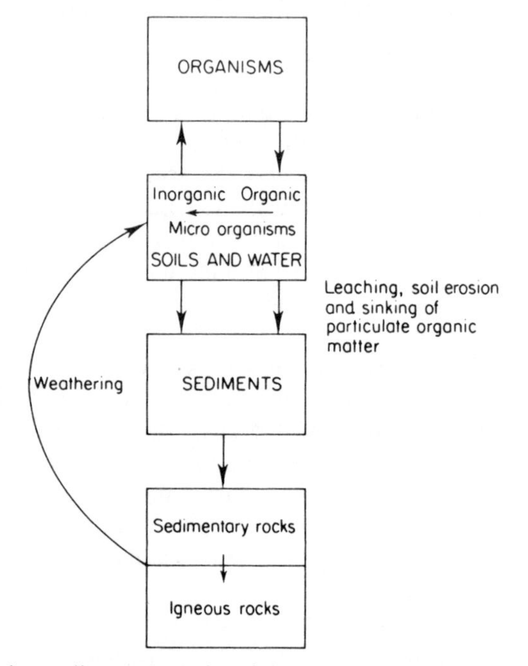

Figure 12.7 Local or sedimentary cycles of elements which are not gaseous and have no gaseous compounds of environmental significance.

Elements such as calcium, magnesium and potassium are retained in soil by cation exchange but despite this, large quantities are lost to solution in high rainfall areas, thus budgets established during the Hubbard Brook watershed study in the north-eastern U.S.A. show that 98% of Ca, 94% of Mg and 78% of K were lost in solution compared with an immobile element such as iron of which the entire loss was particulate. The output of calcium from the system represented 13·9 kg ha^{-1} y^{-1} compared with an iron loss of only 0·6 kg ha^{-1} y^{-1}. These outputs represent mainly the products of weathering of soil mineral materials. Some elements enter ecosystems dissolved in rainwater or as atmospheric dust in quantities as large or larger than the annual leaching loss; thus chloride of oceanic origin enters the Hubbard Brook system at 6·2 kg $ha^{-1} y^{-1}$ compared with a loss of 4·6 kg $ha^{-1} y^{-1}$.

Phosphorus is of interest as a major plant nutrient which is very immobile in soil. Many animal tissues and products, bone, urine and faeces are P-rich and in areas where they become concentrated such as human habitations the large content of soil phosphorus may persist for very long periods of time, for example soil P-analysis has been used in archaeology to identify habitation sites several thousand years old. Similarly in agricultural fertilization much of the applied P remains in the soil and does not cause the major eutrophication and nitrate problem which is associated with N-fertilizers. The major source of P-eutrophication is sewage containing P-compounds from domestic and industrial detergents.

CYCLING OF TOXIC ELEMENTS

Several non-essential elements, despite their substantial toxicity, are freely cycled through biological systems and may become concentrated in the biosphere. Examples are mercury, lead, cadmium and fluorine. In addition to these elements a new problem appeared with the artificial production of radionuclides in nuclear reactions to which the main contributor has been weapons testing. Some radioisotopes of essential elements such as ^{14}C have been released causing both biological hazard and, in this case, interference with the ^{14}C dating process. Many radioisotopes of non-essential elements behave as analogues of essential elements, thus ^{90}Sr, a Ca analogue, is strongly accumulated by plants in Ca-deficient environments while ^{137}Cs behaves similarly to K. The isotopes of plutonium, ^{239}Pu, ^{240}Pu and ^{241}Pu, though a man-made element, are taken up by and with extreme hazard to organisms.

Several toxic elements, for example lead and mercury, have been dissipated locally by mining operations since early times but the growing industrial use of these and other elements during the past century has now created a more extensive problem. Lead enters the environment in enormous quantities and is particularly efficiently dispersed to the atmosphere by the use of tetraethyl and tetramethyl lead as antiknock additives to petrol (gasolene), which may contain almost 1 g Pb dm^3! About $2{\cdot}5 \times 10^5$ t y^{-1} Pb enters the oceans from this source and the mean sea-water concentration has increased almost sevenfold during the past 50 years and is now about 0·07 $\mu g\,kg^{-1}$ (Goldberg, 1971).

Many natural soils, peats and even polar-ice show Pb-enrichment dating from the beginnings of industrial and vehicular input (Murozumi *et al.* 1969; Livett *et al.* 1979). The degree of enrichment is similar to that cited for ocean water, for example Ruhling and Tyler (1968) analysed herbarium specimens of mosses collected in Sweden during the 19th and 20th centuries and found an original mean Pb content of 20 $\mu g\,g^{-1}$ rising to 90 $\mu g\,g^{-1}$ by the late 1960s, with a particularly sharp rise after 1950 coinciding with the increase in petrol driven vehicles on Swedish roads.

It is fortunate for man that lead is not strongly absorbed from soil by plants (e.g. Ruhling and Tyler, 1968) so that the main toxicity hazard is from inhalation of dust or ingestion of surface-contaminated food. Plants on heavily contaminated soil such as mining spoil do however absorb large quantities of Pb up to several thousand $\mu g\,g^{-1}$ compared with normal plant content of between 1 and 15 $\mu g\,g^{-1}$ (e.g. Johnston and Proctor, 1977).

Mercury is used industrially as an electrolytic electrode in the production of chlorine, from which process about 25 % of world Hg production escapes in flue-gases! Little attention was given to mercury as an environmental toxin as it was formerly believed that its inorganic compounds would be so insoluble as to pose no problem. It is now known that bacteria such as *Methanobacterium amelanskis* can form methyl mercury compounds, particularly in sulphide-rich sediments, and these are then mobile either in solution or as atmospheric volatiles. Kozuchowski and Johnston (1978) showed mercury loss from the leaves of *Phragmites australis* (reed grass) by leakage of volatiles from bottom sediments through the plant's intercellular space. Methyl mercury chloride (CH_3HgCl) is particularly toxic to animals as it is easily passed across cell membranes.

Cadmium behaves in the environment similarly to zinc (same group, next period) and is geochemically associated with it; thus a major source of Cd-pollution is Zn mining and smelting in addition to its release by other industries such as metal-plating. Cadmium is quite mobile in soils and water and thus freely taken up by plants and passed on to animals (Coughtrey and Martin, 1976).

Fluorine occurs naturally as calcium fluoride in a variety of minerals but the main sources of atmospheric pollution are aluminium smelting using cryolite (Na_2AlF_6) as a flux, coal burning and the firing of clays in brick manufacture (many clays and shales contain between 500 and 800 $\mu g\,g^{-1}$ F and coal often up to 500 $\mu g\,g^{-1}$, all of which may be liberated in high temperature combustion). Fluorine is freely mobile in the atmosphere and ultimately appears in rainfall as fluoride. Gaseous F_2 enters open stomata, causes collapse of mesophyll cells, loss of photosynthetic capacity and necrosis. Herbivores may accumulate, at worst a lethal dose, but more often suffer minor fluorosis symptoms and fail to thrive satisfactorily.

Fluorine shares with SO_2 a rather specific toxic effect on lichens, of which the distribution, performance and F or S content may be used to monitor pollution (Gilbert, 1970a, b; Leblanc and de Sloover, 1970). The topic has attracted so much research that a complete review volume appeared in 1973 (Ferry *et al.*) from which it is apparent that the air pollution history of a country such as Britain becomes preserved in the regional diversity of corticolous lichen communities: corticolous species are stressed as they are not subject to geological variation as are saxicolous species.

It is not only inessential elements which cause toxicity; many essential elements may exceed threshold concentrations beyond which they are harmful or they may occur as specifically toxic compounds. B, Cu, Mn, Zn and sometimes Mo and Se are toxic at concentrations within less than an order of magnitude of required concentrations. Deficiency and toxicity of these elements may thus be encountered even in the same geographical area and often on the same geology (Applied Geochemistry Research Group, 1978; Davies, 1980). The major elements N, P and K occasionally occur at sufficient concentration to cause toxicity, particularly in agricultural and horticultural soils. Oxygen in the form of trioxygen (ozone: O_3), nitrogen and sulphur oxides are all examples of essential elements which may be harmful air pollutants in a specific molecular form.

The adverse disruptions of biogeochemical cycles, often termed environmental pollution, cannot be detailed further here but papers and reviews may be consulted under the following headings: general environmental pollution: Chadwick and Goodman (1975) and Holdgate (1979); air pollution: Mudd and Kozlowski (1975) and Mansfield (1976); soils: Lisk (1972), Leeper (1978) and Davies (1980); water: Cronin (1975) and Burton and Liss (1976); specific elements and their cycling: Nriagu (1976; 1978; 1979a, b) and Oehme (1978); environmental reclamation: Hutnik and Davies (1973), Holdgate and Woodman (1978) and Bradshaw and Chadwick (1980).

SOIL AND ECOSYSTEM CYCLES

The cycling of elements through ecosystems results in surface concentrations because organic litter deposited on the soil surface contains elements which plants have 'mined' from deeper horizons of the soil. This process commences with pioneer plants and reaches some sort of equilibrium at climax according to the seral relationships of biomass and productivity (p. 329). The quantities of elements cycled become larger as energy flow increases (Figure 12.2) but may also limit energy flow by their deficiency: a 'chicken and egg' problem of causation and consequence.

The elemental profiles in soil vary with site and plant association, and the long-term stability of the cycle is a function of the nutrient-richness of the parent material, nutrient conservation by the ecosystem and the balance between losses due to leaching, erosion, blown litter and migrating animals and gains from precipitation and dust.

Several workers have detailed the changes in soil chemical and physical composition caused by individual trees (Zinke, 1962; Gersper and Hollowaychuk, 1970; Kellman, 1979). The study by Kellman showed substantial enrichment of savanna soils beneath tree canopies, particularly with cationic elements. In this case the enrichment was explained as enhanced capture from precipitation. Late-successional species are better adapted to intercept atmospheric inputs, a point made by Gorham *et al.* (1979) in their review of successional effects on chemical budgets.

Commercial afforestation has made it possible to compare the effects of sudden changes in vegetation composition on nutrient distribution within profiles. Wright (1956) investigated dune soils, in Scotland, under different-aged plantations of *Pinus nigra* var. *calabrica*, (Corsican pine) and showed, for several elements, that absorption of nutrients reduced availability in the deeper parts of the profile, but the return in litterfall increased availability near the surface

(Figure 12.8). The effect was particularly marked for magnesium and calcium but potassium initially decreased in the surface soil and the loss was never quite made up by the litter input. Leaching, superimposed on the biological cycle, may have been responsible for the overall lowering of potassium availability.

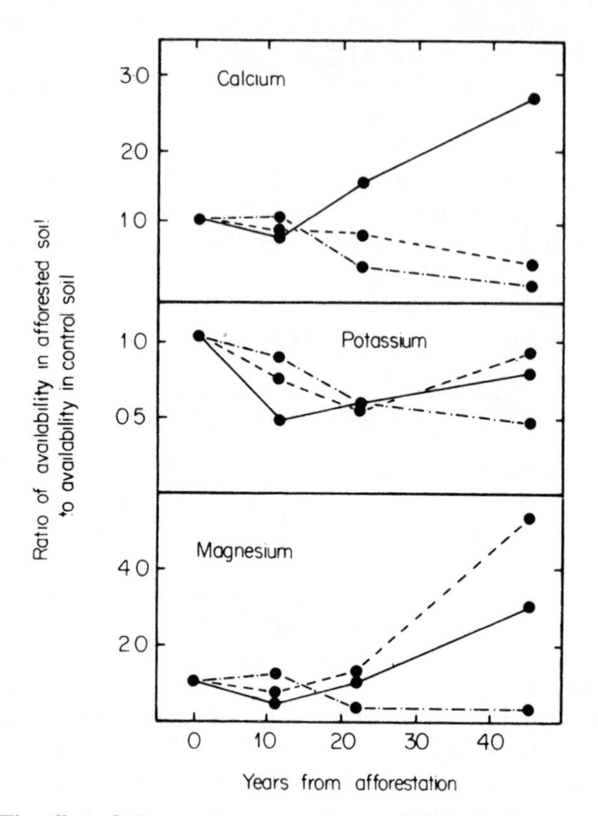

Figure 12.8 The affect of afforestation on nutrient availability in the upper 60 cm of a soil profile. The data for 0 years are from control sites adjacent to the afforested plots. Soil depth ——— 0 cm; ----- 15 cm; -·-·-·-·- 60 cm. Data of Wright (1956).

Conifers, and in particular pines, are less active in nutrient biocycling than deciduous species. Rodin and Bazilevich (1967) cite values of 50–100 kg ha^{-1} of ash elements annually returned in the litter of conifers and 200–270 kg ha^{-1} for deciduous species. This difference is particularly marked for calcium, which is much more strongly cycled by deciduous species than by conifers, and is probably responsible for maintaining the cation-saturated status of most Brown Forest Soils by counteracting the leaching process, returning nutrients from the deeper horizons to the surface (Figure 12.9).

In productive ecosystems a large quantity of nutrient may become immobilized in the standing crop and litter. In nutrient-rich systems, though the amounts immobilized are large, they form only a small proportion of the total soil resource

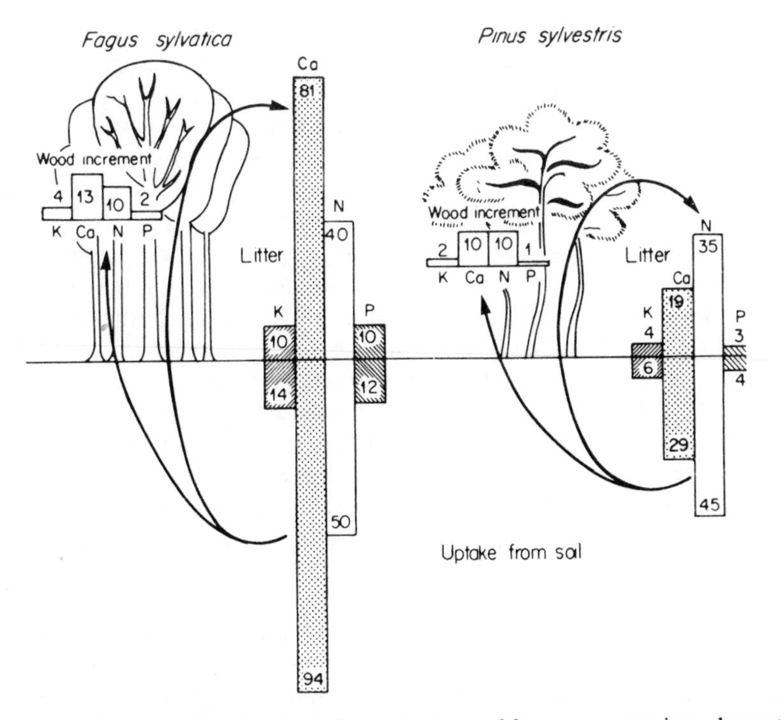

Figure 12.9 The annual cycle of uptake and return of four macronutrient elements in deciduous (hardwood) and coniferous (softwood) forest. Figures are kg ha^{-1} y^{-1}; those in the tree crowns represent annual retention in wood increment and those on the soil surface, return in litter. Note that the return represents a large proportion of uptake. After Duvigneaud and Denaeyer de Smet (1970).

but it has been postulated that the vegetation of oligotrophic soils may immobilize a considerable proportion of the ecosystem nutrient. Rennie (1955) suggested that afforestation of nutrient-poor soils and removal of timber may cause a nutrient loss which will degrade the habitat and prevent more than one timber crop from being grown. This conclusion is of great silvicultural significance but is still in dispute, for example Miller (1963) gave estimates of nutrient removal in the timber of 100-year-old *Nothofagus truncata* (southern beech) which, though large, were of the same order as the annual additions from rainfall. Afforestation with conifers may, however, be unwise in some marginal soil areas as their more limited activity in biocycling of nutrient elements may increase rates of leaching loss; a subject which is in need of further research.

Despite the large quantities of nutrients held in the living plant cover, the annual turnover is considerable. It mainly derives from litterfall but there is leaching of some elements from leaves by precipitation throughfall and stemflow. The turnover is a closed biological cycle and is responsible for establishing the nutrient distribution patterns described above. The cycle is not completely closed, a geochemically open component arising from the leaching of nutrients to the groundwater, soil erosion and input of nutrients in precipitation. As noted earlier in this chapter, the geochemical part of the nutrient cycle becomes closed only with the passage of extended periods of geological time.

Estimates of input in precipitation vary widely, since sample collection and the measurement of low concentrations of nutrients in precipitation are difficult. Levels also vary with geographical situations such as proximity to natural and industrial sources of airborne materials, and with distance from the sea, which forms a prolific source of aerosols. For the temperate zones Ovington (1968) suggests annual ranges of K, 1–10; Ca, 3–19; Mg, 4–11; N, 0·8–4·9; and P, 0·2–0·6 (kg ha^{-1}). Additional income may originate from airborne dust which is deposited by impaction on the canopy.

Leaching losses are very difficult to measure and there are few data on this topic. Miller (1963) tentatively suggests, for a *Nothofagus truncata* forest stand in New Zealand, an annual loss in kg ha^{-1} of: K, 15; Ca, 30; Mg, 15; S, 15; N, 2 and P, 0·03. These values may be high by comparison with other temperate forests on eutrophic soils. Crisp (1966) by means of a catchment drainage study, established a nutrient budget for a moorland area of blanket bog, eroding peat and some grassland. He found net annual losses, in kg ha^{-1}, of K, 8·0; Ca, 50; P, 0·2–0·4 and N, 9·5; but a large proportion of the N and P was contributed from peat erosion and does not represent direct loss from the available plant nutrient pool. The comparison of income and loss provided by these data certainly suggests that a high P/E ratio, combined with nutrient removal by timber cropping, would deplete the resources of the ecosystem. With an originally low supply the depletion may be of considerable significance as elements such as K, Ca and Mg are often limiting under natural conditions.

The idea that nutrient cycling and biomass reach equilibrium under climax conditions must be modified if the long-term balance between income and leaching loss is not compensated by release from parent minerals. The time scale of the sere is thus relevant to the stability of the climax and, given low rates of weathering, coupled with inefficient recycling of nutrients, an apparent climax may fall into decay (Goreham *et al.*, 1979). External interference may also break the cycle and cause environmental deterioration; for example, the genesis of many North European heathlands was initiated by human destruction of the climax forest, which enhanced nutrient loss and caused soil deterioration. In northern temperate areas it is not possible to judge the stability of the apparent climax vegetation, as the time period from the early Postglacial until the late Stone Age or Bronze Age, when man began significantly to interfere with vegetation, was probably insufficient for a climax to establish and prove its long-term stability.

Little experimental data existed until the early 1970s concerning the effects of cutting, fire or other major disturbance on nutrient cycling and conservation within ecosystems. A number of long-term catchment studies have now come to fruition in which geologically 'waterproof' watersheds have been used to construct nutrient budgets by measuring and analysing rainfall and streamflow.

The work of Likens *et al.* (1970) compared nutrient losses from an undisturbed and a clear-felled watershed in the Hubbard Brook catchment, New Hampshire,

U.S.A. and showed enormous differences in composition of the outflow waters (Figure 12.10).

Undisturbed ecosystems are remarkably conservative of nutrients, thus, despite an excess of precipitation over evaporation of some 80 cm, undisturbed forest at Hubbard Brook shows leaching losses of only 4 kg N ha^{-1} y^{-1} and 2·4 kg K ha^{-1} y^{-1}, though calcium output is somewhat larger at 13·9 kg ha^{-1} y^{-1} (Likens *et al.* 1977).

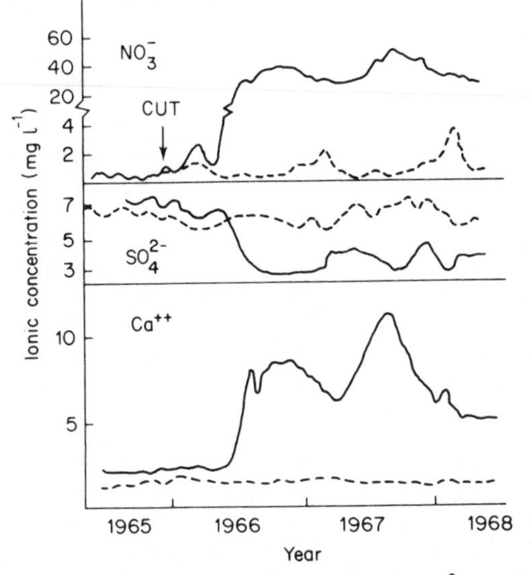

Figure 12.10 The change in concentration of NO_3^-, SO_4^{2-} and Ca^{2+} dissolved in streamwater following clear-felling of a watershed in New Hampshire, USA. The forest was cut at the beginning of 1966 after a calibration period; regrowth was prevented. Clear cut watershed ———; forested control watershed - - - - -. After Likens *et al.* (1970).

Disafforestation causes a major increase in losses mediated through its effect on the nitrogen cycle. Figure 12.10 shows the enormous increase in dissolved nitrate which appeared in streamwater after cutting and, in 1 year caused sufficient eutrophication to promote an algal bloom in addition to maintaining the NO_3^- content of the water in excess of health standards for drinking water. The loss represents an increase from 4 to 142 kg N ha^{-1} y^{-1}. Most of this could be accounted for by nitrification of soil organic nitrogen which would normally then be taken up by trees and cycled through litter-fall. Rodin and Bazilevich (1967) show amounts of 100 kg N ha^{-1} y^{-1} as a possible contribution from deciduous forest litterfall. Increased summer soil temperatures in the absence of shading may also have increased the nitrification rate.

The production of excess NO_3^- and H^+ by nitrification causes displacement of metal cations from the exchange complex, for example Ca increased by about nine times in the streamwater (Figure 12.10). This, with other cations such as K and Mg, was about equivalent to the nitrate anion loss.

Sulphate was the only ion to decrease, loss being halved. Likens *et al.* (1977) show that the SO_4^{2-} loss from the undisturbed forest is almost exactly balanced by atmospheric input, some of which will be in precipitation and some impacted on the canopy from aerosols. Loss of canopy alone may be sufficient to explain the reduction of SO_4^{2-} output but as water-yield was increased 39 % in the first and 28 % in the second year after cutting it may be that reduced interception and transpiration increased soil-wetness enough to cause sulphide formation and immobilization in the deeper parts of the soil profiles.

Chapter 13

Plants in ecosystems

INTRODUCTION

The main theme of this text has been the interaction between plant and environment mediated through plant function, a relationship often described as 'physiological ecology'. It is unrealistic to consider the plant physiological response in the absence of the biotic pressures to which it will normally be exposed. These comprise competition between plants, effects of grazing or other predation, the relationship between plants and microorganisms both through disease or symbiosis and through decomposition effects which close elemental cycles, in addition to the more tenuous relationships such as those between plant and disease vector, plant and pollinator, plant and carnivores as controllers of herbivore populations.

PLANTS AND ANIMALS

Plant predators

Plants and animals manifest very strong co-evolved relationships which embrace not only the reciprocal effects of grazing but also those of pollination, dispersal and other interactions. It has been suggested that predation pressure has been instrumental in developing Angiosperm subgroups, or even the Angiosperms themselves, with their diversity of secondary metabolites which are often grazing-deterrents (Erlich and Raven, 1964). These workers suggest that plant–herbivore co-evolution is paramount in the development of terrestrial diversity and certainly Janzen (1970) has argued convincingly that tropical forest trees which are under extreme seed-predator pressure owe their very survival to the large interspecific-distance which high diversity produces.

The plant predator may be a grazing herbivore or it may be more akin to an animal parasite, tapping the contents of individual cells as do aphids, plant bugs, nematodes and others. The impact of predation may be:

(1) Uniform and non-lethal removal of organs or metabolites which change the physiological status of the plant in relation to the environment, competition and entry of pathogens.
(2) Death or competitive elimination of the individual.

(3) Production of regeneration niches, often bare soil.

(4) Modification of the soil by input of faeces or urine, creating new niches or modifying competitive status.

(5) Benefit, such as pollination or dispersal. Predation may be co-evolutionarily 'encouraged' for this advantage.

The consequences of steady grazing pressure have been observed for many years. Tansley and Adamson (1925) used grazing exclosures to protect English chalk grassland from the rabbit (*Lepus cuniculus*): the characteristic rosette and creeping species such as *Cirsium acaulon* (stemless thistle) and *Helianthemum nummularium* (rockrose) rapidly succumbed to competition from robust grasses such as *Helictotrichon pratensis* (meadow oat) once they were not suppressed by grazing. It is interesting to note the same type of competitive inhibition by robust grasses which Willis (1963) found after NPK fertilization of dune grassland. Ecologists are familiar with these effects of rabbits and other mammals throughout the world: in North America Egler (1977, unpublished) has most aptly described the concentric areas of devastation around warrens as 'bunny belts', a term which might be less amusing to the small farmer.

Grazing 'haloes' have tempted ecologists into other interpretations, for example, Muller *et al.* (1964) attributed herb-free areas around desert shrub thickets in California to allelopathic growth limitation by volatile terpenes (p. 405) but the use of rabbit-exclosures permitted free herb growth within the haloes (Bartholomew, 1970).

If grazing is non-selective, like mowing, it automatically selects against tall, fast-growing species and favours prostrate life-forms which would otherwise succumb to shading. Within species, extreme polymorphism may develop: in *Arrhenatherum elatius* (false oat grass) this ranges from tall upright individuals in deep swards to completely prostrate ones under grazing (Mahmoud *et al.* 1975). Selection is rapid, grazing detectably favouring a prostrate narrow-leaved cultivar of *Lolium perenne* (rye grass) within 4 months and permitting its increase to 85% in competition with an upright cultivar within $2\frac{1}{2}$ years (Brougham and Harris, 1967).

If grazing is selective its effects may be even more dramatic, sometimes reducing an otherwise vigorous species to minor representation in the sward though rarely causing extinction, as the search-effort by the predator becomes unrewarding at low densities. This effect may promote greater plant species diversity, the creation of 'grazing niches' permitting otherwise physiologically similar plant species to co-exist without exclusive competition.

Mammalian herbivores may show strong food preferences. Given grazing choice horses will 'vote with their feet', preferring a mixture of palatable cultivated grasses to a single species sward of *Agrostis tenuis* (browntop) by an intensity factor of seven times (Archer, 1973). Food availability plays a part: rabbits are very critical in food choice at low population levels but, as grazing pressure increases, the protected territories of dominant males become smaller until the area is heavily grazed throughout and very few species are avoided. Moderate rabbit grazing thus creates a diversity of habitat; a mosaic of heavily and lightly grazed grassland of which the precise pattern depends on animal behaviour.

The effects of grazing in limiting more palatable species have also been shown in long-term experiments, for example the grazing exclosure and fertilization of Welsh hill-pasture described by Milton and Davies (1947) and Jones (1967). Protected from grazing, the upland pasture was converted by fertilization into a sward closely resembling a lowland meadow but, without protection from grazing, the lowland species were unable to flourish (Figure 13.1). Despite its relatively poor pasture quality the *Festuca–Agrostis* sward without fencing was further degraded by overgrazing, permitting the invasion of agriculturally undesirable species.

Plants contain an enormous number of secondary metabolite biochemicals of unknown function (Harborne, 1972), some of which have been linked to grazing deterrence, particularly amongst invertebrates. Young leaves which are of low secondary metabolite content are more at risk than old leaves, for example the seasonal pattern of feeding by *Operophtera brumata* (winter moth) on *Quercus spp* (oak) leaves is mainly determined by accumulation of tannins (Feeny, 1970). Selection would favour plants armed with grazing deterrents and would be stronger in plants with a long pre-reproductive phase, a speculation supported by Cates and Orians' (1975) study of 100 species from different stages of succession in which they showed that slugs markedly preferred early successional species which also have a short life-cycle. It is noteworthy that the leaves of tree species found to be unpalatable to earthworms (Satchell, 1967) were also late successional species.

A few species toxic to farm animals have been studied in more detail, for example the genus *Senecio* which produces liver-damaging alkaloids. A few organisms have evolved immunity and, in the remarkable case of *Tyria jacobaea* (cinnabar moth), the grazing larva stores the alkaloid and passes it on to the adult moth which is thus distastefully toxic to predatory birds, a defence coupled with warning colouration.

Another detailed study has been the ecology and genetics of cyanogenesis (Jones, 1973). Some 2000 species of plant have been found which contain one or more of 23 cyanogenic glucosides and the enzyme B-glucosidase. When cellular compartmentation is damaged by crushing or freezing, the enzyme mediates liberation of HCN from the glucoside, possibly acting as a grazing deterrent. This is certainly the case in laboratory feeding experiments with snails which prefer the acyanogenic forms of *Lotus corniculatus* (birdsfoot trefoil).

Genetic studies have been possible because several species, for example *L. corniculatus* and *Trifolium repens* (white clover) are polymorphic for cyanogenesis: plants may be allelic for presence or absence of the glucoside and separately allelic for presence or absence of the enzyme thus cyanogens must carry both glucoside and enzyme while acyanogens may lack both, one or the other. The interaction with grazing and other environmental factors is multifactorial and difficult to characterize but it seems that heavy rabbit grazing selects against the acyanogenic *T. repens* and at least one mammal (*Microtus agrestis*—common vole) discriminates against the cyanogen if offered alternative

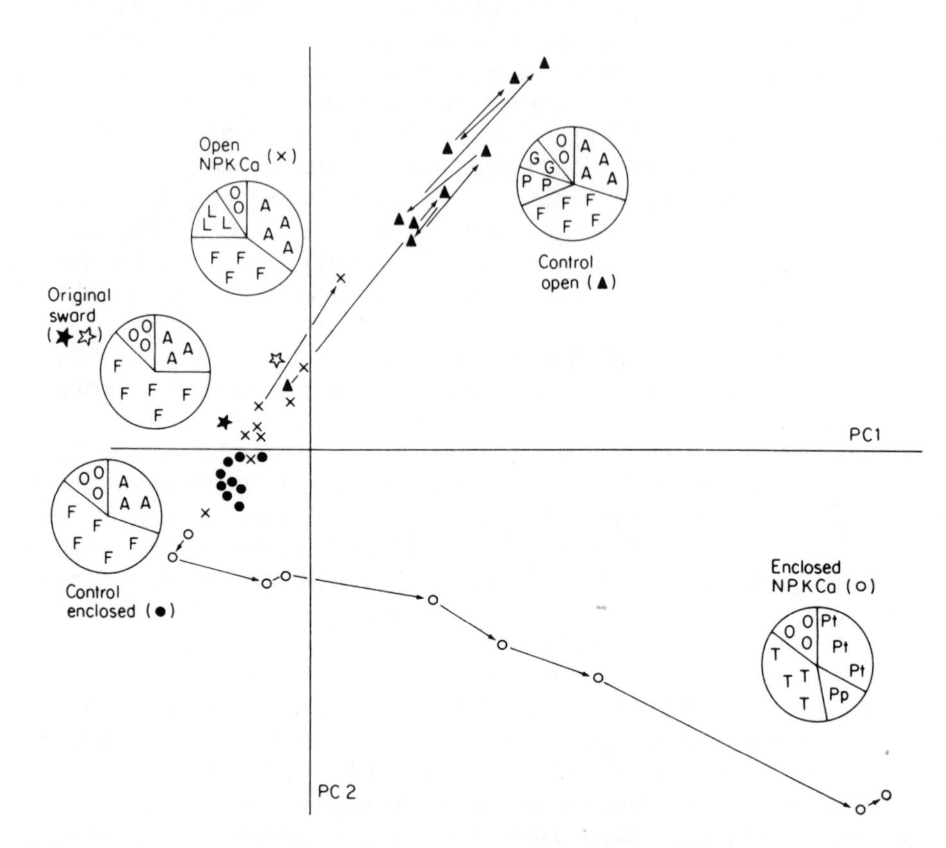

Figure 13.1 The effect of fertilization, grazing and time on the species composition of a *Festuca–Agrostis* hill-grassland in mid-Wales. The diagram shows part of a principal component ordination of the correlation matrix between species and elegantly illustrates the drift in species composition during the 11-year experimental period. The original open and enclosed swards are indicated, respectively, by the symbols ☆★. Each plotted point represents the species-composition data-set for one year, and where significant changes have occurred between years the linking arrows indicate the time sequence. The last plotted point is for 1944. The ordination was computed by R. S. Smith (1975) from data of Milton and Davies (1947) which were also used to construct the sector diagrams for species composition in 1946. The enclosed control plots with moderate grazing (●) show no appreciable change from the original species composition but fertilization of these enclosed plots (○) caused an enormous change, the original *Festuca–Agrostis* sward being entirely superseded by a *Poa* spp.–*Trifolium repens* mixture resembling a lowland pasture. Grazing was totally uncontrolled in the open plots and overgrazing has caused the control plot (▲) to change considerably to a species composition resembling many heavily grazed hill pastures in which *Galium saxatile* and *Potentilla erecta* are common. Attraction of sheep to the adjacent fertilized plots may have exacerbated this over-grazing. The open, fertilized plot (×) was prevented from great change by grazing and differs from the original sward mainly in the advent of *Luzula sylvatica.* Key: A, *Agrostis* spp.; F, *Festuca ovina*; P, *Potentilla erecta*; G, *Galium saxatile*; Pt, *Poa trivialis* PP, *Poa pratensis*; T, *Trifolium repens*; L, *Luzula sylvatica*; O, other species.

food (Jones, 1973). The distribution of cyanogenic *T. repens* is negatively correlated with incidence of frost and that of *L. corniculatus* with either droughting (Jones, 1973) or salt-spray damage (Keymer and Ellis, 1978) all of which would be expected to damage compartmentation.

The chemical interactions between plant predators and their hosts may far exceed simple effects of metabolite loss, a point illustrated to extreme by gall-forming insects. Aphids may influence tree growth more by injecting growth-inhibiting materials than by removing metabolites (Dixon, 1971a): certainly this is the case for leaf-size in *Acer pseudoplatanus* (sycamore). Not only is wood-formation massively inhibited in this species but nutrient cycling is also affected as leaves fall early with a larger than normal nitrogen content. Interspecific differences are also substantial, *Tilia* × *vulgaris* (lime), for example being much less affected by infestation despite inhibition of root growth (Dixon, 1971b). Van Emden (1972) describes insects as remarkably good phytochemists and this is indeed true of the aphids which not only locate a suitably aged host of the correct species but also find the best position for tapping phloem contents of a required amino acid composition.

Catastrophic predation may occur amongst adult woody-plants, sheep, goat, grey squirrel, rabbit and porcupine all being recorded as tree-girdlers, while larger mammals such as the elephant may totally uproot or break-off adult trees. A few mammals are tree-felling specialists, the beaver for example adding such activity to its great ecological impact as a paludifier. The effects of such predation are little studied except for the general observation that it contributes to the cycle of death and regeneration making available temporary niches for the establishment of tree seedlings and for the more transient subordinate species associated with open areas in woodland. Plants which are delicately poised in relation to the light-climate of the woodland floor may be reliant on the occasional opening of the canopy in mature woodland. The drifts of spring blossom associated with European hedgerow species such as *Crataegus monogyna* (hawthorn) are a man-made feature of the countryside, such species, originally woodland undershrubs, probably flowered only when exposed by a canopy gap.

Control of forest regeneration by predation is much more effective at the seedling stage. Many tree seedlings, especially those with epigeal germination, are killed by loss of growing point and cotyledons. Slugs, snails, mice and voles are well able to destroy enormous numbers of seedlings in this way. The occurrence of mast-years in so many temperate forest trees is probably of survival value as sufficient seedlings may be produced to saturate the predator demand. Dispersal mechanisms add to the chances of survival by removing the seedlings from the predator 'rain' below the parent canopy. The long-term development and cyclic shifting of woodland mosaics with all of their varying environmental interactions may very well be an evolutionary consequence of the plant–predator interaction.

Catastrophic predation usually affects only individuals in a population but infrequently, herbivore damage may cause very severe damage even to the extent of starving the predator. Such events are usually identifiable with failure of feedback control of grazers by carnivores. A classic example was the invasion of

Isle Royale in Lake Superior by moose (*Alces alces andersonii*) in 1908 leading to destruction of forest and pond vegetation by 1930. Wolves (*Canis lupus*) arrived across winter ice in 1948 and rapidly reduced the moose population to about $2\,\text{km}^{-2}$, permitting regeneration of the vegetation. The entire competitive balance and physiological ecology of the vegetation is thus under a feedback control involving much less than 1 % of the photosynthetically fixed energy! (Figure 13.2).

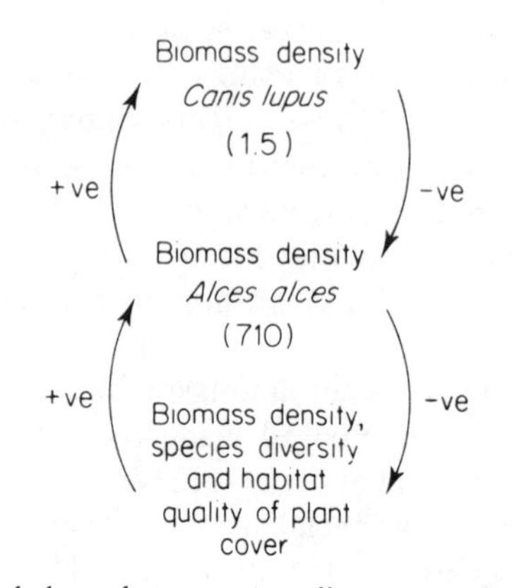

Figure 13.2 The feedback loop between standing crop of wolf (*Canis* lupus), moose (*Alces alces*) and vegetation in the Isle Royale ecosystem. Bracketed figures are standing crop $\text{kg}\,\text{km}^{-2}$. Data: Jordan *et al* (1971).

Violent pulsations of plant–herbivore systems are often cyclic and appear most prominently, or at least are most easily studied, in simple, stressed ecosystems. Schultz (1969) has described the sequence of events in the approximate 4-year cycle of lemming (*Lemmus mucronatus*) numbers in Alaskan tundra: in spring of the population-explosion year the grass-sedge forage has a high P and Ca content, and further stores of P and Ca exist in dead organic matter (Figure 13.3). The sward is heavily grazed giving it a low albedo which, coupled with the effects of burrowing, permits deep thawing. By late summer the population peaks and then crashes as a consequence of migration and predation.

During the following year the overgrazed sward is unproductive and deficient in P and Ca but by the third year some P and Ca has been mineralized, primary production is increased but the forage is still low in P and Ca. In the fourth year this trend reaches its limit; primary production is high and the forage is rich, not only in P and Ca but also N and K. Jordan and Kline (1972) used a mathematical model to investigate the suggestion that P-deficiency and the mineralization time-lag might be a prime causal factor in this particular microtine cycle. Their model, programmed with Schultz's data, simulates the 4-year cycle of forage-P content, but the single change of including a P-storage compartment equivalent to tree-boles in a forest system completely damped the oscillation. This result

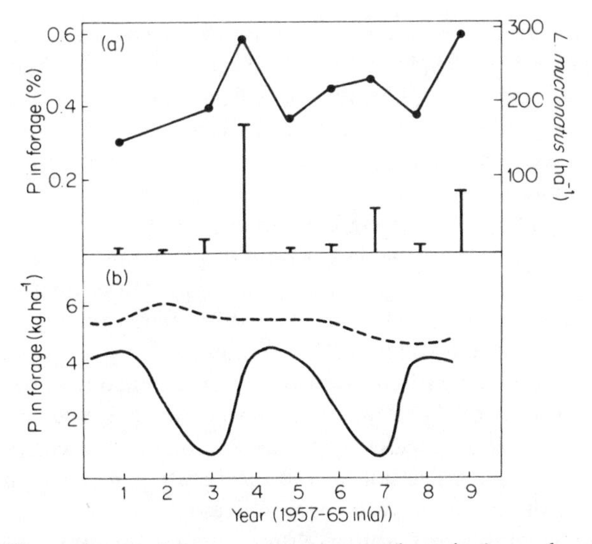

Figure 13.3 Fluctuations of *Lemmus mucronatus* (lemming) numbers with time in relation to phosphorus status of herbage in the Point Barrow, Alaska tundra ecosystem. (a) Field measurements of forage P-content (○) and lemming numbers (bars). After Schultz (1969). (b) Mathematical model prediction of P-status using Schultz's data (——) and using the same data plus a P-storage compartment equivalent to the standing wood biomass of a tropical forest (-----). After Jordan and Kline (1972).

suggests that the role of P is crucial but Schultz warned that it is misleading to identify causality with a single factor even in a simple ecosystem and it must be remembered that a model only includes those factors which its creators have thought important.

Catastrophic predation is the underlying principle of biological control which has been applied successfully to various weed species. The best-known example is control of prickly pear cactus (mainly *Opuntia stricta*) in Australia by the moth *Cactoblastis cactorum* after 1925. Another success story is control of perforate St John's wort (*Hypericum perforatum*) in California by a chrysomelid beetle. Harper (1977) notes that the chrysomelids prefer to oviposit in sunlit sites and *H. perforatum* is now confined to shade conditions–a salutary warning to the physiological ecologist who forgets his animals! The moth *Tyria jacobaea*, mentioned above in relation to phytotoxins, has also been used but with more variable results in control of *Senecio* spp.

In almost all cases of biological control the pest has not been eliminated but reduced in population, controlled either by the search range of the predator (e.g. *C. cactorum*) or by limitation to a specific environmental or competitive niche (e.g. *H. perforatum*). In most cases the introduced predator is chosen to be monophagous, as polyphagous organisms are considered to threaten other species. For this reason the predator population has always declined before the prey plant has become extinct. A polyphage with a preference for a specific prey might be expected to cause extinction and it is not possible to say how many plant distributions are controlled in this way in natural ecosystems where nature has already completed her 'experiments'. The information we have concerning perturbations of species equilibrium come from agricultural disturbance and introduction and the most spectacular examples involve transoceanic introductions which remove both predator and prey species from the stabilizing effects of their natural enemies.

The secondary effects of grazing herbivores include trampling and localized enrichment of soil with urine and faeces both of which may compound the effect of heavy grazing in providing patches of bare soil. Many burrowing animals, including some carnivores, may also contribute bare soil regeneration sites, ranging from the transient soil mounds of moles (*Talpa* spp) to the longer-lived 'hills' of ants and termites. Mole and ant-hills in grassland often provide a focus of seedling regeneration on bare soil and also create species diversity, ant-hills permitting the survival of species which are outcompeted elsewhere (King, 1977). Grubb's (1977) critical discussion of the regeneration niche suggests that the regular or sporadic appearance of bare soil patches within otherwise closed vegetation will inevitably be a source of diversity and, for some species, the only means of replacing themselves, thus King (1977) lists over 30 species which occurred only on ant-hills in his sample sites: these included a number of winter annuals which usually occur on bare soil, for example the grasses *Aira caryophyllea*, *A. praecox* and *Vulpia bromoides* together with species which are now mainly ruderal such as *Stellaria media* (chickweed) and *Senecio vulgaris* (groundsel).

Many mammalian herbivores either use specific latrine sites or congregate at night and in shelter from bad weather. The specialized flora of rabbit latrines has often been noted but little studied in relation to soil conditions. Victorian naturalists referred to 'lair flora' species, for example *Urtica dioica* (nettle), *Galium cruciata* (crosswort) and *Rumex acetosa* (sorrel) are often associated with sheep-shelters in Britain and *U. dioica* is world-wide associated with guano soils of sea-bird colonies. All of these soils are P-eutrophicated as indicated by *U. dioica* (p. 275) and such effects may persist for long after the land-use has changed; formerly grazed woodlands may often by identified by such species as *U. dioica* and *Sambucus nigra* (elder) and even prehistoric midden sites several thousand years in age may retain *U. dioica* in Britain.

The role of seed-predating animals in relation to plants is a fascinating study in cost benefit when it is realized that predation often approaches 100% despite the reliance of the plant on predators to disperse its seeds. Janzen (1971) recorded complete loss of beans from the leguminous tree *Cassia grandis* to bruchid beetles unless vertebrate seed-predators carried some away from the parent tree and failed to eat them! The strategy of tempting both beneficial and lethal predator seems incredible until it is realized that the locally fallen seed, even if it germinated, would be unlikely to survive the rain of invertebrate predators falling from the parent canopy (Janzen, 1971). The remarkable synchronized flowering of monocarpic bamboos at intervals of several decades may be interpreted as an adaptive mechanism which saturates the demands of seed predators (Evans,

1976) and ensures establishment of a new generation. Many tropical trees and some temperate trees show similar 'mast' years which may also be related to predator satiation.

The overall effects of plant predation are literally shaping the world: grassland versus forest and species survival within an ecosystem often reflecting the balance of herbivores rather than plant physiological–environmental interactions. The coarse grain of temperate habitats which so contrasts with the fine-grained species-diverse systems of the moist tropics is a reflection of a favourable environment at equatorial latitudes but much of it must be seen through the distorting mirror of herbivore behaviour. In some parts of the world man has created a herbivore-pressured environment while, in some areas with wild large herbivores, he has destroyed the grazing pressure. In some cases there is a rhetorical question: did herbivores create grasslands or *vice versa*?

The magnitude of the biotic effect is in no way a reflection of the scale of the predator, for example root-feeding *Typula* spp (crane fly) larvae may cause a productivity loss of 20–30% in grassland (Harper, 1977) while Varley and Gradwell show oak (*Quercus robur*) timber growth to range as much as 40 to 50% around the mean in correlation with caterpillar infestation, a correlation which they assumed to be causal. Thus small organisms may, by their sheer numbers, have similar impact to that of large predators and, because of their short life-cycles and rapid population changes, a much more variable effect.

The consequence of heavy grazing which is so obvious in grassland is the favouring of plants with effective vegetative reproduction such as the grasses in which the meristem is basal rather than apical. At the same time sexual reproduction may be suppressed by loss of inflorescences or their failure to form. Similar effects may be traced in woody plant ecosystems, perhaps reaching their extreme in human management of woodland and hedgerow in which coppicing imposes repeated loss of above ground parts, supresses flowering, encourages adventitious new growth and confers effective immortality on some individuals. Rackham (1980) for example records coppiced ash (*Fraxinus excelsior*) of perhaps 1000 y in age.

Pollination

Pollination by animals is usually related to predation, nectar or various plant organs being offered as bait. The evolutionary advantage of successful pollination outweighs substantial energy expenditure in advertising (floral parts and scent) and payment (nectar or tissues). In this context, Whitehead (1969) has given an interesting account of the factors favouring a decrease in anemophilly and increasing animal pollination as low latitudes are approached. The conditions for successful wind pollination are those which favour arrival of pollen at a stigma without the necessity for too high a pollen production: the conditions are large stands of single species, low vegetation density coupled with deciduous habit, relatively dry climatic conditions, good climatic signals to synchronize flowering and too few animals for efficient pollination. In the tropics none of these conditions are met and the evolutionary cost–benefit analysis is weighted in favour of the search and find behaviour of the animal pollinater rather than the random dispersal of anemophilly.

Once the 'payoff' becomes substantial, animal pollination may be coevolved to extremes. The tropical figs (*Ficus* spp) have a remarkable relationship with a pollinating wasp whose larvae eat more than 50 % of the fig ovules in exchange for pollinating services (Janzen, 1979)! Similarly various species of *Yucca* (Joshua tree, etc.) have an obligate pollination relationship with a moth. In any such case the distribution and survival of the plant may be at the mercy of the climatic and biotic controls of the pollinator, altering its range in a subtle way which plant physiology alone could not reveal.

Competition between plant species for the service of pollinators may occur, for example Free (1968) cites the problem of *Taraxacum officinale* in apple orchards and Levin (1970) experimentally showed the effects of both flower colour and commonness in influencing insect choice between otherwise similar *Phlox* spp flowers. Any minority species in competition for a pollinator is likely to receive a majority of hetero-pollinations and will decrease in a cascade effect from generation to generation (Levin and Anderson, 1970). Some species are remarkably effective in attracting pollinators, for example Beattie *et al.* (1973) found 44 % to 82 % of all available insects preferentially visited *Frasera speciosa* (monument plant) in the Rocky Mountains. This plant has spectacular blossoms and the further advantage of strongly synchronized flowering. If two species are very similar there is likely to be an evolutionary advantage to desynchronization of their flowering and to aquisition of species specific pollinators.

Seed predators and dispersal

The effects of animals on seed dispersal are sufficiently well known to be treated in some elementary botany texts. Seeds of fleshy fruits are often dispersed by passage through animal guts and may germinate more freely after such an experience. Temple (1977) cited the remarkable case of the tree *Calvaria major* which used to be common in Mauritius. The dodo (*Raphus cucullatus*), now extinct, ate the seeds for their pulpy, succulent mesocarp tissue and the consequent broke the seed dormancy by abrading the hard endocarp in the bird's gizzard. The tree is now almost extinct as seeds have not germinated since the demise of the dodo, but they can be made to germinate artificially by feeding to turkeys, birds of similar size!

A less frequently realized effect is the interaction between seed predation and germination when animals bury caches of seed and then fail to rediscover them during winter. In North America the magpie (*Garrulus glandarius*) buries enormous numbers of acorns (*Quercus* spp), while small mammals such as squirrels (e.g. *Sciurus* and *Neosciurus* spp), mice (*Peromyscus* spp) and voles (*Clethryonomys* spp) show similar behaviour throughout the world.

PLANT PATHOGENS

The analysis of pathogen resistance into horizontal and vertical components has significantly changed the approaches of the plant breeder to epidemic disease (Van der Plank, 1975; Scott and Bainbridge, 1978) but it is also a useful ecological concept as endemic disease is common in natural ecosystems (endemic disease is that which is always present, often prevalent). Horizontal resistance may be defined as the broad, generalized resistance of a host species to a pathogen species in which different varieties of the host do not have their 'own' strain of the pathogen. Vertical resistance, however, involves a strong interaction between host variety and pathogen race, implying a gene-for-gene relationship. Both types of resistance can occur together but horizontal resistance can never be zero (Van der Plank, 1975).

Horizontal resistance is polygenically controlled and has evolved as an 'old' mechanism derived from generations of selective contact between host and pathogen. Surplus horizontal resistance is probably harmful as the gene complex controlling it is also responsible for a range of healthy plant metabolic functions and it is hardly likely that the two could be optimized simultaneously. For this reason horizontal resistance is selected-for only when host meets pathogen. Vertical resistance may be negligible when a host meets a new strain of pathogen but, involving gene-for-gene selection, may be rapidly bred into the population.

Species of undisturbed ecosystems will have fairly strong horizontal resistance to the endemic pathogens combined with a shifting mosaic of species-populations differentiated to give vertical resistance to the endemic range of pathogen strains. In this situation a major epidemic would be unlikely unless a susceptible host was introduced, or a virulent strain arises *de novo* or is introduced from outside the system. It is for this reason that epidemics are most usually confined to agro-ecosystems or disturbed habitats. Little is known of endemic disease in the wild but Van der Plank (1975) gives useful discussion of endemism and several writers in Scott and Bainbridge (1978) have investigated the use of crop varietal mixes as a buffer against epidemic disease.

It is difficult to assess the role of endemic disease in controlling eco-physiological interactions between plants but a few examales of the consequences of perturbation suggest that it may be substantial. The Dutch elm disease, caused by the fungus *Ceratocystis ulmi*, first flared up in Europe after the First World War and then in the 1920s reached the U.S.A. on logs exported from France (Brasier and Gibbs, 1978). The beetle *Hylurgopinus rutipes*, a bark-boring vector, quickly caused a disastrous epidemic amongst the non-resident American *Ulmus* spp. During the 1960s a very aggressive strain of the disease was re-imported to Britain on Canadian rock-elm (*U. thomasii*) logs and the British vector, *Scolytus scolytus*, rapidly spread the disease lethally to over 60 % of Southern British elms by 1976 (Brasier and Gibbs, 1978).

Horizontal resistance to the aggressive strain is very slight and there is only small vertical resistance in one species, *U. carpinifolia*. However, the beetles do not attack very young trees and some root-sucker populations are arising from killed *U. procera*: it is very difficult to predict the future if the aggressive strain becomes endemic as the impact of massive tree death on the *S. scolytus* populations cannot be assessed and the disease can only re-invade in the presence of the vector. This example is a single illustration of the difficulty of the problem and our inability to predict pathogenic events in wild populations.

The composition of any ecosystem is the consequence of the fortuitous arrival of organisms, environmental and competitive limits compounded by predation and pathogenicity: we are looking at the end of 'natural experiments' in which some treatments were lethal. There is no possibility of assessing alternative consequences in the absence of specific chosen organisms unless a great deal is known about the system and the physiological ecology of the organism, in which case mathematical modelling *may* help. Harper (1977) notes that the only way to obtain such data is to purturb the system by either eliminating pathogens and predators, for example the application of molluscicide, insecticide and fungicide to permanent grassland plots, which released *Bellis perennis* (daisy) from population control but the cause was not identified (Foster, cited by Harper, 1977).

SUCCESSION AND REGENERATION

The nature of succession

Bare substrata *are* colonized by organisms and this *is* followed by a gradual change of species composition and environmental conditions with time. Surprisingly, these trite observations have been the centre of hot dispute for almost the whole of the past century, particularly in relation to the mechanism of species replacement and concerning the nature of the species-mixture which exists once replacement has slowed to an imperceptible rate.

It is not possible to detail the history of the dispute in this short chapter, and recent reviews have been given by Drury and Nisbet (1973), Horn (1974) and Connel and Slatyer (1977). Three main points are controversial:

(i) Do early successional species so 'damage' their own habitat that they fail to regenerate, thus enforcing replacement by later-successional species? An affirmative answer implies truly autogenic succession.

(ii) Does species' composition converge during succession to a single or limited number of 'climax' states?

(iii) As succession advances do the properties of diversity and stability develop as supra-organismic characteristics of the ecosystem? This would imply that the interactions between species are so positive and intense that the ecosystem has an integrity which is beyond the simple physiological sum of its parts in organization and homeostasis.

There is no doubt that the development of individual plants or a vegetation canopy may substantially modify the edaphic surroundings (e.g. Zinke, 1962; Gersper and Holowaychuk, 1970 and pp. 103, 371) and obviously causes major changes of microclimate at the soil surface, particularly in relation to radiant heating and water evaporation (e.g. Barclay-Estrup 1971). It is not quite so certain that these changes cause autogenic succession by preventing regeneration of the early arrivals or even 'prepare the way' for latecomers.

The early invaders of bare surfaces are often 'quick buck' strategists with good seed-dispersal and fast growth but short life. The last characteristic prevents them from maintaining a permanently closed canopy or limits the amount of shade which they cast and this opens the door to the 'banker' strategist which, though slower growing, has a much longer life and hoards environmental resources until the early-succession species are competitively bankrupt. This sequential competitive explanation of successional replacement does not require that pioneer species should so far alter their environment that they cannot reproduce and, as such behaviour would be an adaptive disadvantage for a species, the replacement by sequential competitive effects would seem a more likely mechanism.

Grime (1979) interprets most ecological interactions in terms of plant strategies of which he identifies three main components: tolerance of disturbance, epitomized by ruderals (R); tolerance of stress (S) and tolerance of competition (C). During succession in a productive habitat there is a period of intense competition between C–R strategists, at first herbaceous annuals and biennials followed by herbaceous perennials and then a sequence of shade-intolerant shrubs and trees. At this stage of ecosystem development the effects of shading and nutrient immobilization favour the emergence of the S-strategists, in this case long-lived trees of large biomass. Grime also shows that plant species may be ordinated in terms of their maximum relative growth rate and a compound index of morphology to produce a triangular ordination which is closely related to subjectively assessed C–S–R balance.

Horn (1975) treats succession as a sequence of stochastic (random) plant by plant replacement steps in which the replacement of any species occurs with a probability defined by the current status of the ecosystem. In the case of a forest this probability is related to the proportions of different tree species seedlings present below the canopy of each existing adult species. Such Markov-chains of replacement have the property of settling to an eventual equilibrium mixture, the 'climax'. In real ecosystems this climax is indistinct and variable from place to place, forming a mosaic of regeneration phases upon which is superimposed a larger-scale edaphic variability.

The Markov models may have built-in to them forbidden or favoured replacements and these would be reflections amongst other things of relative competitive status. Horn (1971) studied secondary forest succession on abandoned fields in New Jersey, U.S.A. and concluded, similarly to Grime (1979), that early successional species are shade-intolerant and are photosynthetically efficient by virtue of their multilayer foliage canopy which is adapted to high illumination. Such canopies are also selected to be water conservative and, for this reason, early succession species may persist on shallow or arid soil. Bazzaz (1979) reviews evidence for the water efficiency and high maximum photosynthesis of early succession herbs and trees.

Late succession trees by contrast have a monolayer canopy and cast such deep shade that sub-canopy regeneration, even of its own species, cannot occur and is forced to await the appearance of a canopy gap to re-initiate the late succession mosaic, an excellent illustration of the importance of the regeneration niche stressed by Grubb (1977). Long ago, Watt (1925) made this same interpretation of regeneration in British beechwoods (*Fagus sylvatica*) where tree death produces a new regeneration mosaic *tessella* in which the multilayer *Fraxinus excelsior* (ash) first emerges from sub-canopy shade suppression (Wardle, 1959) and the monolayered *F. sylvatica* then regenerates by seed, ultimately overtopping the *F. excelsior* 'nurses'. The similarity between this monolayer species' climax and that described by Horn (1971) is further shown by the climax dominant, *Fagus grandifolia*, belonging to the same genus.

Connel and Slatyer (1977) discussed three mechanisms of succession which they named, respectively, facilitation, tolerance and inhibition. The facilitation mechanism involves the classic concept of early invaders 'paving the way' and is now discredited. The tolerance mechanism involves sequential niche-filling and the gradual gaining of competitive advantage, while inhibition has the central concept of pre-emption of space, possession being the major portion of the law. Connel and Slatyer support this third mechanism and suggest that succession can only occur as canopy gaps appear to form regeneration niches. Considering the overwhelming role of pre-emption in competitive interaction (p. 407 ff), this seems a convincing interpretation. They also point out that many secondary forest successions may terminate with the same dominant species for the simple reason that it persists in vegetative form as basal sprouts and thus automatically pre-empts possession. The reproduction of Horn's (1971) *F. grandifolia* certainly can take place in this way, similarly to the regeneration of cut stumps in coppice-managed woodland, and the individual trees may thus be considered almost immortal.

Grime's (1979) suggestion that nutrient immobilization and consequently nutrient stress occurs in late succession is circumstantially supported by studies of immobilization in standing biomass (Rennie, 1955; R. B. Miller, 1963; H. G. Miller, 1979) and also by the suggestion of Swift *et al.* (1979) that the mineralization of other elements lags behind that of carbon, particularly on infertile soils, and, as organic litter accumulates, it acts as a sink for nutrient elements.

The behaviour of some opportunist early-succession species seems adapted to the pulse of nutrients which become available after windthrow, felling or fire (Likens *et al.* 1977). *Prunus pensylvanica* (pin cherry) regenerates freely from a soil seed bank after felling or burning in New England (U.S.A.) hardwood stands and, within 4 years, has a canopy leaf area index of 4 to 6! During this period it scavenges large proportions of the N, Ca and K which would otherwise be lost to groundwater (Marks and Bormann, 1972). *P. pennsylvanica* is a multilayer species and is successionally shaded-out by monolayer species, climaxing with *Fagus grandifolia*, the scavenged nutrients returning to circulation *via* decomposer mineralization.

The interpretation of succession as an autogenic process, or otherwise, is complicated by the fact that most data concern systems in which presumedly different stages are present as spatial zones, for example those related to watertable depth in subaquatic (hydrosere) and salt marsh (halosere) systems. These are the very cases where the allogenic process of sedimentation alters relative levels of water and substrate. It is not easy to determine whether the spatial zones are genuinely autogenous or simply a reflection of species adaptations to different water levels. One exception to this dilemma is the peat-pool system in ombrogenous bog where changes of substrate level are *only* caused by accumulation of organic materials and consequent zonation must be considered autogenic. It is also possible in the peat ecosystem to identify the subfossil remains of plants and confirm the species-changes which have occurred.

Walker (1970) has used this stratigraphic approach to various British lowland hydroseres and found a diversity of sequences which not only suggest that the concept of uniform and convergent succession is either an oversimplification or incorrect but also shows that analogy between spatial zones and time sequences may be misleading as the present-day surface does not present all possible transitions.

Many studies of mature vegetation have been cited in support of the concept that succession is convergent to a limited number of climax states and, at least in early work, it is hard to escape the implication that the species composition of the climax has supra-organismal significance and is a community characteristic rather than a consequence of the interaction of species characteristics. An example may be taken from the North American mixed deciduous forests already referred to above in the work of Horn (1971) and Likens *et al.* (1977). Old-field studies have shown the *Fagus grandifolia–Acer saccharum* (beech–sugar maple) climax to be preceded by earlier successional species but in their review of evidence of sequences Drury and Nisbet (1973) conclude that both early and late successional species may invade simultaneously and it is simply differentials of growth rate, shade tolerance and other competitive interactions which delimit the surviving species-mix. As time passes and the canopy matures, so different niches develop, including the regeneration niche of senescence, fire or windthrow. The final outcome is defined not so much by some magical convergence of community characteristics but by the upper limit to the number of species available to fill the potential niches. In a relatively species-poor temperate environment it may simply be that there are not sufficient late succession species available for a diversity of climax types to be developed.

In a species-rich environment it would be less easy to mistake a physiognomic climax type for a species-defined climax and this is indeed true, judging from the much greater difficulty of classifying stands of humid-tropical forest (Richards, 1952; Janzen, 1975; Flenley, 1979). All of these workers conclude that the early monoclimax or limited climax theories developed in temperate vegetation are inapplicable to the polyclimax situation of the tropics. The diversity of terminations of succession is likely to reflect the much wider range of late successional tropical species available for niche-filling.

Walker's (1970) revelation of multiple pathways of hydrarch succession may again be related to the relatively large number of species available for colonization of the waterside habitat, in particular during the early stages of succession. Relatively short-term oscillations of water-table may also play a part by diversifying environmental conditions.

Succession and ecosystem properties

A great deal has been written concerning the changes of diversity, stability and other community characteristics during succession (e.g. Diamond and Cody, 1975) and it is difficult in a short space to present any meaningful synopsis. It is obvious and trivial that diversity must increase during early succession: a sample with one species is less diverse than one with two species! It is less easy to analyse diversity of late successional vegetation when relatively small changes of a large diversity index (see E. P. Odum, 1971) may have significance. Horn (1974) argued that a slight perturbation of climax vegetation will permit either invasion or recovery of a suppressed species, both changes increasing diversity indicating that the last stages of succession probably involve some decrease in diversity. Horn described the climax condition as 'a blurred successional patchwork' which would arise, amongst other causes, because of the need for a regeneration niche in a stable vegetation.

Stability, defined as absence of change, must increase as succession proceeds but this is unimportant (Horn, 1974). Dynamic stability, the rate of recovery from disturbance, inevitably becomes less as the ecosystem matures and contains more species. In a stable environment such as the wet tropics diversity ultimately becomes very great because the stress-free environment permits the evolution of very fine-grained niche differentiation (p. 397). At the same time such systems become very fragile as the individuals, by definition, are ill-fitted to cope with the habitat changes of disturbance. Contrasting with this, stressed environments often provide competitive refugia for a limited number of tolerant species (p. 279). Such ecosystems are of low diversity, have few potential niches, but are very resilient to damage. Thus heathland and chaparral, for example, are maintained by fire and show virtually no species changes following fire because their species-limited vegetation is already fire adapted to resist this particular selection pressure.

The relationship of both primary productivity and biomass density with succession has also been debated at some length. In early succession both must increase but the significant changes again are those which occur as the ecosystem approaches equilibrium. The accumulation of living biomass increases the amount of respiring tissue and there comes a time when, statistically, the ecosystem respiration is equal to its photosynthetic production and an equilibrium will be established. At this stage the biomass will contain a large proportion of the nutrients available from the habitat and, if these fail to recycle sufficiently fast or if they are lost by export, this may be a reason why climax-equilibrium cannot be maintained. External disturbance, such as fire or felling, may catastrophically initiate such a disequilibrium.

Various workers have reported reduced productivity and declining biomass at climax (Horn, 1974) and at least for forest ecosystems Horn (1971) has shown that maximum photosynthesis and self-regeneration cannot both be optimized with the same foliage distribution pattern. If a monolayer foliage species cannot regenerate in its own shade the only mechanism for regeneration is the development of a mosaic of regeneration niches provided by senescence of individuals in the monolayer canopy. Some of these will initially become occupied by early succession species which have a potentially higher photosynthetic rate (Bazzaz, 1979). A pure monolayer climax cannot have so high a net primary production as an earlier successional canopy of multilayer species despite the fact that these are competitively ephemeral (Horn, 1974).

To summarize, it seems that succession must be interpreted through the physiological–ecological characteristics of individual species and that many of its patterns are the result of stochastic replacement of one species by another under the constraints of competitive interaction. Changes of diversity, dynamic stability, productivity and biomass are simply interpreted and of no deep significance in early succession but as the elusive equilibrium climax is approached begin to reflect properties of the system which has assembled itself.

In the context of convergence as a property of succession it is interesting that any form of vegetation ordination (Whittaker, 1978) usually results in separation of individual stands despite visual uniformity of the sampled plant-cover. If the geographical area from which the samples are taken is increased, the degree of stand separation also tends to increase. This suggests that no two vegetation stands are ever identical and provides further circumstantial evidence against major convergence and against the now discredited Clements' (1916) organismic interpretation of succession.

Plants and catastrophe

Most disasters which befall plant communities are extensions of normal events, thus drought, frost and flood are factors for which selection has built in some degree of tolerance and, for this reason it is rare to find a total kill and, if this does occur it is often in already disturbed and often species-poor ecosystems. During the major drought of summer 1976 in Northern Europe very large areas of monospecifically dominated *Calluna vulgaris* (heather) moorland were killed by either water shortage and/or thermal overloading on shallow mineral soils, and Rymer (1979) records a frost-kill of over 60 % of gorse (*Ulex europaeus*) in mid-Wales in 1962–63. Both of these events occurred in already fire- and grazing-stressed systems. Watt (1954) found that *Pteridium aquilinum* (bracken) rhizomes in Eastern England were killed to a depth of 19·5 cm by the severe winter of 1940: again, bracken is a plant which, without human interference, would be sheltered by an open woodland canopy.

At first sight, fire does not appear to be in the same category of disasters as it is an all or nothing event but, of course some perennial plants survive fire, as do some seeds, and though there are some humid ecosystems which rarely, if ever, experienced fire before man began his work, almost any terrestrial soil in regions with a dry season will reveal some charcoal fragments, elemental carbon being resistant to microbial attack and, despite the apparent unpredictability of fire the ecosystems of such areas have some in-built tolerance. In some cases fire is essential for maintenance of stability: Blackburn and Tueller (1970) describe the invasion of arid sagebrush (*Artemesia nova*) communities in Nevada by pinyon and juniper (*Pinus monophylla* and *Juniperus monophylla*) once overgrazing reduced the vegetation to a level where it could no longer fuel cyclic fires.

These fire-obligate ecosystems are more widespread than might be suspected, thus fire is an integral part of most sclerophyll shrub communities ranging from the heathlands of Northern Europe, South America, South Australia and New Zealand together with the Mediterranean climate garigues, chapparalles and their formerly more extensive maquis forest cover (Specht, 1979). Similarly much of the high latitude boreal conifier forest with its slowly decomposing, resin rich litters is now a fire co-evolved system in which some trees retain their cones until fire unseals the resin-bound ovuliferous scales and the seed is dispersed to germinate in the ashes of destruction. Examples are *Pinus banksiana* (jackpine) of North America and *P. halepensis* (aleppo pine) of the European Mediterranean. Both species occur on shallow or sandy, drought-prone soils of high fire risk and, at least in the case of *P. halepensis*, the plant is so inflammable that there is a positive feedback enhancement of the fire risk which then places the pine in a competitively advantageous situation (Kozlowski and Ahlgren, 1974). Many Mediterranean species are also insulated by very thick bark, a European example being the cork oak (*Quercus suber*) and most extreme of all, in the North American Sierra Nevada, *Sequoiadendron gigantea* (Big tree) may be encased in 15 to 60 cm of soft bark. *S. gigantea* is the world's largest (not tallest) tree and may live to 3000 y or more, by which age all individuals are distorted by fire scars: modern man is not the only agency of fire. Again in this species the mature cones may remain on the trees for over 20 years, seed shedding increases after wildfire and seedlings often establish in long rows in the trenches left by combustion of fallen trunks (Hartesveld *et al.* 1975).

Burning in fire-adapted systems initiates a succession which Tansley (1939) termed the burn subsere. It was first described for English lowland heath by Fritsch and Parker (1913) who realized that the heathland vegetation mosaic was a reflection of time elapsed since last burning. Seedlings colonize bare soil together with algae, bryophytes and lichens while the shrub component regenerates adventitiously from burned stem bases. Gorses and ling (*Ulex* spp and *Calluna vulgaris*) regenerate freely in this way, while geophytes such as *Pteridium aquilinum* (bracken) are positively favoured by the temporary removal of surface-rooted competitors.

Geophytes are not damaged as soil is of low thermal conductivity and cannot in any case exceed 100 °C until it has dried. At 2·5 cm depth, temperatures rarely exceed 50 °–60 °C and in normal wildfires fall again within less than 1 hour (Raison, 1979). At 5 cm heating is often insignificant, permitting tubers, rhizomes and seedbank to survive. Under a fuel source such as fallen logs or peat, 50–70 °C may be reached, even at 30 cm depth, totally killing all soil organisms and modifying the physical and chemical nature of the soil (Raison, 1979). Temperatures above the surface are much higher, often exceeding 300–500 °C, and with plenty of dry fuel rising to maxima in excess of 1000 °C.

A few plants have become obligately dependent on fire at least to emerge as dominant or common species. The Australian *Eucalyptus regnans* (gumtree) rarely produces young plants, succumbing to predation, leaf-spot fungus attack or competition from *Notofagus cunninghamii* (southern beech), a late succession deep-shade casting species. Despite this, mature, even aged, stands of eucalypyts exist and include some of the world's tallest trees. The establishment of these groves is only possible when sporadic fire removes the beech competition and provides a microclimate which is adverse to the leaf-spot fungus! (Evans, 1976).

Many fire-maintained vegetations such as temperate heathland, Mediterranean chaparralle and garigues contain species which are very common or dominant but whose ecological role would be much more limited without fire control of tree cover. *Calluna vulgaris* (heather) and *Erica* spp in heathlands, and shrub-oaks such as *Quercus coccifera* (Kermes oak) of the European Mediterranean, are cases in point. These plants quite literally succeed themselves after fire because they either redevelop vegetatively from basal sprouts or because there is a substantial soil seed bank of which dormancy is often broken by fire, certainly amongst some of the Ericaceae.

COMPETITION

Almost 200 years ago Malthus (1798) wrote: 'Population, when unchecked, increases in a geometrical ratio... This implies a strong and constantly operating check on population from the difficulty of subsistence.' The full implication of Malthus' words passed unheeded for almost a century as the realization of his predictions was delayed by the Industrial Revolution which permitted the agricultural exploitation of new land areas and the establishment of an international food market. Today the human population again faces the dilemma of numbers versus subsistence without any prospect of relief; any increase in agricultural efficiency deriving from plant breeding and cropping advances now being met by immediate increases in population level.

Population crisis has been an ever-present biological fact since, without restraint, a population will always expand to fill and overfill available space. The subject of competition has attracted the attention of plant ecologists for over 150 years. Clements *et al.* (1929) reviewed early work and cited De Candolle (1820) as the first to characterize plant competition: '... all the plants of a given place are in a state of war with respect to each other'. In 1907, Clements, who pioneered so much ecological thought, defined competition in terms which have since been improved only in detail: 'When the immediate supply of a single necessary factor falls below the combined demands of the plants then competition begins.'

Plants may interact by competition for the supply of some particular factor but may also influence each other directly, for example by secretion of toxic metabolic products into the environment. De Candolle's concept of plants at war is, perhaps, a more fitting analogy for the latter type of interaction. The concept of warfare implies a direct and purposive struggle which hardly fits the situation in which a nutrient is removed from the environment before it can be reached by a competitor. As the word 'competition' carries overtones derived from human activities, Harper (1961) suggested its replacement by 'interference' but there is no sign that this has been generally adopted as an alternative.

Another possible approach to competition may be seen in the writings of Margalef (1968), who placed the concept in a cybernetic frame work by describing the competition between two species in terms of their respective feedback relationships with a jointly used resource. Each species has its own, stabilizing (negative) feedback relationship with the resource *A* (Figure 13.4a). If *A* becomes overtaxed then the feedback loop leads to a reduction in demand and the steady state is maintained. However, if a second species (*C*) is introduced which utilizes the same resources it is possible for a resultant positive feedback pathway to be established between species *B* and *C* (Figure 13.4b). If, for example, *B* overdraws the resource *A* then *C* must suffer a deficiency; some of the *C* population may consequently die and return part of resource *A* to the environment. If *B* immediately takes up the returned portion of *A* there is nothing to prevent a continuing *A* deficiency from killing the whole of the population of *C*. In this case the establishment of the positive feedback pathway between *B* and *C* leads to the catastrophic collapse of the *C* population and a 100% success of *B* as a competitor. The potential establishment of such feedback loops between all

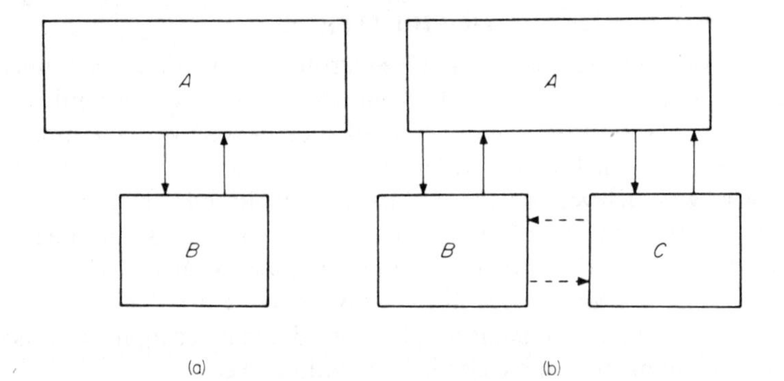

Figure 13.4 Diagram (a) represents the negative feedback relationship between a resource *A* and a consuming species *B*. This will be a stable relationship, the magnitude of the resource pools and the species pools being determined by relative rates of supply, demand and return. Diagram (b) shows the introduction of a second, competing species *C*. Each species now has comparable negative feedback relationships with the resource *A* but the resultant is a possible positive feedback relationship between the two species leading to instability and potential extinction of one species. Reproduced with permission from R. Margalef, *Perspectives in Ecological Theory*, University of Chicago Press, Illinois, 1968, Figure 2, p. 8.

species and individuals of a complex ecosystem draws attention to the immense intricacy of such networks and to the threat that a minor unbalancing of environmental or species relationships could initiate self-reinforcing population crashes, perhaps leading to species extinction.

This model raises a problem which is discussed by Harper (1967) and has recurred frequently as a difficulty for other workers. If competition is so important as a regulator of ecosystems why should species co-occur?

Margalef's model suggests that one species will always be outcompeted by another while Horn's (1974) plant-by-plant successional model predicts a monospecific climax when this is the case.

Niche occupancy

The general self-stabilization of multi-species ecosystems is most easily discussed in terms of niche occupancy. The utilization of the environment by organisms is divisible into three main components: space, resources and time. Any organism which utilizes exactly the same combination of these components as another will be in exclusive competition as suggested by the Margalef model. If one species differs in even one sub-division, for example nitrogen requirement as part of the resource component, then two co-existing species will at first be in unlimited competition but, if the more successful species also has the greater nitrogen requirement it will become resource limited while its competitor is still able to gain nitrogen from the environment and an equilibrium population mixture may be established.

The specific combination of space, resource and time requirements may be defined as a species' niche occupancy and, by analogy, could be a man's address, occupation and working-hours. Because there are so many combinations of resource requirements; a diversity of space occupancies, particularly in soil and a variety of phenologies coupled with microorganism and predator–pathogen effects, the niche can only be defined in terms of a mathematical abstraction, the niche hypervolume, of which the ordinates are the numerical values of the different components.

Definition of the niche hypervolume begins the building of a bridge linking the work of the population eco-physiologist with the analyses of the quantitative ecologist. The multifactorial axes of a species ordination (Whittaker, 1973; 1978) may be interpreted environmentally and, to some extent, model the axes which define species niche hypervolumes.

Lieth (1960) showed for *Trifolium repens* (white clover) and *Lolium perenne* (rye grass) that the two species form a mobile mosaic in which low clover density areas are invaded by grass and vice versa. As specialization for difference in niche occupancy is a means of avoiding annihilating competition (Harper, 1967) it is tempting to suggest, in cases such as this clover: grass mosaic, that the plants themselves create transient micro-niches in the soil environment. Snaydon (1962) has shown very steep gradients of pH, calcium, phosphorus and potassium in clover associations while the work of Goodman and Perkins (1959) with

Eriophorum angustifolium (cotton grass) and Zinke (1962) with various tree species shows that the plants are capable of establishing considerable microenvironmental gradients of nutrients. As Harper (1967) has suggested, intense niche differentiation would prevent an exclusive struggle by focusing the intense battles within, rather than between, species.

THE STRUCTURE OF COMPETITION

One approach to plant competition is to analyse it in terms of the factors for which plants compete or which are related to competition between plants. These may be tabulated as follows:

<table>
<tr><td rowspan="5">A. Factors for which plants compete</td><td>(a) Space</td><td>Above and below ground</td></tr>
<tr><td>(b) Light</td><td rowspan="2">Above ground</td></tr>
<tr><td>(c) Carbon dioxide</td></tr>
<tr><td>(d) Nutrients</td><td rowspan="2">Below ground</td></tr>
<tr><td>(e) Water</td></tr>
<tr><td rowspan="2">B. Plant characteristics which cause interaction</td><td colspan="2">(a) Passive root interactions such as the normal production of respiratory CO_2</td></tr>
<tr><td colspan="2">(b) Direct interaction due to the secretion of specific toxins into the environment (Allelopathy)</td></tr>
<tr><td rowspan="6">C. Interactions with external factors which influence or cause competition</td><td colspan="2">(a) Competition for pollinators</td></tr>
<tr><td colspan="2">(b) Competition for agents of dispersal</td></tr>
<tr><td colspan="2">(c) Selective pressure or disturbance of ecological equilibria by animals and man</td></tr>
<tr><td colspan="2">(d) Disturbance of the environment which provides bare soil or seedling niches</td></tr>
<tr><td colspan="2">(e) Influence of temperature, humidity, exposure, wind etc. on other competitive factors</td></tr>
<tr><td colspan="2">(f) Adverse soil conditions such as toxic solute content, heavy metals, excess calcium carbonate etc.</td></tr>
</table>

FACTORS FOR WHICH PLANTS COMPETE

Space

Donald (1963) noted that plant competition for space, in the sense of physical interaction, rarely arises. He cites the example suggested by Clements *et al.* (1929) of competition amongst close-sown tuberous crops where individual tubers may become polygonal or even be lifted from the soil by their neighbours. Further examples may be seen when clusters of seeds have germinated in the same spot and, in woody plants, when neighbouring branches actually clash, leading to damage, deformity and sometimes infection. Seed establishment is probably the commonest example, most soils presenting only a limited number of niches in which the water supply requirements for germination are satisfied. Harper's (1961) experiments with brome grasses *Bromus* spp (Figure 13.5) illustrate this point: when sowed on to a compacted soil surface the number of emergent seedlings was limited in relation to seeding-rate by the small number of surface crack sites available for germination. By contrast in the rough-surfaced soil no limitation occurred and the seedling number was proportional to the seeding rate. These thus represent the fairly limited number of examples of competition for space. In the light of the discussion of soil niche differentiation above it may, however, be necessary to reassess general views on competition for space in the soil matrix.

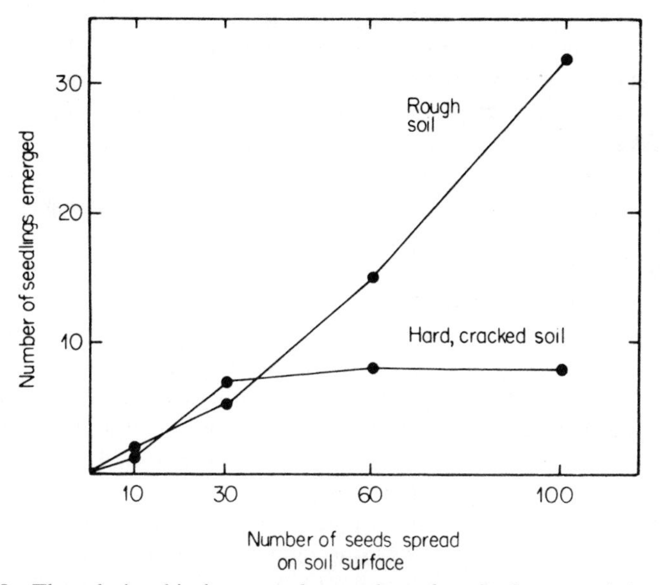

Figure 13.5 The relationship between the number of seeds shown and the number of seedlings emerging for a mixed *Bromus* spp population. Reproduced with permission from J. L. Harper, Approaches to the study of plant competition, *SEB Symp. XV*, 1961, 1–39, Figure 12.

Light

Competition for light is unique in that there is no storage: incoming light must be used or lost. Leaves behave toward light as individual units: when they remain for long periods below compensation point they are not supported by export of assimilates from other parts of the plant and quickly die. For this reason competition for light is between individual leaves rather than between plants. The canopy architecture and its relationship to leaf display and insolation is, thus, a most important factor in determining peak photosynthetic rates and light competitive abilities of plants.

Competition for light is often the ultimate interaction between plants because competition caused by any other factor will reduce growth rate and leaf area, resulting in the replacement of the primary competitive impact by the light-starvation of the suppressed species. Jennings and Aquino (1968) showed that strongly competitive rice varieties received more light than weaker varieties and showed evidence that competition for water and nutrients operated through the reduction of light-intercepting foliage.

The relationship of photosynthetic to respiration balance in sun–shade strategies has been extensively discussed in chapter 5: the general conclusion could be for most mature ecosystems that lethal competition for light is prevented by a wide range of niche specialization. Cohabiting plants may have their growth suppressed by competitive shading but in well adapted systems not to the extent of failing to regenerate.

The increasing burden of respiratory tissue in late succession was discussed on p. 131 but relatively few studies have been made on individual wild plant species. Evans (1972) presents some measurements made on tropical African tree species which show differences between species and age of trunk sample, the latter because, presumably, young trunks have a larger proportion of superficial living tissues to inactive xylem. Evans' Ivory Coast data show a total ecosystem respiration of 18·5 $t\ ha^{-1}\ y^{-1}$, about twice the annual dry matter accumulation! He compares this with Danish beechwood (*Fagus sylvatica*) respiration of 4·5 $t\ ha^{-1}\ y^{-1}$, only half the annual dry matter increment, the difference probably attributable to temperature difference.

Even herbaceous species show the same effect, for example Went (1957) found that the compensating light intensity for *Fragaria vesca* (strawberry) became very high (about 25 % full sun) because non-photosynthetic tissues had accumulated. In a natural ecosystem where primary production has to support predation and pathogens such effects may be very important and it is notable that final loss of shaded leaves is often through fungal attack.

Many plants adjust to increasing canopy shade by 'self-pruning', for example the death of lower branches in competing conifers is intentionally exploited by the forester to produce knot-free timber and most gardeners are familiar with the yellowing and death of lower leaves in an uncut lawn.

Carbon dioxide

Despite the fact that carbon dioxide is limiting to high-light photosynthesis, plants do not usually compete for their carbon source, as atomospheric turbulence is so effective in replenishing the supply. This may not be the case in some still-water systems where diffusive transport may be limiting and a first come, first served competition may exist. There is an interesting but ecologically unimportant demonstration that a C-4 leaf may continuously gain carbon from a C-3 leaf in an illuminated, sealed chamber.

Nutrients

Bradshaw (1969) indicates the range of habitats in which nutrient deficiency, and consequently competition, must occur by suggesting that simple NPKCa fertilization experiments will cause extensive changes in most plant associations. This is exemplified by Willis' (1963) work with sand dune calcareous grassland in which an NPK addition caused a surge of growth in the dominant but normally stunted grasses *Festuca rubra* (red fescue), *Agrostis stolonifera* (creeping bent) and *Poa pratensis* (meadow grass). The sward depth was much increased and the whole range of rosette and creeping perennials so characteristic of calcareous dune and limestone grassland was quickly ousted by competition for light. The conclusion may be drawn that the physiognomy of such associations is sculpted by the limiting effects of competition for the three major nutrients.

The greatest problem in studying nutritional competition is the complexity of the interaction for so many major, minor and trace elements coupled with the range of competing species, some of which may also show strong differentiation of nutrition ecotypes. The latter point imposes rigorous requirements on the choice of experimental material for work in this field.

Donald (1963) notes that a plant's success in gaining a greater share of the limiting nutrient may cause such an increase in growth that a competing species may be suppressed secondarily by shading. He described several experimental devices by which above and below ground competition effects could be separated in culture. Partitions were used to separate root or shoot systems, permitting factorial investigation of soil and aerial effects. Table 13.1 shows the results of

Table 13.1 The interaction of competition for nitrogen and for light in *Lolium perenne* and *Phalaris tuberosa* (Data of Donald, 1958)

	Yield dry wt (g per treatment)			
	No competition	Competition for light	Competition for nitrogen	Competition for both
L. perenne	4·71	4·19	4·31	4·72
P. tuberosa	4·67	3·19	1·17	0·32

such an experiment using the grasses *Phalaris tuberosa* and *Lolium perenne* (tuberous canary grass and rye grass). The depression of *P. tuberosa* yield caused by competition for nitrogen was strongly reinforced when the plants were also allowed to compete for light and supports the suggestion by Jennings and Aquino (1968), that competition for many factors may ultimately operate through modified light relationships.

A great deal of the competition literature concerns agricultural plants which are grown either as single species or as mixtures of a few grassland species. The majority have been selected for their high vegetative or reproductive yields and consequently tend to be demanding nutritionally. The situation in natural ecosystems is very different, the habitats forming large- and small-scale mosaics of varying nutrient status which interact with the distribution of species and ecotypes having a very wide range of nutritional requirements.

Tansley (1917), over 50 years ago, grew the calcifuge, *Galium saxatile* (heath bedstraw), in competition with the calcicole, *Galium sterneri* (sterner's bedstraw), on soils of different pH. Each was capable of growing alone on either an acid peat or a calcareous soil but in competition *G. saxatile* was handicapped on the calcareous soil and *G. sterneri* on the peat. It may be inferred that both species have a wider range of physiological tolerance than is manifested ecologically, competition limiting the potentially wide distribution.

Rorison (1969) illustrates this for several species (Figure 13.6). The physiological response curves of most plants show that they grow best at

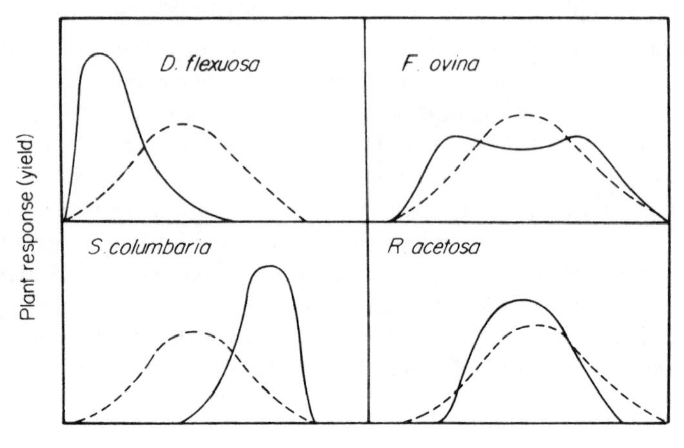

Figure 13.6 Diagrammatic representation of the physiological response curve (-----) and the ecological response curve (———) to soil pH of *Deschampsia flexuosa*; *Festuca ovina*; *Scabiosa columbaria* and *Rumex acetosa*. Reproduced with permission from I. H. Rorison, *Ecological Aspects of the Mineral Nutrition of Plants*, Blackwell, Oxford, 1969, p. 159, Figure 2.

intermediate pH values and in fertile soils; but their ecological responses may be very different. The work of Hackett (1967) and others shows that *Deschampsia flexuosa* (wavy hairgrass), for example, has a rather low growth rate, a low phosphorus requirement and tolerance of high aluminium levels. When grown alone it performs best in fertile soils but in competition is swamped by other, faster growing, species. Hackett suggested that 'acid soils confer no specific nutritional benefits on *D. flexuosa* but are a refuge for it from competition from more vigorous but less acid tolerant species'.

Another striking example is the competitive relationship between *Mercurialis perennis* and *Urtica dioica* (dog's mercury and nettle), both rhizomatous plants of the field layer in some types of British woodland. In culture *M. perennis* grows well in soils of such low phosphate status that *U. dioica* cannot grow beyond the first leaf pair, being limited by P deficiency (Pigott and Taylor, 1964). The authors suggest that *U. dioica* is excluded from many otherwise suitable habitats by low P availability and showed that P fertilization permitted the establishment of *U. dioica* seedlings which could then outcompete *M. perennis* because of their greater height. On a natural gradient of soil phosphorus one would expect to find: first *Mercurialis* alone; then an *Urtica/Mercurialis* mixture in which *Urtica* could not become overcompetitive because of P deficiency size limitation; finally, with high soil P, *Urtica* alone, excluding *Mercurialis* by competition for light.

Water

Competition for water can be compared with competition for nutrients in that plants which produce more dry weight and leaf area for the same uptake are usually better competitors. The further analogy with tolerance of nutrient deficits is the plant which is inherently drought-tolerant.

Few studies of competition for water have been made in natural ecosystems, most investigations having been of agricultural species. Goode (1956) and McWilliams and Kramer (1968) for example both competed a shallow-rooted annual grass with a deep-rooted perennial. In both cases, *Poa annua* versus *Lolium perenne* and *Phalaris minor* (annual canary grass) versus *P. tuberosa*, the plants escape competition for water by this niche differentiation of root stratification but, in the case of *Lolium perenne* this advantage is lost if it is overgrazed, as the deep roots are no longer formed. If water is taken from soil-storage the rate of root extension may be of competitive importance. The undesirable range grass *Bromus tectorum* (cheatgrass) for example, is able to start root extension in very cold soil early in the growing season and outcompetes the desirable dominant *Agropyron spicatum* (bluebunch wheatgrass) (Harris and Wilson, 1970).

An elegant field study of a two-species desert-shrub community was recently made by Fonteyn and Marshall (1978). They selectively removed either *Larrea divaricata*, (creosote bush), *Ambrosia trifida* (ragweed) or both shrubs in a 100 m^2 circle around one or other species and then measured minimum daytime leaf water potential (Figure 13.7). Removal of all plants increased the water potential of *A. trifida* by 1·4 MPa but removal of *L. divaricata* had little effect on either species, thus suggesting that *A. trifida* is in strong inter- and intra-specific competition for water.

FACTORS CAUSING COMPETITION

Passive root interaction

Under normal conditions the soil atmosphere contains less oxygen and more CO_2 than the air above it: a consequence of root and microorganism respiration. There may be a passive interaction between the roots of different species if one is more sensitive than the other to high soil CO_2 or low O_2 concentrations. This interaction is most likely to arise or contribute to competitive effects in wet soils where there is a shortage of air-filled pores and a consequently reduced oxygen diffusion rate. Sheikh (1970), for example, found that *Molinia caerulea* was less sensitive to high CO_2 levels than *Erica tetralix* and, as a result, on soils of

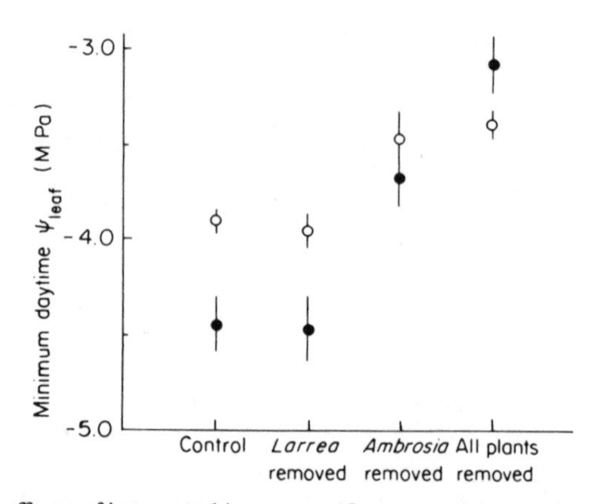

Figure 13.7 The effects of inter- and intra-specific competition on leaf water potential of the desert shrubs *Ambrosia trifida* and *Larrea divaricata.* The experimental treatments involved the removal of either or both species in a 100 m^2 circle around a central bush of each species. No plants were cleared from the control treatment. Minimum daytime leaf water potential of the central bush was measured. *L. divaricata* (○); *A. trifida* (●). Data of Fonteyn and Mahall (1978).

moderate nutrient content under waterlogged conditions, *E. tetralix* is outcompeted by *M. caerulea.* The root réspiration of the two species must make a passive contribution to this interaction in the sense that the normal respiratory production of CO_2 has become involved in the competitive balance.

Allelopathy

De Candolle (1832) suspected that plants released toxic materials into soils and that these lasted long enough to necessitate the rotation of crops. However, at the time, there was little real evidence for this, or knowledge of other soil effects such as nutrient deficiency and pathogen accumulation. From the beginning of this century evidence has accumulated that plants may, directly or indirectly, harm each other through release of chemicals to the environment, the phenomenon of allelopathy (Rice, 1974, 1979).

Plants are known to produce many hundreds of different chemicals which 'leak' to the environment and which can *in vitro* show phytotoxic effects (Rice, 1974). It has proved remarkably difficult to demonstrate any ecological significance for these effects, firstly because of the experimental difficulty of separating competition for a soil resource from the effect of toxin secretion, secondly because most of the toxic secretions are very rapidly metabolized by soil microorganisms, and thirdly, for the theoretical reason that plants very rapidly show selection of tolerant populations when faced with most other known toxicity situations (Harper, 1977, and p. 282).

Rice's two reviews (1974, 1979) are filled with examples of experiments in which potential phytotoxins have been identified in plant leachates and this has then been correlated with the effect of the plant in supressing associated species. More or less attention has been devoted to the almost impossible task of eliminating other interactive effects such as competition for nutrients, water and light, differential modification of the physical environment and selective predation.

The problem of causal interpretation is illustrated by Muller's (1966) assertion that herb-free haloes around the chaparral shrubs *Salvia leucophylla* (purple sage) and *Artemesia californica* (sagebrush) were caused by their releasing volatile phytotoxic terpenes. A mechanism of mitochondrial respiration inhibition was subsequently demonstrated (Muller *et al.* 1969) and, in the first edition of this text the conclusions were accepted as fairly convincing. However, Bartholemew (1970), using grazing exclosures (p. 378) has shown the herb-free areas to be caused by small mammal predation without need to invoke allelopathy. Many more such examples may be found in Rice's reviews (1974, 1979) and in Harper (1977).

Some very early workers attempted to eliminate direct competitive effects by watering one species with the leachates from watering another, for example Bedford and Pickering's (1919) investigation of fruit tree inhibition by grass. This approach has been repeated many times and, in more modern work, nutrient addition and plant analysis used to eliminate nutrient depletion as a source of interaction. A recent example is the work of Newman and Rovira (1975) with grassland herbs: reciprocal experiments with eight species showed some inhibitory effect in all cases. Four species were more sensitive to auto- than to allo-inhibition and four *vice versa*. The allo-inhibited species were generally carpet-formers (e.g. *Holcus lanatus*—Yorkshire fog grass) and the auto-inhibited species those which occur as isolated individuals or small patches (e.g. *Plantago lanceolata*—ribwort plantain).

The fact that this experiment showed results which correlate with ecological behaviour suggests that allelopathy may genuinely have been shown here, though it is unfortunate that the work was done in sand culture rather than soil, which might be more potent in detoxifying allelochemicals.

Harper (1977) has made the useful suggestion that the demands of Koch's Postulates in microbiology should be met in experimental assessment of allelopathic effects: the chemical should be extracted, identified and it should cause 'disease' symptoms in the suppressed plant comparable with those seen in normal field interaction. Much of the biochemical work has already been done (Rice, 1974) but it needs consolidating by this ecological, 'disease causation' approach.

INTERACTION AND EXTERNAL FACTORS

Competition for pollinators and agents of seed dispersal is generally related to the 'attractiveness' of the flower or fruit to another organism and has already been discussed (pp. 385–6). Similarly, plants which rely upon the palatability of their fruit for seed dispersal must encounter agent preference as a selective pressure. Posing considerable sampling problems, these aspects of competition have not been extensively worked.

Disturbance of ecosystems and selective pressure by man and animals cause extensive variation in competitive status of plants in natural ecosystems. Such pressures may be visualized as deforming or displacing niche hyperspaces, thus permitting some species to increase in numbers and others to regress. Niche space which is not efficiently occupied may also be taken over by entirely new species which have been introduced by man or animals. The latter change is epitomized by the recent behaviour of the hybrid of *Spartina maritima* and *S. alterniflora* (cord grasses in Britain subsequent to its origin in Southampton Water shortly before 1870 (Goodman *et al.* 1969). The hybrid originated as a result of the importation of *S. alterniflora* from North America in the early nineteenth century. Present day populations consist of a sterile F_1 hybrid (*S.* × *townsendii*) and a fertile amphiploid (*S. anglica*) which is believed to have arisen in *c.* 1890, accounting for the very sudden spread of the plant after that date.
S. anglica is a colonizer of bare, tidal mud at the lower end of the salt-marsh association, entering the niche which was formerly occupied by various annual *Salicornia* spp. It is a most successful competitor as it spreads vegetatively with great vigour from tussocks established from seed or vegetative fragments. Its perennial habit permits rapid invasion of the *Salicornia* zone and subsequent entry into other parts of the marsh. In the century since its appearance it has spread to the majority of British salt-marshes, with a consequent alteration in their physiognomy and reduction in species diversity.

Similar consequences have followed the introduction of many species to habitats outside their normal geographical range, sometimes with disastrous effects on land-use. *Opuntia* spp (prickly pear), for example, spread unchecked over many square miles of Australian grazing land where they were free from serious competition or predation and were brought under control only when a predator, the moth *Cactoblastis cactorum*, was introduced as a biological controller. Many examples are cited in great detail by Salisbury (1961).

Plants of specialized habitats which require bare soil for their establishment have been allowed to enter other ecosystems through man's agricultural activities since the late Stone Age or early Bronze Age, the pollen record showing a sudden increase of 'weed' species associated with the appearance of cereals. A good example is *Poa annua* (annual meadow grass), which is very common in cultivated and trampled ground in Britain but rather rare in other habitats.

ECOSYSTEMS, COMPETITION AND SELF-REGULATION

A great deal of the literature on plant competition is derived from agricultural experiments in which the interspecific effects of 'weeds' or the intraspecific effects of crop density have been investigated in two-species or mono-specific stands with very small spacing. Much of this work has little relevance to the role of competition in stabilizing natural ecosystems and in controlling habitat specificity.

The reason for this is that, except for some early succession stages, niche differentiation limits competitive interaction in the mature ecosystem though its effects are always visible and often resurgent during regeneration. The most valuable experiments are those which have given insight to the processes which control plant density and reproductive success in multispecies ecosystems coupled with a study of the strategies by which species or populations escape competition either by niche occupancy or by retreat to refugium ecosystems.

During early succession, as high densities of short-lived plants develop from seed, for example in post-fire forest stands, it quickly becomes obvious that a relatively small number of individuals mature and many others become suppressed, often to die of pathogenicity as shading increases. This process seems to follow the biblical maxim: 'Unto everyone that hath shall be given.' Larger plants tend to become larger and smaller plants relatively smaller until some are eliminated. Such trends may often be traced back to the influence of seed weight, individuals with larger than average food resources giving seedlings which are most likely to become the dominants of the population (Black, 1958). The consequence is that weight variation, which at the seedling stage is normally distributed, becomes steadily more skewed until the population is dominated by a few large plants and contains large numbers of small plants some of which are destined to be eliminated.

This process of increasing dominance of large individuals is also deducible from the work of Hozumi *et al.* (1955) who grew *Zea mais* (sweetcorn) in single rows at two different spacings. Autocorrelation analysis between the weights of the *n*th plant in a row and the $(n + 1)$th to $(n + 5)$th showed the markedly

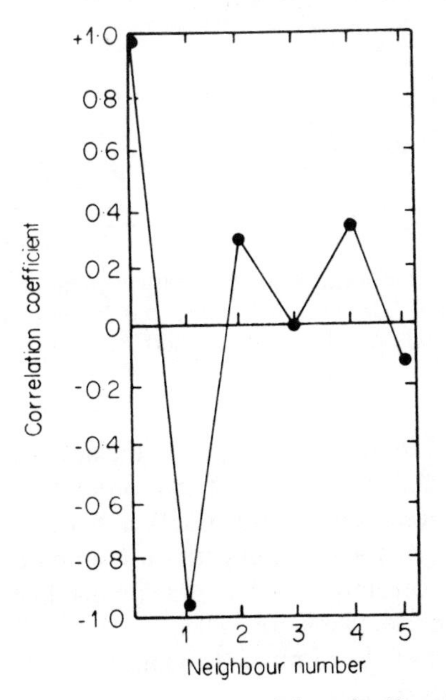

Figure 13.8 The correlation between shoot weights of a *Zea mays* plant and its first to fifth neighbours. After K. Hozumi *et al.*, *J. Inst. Polytech. Osaka City Univ. Series D*, **6**, 121–30 (1955).

alternating effect seen in Figure 13.8. The first $(n + n + 1)$ correlation was always negative and the subsequent values then fluctuated alternately between positive and negative, suggesting that a large plant tended to depress the weight of its neighbours while a small plant caused enhanced neighbour weight. It should be mentioned that the authors also detected another effect in these experiments which would tend to cancel the consequences of this alternation. This was the negative correlation between shoot length and elongation rate which, of course, would tend to equalize plant size. The behaviour of the crop as a whole must be governed by the balance between these two effects.

Figure 13.9 is a striking example of the accumulation of suppressed survivors and a very few large successful plants in a densely sown mono-specific stand. The relative frequency distribution at maturity is so skewed that over 70% of the

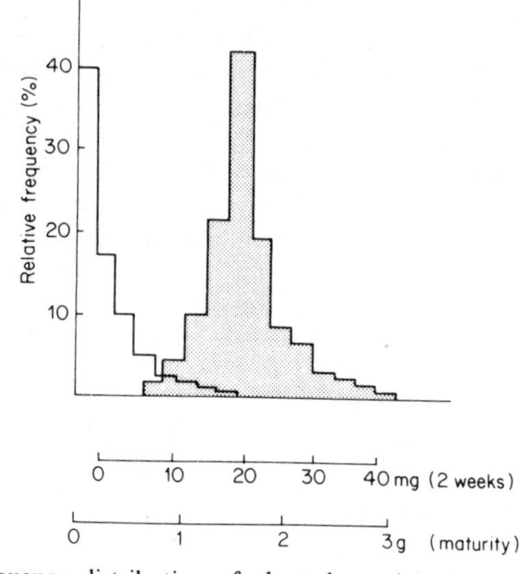

Figure 13.9 Frequency distribution of plant dry weight in a population of *Linum usitatissimum* (flax) sown at 3600 seeds m^{-2} measured 2 weeks after emergence (stippled) and at maturity (unshaded). Size-classes at 2 weeks: 3 mg, and at maturity: 0·2 g. Data of Obeid *et al.* (1976).

plants have a weight less than one-third that of the heaviest individuals. This can be compared with an almost normal distribution at 14 days which reflects the normal distribution of seed weight frequency. Weight distributions of this sort are the rule in natural systems where competition is intense; early successions or constantly rejuvenated vegetation such as grazing plagioclimax grassland. More mature ecosystems do not show the effect so strongly because suppressed individuals have had time to be eliminated and niche differentiation for avoidance of competition has occurred.

The occurrence of this skewing in the frequency of plant weights raises questions concerning the pre-emption of space, plastic responses of individuals and the process of density-dependent self-thinning. Suppressed individuals are usually in ultimate competition for light and the successful few are those which, through greater seed weight, favourable germination niche, a patch of moist, deep soil or some other advantage, win the race for pre-emption of a larger than average portion of the canopy area.

Plastic responses are most often seen in dense juvenile systems or amongst closely sown crops. In late successional systems the suppressed individuals have already joined the ranks of the dead. Plastic response is seen to extreme in the Clements *et al.* (1929) classic experiments with *Helianthus annuus* (sunflower). Their results and spectacular photographs show that isolated plants (160 cm spacing) produced a total leaf area which was a thousand times more than plants at 5 cm spacing, and stem bases of eighty times the cross-sectional area! These huge differences were reflected in a height reduction of only 56 %. Such spindly, poorly leaved individuals (the 'whips' of foresters) are susceptible to wind damage, infection with pathogens and, amongst herbaceous plants, often succumb to basal damage by snails and other organisms which larger plants might tolerate. The enormous reduction of individual leaf area is partly a consequence of reduced growth but the main loss is a consequence of shading in the lower canopy which causes leaf senescence and self-pruning.

Self-thinning involves processes related to the behaviour of the individual which have the statistical result that there is an upper limit to maintainable density and that this drifts downward with time from establishment of the juvenile plant. At any time during this process there is a linear relationship between the reciprocal of mean plant weight and prevailing density. This would be expected if the consequence of competition is the simple partitioning of available habitat-volume between individuals. The relationship has been named the reciprocal yield law but care must be taken with its interpretation, as the reciprocal weights of very small plants are very large numbers which distort the significance of calculated regressions unless suitably transformed (Harper, 1977).

The maximum density shifts downward as plant weight increases, following a $\frac{3}{2}$ power-law in which:

$$W = x(D_{en})^{-1.5} \tag{13.1}$$

where W = the mean weight of surviving plant (g); D_{en} = the density of survivors (plants m^{-2}); and x is a constant ranging from about 10^4 to 10^5.

The value of x is partly a function of canopy geometry and the capacity of a species to pre-empt the trapping of light. This is of particular interest if it is related to Horn's (1971) discussion of multilayer and monolayer canopies in governing successional ability and Grime's ordination of competitive strategies as a part-function of a morphological index which reflects the space-occupying ability of a species.

Figure 13.10 shows some examples of survivor density versus mean weight curves in which it may be noted that species with a deep canopy of broad, near-horizontal leaves (*Chenopodium album*—goosefoot) have a much greater value of

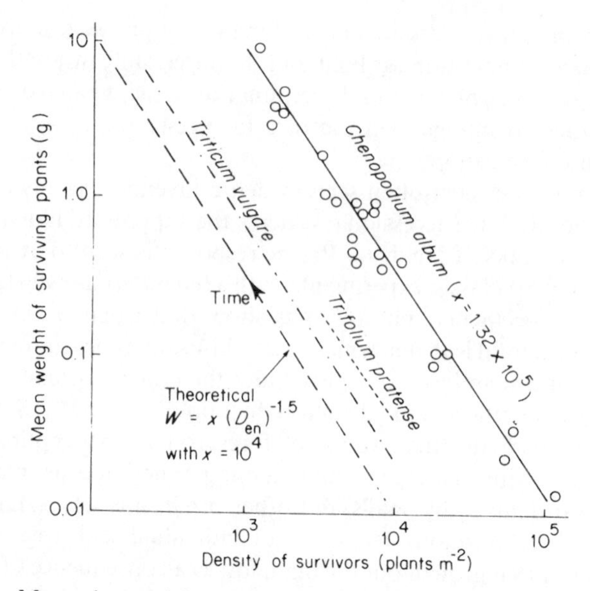

Figure 13.10 Mean plant weight and its relationship to the change of plant density with time: the 3/2 power law. The theoretical line for $x = 10^4$ is shown carrying an arrow which indicates change with passage of time. The original data points for *C. album* are plotted to indicate the degree of experimental variation, but they are omitted from other species for clarity. Data from Harper (1977).

x than either decumbent broad-leaved species (*Trifolium pratense*—clover) or near-upright grass-leaved plants such as wheat (*Triticum vulgare*). This also reflects competitive ability as *C. album* is a highly competitive ruderal (Grime, 1979), *T. repens* only survives in grassland swards when grazing or cutting protect it from competition and *T. vulgare* is a very uncompetitive agricultural species as farmers know to their cost.

Population growth and interaction

Patterns of population increase have often been modelled in simplistic form by the logistic equation:

$$\frac{dN}{dT} = r_{ep}N\left(\frac{K - N}{K}\right) \qquad (13.2)$$

where dN/dT = change of number (N) with time (T), r_{ep} = reproductive rate and K is a limiting constant.

Rate of population growth	=	Intrinsic rate of increase	×	Actual realization of potential increase

In stable populations N is likely to remain nearly equal to K for long periods, density-dependent mortality and limited reproduction dominating the regulatory process. Harper (1967) points out that many plant populations spend a large proportion of their time in recovering from environmental catastrophes. These populations will generally be increasing at something near to their intrinsic rates and population size will be a function of the magnitude of the last catastrophe and the time available for regrowth.

The value of r_{ep} is twofold, being represented in local vegetative multiplication and also in the further flung effects of seed production. The first may be looked upon as a large capital investment producing a small but secure return in increasing population size while the second represents a small capital investment with a potentially high return in increased numbers but a much greater risk of total loss. Harper discussed the strategy of the plant life-cycle in governing its potential competitive relationships. In terms of seed production the rate of population increase is critically governed by precocity of reproduction, for example an individual producing two offspring in the first year, then dying, has the same potential rate of increase as an individual producing one offspring per year for ever. Harper contrasted the biennial *Digitalis purpurea* (foxglove) producing *c.* 100,000 seeds every 2 years with an annual counterpart which would achieve the same population growth rate with only 330 seeds p.a. However, high risk of seed mortality must affect the economy of these two types of plant; the annual risks total seed loss but this is much less likely with the biennial.

The value of K is a measure of the limiting factors of the environment and when this includes competition from one or more other species the values of K in the equations for each species may be made self-adjusting for the population density of the other species (Hutchinson, 1965):

$$\frac{dN_1}{dT} = r_{ep1}N_1\frac{(K_1 - N_1 - \alpha N_2)}{K_1}$$
$$\frac{dN_2}{dT} = r_{ep2}N_2\frac{(K_2 - N_2 - \beta N_1)}{K_2} \tag{13.3}$$

where N_1 and N_2 are the respective numbers of species 1 and 2, r_{ep1} and r_{ep2} are their relative growth rates and K_1 and K_2 are constants for each species. The condition for continued growth of one species will inhibit its own further growth more than that of the other species, thus preventing competition from culminating in an exclusive struggle.

Returning to the Margalef (1968) feedback concept discussed at the beginning of the chapter, it was postulated that a positive feedback could be established between a pair of species competing for a single limiting resource. This feedback could result in one species 'mopping up' the resource to such an extent that the second species would be driven to extinction. Hutchinson's model, however, averts this catastrophe, as a density-dependent negative feedback within each species prevents the population explosion which would be necessary to remove all of the resource from the environment. The equilibrium proportions of the two species in the population will depend on the rate of this internal feedback mechanism and its critical density dependence.

One of the most satisfactory experimental approaches to competition between two species (or populations) was first described by de Wit (1960). Two species are sown or planted together as a replacement series in different ratios, but the overall plant density is maintained constant. If there is no competition and growth rates are equal then there is no change in ratio during the experimental period (Figure 13.11a) but if one grows more rapidly than the other, the relationship becomes distorted as in Figure 13.11b. There is still, however, no competitive effect of one upon the other. If competition occurs then the successful plant increases its yield at the expense of the unsuccessful competitor (Figure 13.11c).

When data from such experiments are expressed as species ratios at sowing, versus the ratios at harvest (input ratio versus output ratio), they may be used to propound four different behavioural situations (Figure 13.12a and b). When there is no drift of species ratio from input to output the association will remain stable and the *relative reproductive rate* may be expressed as unity:

$$RRR = \frac{A_{\text{harvest}}/A_{\text{sown}}}{B_{\text{harvest}}/B_{\text{sown}}} = 1{\cdot}0 \qquad (13.4)$$

The slope of the input/output relationship will thus be 45° (Figure 13.12a). If, however, one species has a reproductive rate which is higher than the other, irrespective of the sowing ratio, then the slope remains at 45° but it is displaced either above or below the unity line and, with the course of time, the species with the lower reproductive rate will become extinct. If the relative reproductive rate of the two species is sensitive to sowing ratio, the slope of the line will be altered. If the slope is greater than 45° one, or the other, species will drift toward extinction according to the initial seed ratio. If the slope is less than 45° the two species will converge toward an intermediate value lying on the unity line and the association will become stable.

The situations which promote stability are, therefore, those in which $RRR = 1$ and there is no competitive interaction, or those in which the increased ratio of one species in a seed mixture is reflected as a decrease of that species in the harvest mix or *vice versa.* Such a response may be interpreted as 'self-competitive limitation' at high initial ratios and 'space exploitation' at low ratios. In the unstable situation with a slope of more than 45° the species which is at a high level in the initial seed mix gains a long-term competitive advantage which compounds with time.

Examination of these four potential deviations from equilibrium suggests that most circumstances are likely to eliminate one species of a competing pair and yet it is notable that natural ecosystems support a great diversity of co-existing species. In fact, it is only the unusually harsh environment, sometimes grossly polluted or damaged, which is characterized by very low species diversity.

Early work with de Wit replacement series experiments was confined to agricultural plants, but the technique is now proving powerful in elucidating the responses of potential competitors in natural ecosystems. Goldsmith (1973), for example, observed that *Armeria maritima* (thrift) was most successful in a sea-cliff sward of *Festuca rubra* (red fescue) where salt-spray salinated the soil. In a

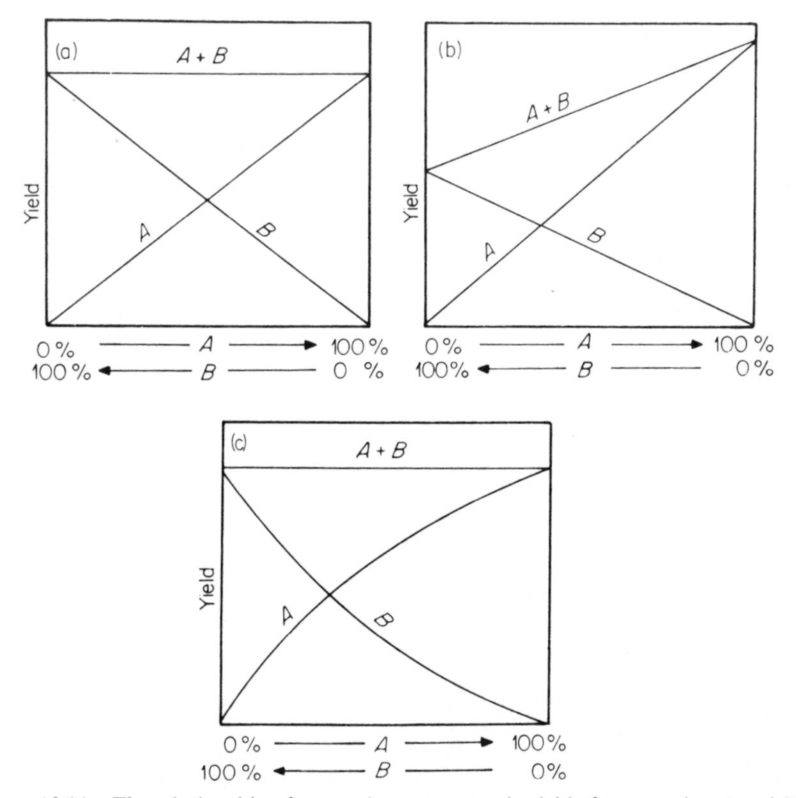

Figure 13.11 The relationship of vegetative or propagule yield of two species, *A* and *B*, to proportions of each in the sowing or planting mixture. (a) No competitive interaction and identical growth or seed production rates; (b) No competitive interaction but different growth or seed production rates; (c) Competitive interaction in which the proportion of *B* in the harvest is suppressed and that of *A* enhanced. Redrawn with permission from C. T. de Wit, *Verslag. Landbouwk. Onderz. Ned.*, **66**, 1–82 (1960).

replacement series *F. rubra* proved to be the better competitor in normal soil but a treatment irrigated with seawater showed *A. maritima* to be the better competitor despite the much reduced growth of both species.

It is possible to use replacement series multifactorially amongst several species, for example Pemedasa and Lovell (1974) competed four dune annuals and one perennial species in all possible combinations and with normal and enhanced nutrient regimes (Figure 13.13). The perennial *Festuca rubra*, as might be expected, outcompeted all of the annuals (Figure 13.13a) and was more successful with high nutrients (Figure 13.13b). Amongst the annuals the slightly more robust *Vulpia membranecea* (dune fescue) suppressed all other annuals (Figure 13.13e and f) while the two hair grasses (*Aira praecox* and *A. caryophyllea*), both of similar morphology, showed little competitive interaction (Figure 13.13c). *Cerastium atrovirens* (mouse-ear chickweed) suppressed the two weakly-growing *Aira* spp (Figure 13.13d) but succumbed to the stronger *V. membranacea* (Figure

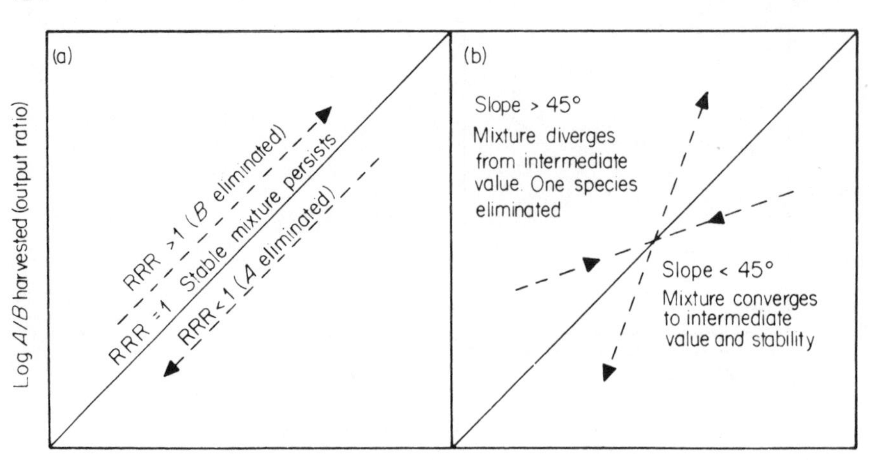

Figure 13.12 The relationships which may exist between input (sown) and output (harvested) ratios of two species grown in competition. (a) The continuous line with a slope of 45° represents an input-output ratio of unity giving a stable situation in which both *A* and *B* will persist. Lying above or below this line, but still with unit slope, are the two pecked lines representing a constant reduction in the proportion of one species at each harvest. Both of these situations are unstable and lead to the ultimate extinction of one species. (b) In this case the two pecked lines represent a situation in which the input-output ratios are density dependent so that the slope is >45° or <45°. With a slope >45° the situation is unstable; above the intersection with the continuous line, *B* will gradually disappear, and below the intersection *A* will gradually disappear. With a slope of <45° the mixture converges to an intermediate value and *A* and *B* persist together. Redrawn from Donald (1963) after de Wit, *Verslag. Landbouwk, Onderz. Ned.*, **66**, 1–82 (1960).

13.13f). These results not only reflect the field behaviour of the plants but are also particularly relevant to these early successional species which often establish on bare soil at incredibly high seedling density.

In replacement series experiments the suppression of one species may be equalled by the increment of the other so that relative yield total (line $A + B$, Figure 13.11) is constant irrespective of proportions. If this is not the case the relative yield total may be greater (or less) than the yield of either species alone. In many cases the data of Pemedasa and Lovell (1974) show much greater than expected yields in the 50:50 mixtures, particularly in the cases of *V. membranacea* versus either *F. rubra* or *A. caryophyllea* (Figure 13.13a, b and e). In both cases most of the extra yield comes from the successful partners; *F. rubra* in the first case and *V. membranacea* in the second, suggesting that these two species perform better in competition with a weaker species than they do in self-competition.

Explanation of such effects is not easy unless one partner can be identified as making a special contribution, thus Hall (1974) gave an excellent account of replacement series competition between a grass (*Chloris guyana*) and a legume

(*Stylothanses humilis*) in which ^{15}N labelling was used to differentiate soil-derived and *Rhizobium*-derived nitrogen in the plants. The two species were in competition for soil-N and the robust *C. guyana* was able to suppress the weaker legume but, despite this, some *Rhizobium*-fixed N became available to *C. guyana* with the result that the relative yield total rose to 1·3, when *S. humilis* comprised 80 % of the mixture but most of the extra growth was contributed by *C. guyana*. It is not so easy to explain the enhanced relative yield totals of Figure 13.13, none of the species possessing an obvious niche-differential strategy such as N-fixation.

The use of mixed species experimentation offers much to the plant ecologist, bridging the gap between physiological ecology and field studies of distribution and population biology. Together with studies of animal-plant and pathogen–plant interaction they form a most exciting and rapidly growing part of biological science.

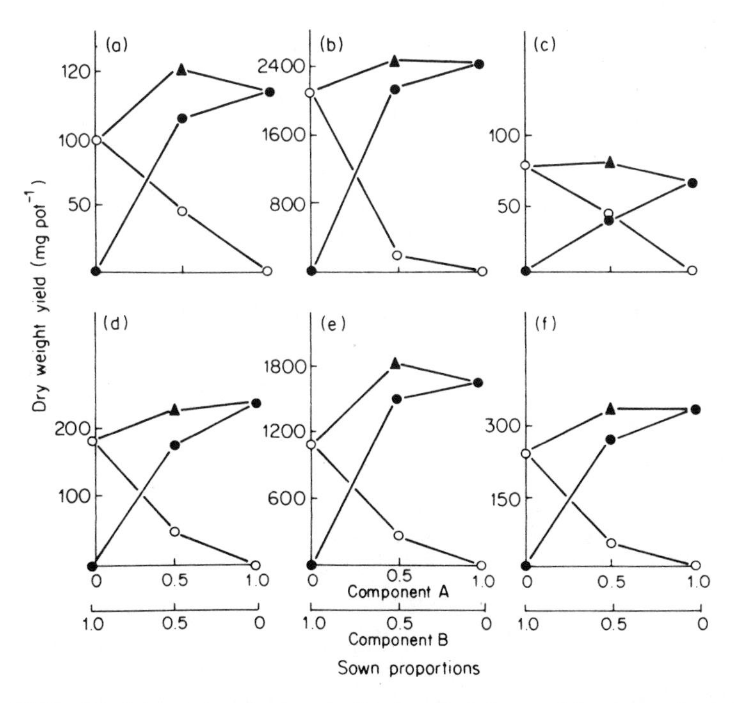

Figure 13.13 Replacement series graphs for some of the combinations of species and treatments used by Pemedasa and Lovell (1974) to show the interrelationship between dry matter production in pure stands and mixed-species stands of various sand dune plants. The first-named species (component A) is always plotted with density increasing toward the right-hand end of the ordinate; +N and −N respectively indicate addition and no addition of nutrients while L, M and H are sowing densities of 6, 18 and 54 plants per pot. (a) *Festuca rubra v. Vulpia membranacea* −N; L. (b) *F. rubra v. V. membranacea* +N; H. (c) *Aira caryophyllea v. A. praecox* −N; M. (d) *A. praecox v. Cerastium atrovirens* +N; M. (e) *A. caryophyllea v. V. membranacea* +N; M. (f) *C. atrovirens v. V. membranacea* +N; H. Note the differing scales of yield which vary according to nutrition.

CONCLUSION

Perhaps the last word might rest with Darwin: 'It is good thus to try in our imagination to give any form some advantage over another. Probably in no single instance should we know what to do, so as to succeed. It will convince us of our ignorance on the mutual relations of all organic beings: a conviction as necessary as it seems to be difficult to acquire.'

Thus, in the final paragraph of 'The Struggle for Existance' in *Origin of Species*, written well over a century ago, he placed his pen delicately on the near-impossible task of the ecologist who attempts to piece together innumerable fragments of knowledge into a mosaic which pictures our understanding of the living world.

There is yet much for the experimental ecologist to do, not only to explain but to educate, for our world is frighteningly fragile and much has already been lost for ever.

Appendix

Appendix Table 1.1a Milestones on the road to ecological thought

Date		Reference	Page (if in text)
40,000 B.C.	Upper Palaeolithic: concepts of species and habitat? Cave paintings.		
6000	Neolithic: beginning of cereal-based agriculture. Plants and water irrigation.		
3100	Sumeria: first syllabic writing.		
1000	Greece: first alphabets.		
427–347	Plato: *Critias*—appreciation of disafforestation and soil loss (*c.* 350 B.C.).	Crowther (1953)	2, 58, 66
384–323	Aristotle: classification of organisms—'*scala naturae*' — influenced scientific thought for 2000 years.	*Encyclopedia Britannica* (1970)	
371–285	Theophrastus: *Enquiry into Plants* — ecological understanding of plant distribution in relation to habitat — '...some love wet and marshy ground...'.	Kormondy (1965)	
95–55	Lucretius: *De Rerum Natura* — 'nothing is created out of nothing.' Nature of atoms, compounds and life.	Latham (1951)	
70–19	Virgil: *Georgics* — crops and environment; soil-testing; crop-management.	Crowther (1953)	
23–79 A.D.	Pliny: 37 books of natural history.		
1400–1500	Printing in Europe. Herbalists descriptions of plants and habitats.	e.g. Gerard (1597)	
1577–1644	Van Helmont first understood relationship between soil and water in plant growth.	Russel (1973)	147 ff

Appendix Table 1.1a (*continued*)

Date		Reference	Page (if in text)
1628–1705	John Ray introduced binomial classification.		
1707–1778	Linnaeus: foundation of modern taxonomy. Observations of phenology and plant geography.	Kormondy (1965)	
	At about this time biological science, including the embryonic ecology, showed a number of divergences of development which resulted in the foundation of the topics which we know as descriptive and quantitative ecology, plant geography, vegetation dynamics, physiological ecology, ecology and applied ecology amongst others. In the remainder of the table the development of these theses is, for convenience, considered separately.		

Appendix Table 1.1b Plant geography, descriptive and quantative ecology, community and population dynamics

Date		Reference	Page (if in text)
1685	King: first concepts of plant succession.	Clements (1928)	388 ff
1729–1810	Early work on plant succession, in particular related to peat formation. (Degner, 1729; Anderson, 1794; De Luc, 1806; Rennie, 1810).	Clements (1928)	388 ff
1798	Population, competition and food-supply.	Malthus (1798)	2, 395 ff
1805	First plant geography	Humboldt (1805)	
1842	Steenstrup: peat stratification and vegetation history.	Clements (1928)	104–5
1850	First use of quadrats by Colniess. Measurement of vegetation.	McLean (1973) and Ivimey-Cook	
1851	*Origin of Species.* Concepts of fitness and survival through adaptation.	Darwin (1859)	
1861	Von Post: description of the role of animals and plants in soil formation.	Clements (1928)	98–9
1866	Use of term: *oecology.*	Haekel (1866)	

Appendix Table 1.1b (*continued*)

Date		Reference	Page (if in text)
1891	Dune succession in Denmark.	Warming (1891)	
1896	Textbook of ecology.	Warming (1896)	
1898	Textbook of physiological plant geography — see Table 1.1c.	Schimper (1898)	
1899	Sand dune succession in North America	Cowles (1899)	
1900	First textbook of soil science: soil classification system and relationships of vegetation types to soil groups.	Dokuchayev (1900)	54
1903	Textbook of soil science: first general British account.	Hall (1903)	Chapters 3 & 4
1905	*Research Methods in Ecology*—classic text. Use of quadrats, instrumental methods for environmental measurement. Experimental physiological approach.	Clements (1905)	3
1911	*Types of British Vegetation*—use of successional relationships and habitat factors in classifying an extensive vegetation.	Tansley (1911)	
1912	*A Study of the Soils of the United States*	Coffey (1912)	53
1913	Braun-Blanquet and Furrer: early report of the phytosociological approach to vegetation classification which has come to dominate European ecology. See also Braun-Blanquet (1932).	Kormondy (1965) McKintosh (1978)	
1916	*Plant Succession* — concepts of habitat relationships and succession in which plant communities behave as pseudo-organisms: the organismic hypothesis. Climatic climax. Climax theory.	Clements (1916) Clements (1936)	308 ff
1917	Refutation of organismic hypothesis. Community composition is a reflection of random arrival and environmental selection. The Gleason–Clements controversy can be traced down to the present day in 'continuous variation' or 'hierarchical unit' classification techniques.	Gleason (1917) Gleason (1926)	392–3

Appendix Table 1.1b (*continued*)

Date		Reference	Page (if in text)
1920	Independent re-derivation of Verhulst's logistic equation for population growth.	Pearl and Reed (1920) Kormondy (1965)	410
1920	*Plant Indicators*—value of species is indication of environmental and biotic conditions	Clements (1920)	
1920	The classification of vegetation	Tansley (1920)	
1926	Tansley and Chipp: *Aims and methods of vegetation study.*	Tansley and Chipp (1926)	
1925–6	Lotka–Volterra equation for growth of interacting populations.	E. P. Odum (1971)	411
1929	*Plant Competition*	Clements, Weaver and Hanson (1929)	395 ff
1932	*Plant Sociology*	Braun-Blanquet (1932)	
1935	First use of term ecosystem.	Tansley (1935)	
1939	*British Isles and their Vegetation.* Developed from Tansley (1911).	Tansley (1939)	
1951	Analysis of continuous variation in vegetation systems — the continuum concept.	Curtis and McIntosh (1951)	
1953	*Fundamentals of Ecology.* First major textbook to use the holistic, ecosystem approach. Second edition 1959; third, much improved 1971.	E. P. Odum (1953; 1959; 1971)	
1954	Use of factor analysis to produce a vegetation ordination in which continuous variation is graphically represented by position relative to pairs of graph axes (ordinates).	Goodall (1954)	
1955, 1957	Odum brothers attempt to quantify a holistic ecosystem study.	Odum and Odum (1955) Odum (1957)	338
1957	Use of similarity coefficient for simple ordination	Bray and Curtis (1957)	
1959	Development of association analysis for objective hierarchical classification of vegetation	Williams and Lambert (1959)	
1960s	Increased efficiency of ordination by principal component analysis and other methods.	e.g. Orloci (1966; 1975)	

Appendix Table 1.1b (*continued*)

Date		Reference	Page (if in text)
	From the 1960s onward so much ecological publication has appeared that it is possible only to list some of the main areas in which advances have occurred, with references to review volumes.		
1960–	Increasing sophistication of approach to theoretical ecology including approaches to species diversity, niche differentiation, and competitive interaction. McArthur and Wilson (1967); E. P. Odum (1971); Colinveaux (1973); Harper (1977); May (1976).		
	Plant strategies and population biology. Harper (1977); Grime (1979).		
	Extensive publication of multifactorial analytical work. Greig Smith (1964); Whittaker (1973); Kershaw (1973); Mueller-Dombois and Ellenberg (1974); Orloci (1975); Pielou (1977).		

Appendix Table 1.1c Physiological ecology, production and systems ecology

Date		Reference	Page (if in text)
1661	*The Sceptical Chymist*: nature of elements, compounds, soil-salts.	Boyle (1661)	
1669	First use of solution culture, establishing the need for dissolved matter.	Woodward (1669)	
1727	*Vegetable Staticks*: first comprehensive whole-plant physiology. Air 'is wrought into the composition of plants'. Role of sunlight.	Hales (1727)	
1776	Priestley: plant life 'restores injured air'.	Rabinowitch and Govindgee (1969)	
1779	Ingen–Housz: established role of light and darkness in the air injuring and restoring properties of animals and plants	Rabinowitch and Govindgee (1969)	

Appendix Table 1.1c (*continued*)

Date		Reference	Page (if in text)
1782	Senebier: increase in plant weight derives from the interaction of 'phlogisticated air' and light. In later publication this statement was re-expressed in the terms of Lavoisier's modern chemistry.	Russel (1973)	
1801	Role of wood in upward movement of water, bark in downward movement of assimilates, nature of ray tissues in transport and movement of leaves in response to light.	Knight (1801)	
1804	De Saussure: first correct equation for photosynthesis.	Lieth and Whittaker (1975)	
1847–59	*Cybele Brittanica* vols. 1–4. Geographical distribution of British plants with reference to climate and altitude.	Watson (1847–59)	
1840	Final acceptance of air as CO_2 source for plants and speculation on the global magnitude of the photosynthetic process.	Liebig (1840) Lieth and Whittaker (1975)	346 ff
1858	Sachs revives the technique of solution culture for nutritional investigation. Major nutrient requirements of plants fairly well known.	Pfeffer (1881) (Translation 1900)	Chapter 9
1861	At this time in general textbooks of botany, physiological knowledge was still comparatively rudimentary and generalized.	e.g. Bently (1861)	
1862	Law of the minimum	Liebig (1862)	
1881	Pfeffer: textbook of plant physiology — a basic knowledge of all fundamental whole-plant physiological processes had accumulated. Photosynthesis, respiration, translocation, water relations, mineral nutrition, nitrogen assimilation and environmental effects were all treated in a fashion which parallels modern thought.	Pfeffer 1881 (Translation 1900)	

From the 1880s onward, many papers were published which would today be considered physiological–ecological in nature. Prior to the establishment of ecological journals in the 1900s they appeared in journals such as: *American Naturalist*; *Annals of Botany*; *Annales des*

Appendix Table 1.1c (*continued*)

Date		Reference	Page (if in text)
	Sciences Naturelles Botaniques; *Bulletin of the Torrey Botanical Club*; *Botanical Gazette*; *Comptes Rendus des Seances de l'Academie des Sciences*; *Flora*; *Jahrbucher für Wissenschaftliche Botanik*; *Journal of the Linnaean Society*; *Nature*; *New phytologist*; *Philosophical transactions of the Royal Society, London*; *Transactions of the Royal Society, London.*		
1882	Textbook of plant physiology.	Sachs (1882)	
1885	Demonstration of N-fixation by soil microorganisms.	Berthelot (1885)	234 ff
1887	Arrhenius: dissociation theory which ultimately led to more critical understanding of soil-plant solute relations.	Clark (1923)	89 ff, 259
1889	Hellrigel's first demonstration of symbiotic N-fixation.	Pfeffer (1900)	235
1893	Winogradsky speculated on energy consuming (carbohydrate) efficiency of N-fixation	Pfeffer (1900)	237
1894	Textbook of botany including well-known section on plant physiology.	Strasburger (1894)	
	Formulation of cohesion–tension theory of water transport	Dixon and Joly (1894)	147ff, 163
	Textbook of practical plant physiology with ecological discussion	Darwin and Acton (1894)	
1884	*Physiological Plant Anatomy*	Haberlandt (1884)	180 ff
1898	Textbook of physiological plant geography	Schimper (1898)	
	Account of environmental effects on stomatal function	Darwin (1898)	
1900	Theoretical treatment of diffusion and leaf structure	Brown and Escombe (1900)	165–6
	Detailed discussion of plant distribution in relation to soil conditions. Plant growth experiments with soils and amended soils	Roux (1900)	
1903	Osmosis, water movement and transpiration	Livingstone (1903)	Chapter 6.
1904	Monograph on transpiration	Burgstein (1904)	Chapter 6.
1905	Energy balance of plant leaves	Brown and Escombe (1905)	15
	Clements' first textbook (see Table 1.1b) commencing a series of	Clements (1905, 1907, 1916, 1920, 1928) and	

Appendix Table 1.1c (*continued*)

Date		Reference	Page (if in text)
	publications which greatly influenced the development of physiological ecology and continued until the late 1930s.	Clements *et al.* (1929)	
1905		Blackman (1905)	
1909	Sorensen's first use of the pH notation which has been of such use to biologists.	Clark (1923)	90 ff
1919–20	Introduction of growth-analytical techniques.	Blackman (1919); West *et al.* (1920); Evans (1972)	137–40
1922	First description of physiological race differentiation.	Turesson (1922; 1931)	see index: population
1925	Introduction of statistical methods: regression, correlation, contingency, analysis of variance. Development of experimental design.	Fisher (1925; 1935)	
1925	Textbook of plant water relationships written with a strong physiological–ecological flavour.	Maximov (1929 English translation)	Chapter 2
1927	*The Climate Near the Ground.*	Geiger (1927) (1965 Englished)	
1931	One of the first 'physiological ecological' textbooks: *Environment and Plant Development.*	Lundegardh (1931)	
1935	First use of term 'ecosystem'.	Tansley (1935)	
1938	First air-conditioned greenhouses for experimental plant physiology.	Went (1957)	
1942	Trophic–dynamic aspects of ecology: a seminal paper for later developments in ecosystem quantification.	Lindemann (1942)	338
1948–9	Pasadena, California. First extensive group of controlled environment chambers permitting some degree of 'real world' simulation. Screening of climatic responses of wild species.	Went (1957)	
1953–7	Odum brothers ecosystem concepts and research which has been the foundation of later attempts to quantify ecosystem function.	E. P. Odum (1953; 1971) Odum and Odum (1955) Odum (1957)	338

Appendix Table 1.1c (*continued*)

Date		Reference	Page (if in text)
1962–	Hubbard Brook ecosystem study of forest Bio-geochemistry	Likens, Bormann *et al.* (1977)	374
1964–74	International Biological Programme. Synthesis reports published by Cambridge University Press.		
1966	First textbook of biological systems analysis	Watt (1966)	
1969	Systems analysis	Van Dyne (1969)	
1972	*Phytochemical ecology*	Harborne (1972; 1977)	379
	Plant ecology	Stalfelt (1972)	
1974	*Plants and Environment*	Daubenmire (1974)	
1975	*Physiological Plant Ecology*	Larcher (1975)	
1976	*Introduction to Physiological Plant Ecology*	Bannister (1976)	
	In addition to the above-listed texts, numerous specialist books in physiological ecology have appeared during the past decade: *Studies in Ecology* vols. 1–32 (1973–9) (Springer-Verlag) and a series of monographs: *Physiological Ecology* (Academic Press), commencing in 1971, are particularly useful. Many are referred to in the text.		

Bibliography

Figures in brackets at the end of each entry are the page number on which the article is cited.

Adams, F. (1974), Soil solution. In *The Plant Root and its Environment*, ed. Carson, E. W. pp. 441–81, Univ. Press of Virginia, Charlottesville. (89)

Ahmad, I., and S. J. Wainwright (1976). Ecotype differences in leaf surface properties of *Agrostis stolonifera* from salt marsh spray zone and inland habitats. *New Phytol.*, **76**, 361–6. (288)

Alexander, M. (1961). *Introduction to Soil Microbiology*. John Wiley, New York. (304)

Allam, A-W, I. (1971). Soluble sulphides in rice fields and their *in vitro* effects on rice seedlings. Ph.D. Dissertation. Louisiana State Univ. (304)

Allison, F. E. (1965). Organic carbon. In *Methods of Soil Analysis*, ed. Black, C. A. pp. 1367–78, Am. Soc. Agron., Wisconsin. (110)

Allison, F. E. (1968). Soil aggregation—some facts and fallacies as seen by a microbiologist. *Soil Sci.*, **106,** 136–43. (115)

Anderson, G. (1975a). Sulfur in soil organic substances. In *Soil Components*. Vol. 2, ed. Gieseking, J. pp. 333–41, Springer, Berlin. (105, 243)

Anderson, G. (1975b). Other organic phosphorus compounds. In *Soil Components*. Vol. 1, ed. Gieseking, J. pp. 305–31, Springer, Berlin. (241)

Anderson, M. (1971). Radiation and crop structure. In *Plant Photosynthetic Production: Manual of Methods*, ed. Sestak, Z., J. Catsky, and P. G. Jarvis. pp. 412–66, Junk, The Hague. (34)

Antonovics, J., A. D. Bradshaw, and R. G. Turner (1971). Heavy metal tolerance in plants. *Adv. ecol. Res.*, **7**, 1–85. (281, 282)

Applied Geochemistry Research Group. (1978). *The Wolfson Geochemical Atlas of England and Wales*, Clarendon Press, Oxford. (250, 371)

Arber, A. (1920) *Water Plants*. Cambridge Univ. Press, Cambridge. (319)

Archer, M. (1973). The species preferences of grazing horses. *J. Brit. Grassl. Soc.*, **28**, 123–8. (378)

Arikado, H., and Y. Aduchi (1955). Anatomical and ecological responses of barley and some forage crops to the flooding treatment. *Bull. Fac. Agric. Mie Univ.*, **11,** 1–29. (320)

Arikado, H. (1959). Comparative studies on the development of the ventilating system between lowland and upland rice plants growing under flooded and upland soil conditions. *Bull. Fac. Agric. Mie Univ.*, **19,** 1–10. (320)

Aristovskaya, T. V., and G. A. Zavarzin (1971). Biochemistry of iron in soil. In *Soil Biochemistry*, Vol. 2, ed. McLaren, A. D., and J. Skujins pp. 385–408, Dekker, New York. (102, 247)

Armstrong, W. (1964). Oxygen diffusion from the roots of some British bog plants. *Nature, Lond.*, **204,** 801–2. (314)

Armstrong, W. (1967a). The relationship between oxidation-reduction potentials and oxygen diffusion levels in some waterlogged organic soils. *J. Soil Sci.*, **18,** 27–34. (296)
Armstrong, W. (1967b). The use of polarography in the assay of oxygen diffusing from roots in anaerobic media. *Physiol. Pl.*, **20,** 540–53. (295, 314, 315)
Armstrong, W. (1967c). The oxidizing activity of roots in waterlogged soils. *Physiol. Pl.*, **20,** 920–6. (318)
Armstrong, W. (1968). Oxygen diffusion from the roots of woody species. *Physiol. Pl.*, **21,** 539–43. (328)
Armstrong, W. (1970). Rhizosphere oxidation in rice and other species: a mathematical model based on the oxygen flux component. *Physiol. Pl.*, **23**, 623–30. (315, 316, 317)
Armstrong, W. (1971a). Oxygen diffusion from the roots of rice grown under non-waterlogged conditions. *Physiol. Pl.*, **24,** 242–7. (320)
Armstrong, W. (1971b). Radial oxygen losses from intact rice roots as affected by distance from the apex, respiration and waterlogging. *Physiol. Pl.*, **25,** 192–7. (320, 322)
Armstrong, W. (1972). A re-examination of the functional significance of aerenchyma. *Physiol. Pl*, **27,** 173–77. (321)
Armstrong, W. (1979). Aeration in higher plants. In *Advances in Botanical Research*, Vol. 7, ed. Woolhouse, H. W., pp. 225–332, Academic Press, London. (292, 315, 321, 322, 326)
Armstrong, W., and D. J. Boatman (1967). Some field observations relating the growth of bog plants to conditions of soil aeration. *J. Ecol.*, **55,** 101–10. (293, 312, 313, 315, 327)
Armstrong, W., and T. J. Gaynard (1976). The critical oxygen pressures for respiration in intact plants. *Physiol. Pl.*, **37,** 200–6. (308, 318–9, 322)
Armstrong, W. and D. J. Read (1972). Some observations on oxygen transport in conifer seedlings. *New Phytol.*, **71,** 55–62
Armstrong, W., and E. J. Wright (1976). An electrical analogue to simulate the oxygen relations of roots in anaerobic media. *Physiol. Pl.*, **36,** 383–7. (321, 322)
Arnon, D. L., and C. M. Johnson (1942). Influence of hydrogen ion concentration on the growth of higher plants under controlled conditions. *Pl. Physiol., Lancaster*, **17,** 525–39. (265)
Ashby, E. (1937). Studies in the inheritence of physiological characteristics III. *Ann. Bot.*, **1,** 11–41. (138)
Ashenden, T. W. (1978). Drought avoidance in sand dune populations of *Dactylis glomerata* L. *J. Ecol.*, **66,** 943–51. (185, 186)
Ashenden, T. W., W. S. Stewart, and W. Williams (1975). Growth responses of sand dune populations of *Dactylis glomerata* L to different levels of water stress. *J. Ecol.*, **63,** 97–107. (139, 184)
Ashton, T. (1956). Effects of a series of cycles of alternating low and high soil water content on the rate of apparent photosynthesis in sugarcane. *Pl. Physiol., Lancaster*, **31,** 266–74 (171)
Atkins, J. G. (1958). *Rice diseases*. U.S. Dept. Agric. Farmers Bull. 2120. (311)
Atkinson, B. W., and P. A. Smithson (1976). Precipitation. In *The Climate of the British Isles*, ed. Chandler, T. H., and S. Gregory pp. 129–82, Longman, London. (225)
Avadhani, P. N., H. Greenway, R. Lefroy, and L. Prior (1978). Alcoholic fermentation and malate metabolism of rice germinating at low oxygen concentrations. *Aust. J. Pl. Physiol.*, **5,** 15–25. (327)
Baba, I., K. Inada, and K. Tajima (1964). Mineral nutrition and the occurrence of physiological diseases. In *The Mineral Nutrition of the Rice Plant*, ed. International Rice Research Instute pp. 173–95, John Hopkins, Baltimore. (311, 312)
Baba, I., I. Iwata, and K. Tajima (1963). Physiological injury. In *Theory and Practice of Growing Rice*, ed. Matsubayashi, M. pp. 149–72, Fuji, Tokyo. (311)
Babalola, O., L. Boersma, and C. T. Youngberg (1968). Photosynthesis and transpiration of Monterey pine seedlings as a function of soil water suction and soil temperature. *Pl. Physiol., Lancaster*, **43,** 515–21. (162)

Babel, V. (1975a). Micromorphology of soil organic matter. In *Soil Components*, Vol. 1, pp. 369–473, Springer, Berlin. (99)
Babel, V. (1975b) Distribution of dropping fabrics in Central European humus forms. In *Progress in Soil Zoology*, ed. Vanek, J. pp. 85–94, Junk, The Hague. (99)
Bacha, R. E., and L. R. Hossner (1977). Characteristics of coatings formed on rice roots as affected by iron and manganese additions. *Soil Sci. Soc. Amer. J.*, **41,** 931–5. (313)
Baker, D. A., and L. L. Hall (1975). *Ion Transport in Plant Cells and Tissues*, North Holland, Amsterdam. (262)
Balandreau, J., G. Rinaudo, F-H. Ibbitssam, and Y. Dommergues (1975). Nitrogen fixation in the rhizosphere of the rice plant. In *Nitrogen Fixation by Free-living Organisms* ed. Stewart, W. D. P., pp. 57–70, Cambridge Univ. Press, Cambridge.
Bange, G. G. J. (1953). On the quantitative explanation of stomatal transpiration. *Acta. bot. neerl.*, **2,** 255–97. (166)
Bannister, P. (1964a,b,c). The water relations of certain heath plants with reference to ecological amplitude. I. Introduction. Germination and establishment. II. Field studies. III. Experimental studies and general conclusions. *J. Ecol.*, **52,** 423–32, 481–97, 499–509. (187, 310)
Bannister, P. (1976). *Introduction to Physiological Plant Ecology*, Blackwell, Oxford. (425)
Barber, D. A. (1969). The influence of the microflora on the accumulation of ions by plants. In *Ecological Aspects of the Mineral Nutrition of Plants*, ed. Rorison, I. H., pp. 191–200, Blackwell, Oxford. (258)
Barber, D. A., M. Ebert, and N. T. S. Evans (1962). The movement of ^{15}O through barley and rice plants. *J. exp. Bot.*, **13,** 397–403. (314)
Barber, S. A., J. M. Walker, and E. H. Vasey (1963). Mechanisms for the movement of plant nutrients from the soil and fertilizer to the plant root. *J. agric. Fd. Chem.*, **11,** 204–7. (255)
Barclay-Estrup, P. (1971). The description and interpretation of cyclical processes in a heath community. III Microclimate in relation to the Calluna cycle. *J. Ecol.*, **59,** 143–66. (389)
Barley, K. P. (1961). The abundance of earthworms in agricultural soils and their possible significance in agriculture. *Agronomy*, **13,** 249–68. (99)
Barley, K. P. (1963). Influence of soil strength on the growth of roots. *Soil Sci.*, **96,** 175–86. ( )
Barrow, N. J., N. Malajczuk, and T. C. Shaw (1977). *New Phytol.*, **78**, 269–76. (257)
Barrs, H. D. (1968). Determination of water deficits in plant tissues. In *Water Deficits and Plant Growth*, ed. Kozlowski, T. T., pp. 236–68, Academic Press, New York. (152)
Bartholemew, B. (1970). Bare zone between Californian shrub and grassland communities: the role of animals. *Science*, **170,** 1210–12. (405)
Bartlett, R. J. (1961). Iron oxidation proximate to plant roots. *Soil Sci.*, **92,** 372–79. (314)
Baskin, J. M., and C. C. Baskin (1976). High temperature requirement for afterripening in seeds of winter annuals. *New Phytol.*, **77,** 619–29. (178, 221–2)
Bauer, H., W. Larcher, and R. B. Walker (1975). Influence of temperature stress on CO_2 gas exchange. In *Photosynthesis and Primary Production in Different Environments*, ed. Cooper, J. P., pp. 557–86, Cambridge Univ. Press, Cambridge. (205)
Bawden, M. G. (1965). A reconnaisance of the land resources of Eastern Bechuanaland. *J. app. Ecol.*, **2,** 357–65. (69)
Bazzaz, F. A. (1979). The physiological ecology of succession. *Ann. Rev. Ecol. Syst.*, **10,** 351–71. (109, 389, 393)
Bear, F. E. (1960). Symposium on bicarbonates. *Soil Sci.*, **89,** 241–302. (265)
Bear, F. E. (1964). *Chemistry of the Soil*, Reinhold, New York. (233, 240, 241, 245)
Beattie, A. J., D. E. Breedlove, and P. R. Erlich (1973). The ecology of the pollination of *Frasera speciosa*. *Ecology*, **54,** 81–91. (386)

Beckel, D. K. B. (1956). Cortical disintegration in the roots of *Bouteloua gracilis* (H.B.K.) Lag. *New Phytol.*, **55,** 183–90. (324)

Bedford, D. of, and S. U. Pickering (1919). *Science and Fruit Growing*, Macmillan, London. (405)

Bently, R. (1861). *A Manual of Botany*, Churchill, London. (422)

Berthelot, M. (1885). Fixation directe de l'azote atmospherique. *C.R. Acad. Sci.*, **101,** 775–84. (423)

Bewley, J. D. (1979). Physiological aspects of desiccation tolerance. *Ann. Rev. Pl. Physiol.*, **30,** 195–238. (175)

Bhat, K. K. S., and P. H. Nye (1973). Diffusion of phosphate to plant roots in soil 1. Quantitative autoradiography of the depletion zone. *Plant and Soil*, **38,** 161–75. (258)

Bierhuizen, J. F. (1976). Irrigation and water use efficiency. In *Water and Plant Life*, ed. Lange, O. L., L. Kappen, and E. D. Schultz, pp. 421–31, Springer, Berlin. (171)

Billings, W. D. (1974). Arctic and alpine vegetation: plant adaptation to cold summer climates. In *Arctic and Alpine Environments*, ed. Ives, J. D., and R. G. Barry, pp. 403–43, Methuen, London. (205)

Binet, P. (1978). *Aspects physiologiques de l'halophilie et de la resistance aux sels.* Soc. Bot. de France, Paris. (174, 285, 288)

Bishop, D. C., and M. L. Reed (1976). The C-4 pathway of photosynthesis: Ein Kranz-typ Wirtschaftswunder. *Photochem. Photobiol. Rev.*, **1,** 1–69. (131)

Bjorkman, O. (1966). The effect of oxygen concentration on photosynthesis in higher plants. *Physiol. Pl.*, **19,** 618–33. (133)

Bjorkman, O. (1968). Carboxydismutase activity in shade-adapted and sun-adapted species of higher plants. *Physiol. Pl.*, **21,** 1–10. (132)

Bjorkman, O. (1971). Comparative photosynthetic gas exchange in higher plants. In *Photosynthesis and Photo-Respiration*, ed. Hatch, M. D. *et al.*, pp. 18–32, Wiley, Chichester. (130)

Bjorkman, O. (1975). Environmental and biological control of photosynthesis. In *Environmental and Biological Control of Photosynthesis*, ed. Marcelle, R., pp. 1–51, Junk, The Hague. (131, 132, 204)

Bjorkman, O., N. K. Boardman, J. M. Anderson, S. W. Thorne, D. J. Goodchilde, and N. A. Pyliotis (1972). Effect of light intensity during growth of *Atriplex patula* on the capacity of photosynthetic reactions, chloroplast components and structure. *Carnegie Inst. Washington Yearbook*, 71, pp. 115–35. (134)

Bjorkman, O., E. Gauhl and M. A. Nobs (1970). Comparative studies of *Atriplex* spp. with and without B-carboxylation photosynthesis and their first generation hybrid. *Carnegie Inst. Washington Yearbook*, 68, pp. 620–33. (206)

Bjorkman, O., and M. M. Ludlow (1972). Characterization of the light climate of the floor of a Queensland rainforest. *Carnegie Inst. Washington Yearbook*, 71, pp. 85–94. (23)

Black, C. A. (1965). *Methods of Soil Analysis*, Vols. I and II, Am. Soc. Agron., Wisconsin. (88, 106, 110, 111, 155)

Black, C. A. (1968). *Soil–Plant Relationships* (2nd edn.), Wiley, New York. (91, 112, 241, 242, 244, 251)

Black, J. D. F., and D. W. West (1969). Solid state reduction at a platinum microelectrode in relation to measurement of oxygen flux in soil. *Aust. J. Soil Res.*, **7,** 67–72. (296)

Black, J. N. (1958). Competition between plants of different initial seed size in swards of subterranean clover (*Trifolium subterraneum*) with particular reference to leaf area and microclimate. *Aust. J. agric. Res.*, **9,** 299–318.

Black, J. N. (1964). An analysis of the potential production of swards of subterranean clover (*Trifolium subterraneum* L.) at Adelaide, South Australia. *J. app. Ecol.*, **1,** 3–18. (348, 349, 353)

Blackburn, W. H., and P. T. Tueller (1970). Pinyon and juniper in black sagebrush communities in east-central Nevada. *Ecology*, **51,** 841–8. (394)

Blackman, F. F. (1905). Optima and limiting factors. *Ann. Bot.*, **19,** 281–95. (423, 424)
Blackman, V. H. (1919). The compound interest law and plant growth. *Ann. Bot.*, **33,** 353–60. (137, 139, 424)
Blake, G. R. (1965). Bulk density. In *Methods of Soil Analysis*, ed. Black, C. A., pp. 374–90, *Amer. Soc. Agron.*, Wisconsin. (113)
Bloomfield, C. (1953a, b). A study of podsolization. I. The mobilization of iron and aluminium by cots Pine needles. II. The mobilization of iron and aluminium by the leaves and bark of *Agathis australis* (Kauri). *J. Soil Sci.*, **4,** 5–16, 17–23. (101)
Bloomfield, C. (1954a, b, c). A study of podsolization. III. The mobilization of iron and aluminium by Rima (*Dicradium cupressinum*). IV. The mobilization of iron and aluminium by picked and fallen larch needles. V. The mobilization of iron and aluminium by aspen and ash leaves. *J. Soil. Sci.*, **5,** 39–45, 46–49, 50–59. (101)
Bloomfield, C. (1965). Some processes of podsolization. In *Experimental Pedology*, ed. Hallsworth, E. G., and D. V. Crawford, pp. 257–66, Butterworth, London. (102, 247)
Boardman, N. K. (1977). Comparative photosynthesis of sun- and shade-plants. *Ann. Rev. Pl. Physiol.*, **28,** 355–(131, 132, 134, 136)
Boatman, D. J., and W. Armstrong (1968). A bog type in N. West Sutherland. *J. Ecol.*, **56,** 129–41.
Bodkin, P. C., D. H. N. Spence, and D. C. Weeks (1980). Photo-reversible control of heterophylly in *Hippuris vulgaris* L. *New Phytol.*, **84,** 533–42. (23, 224)
Boeke, J. E. (1940). On the origin of the intercellulary channels and cavities in the rice root. *Ann. Jardin Bot. Buitenzorg*, **50,** 199–208. (323)
Boggie, R. (1972). Effect of water table height on root development of *Pinus contorta* on deep peat in Scotland. *Oikos*, **23,** 304–12. (293, 327)
Bohm, W. (1979). *Methods of Studying Root Systems*, Springer, Berlin. (119, 120, 121)
Bohn, H. L., and B. C. McNeal and G. A. O'Connor (1979). *Soil Chemistry*. Wiley, Chichester. (96)
Bond, G. (1952). Some features of root growth in nodulated plants of *Myrica gale* L. *Ann. Bot.*, **16,** 467–75. (328)
Boodle, L. A. (1934). The scorching of foliage by sea-winds. *J. Min. Agric.*, **27,** 479–86. (229)
Bould, C. (1963). Mineral nutrition of plants in soils and culture media. Part 1. Mineral nutrition of plants in soils. In *Plant Physiology*, ed. Steward, F. C., Vol. 3, pp. 15–96, Academic Press, New York. (241, 242–3, 248)
Boulter, D., D. A. Coult, and G. G. Henshaw (1963). Some effects of gas concentrations on metabolism of the rhizome of *Irish pseudacorus* L. *Physiol. Pl.*, **16,** 541–48. (324, 325)
Bowen, H. J. M. (1979). *Environmental Chemistry of the Elements*, Academic Press, New York. (286, 360, 362, 364, 365, 366)
Boyer, J. S. (1976a). Photosynthesis at low water potentials. *Phil. Trans. R. Soc. Lond.* B, **273,** 501–12. (154, 173, 175)
Boyer, J. S. (1976b). Water deficits and photosynthesis. In *Water Deficits and Plant Growth*, ed. Kozlowski, T. T., Vol. 4, pp. 103–90, Academic Press, New York. (173)
Boyle, R. (1661). *The Sceptical Chymist*. London. (421)
Bradford, K. J., and D. R. Dilley (1978). The effects of root anaerobiosis on ethylene production, epinasty and growth of tomato plants. *Pl. Physiol., Lancaster*, **61,** 506–09. (309)
Bradshaw, A. D. (1969). An ecologists viewpoint. In *Ecological aspects of the Mineral Nutrition of Plants*, ed. Rorison, I. H., pp. 415–27, Blackwell, Oxford. (231, 271, 350, 400)
Bradshaw, A. D., and M. J. Chadwick (1980). *The Restoration of Land*, Blackwell, Oxford. (284, 371)
Bradshaw, A. D., M. J. Chadwick, D. Jowett, R. W. Lodge, and R. W. Snaydon (1960). Experimental investigations into the mineral nutrition of several grass species. III. Phosphate level. *J. Ecol.*, **48,** 631–7. (276)

Bradshaw, A. D., M. J. Chadwick, D. Jowett, and R. W. Snaydon (1964). Experimental investigation into the mineral nutrition of several grass species. IV. Nitrogen level. *J. Ecol.*, **52,** 665–76. (276)
Bradshaw, A. D., R. W. Lodge, D. Jowett, and M. J. Chadwick (1958). Experimental investigations into the mineral nutrition of several grass species. I. Calcium level. *J. Ecol.*, **46,** 749–57. (276)
Bradshaw, A. D., R. W. Lodge, D. Jowett and M. J. Chadwick (1960). Experimental investigations into the mineral nutrition of several grass species. II. Calcium and pH. *J. Ecol.*, **48,** 143–50. (276)
Bradshaw, A. D., T. S. McNeilly, and R. P. G. Gregory (1965). Industrialization, evolution and the development of heavy metal tolerance in plants. In *Ecology and the Industrial Society*, ed. Goodman, G. T., R. W. Edwards, and M. Lambert, pp. 327–43, Blackwell, Oxford. (282)
Brasier, C. M., and J. N. Gibbs (1978). Origin and development of the current Dutch elm disease epidemic. In *Plant Disease Epidemiology*, ed. Scott, P. R., and A. Bainbridge, Blackwell, Oxford. (387)
Braun-Blanquet, J. (1932). *Plant Sociology: the Study of Plant Communities.* McGraw-Hill, New York (translation). (420)
Bray, R. J., and J. T. Curtis (1957). An ordination of the upland forest communities of southern Wisconsin. *Ecol. Monogr.*, **27,** 325–49. (420)
Bremner, J. M., and K. Shaw (1958). Denitrification in soil. *J. agric. Sci.*, **51,** 22–52. (300)
Bremner, J. M. (1967). Nitrogenous compounds. In *Soil Biochemistry*, Vol. 1, ed. McLaren, A. D., and G. H. Petersen, pp. 19–66, Dekker, New York. (233)
Brenchley, W. E. (revised by K. Warrington) (1958). *The Park Grass Plots at Rothamsted Experimental Station 1856–1949. Rothamsted Expt. Sta.*, Harpenden. (272)
Bridges, E. M. (1978). *World Soils*, 2nd edn., Cambridge Univ. Press, Cambridge. (82)
Briggs, L. J. (1897). The mechanics of soil moisture. *U.S. Dept. Agric. Soils Bull.*, 10. (157)
Bright, D. N. E. (1928). The effects of exposure upon the structure of certain heath plants. *J. Ecol.*, **16,** 323–65. (228)
Brix, H. (1962). The effect of water stress on the rates of photosynthesis and respiration in Tomato plants and Loblolly Pine. *Physiol. Pl.*, **15,** 10–20. (171, 173)
Brooks, J., and G. Shaw (1973). *Origin and Development of Living Systems*, Academic Press, New York. (360, 362)
Brooks, J., P. R. Grant *et al.* (1971). *Sporopollenin*, Academic Press, New York. (105)
Brougham, R. W., and W. Harris (1967). Rapidity and extent of changes in geno typestructure induced by grazing in a ryegrass population. *N.Z. J. agric. Res.*, **10,** 56–65. (378)
Brown, H. (1970). Human materials production as a process in the biosphere. *Sci. Amer.*, **223** (3), 194–208. (350)
Brown, H. T., and F. Escombe (1900). Static diffusion of gases and liquids in relation to the assimilation of carbon and translocation in plants. *Phil. Trans. R. Soc. Lond.* B., **193,** 223–91. (423)
Brown, H. T., and F. Escombe (1905). Researches on the physiological processes of green plants with especial references to the interchange of energy between the leaf and its surroundings. *Proc. R. Soc. Lond.* B., **76,** 29–137. (15, 16, 423)
Brown, J. C. (1961). Iron chlorosis in plants. *Adv. Agron.*, **13,** 329–69. (268)
Brown, J. C., and J. E. Ambler (1970). Further characteristics of iron uptake in two genotypes of corn. *Soil Sci. Soc. Amer. Proc.*, **34,** 249–52. (268)
Brown, J. M. A., H. A. Outred, and C. F. Hill (1969). Respiratory metabolism in mangrove seedlings. *Pl. Physiol., Lancaster*, **44,** 287–94. (324)
Bryan, R. B. (1971). The efficacy of aggregation indices in the comparison of some English and Canadian soils. *J. Soil Sci.*, **22,** 166–78. (114, 117)
Buckingham, E. (1907). Studies on the movement of soil moisture. *U.S. Dept. Agr. Bur. Soils Bull.*, **38**. (158)

Budyko, M. I. (1974). *Climate and Life*. Academic Press, New York. (225, 355)

Bunnel, F. C., and K. A. Scoular (1975). ABISKO II A computer simulation model of carbon flux in tundra ecosystems. In *Structure and Function of Tundra Ecosystems*, ed. Rosswall, T., and O. W. Heat, pp. 425–48. Swedish Nat. Sci. Res. Council, Stockholm. (145)

Buol, S. W., F. D. Hole, and R. J. McCracken (1973). *Soil Genesis and Classification*, Iowa State Univ. Press, Ames. (52, 78)

Bunting, B. T. (1965). *The Geography of the Soil*, Hutchinson, London. (59, 66)

Burg, S. P., and E. A. Burg (1965). Ethylene action and the ripening of fruits. *Science*, **148,** 1190–6. (312)

Burges, A. (1967). The decomposition of organic matter in the soil. In *Soil Biology*, ed. Burges, A., and F. Raw, pp. 479–92, Academic Press, London. (110)

Burgess, P. S. (1922). The soil solution extracted by Lipmann's pressure method compared with 1:5 water extracts. *Soil Sci.*, **14,** 191–216. (254)

Burgstein, A. (1904). *Die Transpiration der Pflanzen*, Fischer, Jena. (423)

Burke, M. J., L. V. Gusta, H. A. Quamm, C. J. Weisser, and C. H. Li (1976). Freezing injury in plants. *Ann. Rev. Pl. Physiol.*, **27,** 507–28. (209)

Burnham, C. P. (1975). The forest environment: soils. In *Tropical Rainforests of the Far East* by T. C. Whitmore, pp. 103–20, Clarendon, Oxford. (68, 70, 72)

Burns, R. C., and R. W. F. Hardy (1975). *Nitrogen Fixation by Bacteria and Higher Plants*, Springer, Berlin. (240)

Burr, G. O., C. E. Hart *et al.* (1957). The sugar cane plant. *Ann. Rev. Pl. Physiol.*, **8,** 275–308. (353)

Burris, R. H. (1975). The acetylene reduction technique. In *Nitrogen Fixation by Free-Living Organisms*, ed. Stewart, W. D. P., pp. 249–57, Cambridge Univ. Press, Cambridge. (234)

Burton, L. D., and P. S. Liss (1976). *Estuarine Chemistry*, Academic Press, New York. (371)

Businger, J. A. (1975). Aerodynamics of vegetation surfaces. In *Heat and Mass Transfer in the Biosphere*, ed. de Vries, D. A., and N. H. Afgan, Scripta Book Co., Washington DC. (25, 26)

Caldwell, M. M., and O. A. Fernandez (1975). Dynamics of Great Basin shrub root systems. In *Environmental Physiology of Desert Organisms*, ed. Hadley, N. F., pp. 38–51, Dowden, Hutchinson and Ross, Stroudsburg. (343)

Callow, P. (1977). Ecology, evolution and energy: a study in metabolic adaptation. *Adv. Ecol. Res.*, **10,** 1–62. (334)

Calvin, M., and A. A. Benson (1948). The path of carbon in photosynthesis. *Science*, **107,** 476–80. (127)

Campbell, N. E. R., and H. Lees (1967). The nitrogen cycle. In *Soil Biochemistry*, ed. Mclaren, A. D., and G. H. Petersen, Vol. 1, pp. 194–215, Dekker, New York. (43, 235, 237)

Cannel, R. Q. (1977). Soil aeration and compaction in relation to root growth and soil management. In *Applied Biology* Vol. 2, ed. Coaker, T. H., pp. 1–85, Academic Press, New York. (307)

Cannon, H. L. (1960). Botanical prospecting for ore bodies. *Science*, **132,** 591–8. (283)

Cannon, H. L. (1971). The use of plant indicators in ground-water surveys, geologic mapping and mineral prospecting. *Taxon*, **20,** 227–56. (250, 284, 285, 286)

Cannon, W. A., and E. E. Free (1925). Physiological features of roots with especial reference to the relation of roots to the aeration of the soil. *Carnegie Inst. Wash. Publ.*, **368,** 1–168. (314)

Carr, N. G., and B. A. Whitton (1973). *The Biology of the Blue-Green Algae*, Blackwell, Oxford. (236)

Carrol, J. C. (1943). Effects of drought, temperature and nitrogen on turf grasses. *Pl. Physiol., Lancaster*, **18,** 19–36. (188)

Cates, R. G., and G. H. Orians (1975). Successional status and the palatability of plants to generalized herbivores. *Ecology*, **56,** 410–8. (379)

Causton, D. R., and J. C. Venus (1981). *The Biometry of Plant Growth*, Arnold, London. (139)

Cavalieri, A. J., and Huang, H. C. (1979). Evaluation of proline accumulation in the adaptation of diverse species of marsh halophytes to the saline environment. *Amer. J. Bot.*, **66,** 307–12. (289)

Chadwick, M. J., and G. T. Goodman (1975). *The Ecology of Resource Renewal and Degradation*, Blackwell, Oxford. (371)

Chamberlain, A. C. (1975). The movement of particles in plant communities. In *Vegetation and the Atmosphere*, Vol. 1, ed. Monteith, J. L., pp. 155–203, Academic Press, New York. (228)

Chandler, G. E., and J. W. Anderson (1976a, b, c). Studies of the nutrition and growth of *Drosera* spp. with reference to the carnivorous habit; Studies on the origin of some hydrolytic enzymes associated with the leaves and tentacles of *Drosera* spp.; Uptake and metabolism of insect metabolites by leaves and tentacles of *Drosera* spp. *New Phytol.*, **76,** 129–41; **77,** 51–62; 625–34. (274)

Chandrasekaran, S., and T. Yoshida (1973). Effect of organic acid transformations in submerged soils on growth of the rice plant. *Soil Sci. Pl. Nutr.*, **19,** 635–9. (306)

Chapman, H. D., and P. F. Pratt (1961). *Methods of Analysis for Soils, Plants and Water.* University of California Press, Los Angeles. (88)

Chapman, S. B. (1976). *Methods in Plant Ecology*, Blackwell, Oxford. (36, 37, 38)

Chapman, V. J. (1976). *Coastal Vegetation* 2nd edn., Pergamon, London. (82)

Chapman, V. J. (1977). *Wet Coastal Ecosystems*, Elsevier, Amsterdam. (82).

Chen, C. C., J. B. Dixon, and F. T. Turner (1980). Iron coatings on rice roots. Mineralogy and quantity influencing factors. *Soil Sci. Soc. Amer. J.*, **44,** 635–9. (313)

Chew, R. M., and A. E. Chew (1965). The primary productivity of a desert shrub (*Larrea tridentata*) community. *Ecol. Monogr.*, **35,** 355–75. (192)

Chilvers, G. A., and J. L. Harley (1980). Visualization of phosphate accumulation in beech mycorrhizas. *New Phytol.*, **84,** 319–36. (241, 257)

Chirkova, T. V. (1968). Oxygen supply to roots of certain woody plants kept under anaerobic conditions. *Soviet Plant Physiol.*, **15,** 475–8. (324)

Clapham, A. R. (1969). Introduction. In *Ecological Aspects of the Mineral Nutrition of Plants*, ed. Rorison, I. H., Blackwell, Oxford. (230)

Clapham, A. R., T. G. Tuttin, and E. F. Warburg (1962). *Flora of the British Isles*, Cambridge Univ. Press, Cambridge. (218)

Clark, J. (1961). Photosynthesis and respiration in White Spruce and Balsam Fir. *Syracuse University State Univ. Coll. For. Tech. Pub.*, **85**. (171)

Clark, N. M. (1923). *The Determination of Hydrogen Ions*, Williams and Wilkins, Baltimore. (423)

Clarkson, D. T. (1965). Calcium uptake by calcicole and calcifuge species in the genus *Agrostis*. *J. Ecol.*, **53,** 427–35. (264)

Clarkson, D. T. (1966). Aluminium tolerance in species within the genus *Agrostis* L. *J. Ecol.*, **54,** 167–78. (279)

Clarkson, D. T. (1967). Phosphorus supply and growth rate in species of *Agrostis* L. *J. Ecol.*, **55,** 111–118.

Clarkson, D. T. (1969). Metabolic aspects of aluminium toxicity and some possible mechanisms for resistance. In *Ecological Aspects of the Mineral Nutrition of Plants*, ed. Rorison, I. H., pp. 381–97, Blackwell, Oxford. (267)

Clarkson, D. T. (1974). *Ion Transport and Cell Structure in Plants*, McGraw-Hill, London. (257, 258, 259, 261, 274, 277)

Clarkson, D. T., and J. B. Hanson (1980). The mineral nutrition of higher plants. *Ann. Rev. Pl. Physiol.*, **31**; 239–98. (232, 233, 259, 277)

Clarkson, D. T., and J. Sanderson (1969). The uptake of polyvalent cation and its distribution in the root apices of *Allium cepa*: tracer and autoradiographic studies. *Planta*, **89,** 136–54. (267)

Clausen, J., D. D. Keck, and W. M. Heisey (1948). Experimental studies on the nature of species. I Environmental responses of climatic races of *Achillea.Carnegie Inst. Wash. Publ.*, **581**. (217)

Clement, C. R., M. J. Hopper, and L. H. P. Jones (1978). The uptake of nitrate by *Lolium perenne* from flowing nutrient solution. *J. exp. Bot.*, **29,** 453–64. (253)

Clements, F. E. (1905). *Research Methods in Ecology*, Nebraska Univ. Publ. Co., Lincoln. (1, 419, 423)

Clements, F. E. (1907). *Plant Physiology and Ecology*, Holt, New York. (395, 423)

Clements, F. E. (1916). *Plant Succession*, Carnegie Inst., Washington. (52, 200, 393, 419, 423)

Clements, F. E. (1920) *Plant Indicators*, Carnegie Inst., Washington. (420, 423)

Clements, F. E. (1928). *Plant Succession and Indicators*, Wilson, New York. (418, 423)

Clements, F. E. (1936). Nature and structure of the climax. *J. Ecol.*, **24,** 252–82. (52, 419)

Clements, F. E., J. E. Weaver, and H. C. Hanson (1929). *Plant Competition*, Carnegie Inst., Washington. (395, 409, 420, 423)

Clowes, F. A. L. (1961). *Apical Meristems*, Blackwell, Oxford. (207)

Clymo, R. S. (1962). An experimental approach to part of the calcicole problem. *J. Ecol.*, **50,** 701–31. (265)

Cody, M. L., and H. A. Mooney (1978). Convergence versus non-convergence in mediterranean climate ecosystems. *Ann. Rev. Ecol. Syst.*, **9**, 265–321. (66)

Coffey, G. N. (1912). A Study of the Soils of the United States. *U.S. Dept. Agr. Bur. Soils Bull.*, **85**. (419)

Coleman, E. A. (1946). A laboratory study of lysimeter drainage under controlled conditions of moisture tension. *Soil Sci.*, **62,** 365–82. (197)

Colinveaux, P. A. (1973). *Introduction to Ecology*, Wiley, Chichester. (337, 421)

Colwell, W. E., and R. W. Cummings (1944). Chemical and biological studies on aqueous solutions of boric acid and of calcium, sodium and potassium metaborate. *Soil Sci.*, **57,** 37–49. (248)

Connel, J. H., and R. O. Slatyer (1977). Mechanisms of succession in natural communities and their role in community stability and organization. *Amer. Nat.*, **111,** 1119–44. (388, 390)

Connell, W. E., and W. H. Patrick (1968). Sulphate reduction in soil: effects of redox potential and pH. *Science*, **159,** 86–7. (304)

Cook, C. W. (1943). A study of the roots of *Bromus inermis* in relation to drought resistance, *Ecology*, **24,** 169–82. (185)

Cooper, J. P. (1965). Climatic adaptation of local varieties of forage grass. In *Biological Significance of Climatic Changes in Britain*, ed. Johnson, C. G., and L. P. Smith, pp. 169–79, Academic Press, New York. (215)

Cooper, J. P. (1975). Control of photosynthetic production in terrestrial systems. In *Photosynthesis and Primary Production in Different Environments*, ed. Cooper, J. P., pp. 593–621, Cambridge Univ. Press, Cambridge. (146, 207, 338)

Cooper, J. P., and J. R. McWilliam (1966). Climatic variation in forage grasses II Germination, flowering and leaf development in mediterranean populations of *Phalaris tuberosa*. *J. app. Ecol.*, **3,** 191–212. (220)

Cosgrove, D. J. (1967). Metabolism of organic phosphates in soil. In *Soil Biochemistry*, ed. McLaren, A. D., and G. H. Petersen, Vol. 1, pp. 216–28, Dekker, New York. (43, 105, 241)

Coughlan, S. J., and R. H. Al-Hasan (1977). Studies on uptake and turnover of glycollic acid in the Menai Straits, North Wales. *J. Ecol.*, **65**; 1–46. (344)

Coughtrey, P. J., and M. H. Martin (1976). The distribution of copper, zinc, cadmium and lead within the pulmonate mollusc *Helix aspersa* Mull. *Oecologia*, **23**, 315–22. (370)
Coutts, M. P., and J. J. Philipson (1978). Tolerance of tree roots to waterlogging II. Adaptation of Sitka Spruce and Lodgepole Pine to waterlogged soil. *New Phytol.*, **80**, 71–7. (329)
Coutts, M. P., and W. Armstrong (1976). Role of oxygen transport in the tolerance of trees to waterlogging. In *Tree Physiology and Yield Improvement*, ed. Cannel, M. G. R., and F. T. Last, pp. 361–85, Academic Press, London. (327)
Cowan, I. R., and F. L. Milthorpe (1968). Plant factors influencing the water status of plant tissue. In *Water Deficits and Plant Growth*, Vol. 1, ed. Kozlowski, T. T., pp. 137–93, Academic Press, New York. (116, 147, 165, 184)
Cowles, H. C. (1899). The ecological relationships of the vegetation on sand dunes of Lake Michigan. *Bot. Gaz.*, **27**, 95–117; 167–202; 281–308; 361–91. (419)
Cowles, H. C. (1911). The cause of vegetation cycles. *Bot. Gaz.*, **51**, 161–83.
Cox, G., K. J. Moran, F. Sanders, C. Nockolds, and P. B. Tinker (1980). Translocation and transfer of nutrients in vesicular-arbuscular mycorrhizas III. *New Phytol.*, **84**, 649–59. (257)
Crafts, A. S. (1968), Water deficits and physiological processes. In *Water Deficits and Plant Growth*, Vol. 2, ed. Kozlowski, T. T., pp. 85–133. Academic Press, New York. (168)
Cram, W. J. (1976). Negative feedback regulation of transport in cells. In *Encyclopaedia of Plant Physiology* Vol. 2 *Transport in Plants*, ed. Lutge, U., and M. G. Pitman, pp. 284–316. Springer, Berlin. (263)
Crampton, C. B. (1963). The development and morphology of iron pan podsols in Mid- and South Wales. *J. Soil Sci.*, **14**, 282–302. (94)
Crampton, C. B. (1968–70). The evolution of soils in Morgannwg. *Proc. Cardiff Nat. Soc.*, **95**, 41–52. (78)
Crawford, R. M. M. (1966). The control of anaerobic respiration as a determining factor in the distribution of the genus *Senecio*. *J. Ecol.*, **54**, 403–13. (324)
Crawford, R. M. M. (1969). The physiological basis of flooding tolerance. *Ber. dt. bot. Ges.*, **82**, 111–14. (324)
Crawford, R. M. M. (1971). Some metabolic aspects of ecology. *Trans. Bot. Soc. Edinb.*, **41**, 309–22. (324)
Crawford, R. M. M., and M. McManmon (1968). Inductive responses of alcohol and malic dehydrogenases in relation to flooding tolerance in roots. *J. Exp. Bot.*, **19**, 435–41. (324)
Crisp, D. T. (1966). Input and output of minerals for an area of Pennine moorland: the importance of precipitation, drainage, peat erosion and animals. *J. app. Ecol.*, **3**, 327–48. (374)
Crocker, R. L., and B. A. Dickinson (1957). Soil development on the recessional moraines of the Herbert and Mendenhall glaciers, south eastern Alaska. *J. Ecol.*, **45**, 169–85. (59)
Crompton, E. (1956). The environmental and pedological relationships of peaty gleyed soils. *Trans. 6th Internat. Congr. Soil Sci.*, **6**, 155–61. (75)
Cronin, L. E. (1975). *Estuarine Research*, Vol. 1, Academic Press, New York. (371)
Crowther, E. M. (1953). The sceptical soil chemist. *J. Soil Sci.*, **4**, 107–22. (2, 417)
Currie, J. A. (1961). Gaseous diffusion in the aeration of aggregated soils. *Soil Sci.*, **92**, 40–45. (292)
Currie, J. A. (1962). The importance of aeration in providing the right conditions for plant growth. *J. Sci. Food Agric.*, **13**, 380–5. (293, 294).
Currie, J. A. (1965). Diffusion within soil microstructure: a structural parameter for soils. *J. Soil Sci.*, **16**, 279–89. (293)

Curtis, J. T., and R. P. McIntosh (1951). An upland forest continuum in the prairie-forest border region of Wisconsin. *Ecology*, **32,** 476–96. (420)

Dainty, J. (1962). Ion transport and electrical potentials in plant cells. *Ann. Rev. Pl. Physiol.*, **13,** 379–402.

Danserau, P. (1957). *Biogeography: an Ecological Perspective*, Ronald, New York. (74)

Darwin, C. (1859). *Origin of Species*, Murray, London. (2, 418)

Darwin, C. (1881). *The Formation of Vegetable Mould Through the Action of Worms*, Murray, London. (102)

Darwin, F. (1898). Observations on stomata. *Phil. Trans. R. Soc. Lond.*, **190,** 531–621. (423)

Darwin, F., and E. H. Acton (1894). *Practical Physiology of Plants*, Cambridge Univ. Press, Cambridge. (423)

Daubenmire, R. F. (1974). *Plant Communities*, Harper and Row, New York. (425)

Davidson, D. T. (1965). Penetrometer measurements. In *Methods of Soil Analysis*, ed. Black, C. A., pp. 472–84. Amer. Soc. Agron., Madison. (114)

Davies, B. E. (1980). *Applied Soil Trace Elements*, Wiley, Chichester. (37)

Davies, C. H. (1940). Absorption of water by maize roots. *Bot. Gaz.*, **101,** 791–805. (162)

Davies, D. D., N. H. Nascimento, and K. D. Patil (1974). The distribution and properties of NADP malic enzyme in flowering plants. *Phytochem.*, **13,** 2417–25. (325)

Davies, M. S. (1975). Physiological differences among populations of *Anthoxanthum odoratum* L. collected from the Park Grass experiment, Rothamsted. IV Response to potassium and magnesium. *J. app. Ecol.*, **12,** 953–64. (277)

Davies, M. S., and R. W. Snaydon (1973a). Physiological differences among populations of *Anthoxanthum oderatum* L. collected from the Park Grass Experiment, Rothamsted. I. Response to calcium *J. app. Ecol.*, **10,** 33–45. (277)

Davies, M. S., and R. W. Snaydon (1973b). Physiological differences among populations of *Anthoxanthum odoratum* L. collected from the Park Grass experiment, Rothamsted. II. Response to aluminium. *J. app. Ecol.*, **10,** 47–55. (277, 284).

Davies, M. S., and R. W. Snaydon (1974). Physiological differences among populations of *Anthoxanthum odoratum* L. collected from the Park Grass Experiment, Rothamsted. III. Response to phosphate, *J. app. Ecol.*, **11**, 699–707. (277)

Davy, A. J., and K. Taylor (1974). Seasonal pattern of nitrogen availability in contrasting soils in the Chiltern Hills. *J. Ecol.*, **62,** 793–807.

Day, J., D. Harris, P. J. Dart, and P. van Berkum (1975). The Broadbalk experiment. An investigation of nitrogen gains from non-symbiotic nitrogen fixation. In *Nitrogen Fixation by Free-Living Organisms*, ed. Stewart, W. D. P., pp. 71–84, Cambridge Univ. Press, Cambridge. (235)

Day, P. R. (1965). Particle fractionation and particle-size analysis. In *Methods of Soil Analysis*, ed. Black, C. A., pp. 545–67, Am. Soc. Agron., Madison. (111)

Deb, B. C. (1949). The movement or precipitation of iron oxide in podzol soils. *J. Soil Sci.*, **1,** 112–22. (101)

De Candolle, A. P. (1820). *Essai elementaire de geographie botanique.* Cited in *Plant Competition*, Carnegie Inst., Washington. (395)

Deely, D. J., and F. Y. Borden (1973). High surface temperatures on strip-mine spoils. In *Ecology and Reclamation of Devastated Land*, ed. Hutnik, R. J., and G. Davies, pp. 69–79, Gordon and Breach, New York. (15, 20)

Delwiche, C. C. (1970). The nitrogen cycle. *Sci. Amer.* **223** (3), 136–46. (351, 364)

Denayer de Smet, S. (1974). Premier aperçu de la distribution du cadmium dans divers écosystèms terrestres non-pollués et pollués. *Oecol. Plant.*, **9,** 169–82. (286)

Denny, C. S., and J. C. Goodlet (1956). Microrelief resulting from fallen trees. *U.S. Geol. Surv. Professional Paper*, **288,** 59–68. (228)

Department of the Environment (1976). *Effects of Airborne Sulphur Compounds on Forests and Freshwaters*, Poll. Paper 7, HMSO, London. (284)

Diamond, J. M., and M. L. Cody (1975). *Ecology and Evolution of Communities*, Belknap Press, Harvard. (392)

Dickinson, C. H., and G. J. F. Pugh (1974). *Biology of Plant Litter Decomposition*, Academic Press, London. (99, 207)

Dickson, R. E., and T. C. Broyer (1972). Effects of aeration, water supply and nitrogen source on growth and development of Tupelo Gum and Bald Cypress. *Ecology*, **53,** 626–34. (327)

Dimbleby, G. W. (1953). Natural regeneration of pine and birch on the heather moors of North East Yorkshire. *Forestry*, **26,** 41–52. (277)

Dimbleby, G. W. (1961). Soil pollen analysis *J. Soil Sci.*, **12,** 1–11. (105)

Dimbleby, G. W. (1962). *The Development of the British Heathlands and their Soils*, Clarendon, Oxford. (47, 58, 59, 104, 197)

Dimbleby, G. W. (1965). Postglacial changes in soil profiles. *Proc. R. Soc. Lond.* B, **161,** 355–62. (104)

Dimbleby, G. W. (1975). Archaeological evidence of environmental change. *Nature, Lond.*, **256,** 265–7. (104)

Dimbleby, G. W. (1976). Climate, soil and man. *Phil. Trans. R. Soc. Lond.* B., **275,** 197–208. (104)

Dixon, A. F. G. (1971a and b). The role of aphids in wood formation I and II. *J. app. Ecol.*, **8,** 165–79; 383–99. (381)

Dixon, A. H. (1914). *Transpiration and Ascent of Sapin Plants*, Macmillan, London. (163)

Dixon, H. H., and J. Joly (1894). On the ascent of sap. *Proc. R. Soc. Lond.*, **57,** 3–5. (163, 423)

Dobereiner, J., R. H. Burris, and A. Hollaender (1978). *Limitation and Potentials for Biological Nitrogen Fixation in the Tropics*, Plenum, New York. (240)

Dobereiner, J., and J. M. Day (1975). Nitrogen fixation in the rhizosphere of tropical grasses. In *Nitrogen Fixation by Free-Living Organisms*, ed. Stewart, W. D. P., Cambridge Univ. Press, Cambridge. (235)

Doi, Y. (1952). Studies on the oxidizing power of roots of crop plants. I. The difference with species of crop plant and wild grass. *Proc. Crop. Sci. Soc. Japan*, **21**, 12–13. (314)

Dokuchayev (1900). *Pektsie o Pohovedenie.* In *Collected Works*, Vol. 7, pp. 257–96, Moscow 1955. (50, 419)

Donald, C. M. (1958). The interaction of competition for light and for nutrients. *Aust. J. agr. Res.*, **9,** 421–35. (401)

Donald, C. M. (1963). Competition among crop and pasture plants. *Adv. Agron.*, **15,** 1–117. (398, 401)

Drew, M. C. (1979). Root development and activity. In *Arid-Land Ecosystems, Structure and Function*, ed. Goodall, D. W., and R. A. Perry, pp. 573–606, Cambridge Univ. Press, Cambridge. (179)

Drew, M. C. (1979). Plant responses to anaerobic conditions in soil and solution culture. *Current Advances in Plant Science*, Commentary 36, pp. 14. (307, 308, 309, 312)

Drew, M. C., E. J. Sisworo, and L. R. Saker (1979). Alleviation of waterlogging damage to young barley plants by application of nitrate and synthetic cytokinin and comparison between the effects of waterlogging, nitrogen deficiency and root excision. *New Phytol.*, **82,** 315–29. (308, 309, 324)

Drew, M. C., and L. R. Saker (1978). Nutrient supply and growth of the seminal root system in barley. III. Compensatory increases and growth of lateral roots and rates of phosphate uptake in response to localized supply of phosphate. *J. exp. Bot.*, **29,** 435–51. (277)

Drew, M. C., M. B. Jackson, and S. Gifford (1979). Ethylene promoted adventitious rooting and development of cortical air spaces (aerenchyma) in roots may be adaptive responses to flooding in Zea mays L. *Planta* **147,** 83–8. (324)

Drift, J. van der (1965). The effect of animal activity in the litter layer. In *Experimental Pedology*, pp. 277–85, Butterworths, Oxford. (98)

Dring, M. J. (1967) Phytochrome in red alga *Porphyra tenella. Nature, Lond.*, **215,** 1411–2. (23)

Drury, W. H., and I. C. T. Nisbett (1973). Succession. *J. Arnold Arboretum*, **54**, 331–53. (388, 391)

Dunn, E. L., F. M. Shropshire, L. C. Song, and H. A. Mooney (1976). The water factor and convergent evolution in mediterranean type vegetation. In *Water and Plant Life*, ed. Lange, O. L., L. Kappen and E. D. Schultz, pp. 492–505. Springer, Berlin. (202)

Duvigneaud, P. (1971). *Productivity of Forest Ecosystems*, UNESCO, Paris. (345)

Duvigneaud, P., and S. Denaeyer-de-Smet (1970). Biological cycling of minerals in temperate deciduous forests. In *Analysis of Temperate Forest Ecosystems*, ed. Reichle, D., pp. 199–225, Chapman and Hall, London. (373)

Dryssen, D. (1972). The changing chemistry of the oceans. *Ambio*, **1,** 21–5. (229)

Eagles, C. F. (1967). Apparent photosynthesis and respiration in populations of *Lolium perenne* from contrasting climatic regions. *Nature, Lond.*, **215,** 100–1. (205, 206)

Eagles, C. F., and O. Ostgard (1971). Variation in growth and development in natural populations of *Dactylis glomerata* from Norway and Portugal. I. Growth analysis. II. Leaf development and tillering. *J. app. Ecol.*, **8,** 367–81; 383–90. (207)

Edlefsen, N. E. (1941). Some thermodynamic aspects of the use of soil moisture by plants. *Trans. Am. geophys Un.*, **22,** 917–40. (149)

Edmondson, W. T., and G. G. Winberg (1971). *A Manual of Methods for the Assessment of Secondary Production in Freshwater*, Blackwell, Oxford. (345)

Edwards, C. A., D. E. Reichle, and D. A. Crosby (1970). The role of soil invertebrates in the turnover of soil organic matter. In *Analysis of Temperate Forest Ecosystems*, ed. Reichle, D. E., pp. 147–72, Chapman and Hall, London. (335)

Edwards, D. P., and G. C. Evans (1975). Problems involved in the design of apparatus for measuring the spectral composition of daylight in the field. In *Light as an Ecological Factor II*, pp. 161–87, Blackwell, Oxford. (35)

Effer, W. R., and S. L. Ranson (1967). Respiratory metabolism in Buckwheat seedlings. *Pl. Physiol. Lancaster*, **42,** 1042–52. (325)

Ehlers, W. (1975). Observations on earthworm channels and infiltration on tilled and untilled loess soil. *Soil Sci.*, **119,** 242–9. (116)

Ehlig, C. F., and W. R. Gardner (1964). Relationship between the transpiration and internal water balance of plants. *Agron. J.*, **56,** 127–30. (157)

Ehrlich, H. L. (1971). Biochemistry of the minor elements in soil. In *Soil Biochemistry*, Vol. 2, ed. McLaren, A. D., and J. Skujins, pp. 361–84, Dekker, New York. (43, 247, 248)

Ellenberg, H. (1963). *Vegetation Mittleeuropas mit den Alpen*, Eugene Ulmer, Stuttgart. (74)

Emerson, W. W., R. D. Bond, and A. R. Dexter (1978). *Modification of Soil Structure*, Wiley, Chichester. (114, 118)

Encyclopedia Britannica (1970). William Benton, Chicago. (417)

Engler, R. M., and W. H. Fitzpatrick, Jr. (1975). Stability of sulphides of manganese, iron, zinc, copper and mercury in flooded soil. *Soil Sci.*, **119,** 21 7–21. (314)

Engstrom, A. (1971). *Air Pollution across National Boundaries: the Impact on the Environment of Sulfur in Air and Precipitation*, Rept. Swedish Preparation Comm. UN Conf. Human Environment, Norstedt et Soner, Stockholm. (229)

Eplee, R. E. (1975). Ethylene: a witchweed germination stimulant. *Weed Sci.*, **24,** 434–6.

Epstein, E. (1972). *Mineral Nutrition of Plants: Principles and Perspectives*, Wiley, Chichester. (254, 259, 261, 289)

Epstein, E., R. W. Kingsbury, J. D. Norlyn, and D. W. Rush (1979). Production of foodcrops and other biomass by seawater culture. In *The Biosaline Concept*, ed. Aller, J. C., E. Epstein *et al.*, pp. 77–79, Plenum, New York. (289)

Erdtman, G. (1937). Pollen grains recovered from the atmosphere over the Atlantic. *Meddel. Goteborgs Bot. Tragard*, **12,** 185–96. (228)
Erlich, P. R., and P. H. Raven (1964). Butterflies and plants: a study in coevolution. *Evolution*, **18,** 586–608. (377)
Etherington, J. R. (1962). The growth of *Alopercurus pratensis* L and *Agrostis tenuis* Sibth in relation to soil moisture conditions. Ph.D. Thesis, University of London. (160, 161, 162, 173)
Etherington, J. R. (1967). Soil water and the growth of grasses. II. Effects of soil water potential on growth and photosynthesis of *Alopecurus pratensis. J. Ecol.*, **55,** 373–80. (160, 172, 185)
Etherington, J. R. (1978). *Plant Physiological Ecology*, Arnold, London. (21, 359)
Etherington, J. R. (1981). Limestone heaths in S.W. Britain: their soils and the maintenance of their calcicole-calcifuge mixtures. *J. Ecol.*, **69,** 277–94. (262)
Etherington, J. R., and A. J. Rutter (1964). Soil water and the growth of grasses. I. The interaction of watertable depth and irrigation amount on the growth of *Agrostistenuis* and *Alopecurus pratensis. J. Ecol.*, **52,** 677–89. (196)
Evans, G. C. (1972). *The Quantitative Analysis of Plant Growth*, Blackwell, Oxford. (131, 139, 207, 400, 424)
Evans, G. C. (1976). A sack of uncut diamonds: a study of ecosystems and the future of mankind. *J. Ecol.*, **64,** 1–39. (384, 385, 395)
Evans, N. T. S., and M. Ebert (1960). Radioactive oxygen in the study of gas transport down the root of *Vicia faba. J. exp. Bot.*, **11,** 246–57. (314)
Fairbridge, R. W., and C. W. Finkle (1979). *The Encyclopedia of Soil Science* Part 1, Dowden, Hutchinson and Ross, Stroudsburg. (263)
Farnsworth, F. B., and E. G. Golley (1974). *Fragile Ecosystems*, Springer, Berlin. (222)
Feeny, P. (1970). Seasonal changes in oak leaf tannins and nutrients as a cause of spring feeding by Winter Moth caterpillars. *Ecology*, **51,** 565–81. (389)
Felbeck, G. T. (1971). Chemical and biological characterization of humic material. In *Soil Biochemistry*, Vol. 2, ed. McLaren, A. D., and J. Skujins, pp. 36–59, Dekker, New York. (106, 110, 233)
Felgan, R. S., and C. H. Low (1967). Clinal variation in the surface-volume relationship of the columnar cactus *Lophocereus schotii* in northwest Mexico, *Ecology*, **48,** 530–6 (193)
Fernandez, J. (1978). A simple system to determine photosynthesis in field conditions by means of $^{14}CO_2$. *Photosynthetica*, **12,** 145–9. (137)
Fernandez, O. A., and M. M. Caldwell (1975). Phenology and dynamics of root growth of three cool semi-desert shrubs under field conditions. *J. Ecol.*, **63,** 703–14. (186).
Ferry, B. W., M. S. Baddeley, and D. L. Hawksworth (1973). *Air Pollution and Lichens*, Athlone Press, London. (370)
Fiscus, E. L., and P. J. Kramer (1970). Radial movement of oxygen in plant roots. *Pl. Physiol., Lancaster*, **45,** 667–69. (318)
Fisher, R. A. (1925). *Statistical Methods for Research Workers*, Oliver and Boyd, Edinburgh. (424)
Fisher, R. A. (1935–1966). *The Design of Experiments*, Oliver and Boyd, Edinburgh. (424)
Fitzpatrick, E. A. (1956). An indurated horizon formed by permafrost. *J. Soil Sci.*, **7,** 248–54. (55)
Flaig, W., H. Beutelspacher, and E. Reitz (1975). Chemical components and properties of humic substances. In *Soil Components* Vol. 2, ed. Gieseking, J., pp. 1–211, Springer, Berlin. (102, 106)
Flenley, J. (1979). *The Equatorial Rainforest: a Geological History*, Butterworth, London. (68, 74, 391)
Flowers, T. J., P. F. Troke, and A. R. Yeo (1977). The mechanism of salt tolerance in halophytes. *Ann. Rev. Pl. Physiol.*, **28,** 89–121.

Focht, D. D. (1978). Methods for Analysis of Denitrification. In *Nitrogen in the Environment* Vol. 2, ed. Nielsen, D. R., and J. G. McDonald, pp. 433–90, Academic Press, New York. (238)

Fonteyn, P. J., and B. E. Marshall (1978). Competition amongst desert perennials. *Nature, Lond.*, **275,** 544–5. (403, 404)

Ford, E. D., and J. D. Deans (1978). The effects of canopy structure on stemflow, throughfall and interception loss in a young Sitka Spruce plantation. *J. app. Ecol.*, **15,** 905–17. (195, 197)

Foster, R. C. (1978). Ultramicromorphology of some South Australian soils. In *Modification of Soil Structure*, ed. Emerson, W. W., R. D. Bond, and A. R. Dexter, pp. 103–9, Wiley, Chichester. (115)

Foy, C. D. (1974). Effects of aluminium on plant growth. In *The Plant Root and its Environment*, ed. Carson, E. W., Univ. Virginia Press, Charlotssville. (284)

Foy, C. D., R. L. Chaney, and M. C. White (1978). The physiology of metal toxicity in plants. *Am. Rev. Pl. Physiol.*, **29,** 511–66. (273, 283, 284)

Franco, C. M., and A. C. Magelhaes (1965). Techniques for the measurement of transpiration of individual plants. In *Methodology of Plant Ecophysiology*, ed. Eckardt, F. E., pp. 211–24, UNESCO, Paris. (195)

Frankland, B. (1976). Phytochrome control of seed germination in relation to the light environment. In *Light and Plant Development*, ed. Smith, H., pp. 477–91, Butterworths, London. (224)

Free, J. B. (1968). Dandelion as a competitor to fruit trees for bee visitors. *J. app. Ecol.*, **5,** 169–78. (386)

Freny, J. R. (1967). Sulfur containing organics. In *Soil Biochemistry*, Vol. 1, ed. McLaren, A. D., and G. H. Petersen, pp. 229–59, Dekker, New York. (43, 105, 243)

Fried, M., and H. Broeshart (1967). *The Soil-Plant System*, Academic Press, New York. (233, 241, 242, 244, 245, 251, 263)

Fripiat, J. J. (1965). Surface chemistry and soil science. In *Experimental Pedology*, ed. Hallsworth, E. G., and D. V. Crawford, pp. 3–13, Butterworths, London. (45)

Fritsch, F. E. (1927). The heath association on Hindhead Common. *J. Ecol.*, **15,** 344–72. (223)

Fritsch, F. E., and W. M. Parker (1913). The heath association at Hindhead Common. *New Phytol.*, **12,** 148–63. (223, 394)

Fritschen, L. J., and L. W. Gay (1979). *Environmental Instrumentation*, Springer, Berlin. (34, 35, 36, 37, 38)

Fritts, H. C. (1976). *Tree Rings and Climate*, Academic Press, New York. (175, 176)

Fukui, J. (1953). Studies on the adaptability of green manure and forage crops to paddy field conditions 1. *Proc. Crop. Sci. Soc. Japan*, **22,** 110–12. (314)

Gaastra, P. (1959). Photosynthesis of crop plants as influenced by light, carbon dioxide, temperature and stomatal diffusion resistance. *Mededel Landbouwhogesch. Wageneningen*, **59,** 1–68. (127)

Gambrell, R. P., and W. H. Patrick Jr. (1978). The chemical and microbiological properties of anaerobic soils and sediments. In *Plant Life in Anaerobic Environments*, ed. Hook, D. D., and R. M. M. Crawford, Ann. Arbor Science Pub., Michigan. (290, 294)

Gardner, W. R., and R. H. Nieman (1964). Lower limit of water availability to plants. *Science*, **143,** 1460–2. (167)

Gartside, D. W., and T. McNeilly (1974). Genetic studies in heavy metal tolerance. II. Zinc tolerance in *Agrostis tenuis*. *Heredity*, **33,** 303–8. (282)

Gates, C. T. (1955a, b). The response of the young tomato plant to a brief period of water shortage I. The whole plant and its principal parts. II. Individual leaves. *Aust. J. biol. Sci.*, **8,** 196–214, 215–30. (190, 191).

Gates, C. T. (1957). The response of the young tomato plant to a brief period of water stress. III. Drifts in nitrogen and phosphorus. *Aust. J. biol. Sci.*, **10,** 125–46. (190, 191)

Gates, C. T. (1968). Water deficits and growth of herbaceous plants. In *Water Deficits and Plant Growth*, Vol. 2, ed. Kozlowski, T. T., pp. 135–90, Academic Press, New York. (16, 192)

Gates, D. M. (1962). *Energy Exchange in the Biosphere*, Harper and Row, New York. (10, 11, 12)

Gates, D. M. (1965). Energy, plants and ecology. *Ecology*, **46,** 1–13. (18)

Gates, D. M. (1968). Transpiration and leaf temperature. *Ann. Rev. Pl. Physiol.*, **19,** 211–38. (16)

Gates, D. M. (1980). *Biophysical Plant Ecology*, Springer, Berlin. (20)

Gates, D. M., and La V. E. Papian (1971). *Atlas of Energy Budgets of Plant Leaves*, Academic Press, New York. (16, 17)

Geiger, R. (1965). *The Climate Near the Ground* translated from the German 4th Edition (1961), Harvard U.P., Cambridge, Mass. (11, 226, 424)

Gerard, J. (1597) *The Herbal* Dover Pub., New York (facsimile). (417)

Gemmel, R. P., and G. T. Goodman (1980). The maintainance of grassland on smelter wastes in the Lower Swansea Valley III. Zinc smelter wastes. *J. app. Ecol.*, **17,** 461–8. (282)

Gerloff, C. C. (1976). Plant efficiencies in the use of nitrogen, phosphorus and potassium. In *Plant Adaptation to Mineral Stress in Problem Soils*, ed. Wright, M. J., pp. 161–73, Cornell Univ. (277)

Gersper, P. L., and N. Hollowaychuck (1970). Effects of stemflow water on a Miami soil under a beech tree. I and II. *Soil Sci. Soc. Amer. Proc.*, **34,** 779–94. (371, 388)

Ghadiri, H. and D. Payne (1977). Raindrop impact stress and the breakdown of soil crumbs. *J. Soil Sci.*, **28,** 247–58. (114)

Gieseking, J. (1975). *Soil Components* Vols. 1 and 2, Springer, Berlin. (108)

Gigon, A. and I. H. Rorison (1972). The response of some ecologically distinct plant species to nitrate and to ammonium nitrogen. *J. Ecol.*, **60,** 93–102. (273)

Gilbert, O. L. (1970a, b). Further studies on the effect of sulphur dioxide on lichens and bryophyte. A biological scale for the estimation of sulphur dioxide pollution. *New Physiol.*, **69,** 605–27, 629–34. (370)

Gleason, H. A. (1917). The structure and development of the plant association. *Bull. Torrey bot. Club*, **43,** 463–81. (419)

Gleason, H. A. (1926). The individualistic concept of the plant association, *Bull. Torrey bot. Club*, **53,** 7: 26. (419)

Gleason, H. A., and A. Cronquist (1964). *The Natural Geography of Plants*, Columbia U.P., New York. (74, 226)

Godwin, H. (1975). *The History of the British Flora*, 2nd edn., Cambridge Univ. Press, Cambridge. (104)

Goldberg, E. D. (1971). Atmospheric dust, the sedimentary cycle and man. *Comm. Earth Sci. Geophys.*, **1,** 117–32. (369)

Goldberg, E. D. (1971). Atmospheric dust, the sedimentary cycle and man. *Comm. Earth Sci. Geophys.*, **1,** 117–32. (369)

Goldsmith, F. B. (1973). The vegetation of exposed cliffs at South Stack, Anglesey. I and II. *J. Ecol.*, **63,** 687–729. (412)

Golley, F. B. (1961). Energy values of ecological materials. *Ecology*, **42,** 581–4. (346)

Golley, F. B., K. Petrusewiczk, and L. Ryszkowski (1975). *Small Mammals, their Productivity and Population Dynamics*, Cambridge Univ. Press, Cambridge. (345)

Good, R. (1974). *Geography of the Flowering Plants*, (4th edn.), Longman, London. (74)

Goodall, D. W. (1954). Objective methods for the classification of vegetation. III. An essay on the use of factor analysis. *Aust. J. Bot.*, **2,** 304–24. (420)

Goodall, D. W., and R. A. Perry (1979). *Arid-land Ecosystems*, Vol. 1, Cambridge Univ. Press, Cambridge. (179)

Goode, J. E. (1956). Soil moisture deficits under swards of different grasses species in an orchard. *Ann. rept. East Malling Res. Sta.*, 69–72. (186, 403)

Goodman, G. T., and D. F. Perkins (1959). Mineral uptake and retention in Cotton Grass (*Eriophorum vaginatum*) *Nature, Lond.*, **184,** 467–8. (120, 278, 397)

Goodman, G. T., and S. A. Bray (1975). *Ecological Aspects of the Reclamation of Derelict Land*, NERC, London. (227)

Goodman, P. J. (1969). Intraspecific variation in mineral nutrition of plants from different habitats. In *Ecological Aspects of the Mineral Nutrition of Plants*, ed. Rorison, I. H., pp. 237–53, Blackwell, Oxford. (277)

Goodman, P. J., E. M. Braybrook, J. M. Lambert, and C. J. Marchant (1969). Biological flora of the British Isles. *Spartina Schreb. J. Ecol.*, **57,** 285–3–3. (406)

Goodman, P. J., and W. T. Williams (1961). Investigations into 'die-back' in *Spartina townsendi* agg. III. Physiological correlates of 'die-back'. *J. Ecol.*, **49,** 391–8. (312)

Gorham, E., P. M. Vitousek, and W. A. Reiners (1979). The regulation of chemical budgets over the course of terrestrial ecosystem succession. *Ann. Rev. Ecol. Syst.*, **10,** 53–84. (374)

Goto, Y., and K. Tai (1957). On differences of oxidizing powers of paddy rice seedling roots among some varieties. *Soil and Plant Food*, **2,** 198–200. (314)

Grable, A. R. (1966). Soil aeration and plant growth. *Adv. Agron.*, **18,** 57–106. (290)

Grace, J. (1977). *Plant Response to Wind*, Academic Press, London. (25, 26, 227, 229)

Granhall, U., and H. Selander (1973). Nitrogen fixation in a subarctic mire. *Oikos*, **24,** 8–15.

Green, M. S., and J. R. Etherington (1977). Oxidation of ferrous iron by rice (*Oryza sativa* L.) roots: a mechanism for waterlogging tolerance? *J. exp. Bot.*, **28**, 678–90. (313)

Greene, H. (1963). Perspectives in soil science. *J. Soil Sci.*, **14,** 1–11. (94, 95)

Greenland, D. J. (1979). The physics and chemistry of the soil-root interface: some comments. In *The Soil-Root Interface*, ed. Harley, J. L., and R. S. Russel, pp. 83–98, Academic Press, London. (115, 116)

Greenway, H., and R. Munns (1980). Mechanisms of salt tolerance in non-halophytes. *Ann. Rev. Pl. Physiol.*, **31,** 149–90. (287, 288)

Greenwood, D. J. (1961). The effect of oxygen concentration on the decomposition of organic materials in soils. *Plant and Soil*, **14,** 360–76. (94)

Greenwood, D. J. (1962). Nitrification and nitrate dissimilation in soil. II. Effects of oxygen concentration. *Plant and Soil*, **17,** 378–91. (301)

Greenwood, D. J. (1967). Studies in oxygen transport through mustard seedlings (*Sinapsis alba* L.). *New Phytol.*, **66,** 597–606.

Greenwood, D. J. (1970). The distribution of carbon dioxide in the aqueous phase of aerobic soils. *J. Soil Sci.*, **21,** 314–29. (297)

Greenwood, D. J., and D. Goodman (1967). Direct measurement of the distribution of oxygen in soil aggregates and in columns of fine soil crumbs. *J. Soil. Sci.*, **18,** 182–96. (116, 293, 294)

Gregory, F. G. (1918). Physiological conditions in cucumber houses. *3rd Ann. Rept. Expt. Res. Sta., Cheshunt*, pp. 19–28. (113)

Gregory, F. G. (1926). The effect of climatic conditions on the growth of barley. *Ann. Bot.*, **40,** 1–26. (113)

Gregory, P. H. (1973). *The Microbiology of the Atmosphere*, Hill, Aylesbury. (229)

Gregory, R. P. G., and A. D. Bradshaw (1965). Heavy metal tolerance in populations of *Agrostis tenuis* Sibth and other grasses. *New Phytol.*, **64,** 131–43. (283)

Greig-Smith, P. (1964). *Quantitative Plant Ecology*, Butterworths, London. (421)

Grim, R. E. (1953). *Clay Mineralogy*, McGraw-Hill, London. (84, 87)

Grime, J. P. (1963a). Factors determining the occurrence of a calcifuge species on shallow soils over calcareous substrata. *J. Ecol.*, **51,** 375–90. (271)

Grime, J. P. (1963b). An ecological investigation of a junction between two plant communities in Coombsdale on the Derbyshire limestone. *J. Ecol.*, **51,** 391–402. (271)

Grime, J. P. (1965). Shade tolerance in flowering plants. *Nature, Lond.*, **208,** 161–2. (133)

Grime, J. P. (1979). *Plant Strategies and Vegetation Processes*, Wiley, Chichester. (389, 410, 421)

Grime, J. P., and J. G. Hodgson (1969). An investigation of the ecological significance of lime chlorosis by means of large scale comparative experiments. In *Ecological Aspects of the Mineral Nutrition of Plants*, ed. Rorison, I. H., pp. 67–99, Blackwell, Oxford. (267, 271)

Grime, J. P., and R. Hunt (1975). Relative growth rate: its range and adaptive significance in a local flora. *J. Ecol.*, **63,** 393–422. (139)

Grime, J. P., and T. C. Hutchinson (1967). The incidence of lime chlorosis in the natural vegetation of England. *J. Ecol.*, **55,** 557–66. (268)

Grodzinski, W., R. Z. Klekowski, and A. Duncan (1975). *Methods for Ecological Energetics*, Blackwell, Oxford. (345)

Grubb, P. J. (1977). The maintainance of species-richness in plant communities: the importance of the regeneration niche. *Biol. Rev.*, **52,** 107–45. (384, 390)

Grubb, P. J., H. E. Green, and R. C. J. Merrifield (1969). The ecology of chalk heath: its relevance to the calcicole–calcifuge and acidification problem. *J. Ecol.*, **57,** 175–210. (59, 103, 269)

Grubb, P. J., and M. B. Suter (1971). The mechanism of acidification of soil by *Calluna* and *Ulex* and their significance for conservation. In *The Scientific Management of Animal and Plant Communities for Conservation*, ed. Duffey, E., and A. S. Watt, pp. 115–33. Blackwell, Oxford. (103)

Gulmon, S. L., and H. A. Mooney (1977). Spatial and temporal relationships between two desert shrubs, *Atriplex hymenelytra* and *Tidestromia oblongifolia* in Death Valley, California. *J. Ecol.*, **65,** 831–8. (179)

Haavinga, A. J. (1974). Problems in the interpretation of pollen diagrams of mineral soil. *Geologie en Mijnbouw*, **53,** 449–53. (105)

Haberlandt, G. (1884). *Physiologische Pflanzenanatomie*, Engelmann, Leipzig. (132, 423)

Hackett, C. (1965). Ecological aspects of the nutrition of *Deschampsia flexuosa* (L.). Trin II. The effects of Al, Ca, Fe, K, Mn, P and pH on the growth of seedlings and established plants. *J. Ecol.*, **53,** 315–33. (267)

Hackett, C. (1967). Ecological aspects of the mineral nutrition of *Deschampsia flexuosa* (C). Trin III. Investigation of phosphorus requirement and response to aluminium in water culture and a study of growth in soil. *J. Ecol.*, **55,** 831–40. (279, 402)

Hadley, N. F. (1975). *Environmental Physiology of Desert Organisms*, Dowden, Hutchinson and Ross, Stroudsburg.

Haekel, E. (1866). *Generelle Morphologie der Organismen*, Jena. (1, 418)

Haekel, E. (1869). Ueber Entwicklungsgung und Aufgabe der Zoologie. *Jenaischer Zeitschr. fur Naturwiss.*, **5,** 353–70.

Haider, K., J. P. Martin, and Z. Filip (1975). Humus biochemistry. In *Soil Biochemistry* Vol. 4, ed. Paul, E. A., and A. D. McLaren, pp. 195–244, Dekker, New York. (108, 109)

Hales, S. (1727). *Vegetable Staticks*, W. and J. Innys, London. (421)

Hall, A. D. (1903). *The Soil: an Introduction to the Scientific Study of the Growth of Crops*, Murray, London. (419)

Hall, A. E., and E. D. Schulze (1976). Current perspectives of steady state stomatal responses to the environment. In *Water and Plant Life*, ed. Lange, O. L., L. Kappen, and E. D. Schultz, pp. 169–88, Springer, Berlin. (166)

Hall, C. A. S., and R. Moll (1975). Methods of assessing aquatic primary productivity. In *Primary Productivity of the Biosphere*, ed. Lieth, H., and R. H. Whittaker, Springer, Berlin. (340, 344)

Hall, R. L. (1974). Analysis of the nature of interference between plants of different species I and II. *Aust. J. agric. Res.*, **25,** 739–56. (414)
Hall, S. M., and J. A. Milburn (1973). Phloem transport in *Ricinus:* its dependence on the water balance of the tissues. *Planta*, **109,** 1–10. (174)
Halstead, R. L., and R. B. McKercher (1975). Biochemistry and cycling of phosphorus. In *Soil Biochemistry* Vol. 4, ed. Paul, E. A., and A. D. McLaren, pp. 31–63, Dekker, New York. (43)
Hamner, K. C. (1940). Interaction of light and darkness in photoperiodic induction. *Bot. Gaz.*, **101,** 658–87. (217)
Harborne, J. B. (1972). *Phytochemical Ecology*, Academic Press, New York. (379, 425)
Harborne, J. B. (1977). *Introduction to Ecological Biochemistry*, Academic Press, New York. (425)
Harley, J. L. (1969a). A physiologist's viewpoint. In *Ecological Aspects of the Mineral Nutrition of Plants*, ed. Rorison, I. H., pp. 437–47, Blackwell, Oxford. (257)
Harley, J. L. (1969b). *The Biology of Mycorrhizas* 2nd edn., Hall, London. (257)
Harley, J. L., and R. S. Russel (1979). *The Soil–Root Interface*, Academic Press, London. (258)
Harper, J. L. (1961). Approaches to the study of plant competition. In *Mechanisms in Biological Competition*, ed. Milthorpe, F. L., pp. 1–39, Cambridge Univ. Press, Cambridge. (396, 399)
Harper, J. L. (1967). A Darwinian approach to plant ecology. *J. Ecol.*, **55,** 247–70. (397, 398, 411)
Harper, J. L. (1977). *The Population Biology of Plants*, Academic Press, London. (2, 218, 220, 223, 383, 385, 388, 404, 405, 409, 410, 421)
Harper, J. L., and R. A. Benton (1966). The behaviour of seeds in soil. II. The germination of seeds on the surface of water supplying substrate. *J. Ecol.*, **54,** 151–66. (169, 170)
Harris, G. A., and A. M. Wilson (1970). Competition for moisture among seedlings of annual and perennial grasses as influenced by root elongation at low temperature. *Ecology*, **51,** 530–4. (186, 207, 403)
Hart, M. H. (1978). The evolution of the atmosphere of the earth. *Icarus*, **33,** 23–39. (356)
Hartesveld, R. J., H. T. Harvey, H. S. Shellhorne, and R. E. Stecker (1975). *The Giant Sequoias of the Sierra Nevada*, U.S. Dept. Int. Nat. Park Service, Washington. (394)
Hartwell, B. L., and F. R. Pember (1918). The presence of aluminium, as a reason for the difference in effect of so-called acid soil on barley and rye. *Soil Sci.*, **6,** 259–77. (265)
Harvey, H. W. (1950). On the production of organic matter in the sea off Plymouth. *J. mar. biol. Ass. U.K.*, **20,** 97–137. (335)
Hassouna, M. G., and P. F. Wareing (1964). Possible role of rhizosphere bacteria in the nitrogen nutrition of *Ammophila areneria. Nature, Lond.*, **202,** 467–9. (235, 258)
Hatch, M. D., and C. R. Slack (1966). Photosynthesis by sugar-cane leaves. A new carboxylation reaction and the pathway of sugar formation. *Biochem. J.*, **101,** 103–11. (127)
Hattori, T. (1973). *Microbial Life in the Soil*, Dekker, New York. (101, 116)
Haynes, R. J., and K. M. Goh (1978). Ammonium and nitrate nutrition of plants. *Biol. Rev.*, **53,** 465–510. (237)
Head, W. S., and G. C. Rogers (1969). Factors affecting the distribution and growth of roots of perennial woody species. In *Root Growth*, ed. Whittington, W. J., pp. 280–95, Butterworths, London. (119–120)
Heal, O. W., and D. F. Perkins (1978). *Production Ecology of British Moors and Montane Grasslands*, Springer, Berlin. (103)
Healy, M. T. (1975) *Oxygen Transport in Pisum sativum* L. Ph.D. Thesis, University of Hull. (322)

Healy, M. T., and W. Armstrong (1972). The effectiveness of internal oxygen transport in a mesophyte (*Pisum sativum* L.). *Planta*, **103**, 302–309. (309, 315)
Heap, A. J., and E. I. Newman (1980). The influence of vesicular-arbuscular mycorrhizas on phosphorus transfer between plants. *New Phytol.*, **85**, 173–9.
Heath, O. V. S. (1967). Resistance to water transport in plants. *Nature, Lond.*, **213**, 741. (163)
Heath, O. V. S., and B. Orchard (1957). Temperature effects on the minimum intercellular space CO_2 concentration. *Nature, Lond.*, **180**, 180–1. (19, 206)
Heber, U., and K. A. Santarius (1973). Cell death by cold and heat and resistance to extreme temperatures. Mechanisms of hardening and dehardening. In *Temperature and Life*, ed. Precht, H., J. Christopherson, and H. Hensel, Springer, Berlin. (208, 209)
Heide, H. van der, M. H. van Raalte, and B. M. de Boer-Bolt (1963). The effect of low oxygen content of the medium on the roots of barley seedlings. *Acta Bot. Neerl.*, **12**, 131–47. (314)
Her Majesty's Stationary Office (1977). *How to Write Metric*, HMSO, London. (Prelim page xvii)
Hesse, P. R. (1971). *A Textbook of Soil Chemical Analysis*, Chemical Pub. Co., New York. (88, 110)
Hewitt, E. J. (1963). The essential nutrient elements: requirements and interactions. In *Plant Physiology*, Vol. 3, ed. Steward, F. C., pp. 137–360, Academic Press, New York. (248, 263)
Hewitt, E. J., and T. A. Smith (1975). *Plant Mineral Nutrition*, English Univ. Press, London. (259)
Hewitt, L. F. (1948). *Oxidation Reduction Potentials in Bacteriology and Biochemistry* (5th edn.), London County Council, London. (298)
Heydecker, W. (1973). *Seed Ecology*, Butterworths, London. (223)
Hibbert, A. R. (1967). Forest treatment effects on water yield. In *Forest Hydrology*. ed. Sopper, W. E., and H. W. Lull, pp. 527–43, Pergamon, Oxford. (197)
Hill, R. (1956). Oxidation-reduction potentials. In *Modern Methods of Plant Analysis*, pp. 393–414, ed. Paech, K., and M. V. Tracey, Springer, Berlin. (298)
Hillel, D. (1972). Soil moisture and seed germination. In *Water Deficits and Plant Growth*, Vol. 3, ed. Kozlowski, T. T., pp. 65–89, Academic Press, New York. (170)
Hiron, R. W. P., and S. T. C. Wright (1973). The role of endogenous abscisic acid in the response of plants to stress. *J. exp. Bot.*, **24**, 769–81. (308)
Hoddinot, J., and J. Bain (1979). The influence of simulated canopy light on the growth of six acrocarpous moss spp. *Can. J. Bot.*, **57**, 1236–42. (224)
Hodgson, J. M. (1974). *Soil Survey Field Handbook*, Soil Survey of England and Wales, Rothamsted. (51)
Hoffman, F. M., and C. J. Hillson (1979). Effects of silicon on the life-cycle of *Equisetum hyemale* L. *Bot. Gaz.*, **140**, 127–32. (249)
Hogg, W. H. (1971). Regional and local environments. In *Potential Crop Production*, ed. Wareing, P. F., and J. P. Cooper, pp. 6–22, Heinemann, London. (32)
Holder, C. B., and K. W. Brown (1980). The relationship between oxygen and water uptake by roots of intact bean plants (*Phaseolus vulgaris* L.). *Soil Sci. Soc. Amer. J.*, **44**, 21–5. (308)
Holdgate, M. W. (1979). *A Perspective on Environmental Pollution*, Cambridge Univ. Press, Cambridge. (364, 371)
Holdgate, M. W. and M. J. Woodman (1978). *The Breakdown and Restoration of Ecosystems*, Plenum, New York. (371)
Holland, H. D. (1978). *The Chemistry of the Atmosphere and the Oceans*, Wiley Interscience, New York. (357, 361, 364, 365, 366)
Hollis, J. P. (1967). Toxicant diseases of rice. *La. Agr. Exp. Sta. Bull.*, 614. (311, 318)

Holmes, M. G., and H. Smith (1977a and b). The function of phytochrome in the natural environment. II. The influence of vegetation canopies on the spectral energy distribution of natural daylight; IV. Light quality and plant development. *Photochem. Photobiol.*, **25,** 539–45; 551–7. (224)

Holmgren, P. P., and P. G. Jarvis (1967); CO_2 efflux from leaves in light and darkness. *Physiol. Pl.*, **20,** 1045–51. (206)

Holroyd, E. W. (1970). Prevailing winds on Whiteface Mountain as indicated by flag-trees. *For. Sci.*, **16,** 222–9. (227)

Holtam, B. W. (1971). *Windblow of Scottish Forests in January 1968*, For. Comm. Bull. 45, HMSO, Edinburgh. (227)

Hook, D. D., and C. L. Brown (1972). Permeability of the cambium to air in trees adapted to wet habitats. *Bot. Gaz.*, **133,** 304–10. (329)

Hook, D. D., C. L. Brown, and P. P. Kormanik (1971). Inductive flood tolerance in Swamp Tupelo (*Nyssa sylvatica var. biflora* (Walt.) Sarg.). *J. exp. Bot.*, **22,** 78–89. (324, 325, 328)

Hook, D. D., C. L. Brown, and R. H. Wetmore (1972). Aeration in trees. *Bot. Gaz.*, **133,** 443–54. (327, 328)

Hook, D. D., and J. R. Scholtens (1978). Adaptation and flood-tolerance of trees. In *Plant Life in Anaerobic Environments*, pp. 299–311, ed. Hook, D. D., and R. M. M. Crawford, Ann Arbor Science, Ann Arbor. (327)

Hopkins, H. T., A. W. Specht, and S. B. Hendricks (1950). Growth and nutrient accumulation as controlled by oxygen supply to the plant roots. *Pl. Physiol. Lancaster*, **25,** 193–209. (308)

Horn, H. S. (1971). *The Adaptive Geometry of Tree Species*, Princeton Univ. Press, Princeton. (389, 390, 391, 393)

Horn, H. S. (1974). The ecology of secondary succession. *Ann. Rev. Ecol. Syst.*, **5,** 25–37. (388, 392, 393)

Horn, H. S. (1975). Markovian properties of forest succession. In *Ecology and Evolution of Communities*, ed. Diamond, J. M., and M. L. Cody, pp. 196–211, Belknap Press, Harvard. (389, 409)

Horne, A. J. (1971). The ecology of nitrogen fixation on Signy Island, South Orkney Islands. *Brit. Antarctic Surv. Bull.*, **27,** 893–902. (334)

Howard, P. J. (1969). The classification of humus types in relation to soil ecosystems. In *The Soil Ecosystem*, ed. Sheals, J. G., pp. 41–54, Systematics Assn. Publ. No. 8. (100)

Howard-Williams, C. (1970). The ecology of *Becium homblei* in Central Africa with special reference to metaliferrous soils. *J. Ecol.*, **58,** 745–63. (283)

Hozumi, K., H. Koyama, and T. Kira (1955). Intraspecific competition among higher plants. IV. A preliminary account of the interaction between adjacent individuals. *J. Inst. Polytech. Osaka City Univ. Series D*, **6,** 121–30. (407)

Hsiao, T. C., E. Acedavo, E. Feres, and D. W. Henderson (1976). Water stress, growth and osmotic adjustment. *Phil. Trans. R. Soc. Lond.* B., **273,** 479–500. (166, 169)

Huber, B. (1935). *Der Wärmhaushalt der Pflanzen Naturwiss. und Landwirtschaft*, Heft 17, Friesing, München. (212)

Huck, M. G. (1970). Variation in tap root elongation rate as influenced by composition of soil air. *Agron. J.*, **62,** 815–18. (308)

Hudson, J. P. (1965). Gauges for the study of evapotranspiration rates. In *Methodology of Plant Ecophysiology*, ed. Eckardt, F. D., pp. 443–51, UNESCO, Paris. (196)

Huikari, O. (1954). Experiments on the effect of anaerobic media upon birch, pine and spruce seedlings. *Commun. Inst. Forest. Fenn.*, **42,** 1–13. (328)

Huiskes, A. H. L. (1979). Biological flora of the British Isles: *Ammophila arenaria* (L.) Link. *J. Ecol.*, **67,** 363–82. (221)

Humboldt, A. von (1805). *Ideen zu einen Geographie der Pflanzen*, Cotta, Tubingen. (418)

Hunt, R. (1978). *Plant Growth Analysis*, Arnold, London. (139)

Hurd, E. A. (1976). Plant breeding for drought resistance. In *Water Deficits and Plant Growth* Vol. 4, ed. Kozlowski, T. T., pp. 317–53, Academic Press, New York. (174)
Hurst, H. M., and A. Burges (1967). Lignin and humic acids. In *Soil Biochemistry*, Vol. 1, ed. McLaren, A. D., and G. H. Petersen, pp. 260–86, Dekker, New York. (106, 107)
Hutchinson, G. E. (1957). *A Treatise on Limnology*, Vol. 1, Wiley, New York. (363)
Hutchinson, G. E. (1965). *The Ecological Theatre and the Evolutionary Play*. Yale Univ. Press, Newhaven. (411)
Hutchinson, T. C., and M. Havas (1980). *Effects of Acid Precipitation on Ecosystems*, Plenum, New York. (229, 367)
Hutnik, R. J., and G. Davies (1973). *Ecology and Reclamation of Devastated Land*, Gordon and Breach, New York. (371)
Hutton, J. T., and K. Norrish (1974). Silica content of wheat husks in relation to water transpired. *Aust. J. agric. Res.*, **25,** 203–12. (170, 299)
Hyde, H. A. (1956). Tree pollen in Great Britain. *Acta allergologia*, **10,** 224–45. (228)
Idle, D. B. (1966). The photography of ice-formation in plant tissue. *Ann. Bot.*, **30,** 199–206. (1966)
Incoll, L. D., S. P. Long, and M. R. Ashmore (1977). SI units in publications in plant science. *Curr. Adv. pl. Sci.*, **28,** 331–43. (Prelim page xvii, 149)
Institute of Hydrology (1976). *Water Balance of the Headwater Catchments of Wye and Severn 1970–74*, Rept. 33, NERC, London. (197)
Isherwood, F. A. (1965). Biosynthesis of lignin. In *Biosynthetic Pathways in Higher Plants*, ed. Pridham, J. B., and T. Swain, pp. 133–46, Academic Press, London. (107)
Itai, C., and A. Benzioni (1976). Water stress and hormonal response. In *Water and Plant Life*, ed. Lange, O. L., L. Kappen, and E. D. Schultz, pp. 225–42, Springer, Berlin. (175)
Ivanov, L. (1928). Zur methodik der transpiration bestimmung am standart. *Ber. dtsch. bot. Ges.*, **46,** 306–10. (195)
Ives, J. D. (1974). Permafrost. In *Arctic and Alpine Environments*, ed. Ives, J. D., and R. G. Barry, pp. 159–94, Methuen, London. (208)
Jacks, J. V. (1965). The role of organisms in the early stages of soil formation. In *Experimental Pedology*, ed. Hallsworth, E. G., and D. V. Crawford, pp. 219–26, Butterworths, Oxford. (41, 115)
Jackson, M. B., and D. J. Campbell (1975a). Movement of ethylene from roots to shoots, a factor in the response of tomato plants to waterlogged conditions. *New Phytol.*, **74,** 397–406. (309)
Jackson, M. B., and D. J. Campbell (1975b). Ethylene and waterlogging effects in tomato. *Ann. app. Biol.*, **81,** 102–5. (309)
Jackson, M. B., and D. J. Campbell (1978). Effects of benzyladenine and gibberellic acid on the responses of tomato plants to anaerobic root environments and to ethylene. *New Phytol.*, **82,** 331–30. (309)
Jackson, M. L. (1958). *Soil Chemical Analysis*. Prentice Hall, New Jersey. (88, 98)
Jackson, W. T. (1955). The role of adventitious roots in recovery of shoot following flooding of the original root systems. *Amer. J. Bot.*, **42,** 816–9. (321)
Janzen, D. H. (1970). Herbivores and the number of tree species in tropical forests. *Amer. Nat.*, **104,** 501–28. (377)
Janzen, D. H. (1971). Escape of *Cassia grandis* L. beans from predators in time and space. *Ecol.*, **52,** 964–79. (384)
Janzen, D. H. (1975). *Ecology of Plants in the Tropics*, Arnold, London. (74, 391)
Janzen, D. H. (1979). How to be a fig. *Ann. Rev. Ecol. Syst.*, **10,** 13–51. (386)
Jarvis, M. C., and H. J. Duncan (1976). Profile distribution of organic C, Fe, Al and Mn in soils under bracken and heather. *Plant and Soil*, **44,** 139–40. (104)
Jarvis, M. S. (1963). A comparison between the water relations of species with contrasting types of geographical distribution in the British Isles. In *The Water Relations of Plants*, ed. Rutter, A. J., and F. H. Whitehead, pp. 289–312, Blackwell, Oxford. (172, 173)

Jarvis, P. G., and M. S. Jarvis (1965). The water relations of tree seedlings. V. Growth and root respiration in relation to the osmotic potential of the medium. In *Water Stress in Plants*, ed. Slavik, B., pp. 167–82. Czech. Acad. Sci. Prague. (174)

Jefferies, R. L., and A. J. Willis (1964). Studies on the calcicole–calcifuge habit II. The influence of calcium on the growth and establishment of four species in soil and sand culture. *J. Ecol.*, **52,** 691–707. (263–4)

Jeffery, D. W. (1964). The formation of polyphosphate in *Banksia ornata*, an Australian heath plant. *Aust. J. biol. Sci.*, **17,** 845–54. (241)

Jenkinson, D. S. (1975). The turnover of organic matter in agricultural soils. In *Soil Organic Matter*, ed. Wyn Jones, R. G., Welsh Soils Discussion Group Rept. 16. Aberystwyth. (99)

Jenkinson, D. S., and J. H. Rayner (1977). The turnover of soil organic matter in some of the Rothamsted classical experiments. *Soil Sci.*, **123,** 298–305. (99)

Jennings, D. H. (1968). Halophytes, succulence and sodium in plants—a unified theory. *New Phytol.*, **67,** 899–911. (289)

Jennings, P. R., and R. C. Aquino (1968). Studies in competition on rice. III. The mechanism of competition among phenotypes. *Evolution, Lancaster, Pa.*, **22,** 529–42. (401)

Jenny, H., and K. Grossenbacher (1963). Root–soil boundary zones as seen in the electron microscope. *Proc. Soil Sci. Soc. Amer.*, **27,** 273–9. (116)

Jenny, H., and C. D. Leonard (1934). Functional relationships between soil properties and rainfall. *Soil Sci.*, **38,** 363–81. (92)

Jensen, M. E. (1968). Water consumption by agricultural plants. In *Water Deficits and Plant Growth*, ed. Kozlowski, T. T., Vol. 2, pp. 1–22, Academic Press, New York.

Jermy, A. C., H. R. Árnold, L. Farrel, and F. H. Perring (1978). *Atlas of Ferns of the British Isles*, Botanical Soc. of the British Isles. (169)

Jernelöv, A. (1975). Cycling of mercury in the environment. In *The Ecology of Resource Renewal and Degradation*, pp. 49–55, Blackwell, Oxford. (43)

John, C. D., and H. Greenway (1976). Alcoholic fermentation and activity of some enzymes in rice roots under anaerobiosis. *Aust. J. Pl. Physiol.*, **3,** 325–36. (325)

Johnston, W. R., and J. Proctor (1977). A comparative study of metal levels in plants from two contrasting lead mine sites. *Plant and Soil*, **46,** 251–7. (286, 370)

Jones, D. A. (1973). Coevolution and cyanogenesis. In *Taxonomy and Ecology*, ed. Heywood, V. H., Academic Press, London. (379)

Jones, H. E. (1971a). Comparative studies of plant growth and distribution in relation to waterlogging. II. An experimental study of the relationship between transpiration and the uptake of iron in *Erica cinerea* L. and *E. tetralix* L. *J. Ecol.*, **59,** 167–78. (310, 326)

Jones, H. E. (1971b). Comparative studies of plant growth and distribution in relation to waterlogging. III. The response of *Erica cinerea* L. to waterlogging in peat soils of differing iron content. *J. Ecol.*, **59,** 583–91. (310)

Jones, H. E., and J. R. Etherington (1970). Comparative studies of plant growth and distribution in relation to waterlogging. I. The survival of *Erica cinerea* L. and *E. tetralix* L. and its apparent relationship to iron and manganese uptake in waterlogged soil. *J. Ecol.*, **58,** 487–96. (310)

Jones, J. B. (1976). Introduction: iron deficiency in plants and its correction. *Comm. Soil. Pl. Anal.*, **7,** i. (268)

Jones, L. H. (1961). Aluminium uptake and toxicity in plants. *Plant and Soil*, **13,** 297–310. (97, 265)

Jones, L. I. (1967). Studies on Hill-Land in Wales. *Tech. Bull. Welsh Pl. Breeding Sta.*, **2,** 1–79. (272)

Jones, R. (1967). The relationship of dune-slack plants to soil moisture and chemical conditions. Ph.D. Thesis, University of Wales. (303, 379)

Jones, R. (1972a). Comparative studies of plant growth and distribution in relation to waterlogging. V. The uptake of iron and manganese by dune and slack plants. *J. Ecol.*, **60,** 131–40. (311)

Jones, R. (1972b). Comparative studies of plant growth and distribution in relation to waterlogging. VI. The effect of manganese on the growth of dune and slack plants. *J. Ecol.*, **60,** 141–46. (311)

Jones, R. (1973). Comparative studies of plant growth and distribution in relation to waterlogging. VII. The influence of watertable fluctuation on iron and manganese availability in dune slacks. *J. Ecol.*, **61,** 107–116. (379, 381)

Jones, R., and J. R. Etherington (1971). Comparative studies of plant growth and distribution in relation to waterlogging. IV. The growth of dune and dune slack plants. *J. Ecol.*, **59,** 793–801. (311)

Jordan, C. F., and J. R. Kline (1972). Mineral cycling: some basic concepts and their application in a tropical rainforest. *Ann. Rev. Ecol. Syst.*, **3,** 33–50. (382, 383)

Jordan, P. A., D. B. Botkin, and M. L. Wolf (1971). Biomass dynamics in a moose population. *Ecology*, **52,** 147–52. (382)

Jorgensen, S. C. (1979). *Handbook of Environmental Data and Ecological Parameters*, Pergamon, New York. (286)

Juday, C. (1940). The annual energy budget of an inland lake. *Ecology*, **21,** 438–50. (338)

Kahn, A. A. (1977). *The physiology and biochemistry of seed dormancy and germination*, North Holland, Amsterdam. (223)

Kaku, S., and M. Iwaya (1978). Low temperature exotherms in xylems of evergreen and deciduous broadleaved trees in Japan with reference to freezing resistance and distribution range. In *Plant Cold Hardiness and Freezing Stress*, ed. Li, P. H., and L. Sakai, pp. 227–39, Academic Press, New York. (209)

Kallio, P., and L. Karenlampi (1975). Photosynthesis in mosses and lichens. In *Photosynthesis and Primary Production in Different Environments*, ed. Cooper, J. P., pp. 393–423, Cambridge Univ. Press, Cambridge. (205, 350, 355)

Kaufman, M. R., and K. J. Ross (1970). Water potential, temperature and kinetin effects on seed germination in soil and solute systems. *Amer. J. Bot.*, **57,** 413–9. (170)

Kawase, M. (1974). Role of ethylene in induction of flooding damage in Sunflower. *Physiol Pl.*, **31,** 29–38. (309)

Kawase, M. (1976). Ethylene accumulation in flooded plants. *Physiol. Pl.*, **36,** 236–41. (324)

Kawase, M. (1979). Role of cellulase in aerenchyma development in Sunflower. *Physiol. Pl.*, **66,** 183–90. (323, 324)

Keely, J. (1978). Malic acid accumulation in roots in response to flooding: evidence contrary to its role as an alternative to ethanol. *J. exp. Bot.*, **29,** 1345–9. (325)

Keely, J. (1979). Population differentiation along a flood frequency gradient: physiological adaptations to flooding in *Nyssa sylvatica*. *Ecol. Monographs*, **49,** 89–008. (315, 325, 328)

Keely, J., and E. H. Franz (1979). Alcoholic fermentation in swamp and upland populations of *Nyssa sylvatica*: temporal changes in adaptive strategy. *Amer. Nat.*, **113,** 587–92. (325)

Kellman, M. (1979). Soil enrichment by neotropical savanna trees. *J. Ecol.*, **67,** 565–77. (371)

Kemper, W. D., and W. S. Chepil (1965). Size distribution of aggregates. In *Methods of Soil Analysis*, ed. Black, C. A., pp. 499–510, Am. Soc. Agron. Wisconsin. (114)

Kendrick, W. B., and A. Burges (1962). Biological aspects of the decay of *Pinus sylvestris* litter. *Nova Hedwigia*, **4,** 313–42. (102)

Kershaw, K. A. (1963). Pattern in vegetation and its causality. *Ecology*, **44,** 377–88. (120)

Kershaw, K. A. (1973). *Quantitative and Dynamic Ecology*, 2nd. edn., Arnold, London. (421)

Keymer, R., and W. M. Ellis (1978). Experimental studies on plants of *Lotus corniculatus*

L. from Anglesey, polymorphic for cyanogenesis. *Heredity*, **40,** 189–206. (381)

Kilbertus, G., O. Reisinger, A. Mourey, and J. A. C. da Fonseca (1971). *Biodegradation and humification*, Pierron, Sarregumines. (99)

Kimball, B. A., and E. R. Lemon (1971). Air turbulence effects upon soil gas exchange. *Proc. Soil Sci. Soc. Amer.*, **35,** 16–21. (119)

King, T. J. (1975). Inhibition of seed germination under leaf canopies in *Arenaria serpylifolia, Veronica arvensis* and *Cerastium holosteoides. New Phytol.*, **75,** 87–90. (224)

King, T. J. (1977). The plant ecology of ant-hills in calcareous grassland I, II and III. *J. Ecol.*, **65,** 235–315. (384)

Klepper, B. (1968). Diurnal pattern of water potential in woody plants. *Pl. Physiol. Lancaster*, **43,** 1931–4. (19, 168)

Klikoff, L. G. (1965). Photosynthetic response to temperature and moisture stress of three timberline meadow species. *Ecology*, **46,** 516–17. (189)

Kluge, M. (1976). Carbon and nitrogen metabolism under water stress. In *Water and Plant Life*, ed. Lange, O. L., L. Kappen, and E. D. Schultz, pp. 243–52, Springer, Berlin. (174)

Kluge, M., and I. P. Ting (1978). *Crassulacean Acid Metabolism*, Springer, Berlin. (130, 204, 289, 362)

Knight, T. A. (1801). Account of some experiments on the ascent of sap in trees. *Phil. Trans. R. Soc. Lond.*, **91,** 333–53. (422)

Koller, D. (1972). Environmental control of seed germination. In *Seed Biology* Vol. 2, ed. Kozlowski, T. T., pp. 1–101, Academic Press, New York. (170, 178)

Konings, H., and G. Verschuren (1980). Formation of aerenchyma in roots of *Zea mays* in aerated solutions and its relation to nutrient supply. *Physiol. Pl.*, **49,** 265–70. (324)

Kononova, M. (1961). *Soil Organic Matter*, Pergamon, London. (110)

Kordan, H. A. (1976a). Adventitious root initiation and growth in relation to oxygen supply in germinating rice seedlings. *New Phytol.*, **76,** 81–6. (326)

Kordan, H. A. (1976b). Oxygen as an environmental factor influencing normal morphogenetic development in germinating rice seedlings. *J. exp. Bot.*, **27,** 947–52. (326)

Kormondy, E. J. (1965). *Readings in Ecology*, Prentice Hall, New Jersey. (417, 418, 419, 420)

Kovda, V. A., E. M. Samilova, J. L. Charles, and J. J. Skujins (1979). Soil processes in arid lands. In *Arid-land Ecosystems* Vol. 1, ed. Goodall, D. W., and R. A. Perry, Cambridge Univ. Press, Cambridge. (171)

Kowal, J. M., and A. H. Kassam (1978). *Agricultural Ecology of Savanna*, Clarendon Press, Oxford. (69)

Kozlokski, T. T. (1964). *Water Metabolism in Plants*, Harper and Row, New York. (168)

Kozlowski, T. T. (1968a, b, 1972a, 1976, 1978). *Water Deficits and a Plant Growth*, Vols. 1–5, Academic Press, New York. (168)

Kozlowski, T. T. (ed.) (1972b). Shrinking and swelling of plant tissues. In *Water Deficits and Plant Growth*, Vol. 3, pp. 1–64, Academic Press, New York. (168)

Kozlowski, T. T. (1972c). *Seed Biology*, Vols. 1, 2, and 3, Academic Press, New York. (223)

Kozlowski, T. T. (1973). *Shedding of Plant Parts*, Academic Press, New York. (180)

Kozlowski, T. T., and C. E. Ahlgren (1974). *Fire and Ecosystems*, Academic Press, New York. (394)

Kozuchowski, J., and D. L. Johnson (1978). Gaseous emission of mercury from an aquatic vascular plant. *Nature, Lond.*, **274,** 468–9. (370)

Kramer, P. J. (1938). Root resistance as a cause of absorption lag. *Am. J. Bot.*, **25,** 110–13. (167)

Kramer, P. J. (1940). Causes of decreased absorption of water by plants in poorly aerated media. *Am. J. Bot.*, **27,** 216–20. (167)

Kramer, P. J. (1951). Injury to plants resulting from flooding the soil. *Pl. Physiol., Lancaster*, **26,** 722–36. (308)
Kramer, P. J. (1969). *Plant and Soil Water Relationships*, McGraw-Hill, New York. (327)
Kramer, P. J., and T. T. Kozlowski (1960). *Physiology of Trees*, McGraw-Hill, New York. (216)
Kramer, P. J., W. S. Riley, and T. T. Bannister (1952). Gas exchange of Cypress knees. *Ecology*, **33,** 117–21. (328)
Kriedman, P. E., E. Torokfalvy, and R. E. Smart (1973). Natural occurrence and utilization of sunflecks by grapevine leaves. *Photosynthetica*, **7,** 18–27. (22)
Kruckberg, A. R. (1954). The ecology of serpentine soils. III. Plant species in relation to serpentine soils. *Ecology*, **35,** 267–87. (281, 284)
Kubiena, W. L. (1953). *Soils of Europe*, Murby, London. (51, 81, 99, 102, 103, 115, 116)
Kubin, S. (1971). Measurement of radiant energy. In *Plant Photosynthetic Production: Manual of Methods*, ed. Sestak, Z., J. Catsky, and P. G. Jarvis, pp. 702–65, Junk, The Hague. (34, 35)
Kutschera, L. (1960). *Wurzelatlas mittleeuropàischer Ackerunkräuter und Kulturpflanzen*, DLG-Verlags, Frankfurt. (119)
Ladd, J. N., and B. H. A. Butler (1975). Humus-enzyme systems and synthetic organic polymer-enzyme analogues. In *Soil Biochemistry* Vol. 4, ed. Paul, E. A., and A. D. McLaren, pp. 143–94, Dekker, New York. (105, 106, 108)
Landsberg, J. J., and D. R. Butler (1980). Stomatal response to humidity: implications for transpiration. *Pl. Cell Env.*, **3,** 29–33. (20, 166)
Lang, A. (1965). Effects of some internal and external conditions on seed germination in *Encyclopedia of Plant Physiology*, Vol. 15, ed. Ruhland, W., pp. 848–93, Springer, Berlin. (170)
Lange, O. L. (1959). Unterschungen uber warmhaushalt unt hitzerresistenz mauretanischer wuster- und savannenpflanzen. *Flora*, **147,** 595–61. (19)
Lange, O. L. (1965). Der CO_2 gasweschel von flechten bei tiefen Temperaturen. *Planta*, **64,** 1–19. (205)
Lange, O. L., L. Kappen, and E. D. Schulze (1976). *Water and Plant Life*, Springer, Berlin. (130)
Lange, O. L., E. D. Schulze, L. Kappen, U. Buschbohm, and M. Evenari (1975a). Adaptation of desert lichens to drought and extreme temperature. In *Environmental Physiology of Desert Organisms*, ed. Hadley, N. F., pp. 20–37, Dowden, Hutchinson and Ross, Stroudsburg. (180)
Lange, O. L., E. D. Schulze, L. Kappen, M. Evenari, and U. Buschbohm (1975b). CO_2 exchange pattern under natural conditions of *Carralluma negevensis*, a CAM plant of the Negev desert. *Photosynthetica*, **9,** 31 8–26. (182, 205).
Lanyi, J. K. (1979). Salt tolerance in microorganisms. In *The Biosaline Concept*, ed. Aller, J. C., E. Epstein *et al.*, pp. 217–33, Plenum, New York. (289)
Larcher, W. (1969). The effect of environmental and physiological variables on the CO_2 gas exchange of trees. *Photosynthetica*, **3,** 167–98. (132, 205)
Larcher, W. (1973). Temperature resistance and survival. In *Temperature and Life*, ed. Precht, H., J. Christopherson, and H. Hensel, pp. 203–31, Springer, Berlin. (209, 210, 213)
Larcher, W. (1975). *Physiological Plant Ecology*, Springer Berlin (2nd edn. 1980). (146, 425)
Larsen, S. (1967). Soil phosphorus. *Adv. Agron.*, **19,** 151–210. (241, 242)
Latham, R. E. (1951). *The Nature of the Universe—Lucretius*, Penguin Books, London. (417)
Lauchli, A. (1976). Genotypic variation in transport. In *Encyclopedia of Plant Physiology*, Vol. 2B *Transport in Plants*, ed. Lutge, U., and M. G. Pitman, pp. 372–93, Springer, Berlin. (264)

Leach, G. J., and D. J. Watson (1968). Photosynthesis in crop profiles measured by phytometers, *J. app. Ecol.*, **5,** 381–408. (137)

LeBlanc, F., and J. de Sloover (1970). Relation between industrialization and the distribution and growth of epiphytic lichens and mosses in Montreal. *Can. J. Bot.*, **48,** 1485–96. (370)

Lee, J. A., and G. R. Stewart (1978). Ecological aspects of nitrogen assimilation. *Adv. Bot. Res.*, **6,** 1–43. (273)

Lee, R. B. (1978). Inorganic metabolism nitrogen metabolism in barley roots under poorly aerated conditions. *J. exp. Bot.*, **29,** 693–708. (309)

Leeper, G. W. (1978). *Managing the Heavy Metals on the Land*, Dekker, New York. (371)

Lemon, E., D. W. Stewart, and R. W. Shawcroft (1971). The sun's work in a cornfield. *Science*, **174,** 371–8. (140, 141, 144)

Lemon, E. R., and A. E. Erickson (1952). The measurement of oxygen diffusion in the soil with a platinum micro-electrode. *Proc. Soil Sci. Soc. Amer.*, **16,** 160–3. (294)

Lemon, E. R., and A. E. Erickson (1955). Principle of the platinum micro-electrode as a method of characterizing soil aeration. *Soil Sci.*, **79,** 382–92. (294)

Levin, D. A. (1970). The exploitation of pollinators by species and hybrids of *Phlox*. *Evolution*, **24,** 367–77. (386)

Levin, D. A., and W. W. Anderson (1970). Competition for pollinators between simultaneously flowering species. *Amer. Nat.*, **104,** 455–67. (386)

Levitt, J. (1972). *Responses of Plants to Environmental Stress*, Academic Press, New York (2nd edn. 1980). (174, 177, 180, 209)

Lewis, B. G. (1976). Selenium in biological systems and pathways for its volatilization in higher plants. In *Environmental Biogeochemistry*, ed. Nriagu, J. O., pp. 389–409, Ann. Arbor, Michigan. (286)

Lewis, D. H. (1980). Boron, lignification and the origin of the vascular plants—a unified hypothesis. *New Phytol.*, **84,** 20 9–29. (233)

Lewis, H., and F. W. Went (1945). Plant growth under controlled conditions. IV. Response of California annuals to photoperiod and temperature. *Amer. J. Bot.*, **32,** 1–12. (217)

Lewis, M. C. (1972). The physiological significance of variation in leaf structure. *Sci. Progr.*, **60,** 25–51. (16)

Lewis, M. C., and T. V. Callaghan (1975). Tundra. In *Vegetation and the Atmosphere*, ed. Monteith, J. C., pp. 399–433, Academic Press, New York. (17)

Leyton, L., and L. Z. Rousseau (1957). Root growth of tree seedlings in relation to aeration. In *Physiology of Forest Trees*, ed. Thimmann, K. V., pp. 467–75, Ronald, New York. (328)

Leyton, L., E. R. C. Reynolds, and F. B. Thompson (1968). Interception of rainfall by trees and moorland vegetation. In *The Measurement of Environmental Factors in Terrestrial Ecology*, ed. Wadsworth, R. M., pp. 97–108, Blackwell, Oxford. (195)

Liddle, M. J., and K. G. Moore (1974). The microclimate of sand dune tracks: the relative contribution of vegetation removal. *J. Ecol.*, **62,** 1057–68. (114)

Liebig, J. (1840). *Chemistry and its Application to Agriculture and Physiology*, Taylor and Walton, London. (337, 422)

Liebig, J. (1862). *Die Naturgesetze dies Feldbaues*, Braunschweig, Vieweg. (422)

Lieth, H. (1960). Patterns of change within grassland communities. In *The Biology of Weeds*, ed. Harper, J. L., pp. 27–39, Blackwell, Oxford. Oxford. (120, 343, 397)

Lieth, H. (1968b). The determination of plant dry matter production with special emphasis on the underground parts. In *Functioning of Terrestrial Ecosystems at the Primary Production Level*, pp. 233–41, UNESCO, Paris. (120)

Lieth, H. (1975). Historical survey of primary productivity research. In *Primary Productivity of the Biosphere*, ed. Lieth, H., and R. H. Whittaker, pp. 7–16, Springer, Berlin. (334, 337, 347, 348)

Lieth, H. (1975). Measurement of caloric values. In *Primary Productivity of the Biosphere*, ed. Lieth, H., and R. H. Whittaker, pp. 119–29, Springer Berlin. (346)

Lieth, H., and R. H. Whittaker (1975). *Primary Productivity of the Biosphere*, Springer, Berlin. (225, 349, 355, 360, 422)

Likens, G. E., A. L. Bormann, N. M. Johnson, D. W. Fisher, and R. S. Pierce (1970). The effect of forest cutting and herbicide treatment on nutrient budgets in the Hubbard Brook watershed ecosystem. *Ecol. Monographs*, **40,** 23–47. (272, 374, 375)

Likens, G. E., F. H. Bormann, R. S. Pierce, J. S. Eaton, and N. M. Johnson (1977). *Biogeochemistry of a Forested Ecosystem*, Springer, Berlin. (197, 365, 375, 376, 390, 391, 425)

Lindeman, R. L. (1942). The trophic dynamic aspect of ecology. *Ecology*, **23,** 399–418. (338, 424)

Lindsay, W. L. (1972). Inorganic phase-equilibria of micro-nutrients in soils. In *Micronutrients in Agriculture*, ed. Mortvedt, W., P. M. Giordano, and W. L. Lindsay, pp. 41–57, Soil Sci. Soc. Amer., Wisconsin. (96)

Lisk, D. J. (1972). Trace metals in soils, plants and animals. *Adv. Agron.*, **24,** 267–325. (371)

Livett, E. A., J. A. Lee, and J. H. Tallis (1979). Lead zinc and copper analyses of British blanket peats. *J. Ecol.*, **67,** 865–91. (369)

Livingstone, B. E. (1903). *The Role of Diffusion and Osmotic Pressure in Plants*, Univ. Chicago Press, Chicago. (423)

Livingstone, D. A. (1975). Late Quaternary climatic change, in Africa. *Ann. Rev. Ecol. Syst.*, **6,** 249–80. (3, 68)

Livingston, R. B., and M. L. Allesio (1968). Buried viable seed in successional field and forest stands in Harvard Forest, Massachusetts. *Bull. Torrey Bot. Club*, **95,** 58–69. (222)

Loftfield, J. V. G. (1921). The behaviour of stomata. *Publ. Carnegie Inst. Washington*, **314,** 1–104. (19)

Long, I. F. (1968). Instrumental techniques for measuring the micrometeorology of crops. In *The Measurement of Environmental Factors in Terrestrial Ecology*, pp. 1–32, ed. Wadsworth, R. M., Blackwell, Oxford. (37)

Longman, K. A., and J. Jenik (1974). *Tropical Forest and its Environment*, Longman, London. (74)

Loomis, R. S., and P. A. Gerakis (1975). Productivity of agricultural ecosystems. In *Photosynthesis and Primary Production in Different Environments*, ed. Cooper, J. P., pp. 145–72, Cambridge Univ. Press, Cambridge. (34)

Loveless, A. R. (1962). Further evidence to support a nutritional interpretation of sclerophylly. *Ann. Bot.*, **26,** 551–61. (278)

Lovelock, J. E. and L. Margulis (1974). Homeostatic tendencies of the earth's atmosphere. *Orig. Life*, **5,** 93–103. (357, 367)

Low, A. J. (1972). The effect of cultivation on the structure and other physical characteristics of grassland and arable soils (1945–70). *J. Soil Sci.*, **23,** 363–80. (116)

Lugo, A. E., and S. C. Snedaker (1974). The ecology of mangroves. *Ann. Rev. Ecol. Syst.*, **5,** 39–64. (82)

Lund, J. W. G. (1965). The ecology of freshwater phytoplankton. *Biol. Rev.*, **40,** 231–93. (351)

Lundegardh, H. (1931). *Environment and Plant Development*. Arnold, London (German edn. 1925). (424)

Lütge, U. (1975). Salt glands. In *Ion Transport in Plant Cells and Tissues*, ed. Baker, D. A., and J. L. Hall, pp. 335–76, North Holland, Amsterdam. (288)

Lütge, U., and M. G. Pitman (1976). *Encyclopedia of Plant Physiology*, Vols. 2A and 2B, *Transport in Plants*, Springer, Berlin. (262)

Luxmoore, R. J., and L. H. Stolzy (1969). Root porosity and growth responses of rice and maize to oxygen supply. *Agron. J.*, **61,** 202–204. (324)

Luxmoore, R. J., and L. H. Stolzy (1972). Oxygen diffusion in the soil-plant system V and VI. *Agron. J.*, **64,** 720–29. (321)

Luxmoore, R. J., L. H. Stolzy, and J. Letey (1970). Oxygen diffusion in the soil-plant system. *Agron. J.*, **62,** 317–32. (321)

McArthur, R., and Wilson, E. O. (1967). *The Theory of Island Biogeography*, Princeton Univ. Press, Princeton. (421)

McCloud, D. E., and L. S. Dunavin (1954). Agrohydric balance studies at Gainsville, Florida. In *Publication in Climatology*, pp. 55–68, Seabrook, New Jersey. (196)

McCreee, K. J. (1973). A rational approach to light measurements in plant ecology. *Curr. Adv. Pl. Sci.*, **5,** 39–43. (35)

McCree, K. J., and J. H. Troughton (1966). Non-existence of an optimum leaf area index for the production rate of white clover grown under constant conditions. *Pl. Physiol. Lancaster*, **41,** 1615–22. (135, 136)

McFarlane, M. (1976). *Laterite and Landscape*, Academic Press, London. (68, 70)

McGregor, A. N., and D. E. Johnson (1971). Capacity of desert algal crusts to fix atmospheric nitrogen. *Proc. Soil Sci. Soc. Amer.*, **35,** 843–4. (234)

McIntyre, D. S. (1970). The platinum microelectrode method of soil aeration measurement. *Adv. Agron.*, **22,** 235–83. (295)

McKintosh, R. P. (1978). *Phytosociology*, Academic Press, New York. (419)

McLean, R. O., and W. R. Ivimey-Cook (1973). *Textbook of Theoretical Botany*, Vol. 4, Longman, London. (418)

Macklon, A. E. S., and P. E. Weatherley (1965). Controlled environment studies of the nature and origins of water deficits in plants. *New Phytol.*, **64,** 414–27. (160)

Mackney, D. (1961). A podzol development sequence in oakwoods and heath in central England. *J. Soil Sci.*, **12,** 23–40. (59)

McLaren, A. D., and G. H. Peterson (1967). *Soil Biochemistry*, Vol. 1, Dekker, New York. (104)

MacMannon, M., and R. M. M. Crawford (1971). A metabolic theory of flooding tolerance: the significance of enzyme distribution and behaviour. *New Phytol.*, **70,** 299–306. (325)

McMichael, B. L., W. R. Jordan, and R. D. Powell (1972). An effect of water stress on ethylene production by intact cotton petioles. *Pl. Physiol., Lancaster*, **49,** 658–60. (324)

McPherson, D. C. (1939). Cortical air spaces in the roots of *Zea mays* L. *New Phytol.*, **38,** 190–202. (323)

McRae, I. C., and T. F. Castro (1967). Nitrogen fixation in some tropical rice soils. *Soil Sci.*, **103,** 277–80. (302)

McWilliams, J. R., and P. J. Kramer (1968). The nature of the perennial response in Mediterranean grasses. I. Water relations and survival in *Phalaris*. *Aust. J. agric. Res.*, **19,** 381–95. (121, 186, 403)

Maggs, D. H., and D. M. Alexander (1970). Tests of a uranyl oxalate light integrator for use in fruit tree canopies. *J. app. Ecol.*, **7,** 639–46. (34)

Mahmoud, A., J. P. Grime, and S. B. Furness (1975). Polymorphism in *Arrhenatherum elatius* Beauv. ex J. and C. Presl. *New Phytol.*, **75,** 769–76. (378)

Mair, B. (1968). Frosthartegradienten entlag der knospenfolge auf eschentrieben. *Planta*, **82,** 164–9. (211)

Majmudar, A. M., and J. P. Hudson (1957). The effect of different water regimes on the growth of plants under glass. II. Experiments with Lettuces (*Lactuca sativa Linn.*). *J. hort. Sci.*, **32,** 201–13. (162)

Malthus, T. R. (1798). *The principles of population.* (2, 395, 418)

Malyuga, D. P. (1964). *Biogeochemical Methods of Prospecting*, Consultants Bureau (translation), New York. (283)

Mandal, L. N. (1961). Transformations of iron and manganese in waterlogged rice soils. *Soil Sci.*, **91,** 121–6. (245)

Mann, K. H., and R. O. Chapman (1975). Primary production of marine macrophytes. In *Photosynthesis and Primary Production in Different Environments*, ed. Cooper, J. P., pp. 207–80, Cambridge Univ. Press, Cambridge. (354)
Mansfield, T. H. (1976). *Effects of Air Pollutants on Plants*, Cambridge Univ. Press, Cambridge. (371)
Margalef, R. (1968). *Perspectives in Ecological Theory*, Univ. Chicago Press, Chicago. (396, 411)
Marks, G. C., and T. T. Kozlowski (1973). *Ectomycorrhizae*, Academic Press, New York. (257)
Marks, M. K., and S. D. Prince (1979). Induction of flowering in wild lettuce (*Lactuca serriola* L.). II. Devernalization. *New Phytol.*, **82,** 265–77. (218, 219)
Marks, P. L., and F. H. Bormann (1972). Revegetation following forest cutting: mechanisms for return to steady state nutrient cycling. *Science*, **176,** 914–5. (390)
Martin, M. H. (1968). Conditions affecting the distribution of *Mercurialis perennis* L. in certain Cambridgeshire woodlands. *J. Ecol.*, **56,** 777–93. (310, 315, 320)
Maschner, H., and S. Azarabadi (1979). Role of the rhizosphere in utilization of inorganic iron-III compounds by corn plants. In *The Soil-Root Interface*, ed. Harley, J. L., and R. S. Russel, pp. 428–9, Academic Press, London. (268)
Mather, J. R. (1954). The measurement of potential evapo-transpiration. *Publication in Climatology* **225,** Seabrook, New Jersey. (196)
Matthews, W. (1886). Navajo names for plants. *Amer. Nat.*, **20,** 767–77. (1)
Mathys, W. (1977). The role of malate, oxalate and mustard-oil glucosides in the evolution of zinc resistance in herbage plants. *Physiol. Pl.*, **40,** 130–6. (283)
Maximov, N. A. (1929). *The Plant in Relation to Water*, Allen and Unwin, London. (15, 177, 183, 424)
May, R. M. (1976). *Theoretical Ecology: Principles and Application*, Blackwell, Oxford. (421)
Maynard-Smith, J. (1978). Optimization theory in evolution. *Ann. Rev. Ecol. Syst.*, **9,** 31–56. (3)
Mazelis, M., and B. Vennesland (1957). Carbon dioxide fixation in oxaloacetate in higher plants. *Pl. Physiol., Lancaster*, **32,** 591–600. (324)
Meidner, H. (1975). Water supply, evaporation and vapour diffusion in leaves. *J. exp. Bot.*, **26,** 666–73. (127, 175)
Meidner, H., and T. A. Mansfield (1968). *Physiology of Stomata.* McGraw-Hill, London. (127)
Messenger, A. S. (1975). Climate, time and organism in relation to podzol development in Michigan II. *Proc. Soil Sci. Soc. Amer.*, **39,** 698–702. (104)
Messenger, A. S., J. R. Kline, and D. Wilderotten (1978). Aluminium biocycles as a factor in soil change. *Plant and Soil*, **49,** 703–9. (104)
Meyer, F. H., and D. Göttsche (1971). Distribution of root tips and tender roots of beech. In *Integrated Experimental Ecology*, ed. Ellenberg, H., Chapman and Hall, London. (120, 121)
Milburn, J. A. (1979). *Water Flow in Plants*, Longman, London. (162, 165)
Milburn, J. A., and R. P. C. Johnson (1966). The conduction of sap. II. Detection of vibrations produced by sap cavitation in *Ricinus* xylem. *Planta*, **69,** 43–52. (163)
Millar, A. A., M. E. Duyser, and G. E. Wilkinson (1968). Internal water balance of barley under soil moisture stress. *Pl. Physiol., Lancaster*, **43,** 968–72. (157)
Miller, H. G. (1979). The nutrient budget of even-aged forests. In *The Ecology of Even-aged Forest Plantations*, ed. Ford, E. D., D. C. Malcom, and J. Atterson, pp. 221–56, Inst. Terrest. Ecol., Edinburgh. (390)
Miller, R. B. (1963). Plant nutrients in Hard Beech. III. The cycle of nutrients. *N.Z. J. Soil. Sci.*, **6,** 388–413. (373, 374, 390)
Milner, C., and R. E. Hughes (1968). *Methods for the Measurement of Primary Productivity of Grassland*, Blackwell, Oxford. (345)

Milthorpe, F. L. (1950). Changes in drought resistance of wheat seedling during germination. *Ann. Bot.*, NS **14,** 79–89. (190)

Milthorpe, F. L., and J. Moorby (1974). *An Introduction to Crop Physiology*, Cambridge Univ. Press, Cambridge. (351, 353, 354).

Milton, W. E. J., and R. O. Davies (1947). The yield, botanical and chemical composition of natural hill herbage under manuring and controlled grazing hay conditions. *J. Ecol.*, **35,** 65–95. (350, 352, 379, 380)

Ministry of Agriculture, Food and Fisheries (1970). *Modern Farming and the Soil*, HMSO, London. (350)

Mitsui, S., S. Aso, K. Kumazawa, and T. Ishiwara (1954). The nutrient uptake of rice plants as influenced by hydrogen sulphide and butyric acid abundantly evolving under waterlogged soil conditions. *Trans. 5th Int. Congr. Soil. Sci.*, **2,** 364 (312).

Mohr, E. C. J., and F. A. van Bahren (1959). *Tropical Soils*, van Hoeve, The Hague. (47, 49, 69, 246)

Molisch, H. (1888). Uber Wurzelausschiedungen und deren Einwerkung auf organische Substanzen. *Sitzungsber. Akad. Wiss. Wien. Math. Nat. Kl.*, **96,** 84. (314)

Monsi, M., and T. Saeki (1953). Uber den Lichtfaktor in den Pflanzengesellschaften und seine Bedeutung fur die Stoffproduction. *Jap. J. Bot.*, **14,** 22–52. (21, 22, 135, 139)

Monteith, J. L. (1963). Gas exchange in plant communities. In *Environmental Control of Plant Growth*, ed. Evans, L. T., pp. 95–112, Academic Press, New York.

Monteith, J. L. (1965). Evaporation and environment. In *The State and Movement of Water in Living Organisms*, ed. Fogg, G. E., pp. 205–34, Cambridge Univ. Press, Cambridge. (32)

Monteith, J. L. (1972). *Survey of Instruments for Micro-meteorology*, Blackwell, Oxford. (34, 36, 37, 38)

Monteith, J. L. (1973). *Principles of Environmental Physics*, Arnold, London. (Prelim page xvii) 12, 13, 26, 27, 32, 33, 165, 204, 346)

Monteith, J. L. (1975). *Vegetation and the Atmosphere*, Academic Press, New York. (20, 32, 344)

Monteith, J. L. (1976). Spectral distribution of light in leaves and foliage. In *Light and Plant Development*, ed. Smith, H., pp. 447–59, Butterworths; London. (224)

Mooney, H. A. (1975). Plant physiological ecology: a synthesis. In *Physiological Adaptation to the Environment*, ed. Vernberg, F. J., pp. 19–36, Intex, New York. (131)

Mooney, H. A., O. Bjorkman, and J. Berry (1975). Photosynthetic adaptation to high temperatures. In *Environmental Physiology of Desert Organisms*, ed. Hadley, N. F., pp. 138–51, Dowden, Hutchinson and Ross, Stroudsberg. (15, 20, 212)

Mooney, H. A., and F. di Castri (1973). *Mediterranean-Type Ecosystems*, Springer, Berlin. (74)

Moore, D. P. (1974). Physiological effects of pH on roots. In *The Plant Root and its Environment*, ed. Carson, E. W., pp. 135–51, Univ. Virginia Press, Charlottsville. (249, 265)

Moore, P. D. (1975). Origin of blanket mires. *Nature, Lond.*, **256,** 267–9. (78, 197)

Moore, P. D., and D. J. Bellamy (1973). *Peatlands*, Elek Science, London. (76, 78, 79, 103).

Moore, P. D., and J. A. Webb (1978). *An Illustrated Guide to Pollen Analysis*, Hodder and Stoughton, London. (103)

Morgan, D. C., and H. Smith (1979). A systematic relationship between phytochrome control, development and specific habitat, for plants grown in simulated natural radiation. *Planta*, **145,** 253–8. (224)

Morriset, C. (1978). Structural and cyto-enzymological aspects of the mitochondrion in excised roots of oxygen-deprived *Lycopersicon* cultivated *in vitro*. In *Plant Life in Anaerobic Environments*, ed. Hook, D. D., and R. M. M. Crawford, pp. 497–537, Ann. Arbor, Michigan. (326)

Mudd, J. B., and T. T. Kozlowski (1975), *Responses of Plants to Air Pollution*, Academic Press, New York. (371)

Mueller-Dombois, D., and H. Ellenberg (1974), *Aims and Methods of Vegetation Ecology*, Wiley, Chichester. (421)

Mulder, E. G. (1975). Physiology and ecology of free-living nitrogen fixing bacteria. In *Nitrogen Fixation by Free Living Organisms*, ed. Stewart, W. D. P., Cambridge Univ. Press, Cambridge. (235)

Mulder, E. G., T. A. Lie, and J. W. Woldendorp (1969). Biology and soil fertility. In *Soil Biology*, ed. Burges, A., and F. Raw, pp. 163–208, UNESCO, Paris. (43, 234)

Muller, C. H. (1966). The role of chemical inhibitors (allelopathy) in vegetational composition. *Bull. Torrey Bot. Club*, **93,** 332–51. (405)

Muller, C. H., W. H. Muller, and B. L. Haines (1964). Volatile growth inhibitors produced by aromatic shrubs. *Science*, **143,** 471–3. (378)

Muller, W. H., P. Lorber, B. Haley, and K. Johnson (1969). Volatile growth inhibitors produced by *Salvia leucophylla*: effect on oxygen uptake by mitochondrial suspensions. *Bull. Torrey Bot. Club*, **96,** 89–95. (405)

Murozumi, M., T. J. Chow, and C. Patterson (1969). Chemical concentration of pollutant lead aerosols, terrestrial dusts and sea-salts in Greenland and Antarctic, snow-strata. *Geochim. et Cosmochim. Acta*, **33,** 1247–94. (369)

Neales, T. F. (1975). The gas exchange patterns of CAM plants. In *Environmental and Biological Control of Photo-synthesis*, ed. Marcelle, R., pp. 299–310, Junk, The Hague. (166)

Newbould, P. (1969). The absorption of nutrients by plants from different zones in the soil. In *Ecological Aspects of the Mineral Nutrition of Plants*, ed. Rorison, I., pp. 177–90, Blackwell, Oxford.

Newbould, P. J. (1963). Production ecology. *Sci. Progr.*, **51**, 91–104. (340)

Newbould, P. J. (1967). *Methods for Estimating the Primary Production of Forests*, Blackwell, Oxford. (342)

Newbould, P. J. (1968). Methods of estimating root production. In *Functioning of Terrestrial Ecosystems at the Primary Production Level*, ed. Eckardt, F. D., pp. 187–90, UNESCO, Paris. (120, 343)

Newman, E. I. (1963). Factors controlling the germination date of winter annuals. *J. Ecol.*, **51,** 625–38. (178, 222)

Newman, E. I. (1969a, b). Resistance to water flow in soil and plant. I. Soil resistance in relation to amounts of roots: theoretical estimates. II. A review of experimental evidence on rhizosphere resistance, *J. app. Ecol.*, **6,** 1–12, 261–72. (160)

Newman, E. I. (1976). Water movement through root systems. *Phil. Trans. R. Soc. Lond.* B, **273,** 463–78. (116, 148, 162)

Newman, E. I. (1978). Root microorganisms: their significance in the ecosystem. *Biol. Rev.*, **53,** 511–54. (257, 258)

Newman, E. I., and A. D. Rovira (1975). Allelopathy amongst some British grassland species. *J. Ecol.*, **63,** 727–37. (405)

Newton, J. E., and G. E. Blackman (1970). The penetration of solar radiation through leaf canopies of different structure, *Ann. Bot.*, **34,** 329–48. (21)

Nguyen, B. C., A. Gandry, B. Bansary, and G. Lambert (1978). Re-evaluation of the role of dimethyl sulphide in the sulphur budget. *Nature, Lond.*, **275,** 637–9. (368)

Nicholson, M. (1970). *The Environmental Revolution.* Hodder and Stoughton, London. (2)

Nielsen, D. R., and J. G. McDonald (1978). *Nitrogen in the Environment* Vols. 1 and 2, Academic Press, New York. (240)

Njoku, É. (1957). The effect of mineral nutrition and temperature on leaf shape in *Ipomoea caerulea*. *New Phytol.*, **56,** 154–71. (280)

Nobel, P. S. (1974). *Biophysical Plant Physiology*, Freeman, San Francisco. (184, 290)

Norman, J. M., C. B. Tanner, and G. W. Thurtell (1969). Photosynthetic light sensor for measurements in plant canopies. *Agron J.*, **61,** 840–3. (35)

Nriagu, J. O. (1976). *Environmental Biogeochemistry*, Vols. 1 and 2, Ann Arbor, Michigan. (371)

Nriagu, J. O. (1978; 1979a and b). *Copper in the environment*; *Zinc in the environment*; *Sulphur in environment*, Wiley, Chichester. (371)

Nutman, P. S. (1976). *Symbiotic Nitrogen Fixation in Plants.* Cambridge Univ. Press, Cambridge. (236, 240)

Nye, P. H. (1954). Some soil forming processes in the humid tropics I. *J. Soil Sci.*, **5,** 7–21. (71)

Nye, P. H., and F. H. C. Marriot (1969). A theoretical study of the distribution of substances around roots resulting from simultaneous diffusion and mass-flow. *Plant and Soil*, **30,** 459–72. (255, 256)

Nye, P. H., and P. B. Tinker (1977). *Solute Movement in the Soil-Root System.* Blackwell, Oxford. (121, 170, 253, 256, 257, 261)

Oades, J. M. (1978). Mucilage at the root surface. *J. Soil Sci.*, **29,** 1–16. (116)

Oates, D., and Oates, J. (1976). Early irrigation agriculture in Mesopotamia. In *Problems in Economic and Social Archaeology*, ed. Sieveking, G., I. Longworth, and K. Wilson, pp. 109–36, Duckworth, London. (147)

Obeid, M., D. Machin, and J. L. Harper (1967). Influence of density on plant variations in fibre-flax, *Linum usitatissimum. Crop. Sci.*, **7,** 471–3. (408)

Oden, S. (1962). Electrometric methods for oxygen studies in water and soil; IV. Fundamental problems involved with the design and use of oxygen diffusion electrodes. C.S.I.R.O. Aust. translation No. 6480 C from *Grundforbattring*, **3,** 117–210 (Swedish). (295)

Odening, W. R., B. R. Strain, and W. C. Oechel (1974). The effect of decreasing waterpotential on net CO_2 exchange of intact desert shrubs. *Ecology*, **55,** 1086–95. (189–90)

Odum, E. P. (1953 and 1959). *Fundamentals of Ecology*, 1st and 2nd edns, Saunders, Philadelphia. (420, 424)

Odum, E. P. (1969). The strategy of ecosystem development. *Science*, **164,** 262–70. (337)

Odum, E. P. (1971). *Fundamentals of Ecology*, 3rd edn., Saunders, Philadelphia. (335, 392, 420, 421, 424)

Odum, H. T. (1957). Trophic structure and productivity of Silver Springs, Florida. *Ecol. Monogr.*, **27,** 55–112. (338, 420, 424)

Odum, H. T. (1971). *Environment, Power and Society*, Wiley, New York. (351, 361)

Odum, H. T., and E. P. Odum (1955). Trophic structure and productivity of a windward coral reef community on Eniwetok atoll. *Ecol. Monogr.*, **25,** 291–320. (338, 420, 424)

Odum, H. T., and C. F. Jordan (1970). Metabolism and evapo-transpiration of the lower forest in a giant plastic cylinder. In *The Tropical Rainforest*, ed. Odum, H. T., and R. Pigeon, pp. 1.165–1.189, AEC Div. Tech. Inf., Oak Ridge. (137, 344)

Odum, W. E., and E. J. Heald (1975). Mangrove forests and aquatic productivity. In *Coupling of Land and Water Systems*, ed. Hasler, A., pp. 129–36, Springer, Berlin. (345)

Oehme, F. W. (1978). *Toxicity of Heavy Metals in the Environment.* Parts 1 and 2, Dekker, New York. (37)

Oinonen, E. (1967). The correlation between Finnish bracken (*Pteridium aquilinum* (L.) Kuhn.) clones and certain periods of site history. *Acta Forest. Fenn.*, **83,** 1–51. (278)

Okudu, A., and E. Takahashi (1964). The role of silicon. In *The Mineral Nutrition of the Rice Plant*, ed. International Rice Research Institute, pp. 123–46, Johns Hopkins, Baltimore. (306)

Olatoye, S. T., and M. A. Hall (1973). Interaction of ethylene and light on dormant weed seeds. In *Seed Ecology*, ed. Heydecker, W., pp. 233–49, Butterworth, London. (222)

Oliver, J. (1960). Winds and vegetation on the Dale peninsula. *Field Studies*, **1,** (2), 37–48. (227)

Oliver, S., and S. A. Barber (1966). An evaluation of the mechanisms governing supply of Ca, Mg, K and Na to Soybean roots (*Glycine max*). *Proc. Soil Sci. Soc. Amer.*, **30,** 82–6. (170, 253)

Olsen, C. (1921). Ecology of *Urtica dioica* L. *J. Ecol.*, **9,** 1–18. (275)

Olsen, C. (1958). Iron uptake in different plant spp. as a function of the pH value of the nutrient solution. *Physiol. Pl.*, **11,** 889–90 5. (265)

Olsen, R. A., and J. E. Robbins (1971). The cause of the suspension effect in resin-water systems. *Proc. Soil Sci. Soc. Amer.*, **35,** 260–5. (91)

Oosting, H. J. (1956). *The Study of Plant Communities: an Introduction to Ecology*, Freeman, San Francisco. (74)

Oppenheimer, H. R., and K. Mendel (1939). Orange leaf transpiration under orchard conditions. I. Soil moisture high. *Palestine J. Bot.*, **2,** 171–250. (139, 157)

Ordin, L. (1958). The effect of water stress on the cell wall metabolism of plant tissue. In *Radio Isotopes in Scientific Research*, Vol. 4, pp. 553–64. Pergamon, London. (173)

Ordin, L. (1960). Effect of water stress on cell wall metabolism of *Avena coleoptile* tissue. *Pl. Physiol., Lancaster*, **35,** 443–50. (173)

Ordway, D. E. (1969). An aerodynamicist's analysis of the Odum cylinder approach to net CO_2 exchange. *Photosynthetica*, **3,** 199–209. (137)

Orians, G. H., and O. T. Solbrig (1977a). *Convergent Evolution in Warm Deserts*, Dowden, Hutchinson and Ross, Stroudsberg. (183)

Orians, G. H., and O. T. Solbrig (1977b). A cost-income model of leaves and roots with special reference to arid and semi-arid areas. *Amer. Nat.*, **111,** 677–90. (180, 181)

Orloci, L. (1966). Geometric models in ecology. I. The theory and application of some ordination methods. *J. Ecol.*, **54,** 193–215. (420)

Orloci, L. (1975). *Multivariate Analysis in Vegetation Research*, Junk, The Hague. (420, 421)

Osmond, C. B. (1979). Ion uptake, transport and excretion. In *Arid-Land Ecosystems* Vol. 1. ed. Goodall, D. W., and R. A. Perry, pp. 607–25 Cambridge Univ. Press, Cambridge. (183, 288)

O'Toole, J. C., and R. T. Cruz (1979). Leaf rolling and transpiration. *Pl. Sci. Letts.*, **16,** 111–4. (181)

Overstreet, R., and H. Jenny (1939). The significance of the suspension effect in the uptake of cations by plants from soil water systems. *Proc. Soil Sci. Soc. Amer.*, **24,** 257–61. (250)

Ovington, J. D. (1953). Studies of the development of woodland conditions under different trees. I. Soil pH. *J. Ecol.*, **41,** 13–34. (103)

Ovington, J. D. (1957). Dry matter production by plantations of *Pinus sylvestris*. *Ann. Bot.*, NS**21,** 287–314. (341)

Ovington, J. D. (1959). The circuation of mineral nutrients in plantations of *Pinus sylvestris*. *Ann. Bot.*, **23,** 229–39. (34)

Ovington, J. D. (1961). Some ápsects of energy flow in plantations of *Pinus sylvestris* L., *Ann. Bot.*, NS**25,** 121–20. (341)

Ovington, J. D. (1968). Some factors affecting nutrient distribution within ecosystems. In *Functioning of the Terrestrial Ecosystem at the Primary Production Level*, ed. Eckardt, F. D., pp. 95–105, UNESCO, Paris.

Owen, P. C. (1952). The relationship of the germination of wheat to water potential. *J. exp. Bot.*, **3,** 276–90. (170)

Paine, R. (1971). The measurement and application of the calorie to plant science. *Ann. Rev. Pl. Physiol.*, **2,** 145–64. (346)

Park, Y. D., and A. Tanaka (1968). Studies of the rice plant on an 'Akiochi' soil in Korea, *Soil. Sci. Plant. Nutr.*, **14,** 27–34. (311)

Parker, J. (1968). Drought resistance mechanisms. In *Water Deficits and Plant Growth*, vol. 1, ed. Kozlowski, T. T., pp. 195–234. Academic Press, New York. (169, 174, 179, 181)

Parker, J. (1972). Protoplasmic resistance to water deficits. In *Water Deficits and Plant Growth*, Vol. 3, ed. Kozlowski, T. T., pp. 125–76, Academic Press, New York. (175)
Parsons, J. W., and J. Tinsley (1975). Nitrogenous substances. In *Soil Components*, Vol. 1, ed. Gieseking, J., pp. 263–304, Springer, Berlin. (105, 253)
Pate, J. S., and D. F. Herridge (1978). Partitioning and utilization of photosynthesate in a nodułated annual legume. *J. exp. Bot.*, **29,** 401–12. (235, 237)
Patel, P. M., A. Wallace, and R. B. Clark (1976). Phosphorus levels on the ability of an iron-inefficient or iron efficient corn inbred to take up iron from nutrient solution. *Comm. Soil. Sci. Pl. Anal.*, **7,** 105–6. (268)
Patrick, W. H., and I. C. Mahapatra (1968). Transformation and availability to rice of nitrogen and phosphorus in waterlogged soil. *Adv. Agron.*, **20,** 223–50. (306)
Patrick, W. H., and F. D. Turner (1968). *9th Int. Congr. Soil Sci. Trans.* Vol. IV (299)
Paul, E. A. (1975). Recent studies using the acetylene reduction technique as an assay for field nitrogen fixation levels. In *Nitrogen Fixation by Free-Living Organisms*, ed. Stewart, W. D. P., pp. 259–69, Cambridge Univ. Press, Cambridge. (234)
Paul, E. A., C. A. Campbell, D. A. Rennie, and K. J. McCallum (1964). Investigations of the dynamics of soil humus utilizing carbon dating techniques, *Trans. 8th Int. Congr. Soil Sci.*, 201–8. (99)
Pauling, L. (1930). The structure of mica and related minerals. *Proc. Nat. Acad. Sci. U.S.*, **16,** 123–9. (84)
Peacock, J. M. (1975). Temperature and leaf growth in *Lolium perenne* II. The site of temperature perception. *J. app. Ecol.*, **12,** 115–23. (207)
Pearcy, R. W., J. A. Berry, and B. Bartholemew (1974). Field photosynthetic performance and leaf temperature of *Phragmites communis* under summer conditions in Death Valley, California. *Photosynthetica*, **8,** 104–8. (19)
Pearl, R., and L. J. Reed (1920). On the rate of growth of the population of the United States since 1790 and its mathematical representation. *Proc. Nat. Acad. Sci. Washington*, **6,** 275–88. (420)
Pearsall, W. H. (1938). The soil complex in relation to plant communities. *J. Ecol.*, **26,** 180–93. (297)
Pearsall, W. H. (1952). The pH of natural soils and its ecological significance. *J. Soil Sci.*, **3,** 41–51. (59, 60, 93, 95)
Pemedasa, M. A., and P. H. Lovell (1974). Interference in populations of some dune annuals. *J. Ecol.*, **62,** 855–68. (413, 414, 415)
Penman, H. L. (1948). Natural evaporation from open water, bare soil and grass. *Proc. R. Soc.*, A, **193,** 120–45. (32)
Penman, H. L. (1963). *Vegetation and Hydrology*, Comm. Bureau Soils, Harpenden. (196)
Pereira, H. C. (1956). A rainfall test for structure of tropical soils. *J. Soil. Sci.*, **7,** 68–74. (114)
Pereira, J. S., and T. T. Kozlowski (1977). Variation amongst woody angiosperms in response to flooding. *Physiol. Pl.*, **41,** 184–92. (308)
Perrier, A. (1971). Leaf temperature measurement. In *Plant Photosynthetic Production: Manual of Methods*, ed. Sestak, Z., J. Catsky, and P. G. Jarvis, pp. 632–71, Junk, The Hague. (37)
Perrin, R. M. S. (1965). The use of drainage water analysis in soil studies. In *Experimental Pedology*, ed. Hallsworth, E. G., and D. V. Crawford, pp. 73–96, Butterworths, London. (244)
Perring, F. H. (1959; 1960). Topographical gradients of chalk grassland; Climatic gradients of chalk grassland. *J. Ecol.*, **47,** 447–81; **48,** 415–42. (226)
Perring, F. H., and S. M. Waters (1968). *Atlas of the British Flora*, Botanical Soc. of the British Isles, Nelson, London. (193, 269)
Petrie, A. H. K., and J. G. Wood (1938a, b). Studies on the nitrogen metabolism of plants. I. The relation between the content of amino acids, protein and water in leaves. III. On the

effect of water content on the relationship between proteins and amino acids. *Ann. Bot.*, NS**2**, 33–60, 881–98. (174)

Pfeffer, W. (1900). *The Physiology of Plants* (translation), Clarendon, Oxford. German first edition 1881. (422, 423)

Phillip, J. R. (1966). Plant water relationships: some physical aspects. *Ann. Rev. Pl. Physiol.*, **17,** 245–68. (147)

Phillips, I. D. J. (1964a). Root-shoot hormone relations. I. The importance of an aerated root system in the regulation of growth hormone levels in the shoot of *Helianthus annuus. Ann. Bot.*, **28,** 17–35. (309)

Phillips, I. D. J. (1964b). Root-shoot hormone relations. II. Changes in endogenous auxin concentration produced by flooding of the root system in *Helianthus annuus. Ann. Bot.*, **28,** 37–45. (309)

Phillipson, J. (1971). *Methods of Study in Quantitative Soil Ecology, Population, Production and Energy Flow*, Blackwell, Oxford. (345)

Pielou, E. C. (1977). *An Introduction to Mathematical Ecology*, Interscience, New York. (421)

Pigott, C. D. (1969). Influence of mineral nutrition on the zonation of flowering plants in coastal salt marshes. In *Ecological Aspects of the Mineral Nutrition of Plants*, ed. Rorison, I. H., pp. 25–35, Blackwell, Oxford. (285)

Pigott, C. D., and K. Taylor (1964). The distribution of some woodland herbs in relation to the supply of nitrogen and phosphorus in the soil. *J. Ecol.*, **52,** (supplement), 175–85. (275, 402)

Piper, C. S. (1944). *Soil and Plant Analysis*, Univ. Adelaide Press, Adelaide. (111)

Pitts, R. G. (1969). Explorations in the chemistry and microbiology of Louisiana Rice plant-soil relations. *Ph.D. Diss.*, Louisiana State Univ. (318)

Polunin, N. (1960). *Introduction to Plant Geography and Some Related Sciences*, Longmans, Oxford. (61, 74)

Pomeroy, L. R. (1970). The strategy of mineral cycling. *Ann. Rev. Ecol. Syst.*, **1,** 171–90. (359)

Ponnamperuma, F. N. (1965). Dynamic aspects of flooded soils and the nutrition of the rice plant. In *The Mineral Nutrition of the Rice Plant*, ed. International Rice Research Institute, John Hopkins, Baltimore. (304, 311)

Ponnamperuma, F. N. (1972). The chemistry of submerged soils. *Adv. Agron.*, **24,** 29–95. (290, 294, 302)

Ponnamperuma, F. N., R. Bradfield, and M. Peech (1955). Physiological disease of rice attributable to iron toxicity. *Nature, Lond.*, **175,** 265. (303)

Ponnamperuma, F. N., T. A. Loy, and F. M. Tianco (1969). Redox equilibria in flooded soils: II. The manganese oxide systems. *Soil Sci.*, **108,** 48–57. (303)

Ponnamperuma, F. N., E. Martinez, and T. Loy (1966). Influence of redox potential and partial pressure of carbon dioxide on pH value and the suspension effect of flooded soils. *Soil Sci.*, **101,** 421–31. (298)

Postan, M. M. (1972). *The Medieval Economy and Society*, Wiedenfeld and Nicholson, London. (351)

Postgate, G. R. (1959). Sulphate reduction in bacteria. *Ann. Rev. Microbiol.*, **13,** 505–20. (304)

Prenzel, J. (1979). Mass flow to the root system and mineral uptake of a beech stand calculated from three-year field data. *Plant and Soil*, **51,** 39–49. (253, 254, 255)

Prince, S. D., M. K. Marks, and R. N. Carter (1978). Induction of flowering in wild lettuce (*Lactuca serriola* L.) I. Vernalization. *New Phytol.*, **81,** 265–77. (218, 219)

Proctor, J. (1971). The plant ecology of serpentine. II. Plant response to serpentine soils. *J. Ecol.*, **59,** 397–410. (284)

Proctor, J., and S. R. J. Woodell (1975). The ecology of serpentine soils. *Adv. ecol. Res.*, **9,** 256–366. (250, 284, 286)

Pruitt, W. O. (1978). *Boreal Ecology*, Arnold, London. (74, 208)

Quastel, J. H. (1963). Microbial activities of soil as they affect plant nutrition. In *Plant Physiology*, Vol. III, pp. 671–756, Academic Press, New York. (233)

Raalte, M. H. van (1941). On the oxygen supply of rice roots. *Ann. Jard. Bot. Buitzenzorg*, **51,** 43–57. (314)

Raalte, M. H. van (1943–4). On the oxidation of the environment by the roots of rice (*Oryza sativa* L.). *Hort. Bot. Bogoriensis, Java. Syokubutu-Iho*, **1,** 15–34. (314)

Rabinowitch, E., and Govindjee (1969). *Photosynthesis*, Wiley, New York. (421)

Raciborski, M. M. (1905). Utleniajace: redukajace właśnosei kómorki żywej. I, II and III. *Bull. Int. de L'Acad. Sciences* (*Cracovie*), 338, 668, 693. (314)

Rackham, O. (1975). Temperatures of plant communities as measured by pyrometric and other methods. In *Light as an Ecological Factor II*, ed. Evans, G. C., R. Bainbridge, and O. Rackham, pp. 423–49, Blackwell, Oxford. (17)

Rackham, O. (1980). *Ancient Woodland*, Arnold, London. (385)

Ragg, J. M., and B. Clayden (1973). *The Classification of some British Soils According to the Comprehensive System of the United States*, Soil Survey, England and Wales, Rothamsted. (51, 52)

Raison, R. J. (1979). Modification of the soil environment by vegetation fires with particular reference to nitrogen transformations: a review. *Plant and Soil*, **51,** 73–108. (395)

Rains, D. W. (1979). Salt tolerance of plants: strategies of biological systems. In *The Biosaline Concept*, ed. Aller, J. C., E. Epstein *et al.*, pp. 47–67, Plenum, New York. (288)

Rajashekar, C., and Burke, M. J. (1978). The occurrence of deep undercooling in the genera *Pyrus*, *Prunus* and *Rosa*: a preliminary report. In *Plant Cold Hardiness and Freezing Stress*, ed. Li, P. H., and A. Sakai, pp. 213–25, Academic Press, New York. (209)

Ramakrishnan, P. S. (1968). Nutritional requirements of the edaphic ecotypes in *Melilotus alba* Medic. I. pH, calcium and phosphorus. *New Phytol.*, **67,** 145–57. (264)

Ramakrishnan, P. S. (1970). Nutritional requirements of the edaphic ecotypes in *Melilotus alba* Medic. III. Interference between the calcareous and acidic populations in the two soil types. *New Phytol.*, **69,** 81–6. (264)

Raschke, K. (1976). How stomata resolve the dilemma of opposing priorities. *Phil. Trans. R. Soc. Lond.* B., **273,** 551–60. (166)

Ratcliffe, D. A. (1961). Adaptation to habitat in a group of annual plants. *J. Ecol.*, **49,** 187–293. (178)

Rauner, Ju. L. (1975). Deciduous forests. In *Vegetation and the Atmosphere*, ed. Monteith, J. L., pp. 241–64, Academic Press, New York. (25)

Raven, J. A. (1977). The evolution of vascular land plants in relation to supracellular transport processes. *Adv. Bot. Res.*, **5,** 153–219. (126, 127, 145, 233)

Raven, J. A., and S. M. Glidewell (1975). Photosynthesis, respiration and growth in the shadenalga *Hydrodiction* africanum. *Photosynthetica*, **9,** 361–71. (134)

Raven, J. A., and F. A. Smith (1977). 'Sun' and 'shade' species of green algae in relation to cell size and environment. *Photosynthetica*, **11,** 48–55. (134)

Rawlins, S. L. (1976). Measurement of water content and the state of water in soils. In *Water Deficits and Plant Growth* Vol. 4, ed. Kozlowski, T. T., pp. 1–55, Academic Press, New York. (155)

Rayner, M. C. (1913). The ecology of *Calluna vulgaris. New Phytol.*, **12,** 59–77. (268)

Raynor, G. S., E. C. Ogden, and J. V. Hayes (1972). Dispersal and deposition of timothy pollen from experimental sources. *Agric. Met.*, **9,** 347–66. (228)

Read, D. J., H. K. Koucheki, and J. Hodgeson (1976). Vesicular arbuscular mycorrhiza in a natural vegetation system. *New Phytol.*, **77,** 641–53. (257)

Read, D. J., and W. Armstrong (1972). A relationship between oxygen transport and the formation of the ecotrophic mycorrhizal sheath in conifer seedlings. *New Phytol.*, **71,** 49–53. (315)

Redman, F. H., and W. H. Patrick (1965). The effect of submergence on several biological and chemical soil properties. *La Agric. Expl. Sta. Bull.*, 592. (306)

Rees, A. R. (1972). *The Growth of Bulbs*, Academic Press, New York. (215)

Reid, D. M., and A. Crozier (1971). Effects of waterlogging on the giberellin content and growth of tomato plants. *J. exp. Bot.*, **22,** 39–44 (309)

Rieger, S. (1974). Arctic soils. In *Arctic and Alpine Environments*, ed. Ives, J. D., and R. G. Barry, pp. 749–69, Methuen, London. (55)

Reimer, T. O., E. S. Barghoorn, and L. Margulis (1979). Primary productivity in an early archaean microbial ecosystem. *Precamb. Res.*, **9,** 93–104. (357)

Reimold, R. J., and W. H. Queen (1974). *Ecology of Halophytes*, Academic Press, New York. (285)

Reinhart, K. G. (1967). Watershed calibration methods. In *Forest Hydrology*, ed. Sopper, W. E., and H. W. Lull, pp. 715–23, Pergamon, Oxford. (197)

Reiter, H. (1865). *Die Consolidation der Physiognomik*, Graz. (2)

Rennie, P. J. (1955). The uptake of nutrients by mature forest growth. *Plant and Soil*, **7,** 49–95. (390)

Rennie, P. J. (1966). The use of micropedology in the study of some Ontario podzolic soil profiles. *J. Soil Sci.*, **17,** 99–106. (93)

Rhykerd, C. L., and C. H. Woller (1973). The role of N in forage productivity. In *Forages*, ed. Metcalf, D. S., and R. F. Barnes, pp. 416–24, Iowa State Univ. Press, Ames. (352)

Rice, E. R. (1974). *Allelopathy*, Academic Press, New York. (404, 405)

Rice, E. R. (1979). Allelopathy—an update. *Bot. Rev.*, **45,** 15–189. (404, 405)

Richards, F. J. (1941). The diagrammatic representation of the results of physiological and other experiments designed factorially. *Ann. Bot.*, NS**5,** 249–61. (280)

Richards, P. W. (1952). *The Tropical Rainforest*, Cambridge Univ. Press, Cambridge. (67, 68, 69, 74, 391)

Richards, L. A. (1965). Physical condition of water in soil. In *Methods of Soil Analysis*, ed. Black, C. A., pp. 128–52, Am. Soc. Agron, Wisconsin. (155)

Richards, L. A., and W. E. Loomis (1942). Limitations of autoirrigators for controlling soil moisture under growing plants. *Pl. Physiol. Lancaster*, **17,** 223–35. (160)

Richards, L. A., and C. H. Wadleigh (1952). Soil water and plant growth. In *Soil Physical Conditions and Plant Growth*, ed. Shaw, B. T., pp. 73–251, Am. Soc. Agron. Wisconsin. (172)

Richter, H. (1976). Water status in the plant. In *Water and Plant Life*, ed. Lange, O. L., L. Kappen, and E. D. Schultz, pp. 42–62, Springer, Berlin. (180)

Ripley, E. A., and R. E. Redmond (1975). Grassland. In *Vegetation and the Atmosphere*, ed. Monteith, J. L., Academic Press, New York. (20, 24)

Roberts, H. A., and G. F. Stokes (1966). Studies on weed seeds of vegetable crops VII. *J. app. Ecol.*, **3,** 181–90. (222)

Robson, M. J., and O. R. Jewiss (1968). A comparison of British and N. African varieties of tall fescue (*Festuca arundinacea*) II and III. *J. app. Ecol.*, **5,** 179–90; 191–204. (211, 212, 215)

Rodin, L. E., and N. I. Bazilevich (1967). *Production and Mineral Cycling in Terrestrial Vegetation* (translation), Oliver and Boyd, London. (272, 335, 373, 379)

Rogers, R. W., R. T. Lang, and D. J. D. Nicholas (1966). Nitrogen fixation by lichens and soil crusts. *Nature, Lond.*, **209,** 96–7. (41)

Rogers, W. S. (1969). The East Malling root-observation laboratories. In *Root Growth*, ed. Whittington, W. J., pp. 361–76. Butterworths, London. (119)

Rogers, W. S., and G. C. Head (1969). Factors affecting the distribution and growth of roots of perennial woody species. In *Root Growth*, ed. Whittington, W. J., pp. 280–95, Butterworths, London. (119, 120)

Rogerson, T. L. (1976). Soil water deficits under forested and clearcut areas in northern Arkansas. *Proc. Soil Sci. Soc. Amer.*, **40,** 802–4. (199)

Roo, H. C. de (1969). Tillage and root growth. In *Root Growth*, ed. Whittington, W. J., pp. 339–58, Butterworths, London. (119)

Rorison, I. H. (1960a). Some experimental aspects of the calcicole–calcifuge problem. I. The effects of competition and mineral nutrition upon seedling growth in the field. *J. Ecol.*, **48,** 585–99. (267)

Rorison, I. H. (1960b). Some experimental aspects of the calcicole–calcifuge problem. II. The effects of mineral nutrition on seedling growth in solution culture. *J. Ecol.*, **48,** 679–88. (267)

Rorison, I. H. (1968). The response to phosphorus of some ecologically distinct plant species I. Growth rates and phosphorus absorption. *New Phytol.*, **67,** 913–23. (275)

Rorison, I. H. (ed.) (1969). Ecological inferences from laboratory experiments on mineral nutrition. In *Ecological Aspects of the Mineral Nutrition of Plants*, pp. 155–75, Blackwell, Oxford. (402)

Rorison, I. H. (1973). The effect of extreme soil acidity on the nutrient uptake and physiology of plants. In *Acid Sulphate Soil*, ed. Dost, H., pp. 223–54, Int. Inst. Land Reclamation and Improvement, Wageningen. (265)

Rosenberg, N. J. (1974). *Microclimate: the Biological Environment*, Wiley, Chichester. (11)

Ross, J. (1975). Radiative transfer in plant communities. In *Vegetation and the Atmosphere*, ed. Monteith, J. C., Academic Press, New York. (23)

Rouschal, E. (1937–8). Eine physiologische studie an *Ceterach officinarum. Flora* (*Jena*), **132,** 305–18. (178)

Roux, J. A. Cl. (1900). *Traité Historique, Critique et Expérimental des Rapports des Plantes avec le Sol*, Masson, Paris. (286, 423)

Rovira, A. D. (1979). Biology of the interface. In *The Soil-Root Interface*, ed. Harley, J. L., and R. S. Russel, pp. 145–60, Academic Press, New York. (256)

Ruhling, A., and G. Tyler (1968). An ecological approach to the lead problem. *Bot. Notiser.*, **121,** 321–42. (369, 370)

Rune, O. (1953). Plant life on serpentine and related rocks in the north of Sweden. *Acta phytogogr. suec.*, **31,** 1–139. (283)

Russel, E. W. (1971). Soil structure: its maintenance and improvement. *J. Soil Sci.*, **22,** 137–51. (115)

Russel, E. W. (1973). *Soil Conditions and Plant Growth*, Longman, London. (118, 230, 417, 422)

Russel, R. S. (1977). *Plant Root Systems*, McGraw-Hill, London. (118)

Russel, R. S., and M. J. Goss (1974). Physical aspects of soil fertility—the response of roots to mechanical impedence. *Neth. J. agric. Sci.*, **22,** 305–18. (116)

Rutter, A. J. (1968). Water consumption by forests. In *Water Deficits and Plant Growth*, vol. 2, ed. Kozlowski, T. T., pp. 23–84, Academic Press, New York. (15, 33, 194)

Ryle, G. J. A., C. E. Powell, and A. J. Gander (1979). The respiratory costs of nitrogen fixation in soybean, cowpea and white clover. I and II. *J. exp. Bot.*, **30,** 135–44; 145–53. (237)

Rymer, L. (1979). Ethnobotany and native distribution of gorse (*Ulex europaeus* L.) in Britain. *Env. Cons.*, **6,** 211–3. (393)

Sachs, J. S. (1882). *Vorlesungen uber Pflanzen-Physiologie*, Englemann, Leipzig. (423)

Salisbury, E. J. (1921). The significance of the calcicolous habit. *J. Ecol.*, **8,** 202–15. (268)

Salisbury, E. J. (1925). Note on the edaphic succession in some dune soils with special reference to the time factor. *J. Ecol.*, **13,** 322–8. (92)

Salisbury, E. J. (1952). *Downs and Dunes*, Bell, London. (277)

Salisbury, E. J. (1961). *Weeds and Aliens*, Collins, London. (406)

Salisbury, F. B. (1963). *The Flowering Process*, Pergamon, Oxford. (217, 218, 219)

Salisbury, F. B., and G. G. Spomer (1964). Leaf temperatures of alpine plants in the field. *Planta*, **60,** 497–505. (16)

Salter, P. J., and J. P. Williams (1967). The influence of texture on the moisture characteristics of soils. *J. Soil Sci.*, **18,** 174–81. (111)
Sanchez, P. A. (1976). *The Properties and Management of Soils in the Tropics*, Wiley, Chichester. (68, 70, 71, 89, 214)
Sanders, F. E., B. Mosse, and P. B. Tinker (1975). *Endomycorrhiza*, Academic Press, New York. (257)
Sanders, F. S., and P. B. Tinker (1973). Phosphate flow into mycorrhizal roots. *Pestic. Sci.*, **4,** 385–95. (257)
Sanderson, P. L. (1977). *On the responses of Sitka Spruce and Lodgepole Pine to conditions associated with waterlogged soil.* Ph.D. Thesis, Univ. of Hull. (306)
Sanderson, P. L., and W. Armstrong (1978). Soil waterlogging, root rot and conifer windthrow: oxygen deficiency or toxicity. *Plant and Soil*, **49,** 185–90. (329)
Satchell, J. E. (1967). Lumbricidae. In *Soil Biology*, ed. Burgess, A., and F. Raw, pp. 259–322, Academic Press, London. (99, 115, 379)
Sauer, R. H. (1978). A simulation model for growth, primary productivity, phenology and biomass dynamics. In *Grassland Simulation Models*, ed. Innis, G. S., pp. 55–87, Springer, Berlin. (145)
Scharpenseel, H. W. (1971). Special methods of chromatographic and radiometric analysis. In *Soil Biochemistry*, Vol. 2, ed. McLaren, A. D., and J. Skujins, pp. 96–128, Dekker, New York. (99, 110)
Scharpenseel, H. W. (1971). Radiocarbon dating of soils—problems, troubles, hopes. In *Paleopedology—Origin, Nature and Dating of Paleosols*, pp. 77–88, University Press, Jerusalem. (99, 110)
Scharpenseel, H. W., and H. Schiffman (1977). Soil radiocarbon analysis and dating. *Geophys. Survs.*, **3,** 143–56. (100)
Scheskel, C. (1975). Silicon and nitrate depletion as related to rate of eutrophication in Lakes Michigan, Huron and Superior. In *Coupling of Land and Water Systems*, ed. Hasler, A., pp. 277–98, Springer, Berlin. (351)
Schimper, A. F. D. (1898). *Pflanzen geographie auf physiologische grundlage.* Fischer, Jena. (English translation, 1903, Clarendon Press, Oxford.) (58, 74, 183, 419)
Schlindler, D. W., and E. J. Fee (1975). The rates of Nutrient cycling and radiant energy in aquatic communities. In *Photosynthesis and Primary Production in Different Environments*, ed. Cooper, J. P., pp. 323–43, Cambridge Univ. Press, Cambridge. (352)
Schlichting, E., and U. Schwertmann (1973). *Gley and Pseudogley*, Verlag Chemie, Weinheim. (75)
Schneider, G. W., and N. F. Childers (1941). Influence of soil moisture on photosynthesis, respiration and transpiration of apple leaves. *Pl. Physiol., Lancaster*, **16,** 565–83. (171)
Schnitzer, M. (1971). Characterization of humic constituents by spectroscopy. In *Soil Biochemistry*, Vol. 2, ed. McLaren, A. D., and J. Skujins, pp. 60–95, Dekker, New York. (110)
Schnitzer, M. (1976). The chemistry of humic substances. In *Environmental Biogeochemistry*, Vol. 1, ed. Nriagu, J. O., pp. 89–107, Ann Arbor, Michigan. (106, 107, 108, 109)
Schofield, R. K. (1935). The pF of water in soil. *Trans. 3rd Int. Congr. Soil Sci.*, **2,** 37. (149)
Schlonder, P. F., L. van Dam, and S. I. Scholander (1955). Gas exchange in the roots of mangroves. *Am. J. Bot.*, **42,** 92–8. (328)
Schrenk, H. (1889). Ueber das Aerenchym ein dem kork homologes Gewebe bei Sumpfpflanzen. *Jahrb. Wiss. Bot.*, **20,** 526–74. (320)
Schultz, A. M. (1969). A study of an ecosystem: the arctic tundra. In *The Ecosystem Concept in Natural Resource Management*, ed. Van Dyne, G. M., pp. 77–93, Academic Press, New York. (383)
Scott, A. D., and D. D. Evans (1955). Dissolved oxygen in saturated soil. *Soil Sci. Soc. Amer. Proc.*, **19,** 7–16. (293)

Scott, P. R., and A. Bainbridge (1978). *Plant Disease Epidemiology*, Blackwell, Oxford. (387)

Sculthorpe, C. D. (1967). *The Biology of Aquatic Vascular Plants*, Arnold, London. (23, 319)

Sears, P. B. (1964). Ecology—a subversive subject. *Bioscience*, **14,** 11–13. (2)

Seehy, J. E., and J. M. Peacock (1975). Canopy photosynthesis and crop growth rate of eight temperate forage grasses. *J. exp. Bot.*, **26,** 679–91. (136)

Sestak, Z., J. Catsky, and P. G. Jarvis (1971). *Plant Photosynthetic Production: Manual of Methods*, Junk, The Hague. (38, 137, 140)

Shanmugam, K. I., F. O'Gara, K. Andersen, and R. C. Valentine (1978). Biological nitrogen fixation. *Ann. Rev. Pl. Physiol.*, **29,** 263–76. (236, 240)

Sheikh, K. H. (1970). The responses of *Molinia caerulea* and *Erica tetralix* to soil aeration and related factors. II. Effects of different gas concentrations on growth in solution culture and general conclusions. *J. Ecol.*, **58,** 141–54. (118, 403)

Sherrat, A. (1980). Water, soil and seasonality in early cereal cultivation. *World Archaeology*, **11,** 313–30. (147)

Shields, L. M. (1950). Leaf xeromorphy as related to physiological and structural influences. *Bot. Rev.*, **16,** 399–447. (132)

Sillen, L. G., and A. L. Martell (1974). *Stability Constants*, Chemical Society, London. (96)

Simpson, B. B., and J. Haffer (1978). Speciation patterns in the Amazonian forest biota. *Ann. Rev. Ecol. Syst.*, **9,** 497–518. (68)

Skujins, J. J. (1967). Enzymes in soil. In *Soil Biochemistry*, Vol. 1, ed. McLaren, A. D., and G. H. Petersen, pp. 371–414, Dekker, New York. (106)

Slatyer, R. O. (1957). Significance of the permanent wilting percentage in studies of plant and soil water relations. *Bot. Rev.*, **23,** 585–636. (153)

Slatyer, R. O. (1961a). Effects of several osmotic substrates on the water relations of tomato plants. *Aust. J. Biol. Sci.*, **14,** 519–40. (172)

Slatyer, R. O. (1961b). Methodology of a water balance study concluded on a desert woodland (*Acacia aneura* F. Muell.) community in central Australia. In *Plant-Water Relationships in Arid and Semiarid Conditions*, ed. UNESCO, pp. 15–26, UNESCO, Paris. (196)

Slatyer, R. O. (1961c) Internal water balance of *Acacia aneura* F. Muell. in relation to environmental conditions. In *Plant Water Relationships in Arid and Semiarid Conditions*, pp. 137–46, UNESCO, Paris. (157)

Slatyer, R. O. (1967). *Plant-Water Relationships*, Academic Press, New York. (Prelim page xvii), 173, 347)

Slatyer, R. O., and H. D. Barrs (1965). Modifications to the relative turgidity technique with notes on its significance as an index of the internal water status of leaves. In *Methodology of Plant Ecophysiology*, ed. Eckardt, F. D., pp. 331–41, UNESCO, Paris. (157)

Slatyer, R. O., and S. A. Taylor (1960). Terminology in soil-plant water relations. *Nature, Lond.*, **187,** 922. (149)

Slavik, B. (1974). *Methods of Studying Plant Water Relations*, Springer, Berlin. (27, 154)

Smith, A. M. (1976). Ethylene in Soil biology. *Ann. Rev. Phytopath.*, **14,** 53–73. (284)

Smith, A. M., and T. ap Rees (1979). Pathways of carbohydrate metabolism fermentation in the roots of marsh plants. *Planta*, **146,** 327–34. (325)

Smith, B. N. (1976). Evolution of C-4 photosynthesis in response to changes in carbon and oxygen concentration in the atmosphere through time. *BioSystems*, **8,** 24–32. (356)

Smith, H. (1973). Light quality and germination: ecological implications. In *Seed Ecology*, ed. Heydecker, W., pp. 219–31, Butterworths, London. (224)

Smith, H. (1975). *Phytochrome and Phytomorphogenesis*, McGraw-Hill, London. (223, 224)

Smith, H. (1976). *Light and Plant Development*, Butterworths, London. (224)
Smith, K. A. (1980). A model of the extent of anaerobic zones in aggregated soils and its potential application to estimates of denitrification. *J. Soil. Sci.*, **31**, 263–77. (294)
Smith, K. A., and S. W. F. Restall (1971). The occurrence of ethylene in anaerobic soil. *J. Soil Sci.*, **22**, 430–43. (312)
Smith, K. A., and R. Scott-Russell (1969). Occurrence of ethylene and its significance in anaerobic soil. *Nature, Lond.*, **222**, 769–71. (312)
Smith, R. A. H., and A. D. Bradshaw (1979). The use of metal tolerant plant populations for the reclamation of metalliferous wastes. *J. app. Ecol.*, **16**, 595–602. (282)
Smith, R. S. (1975). *An Ecological Basis for Land-use Decisions in the South Wales Coalfield*, Ph.D. Thesis, University College Cardiff. (103, 380)
Snaydon, R. W., and A. D. Bradshaw (1961). Different responses to calcium within the species *Festuca ovina* L. *New Phytol.*, **60**, 219–34. (276)
Snaydon, R. W. (1962). Micro-distribution of *Trifolium repens* and its relation to soil factors. *J. Ecol.*, **50**, 133–43. (397)
Soil Survey Staff (1960). *Soil Classification: a Comprehensive System. 7th Approximation*, U.S. Dept. Agric., Washington DC. (49, 113)
Soil Survey Staff (1975). *Soil Taxonomy*, U.S. Govt. Printing Office, Washington DC. (49, 50, 52, 53, 60, 72, 78, 79)
Solbrig, O. T. *et al.* (1977). The biota: the dependent variable. The strategies and community patterns of desert plants. In *Convergent Evolution in Warm Deserts*, ed. Orians, G. H., and O. T. Solbrig, pp. 50–106, Dowden, Hutchinson and Ross, Stroudsberg. (180)
Southwood, T. R. E. (1978). *Ecological Methods with Particular Reference to the Study of Insect Populations*, Chapman and Hall, London. (345)
Sparling, J. H. (1967). The occurrence of *Schoenus nigricans* L. in blanket bogs. I. Environmental conditions. *J. Ecol.*, **55**, 1–13. (254)
Sparling, G. P., and P. B. Tinker (1978). Mycorrhizal infection in Pennine grassland II. Effects of mycorrhizal infection. *J. app. Ecol.*, **15**, 951–8. (257)
Specht, R. L. (1957). Dark Island Heath (Ninety Mile Plain, South Australia). IV. Soil moisture patterns produced by rainfall interception and stemflow. *Aust. J. Bot.*, **5**, 137–50. (160, 161)
Specht, R. L. (1979). *Heathlands and Related Shrublands*, Elsevier, Amsterdam. (394)
Spedding, C. R. W. (1975). *The Biology of Agricultural Systems*, Academic Press, New York. (352)
Spence, D. H. N. (1975). Light and plant response in freshwater. In *Light as an Ecological Factor II*, ed. Evans, G. C., R. Bainbridge, and O. Rackham, pp. 93–133, Blackwell, Oxford. (23)
Sprent, J. I. (1976). Water deficits and nitrogen-fixing root nodules. In *Water Deficits and Plant Growth*, Vol. 4, ed. Kozlowski, T. T., pp. 291–315, Academic Press, New York. (171)
Spurr, S. H. (1969). The natural resource ecosystem. In *The Ecosystem Concept in Natural Resource Management*, ed. Van Dyne, G. M., pp. 3–7. Academic Press, New York.
Stace, H. C. T. (1956). Chemical characteristics of terra rossas and rendzinas of southern Australia. *J. Soil Sci.*, **7**, 280–93. (79)
Stalfelt, M. G. (1972). *Stalfelt's Plant Ecology*, Translated by M. S. and P. G. Jarvis, Longman, London. (425)
Stanhill, G. (1957). The effect of differences in soil moisture status on plant growth. A review and analysis of soil moisture regime experiments. *Soil Sci.*, **84**, 205–14. (172)
Stanhill, G. (1965). The concept of potential evapo-transpiration in arid zone agriculture. In *Methodology of Plant Ecophysiology*, ed. Eckhardt, F. D., pp. 109–17, UNESCO, Paris. (196)

Stanhill, G. (1972). Recent developments in water relations studies. *Proc. 18th Int. Hort. Congr., IV*, 367–85. (172)
Stark, N. M., and Jordan, C. F. (1978). Nutrient retention by the root mat of an Amazonian rainforest. *Ecology*, **59**, 434–7. (258)
Starkey, R. L. (1966). Oxidation and reduction of sulphur compounds in soils. *Soil Sci.*, **101**, 297–306. (304)
Stern, W. R., and C. M. Donald (1962). Light relationships in a grass-clover canopy. *Aust. J. agric. Res.* **13**, 599–623. (21, 135)
Steven, H. M. (1953). Storm damage to woodlands in Scotland on January 31 1953. *Nature, Lond.*, 171, 454–6 (227)
Stevenson, F. J. (1967). Organic acids in soil. In *Soil Biochemistry*, Vol. 1, ed. McLaren, A. D., and G. H. Petersen, pp. 119–46, Dekker, New York. (106, 244)
Steward, F. C. (1963). *Plant Physiology*, Vol. 3, Academic Press, New York.
Stewart, G., and J. A. Lee (1974). The role of proline accumulation in halophytes. *Planta*, **120**, 279–89. (174, 287)
Stewart, W. D. P. (1966). *Nitrogen Fixation in Plants*, University of London Press, London. (234, 240)
Stewart, W. D. P. (1973). Nitrogen fixation. In *The Biology of the Blue Green Algae*, ed. Carr, N. G., and B. A. Whitton, pp. 260–78, Blackwell, Oxford. (41)
Stewart, W. D. P. (1975). *Nitrogen Fixation by Free-Living Microorganisms*, Cambridge Univ. Press, Cambridge. (234, 235, 240)
Stewart, W. D. P., M. J. Sampio, A. O. Isichei, and R. Sylvester-Bradley (1978). Nitrogen fixing by soil algae of temperate and tropical soils. In *Limitations and Potentials for Biological Nitrogen Fixation in the Tropics*, ed. Dobereinen, U., R. H. Burris, and A. Hollaender, pp. 41–63, Plenum, New York. (237)
Stocker, O. (1929). Eine Feldmethode zur Bestimmung der momentanen Transpirations und Evaporationsgroesse. *Ber. dt. bot. Ges.*, **47**, 126–36. (157)
Stout, J. D., K. M. Goh, and T. A. Rafter (1981). Chemistry and turnover of naturally occurring resistant organic compounds in soil. In *Soil Biochemistry*, Vol. 5, ed. Paul, E. A., and J. N. Ladd, Dekker, New York. (100, 106)
Strasburger, E. (1894). *A Text-book of Botany*, (translated into English 1898) Macmillan, London. (423)
Stringer, R. N. (1974). The decomposition of *Ulex europaeus* needles. In *Biodegradation et Humification*, ed. Kilbertus, G., O. Reisinger *et al.*, pp. 247–54, Pierron, Paris. (103)
Sutton, C. D., and D. Gunary (1969). Phosphate equilibria in soil. In *Ecological Aspects of the Mineral Nutrition of Plants*, pp. 127–34, Blackwell, Oxford. (251, 252)
Swank, W. T., and J. E. Douglas (1974). Streamflow greatly reduced by converting deciduous hardwood stand to pine. *Science*, **185**, 857–9. (197, 198)
Swift, M. J., O. W. Heal, and J. M. Anderson (1979). *Decomposition in Terrestrial Ecosystems*, Blackwell, Oxford. (99, 390)
Szeicz, G. (1975). Instruments and Their Exposure. In *Vegetation and the Atmosphere*, ed. Monteith, J. L., Academic Press, New York. (34, 36, 38)
Taft, J. L., and W. R. Taylor (1976). Phosphorus dynamics of some coastal plain estuaries. In *Estuarine Processes*, pp. 79–89, Wiley, Chichester. (274)
Takahashi, J. (1960a). Review of investigations on physiological diseases of rice. Part 1. *Inter. Rice Commis. N. L.*, **9**, 1–6. (311)
Takahashi, J. (1960b). Review of investigations on physiological diseases of rice. *Inter. Rice. Commis. N. L.*, **9**, 17–24. (311)
Takai, Y., and T. Kamura (1966). The mechanism of reduction in waterlogged paddy soil. *Folia Microbiol. (Prague)*, **11**, 304–13. (303, 304)
Takai, Y., and T. Asami (1962). Formation of methyl mercaptan in paddy soils I. *Soil Sci. Pl. Nutr.*, **8**, 132–6. (306)

Takijima, T. (1963). Studies on the behaviour of the growth inhibiting substances in paddy soils with special reference to the occurrence of root damage in paddy fields. *Bull. Natl. Inst. agr. Sci. Japan Ser. B.*, **13,** 117–252. (306)

Talling, J. F. (1975). Primary production of freshwater microphytes. In *Photosynthesis and Primary Production in Different Environments*, ed. Cooper, J. L., Cambridge Univ. Press, Cambridge. (353)

Tanaka, A., R. P. Mulleriyawa, and T. Yasu (1968). Possibility of hydrogen sulphide induced toxicity of the rice plant. *Soil Sci. Plant Nutr.*, **14,** 1–6. (311)

Tansey, M. R., and T. D. Brock (1978). Microbial Life at high temperatures: ecological aspects. In *Microbial Life in Extreme Environments*, ed. Kushner, D. J., pp. 159–216, Academic Press, New York. (213)

Tansley, A. G. (1911). *Types of British Vegetation*, Cambridge Univ. Press, Cambridge. (419)

Tansley, A. G. (1917). On competition between *Galium saxatile* L. (*G. hercynicum* Weig.) and *G. sylvestre* Poll. (*G. asperum* Schreb.) on different types of soil. *J. Ecol.*, **7,** 173–9. (1, 263, 401)

Tansley, A. G. (1920). The classification of vegetation and the concept of development. *J. Ecol.*, **8,** 118–49. (420)

Tansley, A. G. (1935). The use and abuse of vegetational concepts and terms. *Ecology*, **16,** 284–307. (420, 424)

Tansley, A. G. (1939). *The British Isles and their Vegetation*, Cambridge Univ. Press, Cambridge. (58, 59, 226, 394, 420)

Tansley, A. G., and M. M. Rankin (1911). The plant formations of calcareous soils. B. The sub-formation of the chalk. In *Types of British Vegetation*, ed. Tansley, A. G., pp. 161–86, Cambridge University Press, Cambridge. (269)

Tansley, A. G., and R. S. Adamson (1925). Studies of the English chalk III. *J. Ecol.*, **13,** 177–223. (378)

Tansley, A. G., and T. F. Chipp (1926). *Aims and Methods of Vegetation Study*, Brit. Emp. Veg. Comm., London. (420)

Tanton, T. W., and S. H. Crowdy (1972a, b). Water pathways in higher plants I and II. *J. exp. Bot.*, **23,** 600–18; 619–25. (162)

Taylor, A. O., and J. O. Rowley (1971). Plants under climatic stress I. Low temperature, high light effects on photosynthesis. *Pl. Physiol., Lancaster*, **47,** 713–8. (205, 208)

Taylor, H. M., and H. R. Gardner (1963). Penetration of cotton seedling taproots as influenced by bulk-density, moisture content and strength of soil. *Soil Sci.*, **96,** 153–6. (117)

Taylor, H. M., and L. F. Ratliff (1969). Root elongation rates of cotton and peanuts as a function of soil strength and water content. *Soil Sci.*, **108,** 113–19. (116)

Taylor, S. A., and R. O. Slatyer (1960). Water–soil–plant relations terminology. *Trans. 7th Intern. Congr. Soil Sci.*, **1,** 395. (149)

Taylorson, R. B., and S. B. Hendricks (1977). Dormancy in seeds. *Ann. Rev. Pl. Physiol.*, **28,** 331–54. (223)

Teal, J. M., and J. W. Kanwisher (1966). Gas transport in the marsh grass *Spartina alterniflora*. *J. exp. Bot.*, **17,** 355–61. (315, 321)

Temple, S. E. (1977). Plant-animal mutualism with Dodo leads to near extinction of plant. *Science*, **197,** 885–6. (386)

Thom, A. S. (1975). Momentum, mass and heat exchange in plants. In *Vegetation and the Atmosphere*, ed. Monteith, J. L., Academic Press, New York. (31, 32, 33)

Thomas, A. S. (1960). Changes in vegetation since the advent of myxomatosis. *J. Ecol.*, **48,** 287–306. (271)

Thomas, W. A., and W. F. Grigal (1976). Phosphorus conservation by evergreenness of mountain laurel. *Oikos*, **27,** 19–26. (278)

Thornton, I., and Webb, J. S. (1980). Regional distribution of trace element problems in Great Britain. In *Applied Soil Trace Elements*, ed. Davies, B., pp. 381–9, Wiley, Chichester. (286)

Thorp, J., and G. D. Smith (1949). Higher categories of soil classification: Order, Suborder and Great Soil Groups. *Soil Sci.*, **67**, 117–26. (51)

Thurston, J. M. (1969). The effect of liming and fertilization on the botanical composition of permanent grassland and on the yield of hay. In *Ecological Aspects of the Mineral Nutrition of Plants*, ed. Rorison, I. H., pp. 3–10, Blackwell, Oxford. (272, 350, 352)

Thut, H. F., and W. E. Loomis (1944). Relation of light to the growth of plants. *Pl. Physiol., Lancaster*, **19**, 117–30. (18)

Tibbits, T. W., and T. T. Kozlowski, (1979). *Controlled Environment Guidelines for Plant Research*, Academic Press, New York. (30)

Tiezen, L. L., and N. K. Wieland (1975). Physiological ecology of arctic and alpine photosynthesis and respiration. In *Physiological Adaptation to the Environment*, ed. Vernberg, F. J., pp. 157–200, Intext, New York. (208, 350)

Tiffin, L. O. (1966). Iron translocation II. *Pl. Physiol., Lancaster*, **41**, 515–8. (268)

Ting, I. P., and S. R. Szarek (1975). Drought adaptation in CAM plants. In *Environmental Physiology of Desert Organisms*, ed. Hadley, N. F., pp. 152–67, Dowden, Hutchinson and Ross, Stroudsberg.

Tinker, P. B. (1975). Soil chemistry of phosphorus and mycorrhizal effects on plant growth. In *Endo mycorrhizas*, ed. Sanders, F. E., B. Mosse, and P. B. Tinker, pp. 353–71, Academic Press, New York. (258, 274)

Todd, G. W., and D. L. Webster (1965). Effects of repeated drought periods on photosynthesis and survival of cereal seedlings. *Agron. J.*, **37**, 399–404. (188)

Torrey, J. G., and D. T. Clarkson (1975). *The Development and Function of Roots*, Academic Press, New York.

Towe, K. M. (1978). Early precambrian oxygen: a case against photosynthesis. *Nature, Lond.*, **275**, 657–61. (356)

Tranquillini, W. (1964). The physiology of plants at high altitude. *Ann. Rev. Pl. Physiol.*, **15**, 345–61. (16)

Tranquillini, W. (1979). *Physiological Ecology of the Alpine Timberline*, Springer, Berlin. (227)

Transeau, E. N. (1926). The accumulation of energy by plants. *Ohio J. Sci.*, **26**, 1–10. (338)

Treshow, M. (1970). *Environment and Plant Response*, McGraw-Hill, New York. (17)

Trlica, M. J., A. J. Dye *et al.* (1973). A field laboratory for gas exchange measurements of grassland swards. *Photosynthetica*, **7**, 257–61. (344)

Troughton, A. (1972). The effect of aeration in the nutrient solution on the growth of *Lolium perenne*. *Plant and Soil*, **36**, 93–108. (309)

Troughton, J. H. (1975). Photosynthetic mechanisms in higher plants. In *Photosynthesis and Primary Production in Different Environments*, ed. Cooper, J. P., pp. 357–91, Cambridge Univ. Press, Cambridge. (350)

Troughton, J. H., H. A. Mooney, J. A. Berry, and D. Verity (1977). Variable isotope ratios of *Dudleya* spp. growing in natural environments. *Oecologia*, **30**, 307–11. (182)

Turesson, G. (1922). The genotypic response of the plant species to the habitat. *Hereditas*, **3**, 211–350. (424)

Turesson, G. (1931). The selective effect of climate upon plant species. *Hereditas*, **15**, 99–152. (424)

Tyler, P. D., and R. M. M. Crawford (1970). The role of shikimic acid in waterlogged roots and rhizomes of *Iris pseudacorus* L. *J. exp. Bot.*, **21**, 677–82. (324)

Ugolini, F. C. (1970). Antarctic soils and their ecology. In *Antarctic Ecology*, ed. Holdgate, M. W., pp. 673–92, Academic Press, New York. (55)

Valentine, I., and K. P. Barley (1976). Effects of soil temperature and phosphorus supply on an annual grass and clover grown in monoculture and in mixed culture. *Plant and Soil*, **44,** 163–77.
Vamos, R. (1959). 'Brusone' disease of rice in Hungary. *Plant and Soil*, **11,** 103–9. (311)
Van Bavel, C. H. M. (1965). Composition of the soil atmosphere. In *Methods of Soil Analysis*, ed. Black, C. A., pp. 315–18, Am. Soc. Agron., Wisconsin. (118)
Van Bavel, C. H. M., and R. J. Reginato (1965). Precision lysimetry for direct measurement of evaporative flux. In *Methodology of Plant Ecophysiology*, ed. Eckhardt, F. D., pp. 129–35, UNESCO, Paris. (196, 197)
Van den Honert, T. H. (1948). Water transport in plants as a catenary process. *Faraday Soc. Disc.*, **3,** 146–53. (163)
Van der Plank, J. E. (1975). *Principles of Plant Infection*, Academic Press, New York. (387)
Van Dyne, G. M. (1969). *The Ecosystem Concept in Natural Resource Management*, Academic Press, New York. (425)
Vanek, J. (1975). *Progress in Soil Zoology*, Junk, The Hague. (99)
Van Emden, H. F. (1972). Aphide as phytochemists. In *Phytochemical Ecology*, ed. Harborne, J. B., pp. 25–43, Academic Press, New York. (381)
Varley, G. C., and G. R. Gradwell (1962). The effect of partial defoliation by caterpillars on the timber production of oak trees in England. *XI Int. Congr. Entomol.* Wien, 2211–4. (385)
Vartapetian, B. B. (1964). Polarographic study of oxygen transport in plants, *Fiziologia Rastenni*, **11,** 774. (314)
Vartapetian, B. B. (1970). The oxygen and ultrastructure of root cells. *Agrochemica*, **15,** 1–19. (326)
Vartapetian, B. B. (1978). Life without oxygen. In *Plant Life in Anaerobic Environments*, pp. 1–11, ed. Hook, D. D., and R. M. M. Crawford, Ann Arbor Science, Ann Arbor. (326)
Vartapetian, B. B., I. N. Andreeva, and G. I. Kozlova (1976). The resistance to anoxia and the mitochondrial fine structure of rice seedlings. *Protoplasma*, **88,** 215–224. (326)
Vartapetian, B. B., I. N. Andreeva, G. I. Kozlova, and L. P. Agapova (1977). Mitochondrial ultrastructure in roots of mesophyte and hydrophyte at anoxia and after glucose feeding. *Protoplasma*, **91,** 243–256. (325, 326)
Vartapetian, B. B., I. N. Andreeva, and N. Nuritdinov (1978). Plant cells under oxygen stress. In *Plant Life in Anaerobic Environments*, ed. Hook, D. D., and R. M. M. Crawford, pp. 13–88, Ann Arbor, Michigan. (308, 326)
Vegis, A. (1963). Climatic control of germination, bud-break and dormancy. In *Environmental Control of Plant Growth*, ed. Evans, L. T., pp. 265–87, Academic Press, New York. (215)
Veihmeyer, F. J. (1927). Some factors affecting the irrigation requirements of deciduous orchards. *Hilgardia*, **2,** 125–288. (160)
Veihmeyer, F. J., and A. H. Hendrickson (1927). Soil moisture conditions in relation to plant growth. *Pl. Physiol., Lancaster*, **2,** 71–82. (159, 172)
Veihmeyer, F. J., and A. H. Hendrickson (1950). Soil moisture in relation to plant growth. *Ann. Rev. Pl. Physiol.*, **7,** 285–304. (172)
Venkataraman, G. S. (1975). The role of blue-green algae in tropical rice cultivation. In *Nitrogen Fixation by Free-Living Microorganisms*, ed. Stewart, W. D. P., pp. 207–18, Cambridge Univ. Press, Cambridge. (235)
Viets, F. G. (1972). Water deficits and nutrient availability. In *Water Deficits and Plant Growth*, Vol. 3, ed. Kozlowski, T. T., pp. 217–39, Academic Press, New York. (170)
Vince-Prue, D. (1975). *Photoperiodism in Plants*. McGraw-Hill, New York. (220)
Vincente-Chandler, J., R. Carlo-Costus *et al.* (1964). *The Intensive Management of Tropical Forages in Puerto Rico*. Univ. Puerto Rico Exp. Sta. Bull. 187. (352, 353)

Virtanen, A. I., and J. K. Meitinen (1963). Biological nitrogen fixation. In *Plant Physiology*, Vol. 2, ed. Steward, F. C., pp. 539–668, Academic Press, New York. (234, 235, 239)
Vollenweider, R. A. (1969); *Methods for Measuring Primary Productivity in Aquatic Environments*, Blackwell, Oxford. (344)
Vomocil, J. A. (1965). Porosity. In *Methods of Soil Analysis*, Vol. 1, ed. Black, C. A., pp. 299–34, Am. Soc. Agron., Wisconsin. (114)
Von Bulow, F. J., and J. Dobereiner (1975). Potential for nitrogen fixation in maize genotypes in Brazil. *Nat. Acad. Sci., Washington, Proc.*, **72,** 2389–93. (236)
Wadleigh, C. H., and H. G. Gauch (1948). Rate of leaf elongation as affected by the intensity of the total soil moisture stress. *Pl. Physiol., Lancaster*, **23,** 485–95. (173)
Waggoner, P. E., R. A. Moss, and J. D. Hesketh (1963). Radiation in the plant environment and photosynthesis. *Agron. J.*, **55,** 36–9. (18)
Wainwright, S. J., and H. W. Woolhouse (1975). Physiological mechanisms of heavy metal tolerance in plants. In *Ecology of Resource Degradation and Renewal*, ed. Chadwick, M. J., and G. T. Goodman, pp. 231–56, Blackwell, Oxford. (283)
Waisel, Y. (1972). *Biology of Halophytes*, Academic Press, New York. (285, 288)
Wagner, H., and G. Michael (1971). Der Einfluss unterschiedlicher Stickstoffversorgung die Cytokininbildung in Wurzeln von Sonnenblumenpflanzen. *Biochemie Physiologie Pflanzen*, **162,** 147 (308)
Walker, D. (1970). Direction and rate in some British postglacial hydroseres. In *Studies in the Vegetational History of the British Isles*, pp. 117–39 ed. Walker, D., and R. G. West, Cambridge University Press, Cambridge. (284, 391, 392)
Walker, R. B. (1954). The ecology of serpentine soils. 11. Factors affecting plant growth on serpentine soils. *Ecology*, **35,** 259–66.
Walker, T. W. (1965). The significance of phosphorus in pedogenesis. In *Experimental Pedology*, ed. Hallsworth, E. G., and D. V. Crawford, pp. 295–316, Butterworths, London. (240)
Wallace, A., S. M. Soufi, J. W. Chai, and E. M. Romni (1976). Iron-phosphorus interaction in buck-beans. *Comm. Soil Sci. Pl. Anal.*, **7,** 101–4. (277)
Walley, K. A., M. S. Kahn, and A. D. Bradshaw (1974). The potential for evolution of heavy metal tolerance in plants. I. Copper and zinc tolerance in *Agrostis tenuis*. *Heredity*, **32,** 309–19. (282)
Wallwork, J. A. (1970). *Ecology of Soil Animals*, McGraw-Hill, New York. (99)
Wallwork, J. A. (1976). *The Distribution and Diversity of Soil Fauna*, Academic Press, New York. (99)
Walter, H. (1971). *Ecology of Tropical and Subtropical Vegetation*, Oliver and Boyd, London. (74)
Walter, H. (1979). *Vegetation of the Earth*, Springer, Berlin. (74, 181, 203)
Wample, R. L., and D. M. Reid (1979). The role of exogenous auxins and ethylene in the formation of adventitious roots and hypercotyl hypertrophy in flooded Sunflower plants (*Helianthus annuus*). *Physiol. Pl.*, **45,** 219–26. (309)
Wang, T. S. C., S. Y. Cheng, and H. Tung (1967). Dynamics of soil organic acids. *Soil Sci.*, **104,** 138–44. (306, 313)
Wang, T. S. C., T.-K. Yong, and Z.-P. Chuang (1967). Soil phenolic acids as plant growth inhibitors. *Soil Sci.*, **103,** 239–46. (306)
Warcup, J. H. (1951). The ecology of soil fungi. *Trans. Brit. Mycol. Soc.*, **34,** 376–99. (101)
Ward, R. C. (1976). Evaporation, humidity and the water-balance. In *The Climate of the British Isles*, ed. Chandler, T. J., and S. Gregory, pp. 183–98, Longman, London. (225)
Wardle, P. (1959). The regeneration of *Fraxinus excelsior* in woods with a field layer of *Mercurialis perennis*. *J. Ecol.*, **47,** 483–97. (390)
Wardle, P. (1974). Alpine timberlines. In *Arctic and Alpine Environments*, ed. Ives, J. D., and R. G. Barry, pp. 371–402, Methuen, London. (228)
Wareing, P. F. (1969). The control of bud dormancy in seed plants. In *Dormancy and Survival*, ed. Woolhouse, H. W., pp. 241–62, Cambridge Univ. Press, Cambridge. (216)

Waring, R. H., and S. W. Running (1976). Water uptake, storage and transpiration by conifers. In *Water and Plant Life*, ed. Lange, O. L., L. Kappen, and E. D. Schultz, pp. 189–202, Springer, Berlin. (168)

Warming, E. (1891). *De psammofile vegetationer in Danmark*, Vid. Medd. Foren. (419)

Warming, E. (1896). *Lerbuch der Oekologischen.* (1, 419)

Wassink, E. C. (1975). Photosynthesis and productivity in different environments: conclusions. In *Photosynthesis and Primary Productivity in Different Environments*, ed..Cooper, J. P., pp. 675–87, Cambridge Univ. Press, Cambridge. (145, 354)

Watanabe, I., K. K. Lee and B. Alimagno (1978). Seasonal change of N_2 fixing rate in a rice field assayed by *in situ* acetylene reduction technique. *Soil Sci. Pl. Nutr.*, **24,** 1–13.

Waters, F. F. (1977). Secondary production in inland waters. *Adv. ecol. Res.*, **10,** 91–164. (345)

Watson, D. J. (1947). Comparative physiological studies on the growth of field crops. I. Variation in net assimilation rate and leaf area between species and varieties and within and between years. *Ann. Bot.*, NS**11,** 41–76. (139)

Watson, H. C. (1847–59). *Cybele Brittanica*, Vols. 2–4, Longman, London. (422)

Watt, A. S. (1925). On the ecology of British beechwoods with special reference to their regeneration. II. Sections II and III. The development and structure of beech communities on the Sussex Downs. *J. Ecol.*, **13,** 27–73. (390)

Watt, A. S. (1947). Pattern and process in the plant community. *J. Ecol.*, **35,** 1–22. (278)

Watt, A. S. (1954). Contributions to the ecology of bracken. VI. Frost and the advance and retreat of bracken. *New Phytol.*, **53,** 117–30. (208, 393)

Watt, A. S. (1955). Bracken versus heather: a study in plant ecology. *J. Ecol.*, **43,** 490–506. (208)

Watt, K. E. F. (1966). *Systems Analysis in Ecology*, Academic Press, New York. (425)

Weatherley, P. E. (1950). Studies in the water relations of the cotton plant. I. The field measurement of water deficits in leaves. *New Phytol.*, **49,** 81–7. (157)

Weatherley, P. E. (1951) Studies in the water relations of the cotton plant. II. Diumal and seasonal fluctuations and environmental factors. *New Phytol.*, **50,** 204–16. (160)

Weatherley, P. E. (1969). Ion movement within the plant and its integration with other physiological processes. In *Ecological Aspects of the Mineral Nutrition of Plants*, ed. Rorison, I. H., pp. 323–40. Blackwell, Oxford. (148, 253)

Weatherley, P. E. (1976). Introduction: water movement through plants. In: A discussion on water relations of plants. *Phil. Trans. R. Soc. Lond.* B, **273,** 435–44. (116, 160)

Weaver, J. E. (1926). *Root Development of Field Crops*, McGraw-Hill, New York. (119, 179, 277)

Weavind, T. E. F., and J. F. Hodgeson (1971). Iron absorption by wheat roots as a function of distance from the root tip. *Plant and Soil*, **34,** 697–705.

Webley, D. M., D. J. Eastwood, and C. H. Gimingham (1952). Development of a soil microflora in relation to plant succession on sand dunes including the 'rhizosphere' flora associated with colonising species. *J. Ecol.*, **40,** 168–78. (101)

Weigolaski, F. E. (1975). *Fennoscandian Tundra Ecosystems*, Part I, Springer, Berlin. (74)

Wellbank, P. J., and E. D. Williams (1968). Root growth of a barley crop estimated by sampling with portable powered soil coring equipment. *J. app. Ecol.*, **5,** 477–81. (342)

Went, F. W. (1943). Effect of the root system on tomato stem growth. *Pl. Physiol., Lancaster*, **18,** 51–65. (309)

Went, F. W. (1944). Plant growth under controlled conditions II. Thermoperiodicity in growth and fruiting of the tomato. *Amer. J. Bot.*, **31,** 135–40. (216)

Went, F. W. (1957). *The Experimental Control of Plant Growth*, Chronica Botanica, Waltham. (131, 177, 215, 217, 400, 424)

Went, F. W. (1958). The physiology of photosynthesis in higher plants. *Presalia*, **30,** 225–49. (19)

Went, F. W. (1975). Water vapour absorption in *Prosopis*. In *Physiological Adaptation to Terrestrial Environments*, ed. Vernberg, F. J., pp. 67–75, Intex, New York. (180)

Went, F. W. (1979). Germination and seedling behaviour of desert plants. In *Arid-land Ecosystems, Structure Functioning and Management*, Vol. 1, ed. Goodall, D. W., and R. A. Perry, pp. 477–89, Cambridge Univ. Press, Cambridge. (178)
Went, F. W., and N. Stark (1968). The biological and mechanical role of soil fungi. *Nat. Acad. Sci., Washington, Proc.*, **60,** 497–504. (257)
Wesson, G., and P. F. Wareing (1969). The role of light in the germination of naturally occurring populations of buried weed seeds. *J. exp. Bot.*, **20,** 402–413. (227)
West, C., G. E. Briggs, and F. Kidd (1920). Methods and significant relations in a quantitative analysis of plant growth. *New Phytol.*, **19,** 200–7. (139, 424)
West, N. E., and J. J. Skujins (1978). *Nitrogen Fixation in Desert Ecosystems*, Dowden, Hutchinson and Ross, Stroudsberg. (171)
Westlake, D. F. (1975). Primary production of freshwater macrophytes. In *Photosynthesis and Primary Production in Different Environments*, ed. Cooper, J. P., pp. 189–206, Cambridge Univ. Press, Cambridge. (354)
Wetzel, R. G. (1975). Primary production. In *River Ecology*, ed. Whitton, B. A., pp. 230–47, Cambridge Univ. Press, Cambridge. (345)
Whitehead, D. R. (1969). Wind pollination in the angiosperms: evolutionary and environmental considerations. *Evol.*, **23,** 28–35. (385)
Whitehead, F. H. (1965). The effect of wind on plant growth and soil moisture relations: a reply to the re-assessment by Humphries and Roberts. *New Phytol.*, **64,** 319–22. (162)
Whitmore, T. C. (1975). *Tropical Rainforests of the Far East*, Clarendon Press, Oxford. 68, 69, 71, 74
Whittaker, R. H. (1973). *Ordination and Classification of Plant Communities*, Junk, The Hague. (397, 421)
Whittaker, R. H. (1978). *Ordination of Plant Communities*. Junk, The Hague. (393, 397)
Whittaker, R. H., and G. E. Likens (1975). The biosphere and man. In *Primary Production of the Biosphere*, ed. Lieth, H., and R. H. Whittaker, pp. 305–28, Springer, Berlin. (336, 337, 346, 353, 354, 355)
Whittaker, R. H., and P. L. Marks (1975). Methods of assessing terrestrial productivity. In *Primary Production of the Biosphere*, ed. Lieth, H., and R. H. Whittaker, pp. 55–118, Springer, Berlin. (340, 342, 343)
Wieringa, P. J., D. R. Nielsen, and R. M. Hagen (1969). Thermal properties of a soil based upon field and laboratory measurements. *Proc. Soil Sci. Soc. Amer.*, **3,** 354–60.
Wiersum, L. K. (1957). The relationship of size and structural rigidity of pores to their penetration by roots. *Plant and Soil*, **9,** 75–85. (116)
Wild, A., P. J. Woodhouse, and M. J. Hopper (1979). A comparison between the uptake of potassium by plants from solution of constant potassium concentration and during depletion. *J. exp. Bot.*, **30,** 697–704. (261)
Wild, H. (1974). Geobotanical anomalies in Rhodesia. 4. The vegetation of arsenical soils. *Kirkia*, **9,** 243–64. (280)
Williams, R. F. (1946). The physiology of plant growth with special reference to the concept of net assimilation rate. *Ann. Bot. London*, **10,** 41–72. (138)
Williams, W. T., and D. A. Barber (1961). The functional significance of aerenchyma in plants. In *Mechanisms in Biological Competition*, ed. Milthorpe, F. L., pp. 132–44, Cambridge Univ. Press, London. (321)
Williams, W. T., and J. M. Lambert (1959). Multivariate methods in plant ecology. I. Association analysis in plant communities. *J. Ecol.*, **47,** 83–101. (420)
Williamson, P. (1976). Above ground primary production of chalk grassland allowing for leaf death. *J. Ecol.*, **64,** 1059–75. (307, 345)
Williamson, R. E. (1964). The effect of root aeration on plant growth. *Soil Sci. Soc. Amer. Proc.*, **28,** 86–90. (307)
Willis, A. J. (1963). Braunton Burrows: the effects on the vegetation of the addition of mineral nutrients to the dune soil. *J. Ecol.*, **51,** 353–74. (271, 272, 378, 401)

Willis, A. J., B. F. Folkes, J. F. Hope-Simpson, and E. W. Yemm (1959a). Braunton Burrows: the dune system and its vegetation. Part I. *J. Ecol.*, **47,** 1–24. (311)

Willis, A. J., B. F. Folkes, J. F. Hope-Simpson, and E. W. Yemm (1959b). Braunton Burrows: the dune system and its vegetation. Part II. *J. Ecol.*, **47,** 249–88. (311)

Winogradsky, S. (1893). Sur l'assimilation de l'azote gazeux de l'atmosphere par les microbes. *C.R. Acad. Sci. Paris*, **116,** 1385–8. (423)

Wit, C. T. de (1959). Potential photosynthesis of crop surfaces. *Neth. J. agric. Sci.*, **7,** 141–9. (140, 145)

Wit, C. T. de (1960). On competition. *Versl. landbouwk. Onderz. Ned.*, **66,** 1–82. (413, 414)

Wit, C. T. de, R. Brouwer, and F. W. T. Penning de Vries (1971). A dynamic model of crop growth. In *Potential Crop Production*, ed. Wareing, P. F., and J. P. Cooper, pp. 117–42, Heinemann, London. (145)

Wolf, D. D., R. B. Pierce, G. E. Carlson, and D. R. Lee (1969). Measuring photosynthesis of attached leaves with air-sealed chambers. *Crop. Sci.*, **9,** 24–7. (137)

Woodell, S. R. J., H. A. Mooney, and A. J. Hill (1969). The behaviour of *Larrea divaricata* (Creosote Bush) in response to rainfall in California. *J. Ecol.*, **57,** 37–44. (191, 194)

Woodward, F. I. (1979a, b). The differential temperature responses of the growth of certain species from different altitudes. I and II. *New Phytol.*, **82,** 385–95; 397–405. (216)

Woodward, F. I., and C. D. Pigott (1975). The climatic control of the altitudinal distribution of *Sedum rosea* (L.) Scop. I and II. *New Phytol.*, **74,** 323–34; 335–48. (216)

Woodward, J. (1699). Thoughts and experiments on vegetation. *Phil. Trans. R. Soc. Lond.*, **21,** 382–98. (421)

Woodwell, G. M. (1978). The carbon dioxide question. *Sci. Amer.*, **238** (1), 34–43. (362)

Woodwell, G. M., R. H. Whittaker, and R. A. Houghton (1975). Nutrient element concentration in plants in the Brookhaven Oak-Pine forest. *Ecology*, **56,** 318–32. (278)

Woolhouse, H. W. (1966). The effect of bicarbonate on the uptake of iron in four related grasses. *New Phytol.*, **65,** 372–5. (265)

Woolhouse, H. W. (1969a). Differences in the properties of the acid phosphatases of plant roots and their significance in the evolution of edaphic ecotypes. In *Ecological Aspects of the Mineral Nutrition of Plants*, ed. Rorison, I. H., pp. 357–80, Blackwell, Oxford. (241, 274)

Woolhouse, H. (1969b), *Dormancy and Survival*, Cambridge Univ. Press, Cambridge. (216)

Wooldridge, D. D. (1965). Soil properties related to erosion of wild-land soils in central Washington. In *Forest Soil Relationships in N. America*, ed. Youngberg, C. T., pp. 141–52, Oregon State Univ. Press, Corvallis. (117)

Worthington, E. B. (1975). *The Evolution of IBP*, Cambridge Univ. Press, Cambridge. (Prelim page xvii 2)

Wright, S. T. C. (1977). The relationship between leaf water potential and the levels of abscisic acid and ethylene in excised wheat leaves. *Planta*, **134,** 183–9. (374)

Wright, T. W. (1956). Profile development in the sand dunes of Culbin Forest, Morayshire. II. Chemical properties. *J. Soil Sci.*, **7,** 33–42. (371, 372)

Yamada, N., and Y. Ota (1958). Study on the respiration of crop plants. 7. Enzymatic oxidation of ferrous iron by root of rice plant. *Proc. Crop. Sci. Soc. Japan*, **26,** 205–10. (317)

Yeo, A. R., and T. J. Flowers (1977). Salt tolerance in the halophyte *Suaeda maritima* (L.) Dun. Interaction between aluminium and salinity. *Ann. Bot.*, **41,** 331–9. (267)

Yong, T. W. (1967). Ecotypic variation in *Larrea divaricata. Am. J. Bot.*, **54,** 1041–44. (192)

Yoshida, T., and R. R. Ancajas (1973). The fixation of atmospheric nitrogen in the rice rhizosphere. *Soil Biol. Biochem.*, **5,** 153–55.

Yoshida, T., and T. Oritani (1974). Studies in nitrogen metabolism in crop plants. 13.

Effects of nitrogen topdressing on cytokinin content of the root exudate of rice plant. *Proc. Crop. Sci. Soc. Japan*, **43,** 47. (308)

Young, A. (1976). *Tropical Soils and Soil Survey*, Cambridge Univ. Press, Cambridge. (62)

Youngs, E. G. (1965). Water movement in soils. In *The State and Movement of Water in Living Organisms*, ed. Fogg, G. E., pp. 89–112, Cambridge Univ. Press, Cambridge. (160)

Yu, P. T., L. H. Stolzy, and J. Letey (1969). Survival of plants under prolonged flooding. *Agron. J.*, **61,** 844–7. (320)

Zahner, R., and J. R. Donelly (1967). Refining correlations of rainfall and radial growth in young red pine. *Ecology*, **48,** 525–30. (193)

Zelitch, I. (1971). *Photosynthesis, Photorespiration and Plant Productivity*, Academic Press, New York. (130, 206)

Zelitch, I. (1975). Environmental and biological control of photosynthesis: general assessment. In *Environmental and Biological Control of Photosynthesis*, ed. Marcelle, R., pp. 25–62, Junk: The Hague. (128, 130)

Zinke, P. J. (1962). The pattern of individual trees on soil properties. *Ecology*, **43,** 130–3. (104, 371, 382, 398)

Zobell, C. E. (1958). Ecology of sulphate reducing bacteria. *Producers Monthly*, **22,** 12–29. (304)

Index